Student Solutions Manual

Cindy Trimble & Associates

Beginning Algebra

Seventh Edition

John Tobey
North Shore Community College

Jeffrey Slater
North Shore Community College

Prentice Hall
is an imprint of

PEARSON

Reproduced by Pearson Prentice Hall from electronic files supplied by the author.

Publishing as Prentice Hall, Upper Saddle River, NJ 07458.

ISBN-13: 978-0-321-57376-6
ISBN-10: 0-321-57376-5

1 2 3 4 5 6 BRR 10 09 08

Prentice Hall
is an imprint of

PEARSON

www.pearsonhighered.com

Contents

3. $\frac{103}{21} = 4\frac{19}{21}$

$$\begin{array}{r} 4 \\ 21\overline{)103} \\ \underline{84} \\ 19 \end{array}$$

4. Answers may vary. Possible solution: First, multiply the whole number by the denominator. Then, add this to the numerator. The result is the new numerator. The denominator does not change.

0.2 Exercises

1. Answers may vary. A sample answer is: 8 is the LCD of $\frac{3}{4}$ and $\frac{5}{8}$ because 8 is exactly divisible by 4.

3. $\frac{4}{5}$ and $\frac{11}{16}$
$5 = 5$
$16 = 2 \cdot 2 \cdot 2 \cdot 2$
$\text{LCD} = 2 \cdot 2 \cdot 2 \cdot 2 \cdot 5 = 80$

5. $\frac{7}{10}$ and $\frac{1}{4}$
$10 = 2 \cdot 5$
$4 = 2 \cdot 2$
$\text{LCD} = 2 \cdot 2 \cdot 5 = 20$

7. $\frac{5}{18}$ and $\frac{7}{54}$
$18 = 2 \cdot 3 \cdot 3$
$54 = 2 \cdot 3 \cdot 3 \cdot 3$
$\text{LCD} = 2 \cdot 3 \cdot 3 \cdot 3 = 54$

9. $\frac{1}{2}, \frac{1}{18}$ and $\frac{13}{30}$
$2 = 2$
$18 = 2 \cdot 3 \cdot 3$
$30 = 2 \cdot 3 \cdot 5$
$\text{LCD} = 2 \cdot 3 \cdot 3 \cdot 5 = 90$

11. $\frac{3}{8} + \frac{2}{8} = \frac{3+2}{8} = \frac{5}{8}$

13. $\frac{5}{14} - \frac{1}{14} = \frac{5-1}{14} = \frac{4}{14} = \frac{2}{7}$

15. $\frac{3}{8} + \frac{5}{6} = \frac{3\times3}{8\times3} + \frac{5\times4}{6\times4} = \frac{9}{24} + \frac{20}{24} = \frac{29}{24} = 1\frac{5}{24}$

17. $\frac{5}{7} - \frac{2}{9} = \frac{5\times9}{7\times9} - \frac{2\times7}{9\times7} = \frac{45}{63} - \frac{14}{63} = \frac{31}{63}$

19. $\frac{1}{3} + \frac{2}{5} = \frac{1\times5}{3\times5} + \frac{2\times3}{5\times3} = \frac{5}{15} + \frac{6}{15} = \frac{11}{15}$

21. $\frac{5}{9} + \frac{5}{12} = \frac{5\times4}{9\times4} + \frac{5\times3}{12\times3} = \frac{20}{36} + \frac{15}{36} = \frac{35}{36}$

23. $\frac{2}{3} + \frac{7}{12} + \frac{1}{4} = \frac{8}{12} + \frac{7}{12} + \frac{3}{12} = \frac{18}{12} = \frac{3}{2} = 1\frac{1}{2}$

25.
$$\begin{aligned} \frac{5}{36} + \frac{7}{9} - \frac{5}{12} &= \frac{5}{36} + \frac{7\times4}{9\times4} - \frac{5\times3}{12\times3} \\ &= \frac{5}{36} + \frac{28}{36} - \frac{15}{36} \\ &= \frac{18}{36} \\ &= \frac{1}{2} \end{aligned}$$

27. $4\frac{1}{3} + 3\frac{2}{5} = \frac{13}{3} + \frac{17}{5} = \frac{65}{15} + \frac{51}{15} = \frac{116}{15} = 7\frac{11}{15}$

29. $1\frac{5}{24} + \frac{5}{18} = \frac{29}{24} + \frac{5}{18} = \frac{87}{72} + \frac{20}{72} = \frac{107}{72} = 1\frac{35}{72}$

31. $7\frac{1}{6} - 2\frac{1}{4} = \frac{43}{6} - \frac{9}{4} = \frac{86}{12} - \frac{27}{12} = \frac{59}{12} = 4\frac{11}{12}$

33. $8\frac{5}{7} - 2\frac{1}{4} = \frac{61}{7} - \frac{9}{4} = \frac{244}{28} - \frac{63}{28} = \frac{181}{28} = 6\frac{13}{28}$

35. $2\frac{1}{8} + 3\frac{2}{3} = \frac{17}{8} + \frac{11}{3} = \frac{51}{24} + \frac{88}{24} = \frac{139}{24} = 5\frac{19}{24}$

37. $11\frac{1}{7} - 6\frac{5}{7} = \frac{78}{7} - \frac{47}{7} = \frac{31}{7} = 4\frac{3}{7}$

39. $3\frac{5}{12} + 5\frac{7}{12} = \frac{41}{12} + \frac{67}{12} = \frac{108}{12} = 9$

41. $\frac{7}{8} + \frac{1}{12} = \frac{42}{48} + \frac{4}{48} = \frac{46}{48} = \frac{23}{24}$

43. $3\frac{3}{16}+4\frac{3}{8}=\frac{51}{16}+\frac{35}{8}=\frac{51}{16}+\frac{70}{16}=\frac{121}{16}=7\frac{9}{16}$

45. $\frac{16}{21}-\frac{2}{7}=\frac{16}{21}-\frac{6}{21}=\frac{10}{21}$

47. $5\frac{1}{5}-2\frac{1}{2}=\frac{26}{5}-\frac{5}{2}=\frac{52}{10}-\frac{25}{10}=\frac{27}{10}=2\frac{7}{10}$

49. $$\begin{aligned}8\frac{1}{4}+10\frac{2}{3}+5\frac{3}{4}&=\frac{33}{4}+\frac{32}{3}+\frac{23}{4}\\&=\frac{99}{12}+\frac{128}{12}+\frac{69}{12}\\&=\frac{296}{12}\\&=24\frac{8}{12}\\&=24\frac{2}{3}\end{aligned}$$

Their total distance was $24\frac{2}{3}$ miles.

51. $$\begin{aligned}8\frac{1}{2}-2\frac{2}{3}-1\frac{3}{4}&=\frac{17}{2}-\frac{8}{3}-\frac{7}{4}\\&=\frac{102}{12}-\frac{32}{12}-\frac{21}{12}\\&=\frac{49}{12}\\&=4\frac{1}{12}\end{aligned}$$

She will have $4\frac{1}{12}$ hours left over.

53. $A=2+\frac{1}{2}+3\frac{1}{2}+\frac{1}{2}+3\frac{1}{2}+\frac{1}{2}+1\frac{1}{2}=12$ in.

$B=\frac{1}{2}+4\frac{5}{8}+\frac{1}{2}+4\frac{5}{8}+\frac{1}{2}+4\frac{5}{8}+\frac{1}{2}=15\frac{7}{8}$ in.

Cumulative Review

55. $\frac{36}{44}=\frac{9\times4}{1\times4}=\frac{9}{11}$

56. $26\frac{3}{5}=\frac{26\times5+3}{5}=\frac{133}{5}$

Quick Quiz 0.2

1. $\frac{3}{4}+\frac{1}{2}+\frac{5}{12}=\frac{9}{12}+\frac{6}{12}+\frac{5}{12}=\frac{20}{12}=\frac{5}{3}$ or $1\frac{2}{3}$

2. $2\frac{3}{5}+4\frac{14}{15}=2\frac{9}{15}+4\frac{14}{15}=6\frac{23}{15}=7\frac{8}{15}$

3. $6\frac{1}{9}-3\frac{5}{6}=\frac{55}{9}-\frac{23}{6}=\frac{110}{18}-\frac{69}{18}=\frac{41}{18}=2\frac{5}{18}$

4. Answers may vary. Possible solution: Write each denominator as the product of prime factors. The LCD is a product containing each different factor. If a factor occurs more than once in any one denominator, the LCD will contain that factor repeated the greatest number of times that it occurs in any one denominator.

0.3 Exercises

1. First change each mixed number to an improper fraction. Look for a common factor in the numerator and denominator to divide by; if one is found, perform the division. Multiply the numerators. Multiply the denominators.

3. $\frac{36}{7}\times\frac{5}{9}=\frac{9\cdot4}{7}\times\frac{5}{9}=\frac{20}{7}=2\frac{6}{7}$

5. $\frac{17}{18}\times\frac{3}{5}=\frac{17}{6\cdot3}\times\frac{3}{5}=\frac{17}{30}$

7. $\frac{4}{5}\times\frac{3}{10}=\frac{2\cdot2}{5}\times\frac{3}{2\cdot5}=\frac{6}{25}$

9. $\frac{24}{25}\times\frac{5}{2}=\frac{12\cdot2}{5\cdot5}\times\frac{5}{2}=\frac{12}{5}=2\frac{2}{5}$

11. $9\times\frac{2}{5}=\frac{9}{1}\times\frac{2}{5}=\frac{18}{5}=3\frac{3}{5}$

13. $\frac{8}{5}\div\frac{8}{3}=\frac{8}{5}\times\frac{3}{8}=\frac{3}{5}$

15. $\frac{3}{7}\div3=\frac{3}{7}\div\frac{3}{1}=\frac{3}{7}\times\frac{1}{3}=\frac{1}{7}$

17. $10\div\frac{5}{7}=\frac{10}{1}\times\frac{7}{5}=\frac{2\cdot5}{1}\times\frac{7}{5}=14$

19. $\frac{\frac{7}{8}}{\frac{3}{4}}=\frac{7}{8}\div\frac{3}{4}=\frac{7}{8}\times\frac{4}{3}=\frac{7}{2\cdot4}\times\frac{2\cdot2}{3}=\frac{14}{12}=\frac{7}{6}=1\frac{1}{6}$

21. $\dfrac{\frac{5}{6}}{\frac{7}{9}}=\dfrac{5}{6}\div\dfrac{7}{9}=\dfrac{5}{6}\times\dfrac{9}{7}=\dfrac{5}{2\cdot 3}\times\dfrac{3\cdot 3}{7}=\dfrac{15}{14}=1\dfrac{1}{14}$

23. $1\dfrac{3}{7}\div 6\dfrac{1}{4}=\dfrac{10}{7}\div\dfrac{25}{4}=\dfrac{10}{7}\times\dfrac{4}{25}=\dfrac{2\cdot 5}{7}\times\dfrac{4}{5\cdot 5}=\dfrac{8}{35}$

25. $3\dfrac{1}{3}\div 2\dfrac{1}{2}=\dfrac{10}{3}\div\dfrac{5}{2}=\dfrac{2\cdot 5}{3}\times\dfrac{2}{5}=\dfrac{4}{3}=1\dfrac{1}{3}$

27.
$$\begin{aligned}\frac{15}{2\frac{2}{5}}&=15\div 2\frac{2}{5}\\&=\frac{15}{1}\div\frac{12}{5}\\&=\frac{15}{1}\times\frac{5}{12}\\&=\frac{3\cdot 5}{1}\times\frac{5}{3\cdot 4}\\&=\frac{25}{4}\\&=6\frac{1}{4}\end{aligned}$$

29. $\dfrac{\frac{2}{3}}{1\frac{1}{4}}=\dfrac{\frac{2}{3}}{\frac{5}{4}}=\dfrac{2}{3}\div\dfrac{5}{4}=\dfrac{2}{3}\times\dfrac{4}{5}=\dfrac{8}{15}$

31. $\dfrac{4}{7}\times\dfrac{21}{2}=\dfrac{2\cdot 2}{7}\times\dfrac{7\cdot 3}{2}=6$

33. $\dfrac{5}{14}\div\dfrac{2}{7}=\dfrac{5}{14}\times\dfrac{7}{2}=\dfrac{5}{2\cdot 7}\times\dfrac{7}{2}=\dfrac{5}{4}=1\dfrac{1}{4}$

35. $10\dfrac{3}{7}\times 5\dfrac{1}{4}=\dfrac{73}{7}\times\dfrac{21}{4}=\dfrac{73}{7}\times\dfrac{3\cdot 7}{4}=\dfrac{219}{4}=54\dfrac{3}{4}$

37. $30\div\dfrac{3}{4}=\dfrac{30}{1}\div\dfrac{3}{4}=\dfrac{30}{1}\times\dfrac{4}{3}=\dfrac{3\cdot 10}{1}\times\dfrac{4}{3}=40$

39. $6\times 4\dfrac{2}{3}=\dfrac{6}{1}\times\dfrac{14}{3}=\dfrac{2\cdot 3}{1}\times\dfrac{14}{3}=28$

41. $2\dfrac{1}{2}\times\dfrac{1}{10}\times\dfrac{3}{4}=\dfrac{5}{2}\times\dfrac{1}{5\cdot 2}\times\dfrac{3}{4}=\dfrac{3}{16}$

43.
$$\begin{aligned}2\frac{3}{4}\times 26&=\frac{11}{4}\times\frac{26}{1}\\&=\frac{11}{2\cdot 2}\times\frac{2\cdot 13}{1}\\&=\frac{143}{2}\\&=71\frac{1}{2}\end{aligned}$$

You would need $71\frac{1}{2}$ yards.

45. $\dfrac{2}{3}\times 4\dfrac{1}{5}=\dfrac{2}{3}\times\dfrac{21}{5}=\dfrac{2}{1}\times\dfrac{7}{5}=\dfrac{14}{5}=2\dfrac{4}{5}$

The mountain bike trail is $2\frac{4}{5}$ miles.

Cumulative Review

47. $\dfrac{11}{15}=\dfrac{11\times 5}{15\times 5}=\dfrac{55}{75}$

48. $\dfrac{7}{9}=\dfrac{7\times 7}{9\times 7}=\dfrac{49}{63}$

Quick Quiz 0.3

1. $\dfrac{7}{15}\times\dfrac{25}{14}=\dfrac{7}{5\cdot 3}\times\dfrac{5\cdot 5}{7\cdot 2}=\dfrac{5}{6}$

2. $3\dfrac{1}{4}\times 4\dfrac{1}{2}=\dfrac{13}{4}\times\dfrac{9}{2}=\dfrac{117}{8}=14\dfrac{5}{8}$

3.
$$\begin{aligned}3\frac{3}{10}\div 2\frac{1}{2}&=\frac{33}{10}\div\frac{5}{2}\\&=\frac{33}{10}\times\frac{2}{5}\\&=\frac{33}{2\cdot 5}\times\frac{2}{5}\\&=\frac{33}{25}\\&=1\frac{8}{25}\end{aligned}$$

4. Answers may vary. Possible solution: Change each mixed number to an improper fraction. Invert the second fraction and multiply the result by the first fraction. Simplify.

0.4 Exercises

1. A decimal is another way of writing a fraction whose denominator is 10; 100; 1000; 10,000; and so on.

3. When dividing 7432.9 by 1000 we move the decimal point <u>3</u> places to the <u>left</u>.

5. $\frac{5}{8} = 8\overset{0.625}{\overline{)5.000}} = 0.625$

$$\begin{array}{r} 4\ 8 \\ \hline 20 \\ 16 \\ \hline 40 \\ 40 \\ \hline 0 \end{array}$$

7. $\frac{3}{15} = 15\overset{0.2}{\overline{)3.0}} = 0.2$

$$\begin{array}{r} 3.0 \\ \hline 0 \end{array}$$

9. $11\overset{0.63}{\overline{)7.000}} = 0.\overline{63}$

$$\begin{array}{r} 6\ 6 \\ \hline 40 \\ 33 \\ \hline 7 \end{array}$$

11. $0.8 = \frac{8}{10} = \frac{4}{5}$

13. $0.625 = \frac{625}{1000} = \frac{5}{8}$

15. $3.4 = \frac{34}{10} = \frac{17\times 2}{5\times 2} = \frac{17}{5}$ or $3\frac{2}{5}$

17.
$$\begin{array}{r} 1.71 \\ +\ 0.38 \\ \hline 2.09 \end{array}$$

19.
$$\begin{array}{r} 2.50 \\ 3.42 \\ +\ 4.90 \\ \hline 10.82 \end{array}$$

21.
$$\begin{array}{r} 46.030 \\ 215.100 \\ +\ \ 0.078 \\ \hline 261.208 \end{array}$$

23.
$$\begin{array}{r} 147.18 \\ -\ 15.39 \\ \hline 131.79 \end{array}$$

25.
$$\begin{array}{r} 7.21 \\ \times\ \ 4.2 \\ \hline 1\ 442 \\ 28\ 84\ \\ \hline 30.282 \end{array}$$

27.
$$\begin{array}{r} 4.23 \\ \times\ 0.025 \\ \hline 2115 \\ 846\ \\ \hline 0.10575 \end{array}$$

29.
$$\begin{array}{r} 58,200 \\ \times\ 0.0015 \\ \hline 29\ 1000 \\ 58\ \ 200\ \\ \hline 87.3000 \end{array}$$
or 87.3

31. $3.02_{\wedge}\overset{2.64}{\overline{)7.97_{\wedge}28}}$

$$\begin{array}{r} 6\ 04\ \ \ \ \\ \hline 1\ 93\ \ 2 \\ 1\ 81\ \ 2 \\ \hline 12\ \ 08 \\ 12\ \ 08 \\ \hline 0 \end{array}$$

33. $0.002_{\wedge}\overset{261.5}{\overline{)0.523_{\wedge}0}}$

$$\begin{array}{r} 4\ \ \ \ \ \ \\ \hline 12\ \ \ \ \\ 12\ \ \ \ \\ \hline 3\ \ \ \\ 2\ \ \ \\ \hline 1\ \ 0 \\ 1\ \ 0 \\ \hline 0 \end{array}$$

35. $0.06_\wedge\overline{)0.03_\wedge048}$ = 0.508

$$\begin{array}{r} 0.508 \\ 0.06_\wedge\overline{)0.03_\wedge048} \\ \underline{3\ \ 0} \\ 48 \\ \underline{48} \\ 0 \end{array}$$

37. $3.45 \times 1000 = 3450$

39. $0.76 \div 100 = 0.0076$

41. $7.36 \times 10{,}000 = 73{,}600$

43. $73{,}892 \div 100{,}000 = 0.73892$

45.
$$\begin{array}{r} 54.8 \\ \underline{\times\ 0.15} \\ 2\ 740 \\ \underline{5\ 48} \\ 8.220 \text{ or } 8.22 \end{array}$$

47.
$$\begin{array}{r} 13.75 \\ 2.55 \\ \underline{+\ \ 0.078} \\ 16.378 \end{array}$$

49.
$$\begin{array}{r} 2.12 \\ 0.027_\wedge\overline{)0.057_\wedge24} \\ \underline{54} \\ 3\ \ 2 \\ \underline{2\ \ 7} \\ 54 \\ \underline{54} \\ 0 \end{array}$$

51. $0.7683 \times 1000 = 768.3$

53.
$$\begin{array}{r} 0.4732 \\ \underline{\times\ \ \ \ 5.5} \\ 23660 \\ \underline{23660} \\ 2.60260 \end{array}$$
The measured data is 2.6026 L.

55. $1.3 \times 5 = 6.5$ milligrams
No; there was 0.3 milligram of copper in excess of the standard for 5 liters of water.

Cumulative Review

57. $3\frac{1}{2} \div 5\frac{1}{4} = \frac{7}{2} \div \frac{21}{4} = \frac{7}{2} \cdot \frac{4}{21} = \frac{1 \cdot 2}{1 \cdot 3} = \frac{2}{3}$

58. $\frac{3}{8} \cdot \frac{12}{27} = \frac{3}{2 \cdot 4} \cdot \frac{3 \cdot 4}{3 \cdot 3 \cdot 3} = \frac{1}{6}$

59. $\frac{12}{25} + \frac{9}{20} = \frac{48}{100} + \frac{45}{100} = \frac{48+45}{100} = \frac{93}{100}$

60. $1\frac{3}{5} - \frac{1}{2} = \frac{8}{5} - \frac{1}{2} = \frac{16}{10} - \frac{5}{10} = \frac{16-5}{10} = \frac{11}{10} = 1\frac{1}{10}$

Quick Quiz 0.4

1.
$$\begin{array}{r} 8.0567 \\ \underline{-\ 2.3489} \\ 5.7078 \end{array}$$

2.
$$\begin{array}{r} 58.7 \\ \underline{\times\ 0.06} \\ 3.522 \end{array}$$

3.
$$\begin{array}{r} 28.8 \\ 0.16_\wedge\overline{)4.60_\wedge8} \\ \underline{3\ 2} \\ 1\ 40 \\ \underline{1\ 28} \\ 12\ \ \ 8 \\ \underline{12\ \ \ 8} \\ 0 \end{array}$$

4. Answers may vary. Possible solution:
Move the decimal point over 4 places to the right on the divisor ($0.0035 \Rightarrow 35$). Then move the decimal the same number of places on the dividend ($0.252 \Rightarrow 2520$).

How Am I Doing? Sections 0.1–0.4

1. $\frac{15}{55} = \frac{3 \cdot 5}{11 \cdot 5} = \frac{3}{11}$

2. $\frac{46}{115} = \frac{2 \cdot 23}{5 \cdot 23} = \frac{2}{5}$

3. $\frac{15}{4} = 4\overline{)15} = 3\frac{3}{4}$

$$\begin{array}{r} 3 \\ 4\overline{)15} \\ \underline{12} \\ 3 \end{array}$$

4. $4\frac{5}{7}=\frac{4\cdot 7+5}{7}=\frac{28+5}{7}=\frac{33}{7}$

5. $\frac{4}{5}=\frac{?}{30}\Rightarrow\frac{4\times 6}{5\times 6}=\frac{24}{30}$

6. $\frac{5}{9}=\frac{?}{81}\Rightarrow\frac{5\times 9}{9\times 9}=\frac{45}{81}$

7. $8=2\cdot 2\cdot 2$
$6=2\cdot 3$
$15=3\cdot 5$
$\text{LCD}=2\cdot 2\cdot 2\cdot 3\cdot 5=120$

8. $\frac{3}{7}+\frac{2}{7}=\frac{3+2}{7}=\frac{5}{7}$

9. $\frac{5}{14}+\frac{2}{21}=\frac{5\cdot 3}{14\cdot 3}+\frac{2\cdot 2}{21\cdot 2}=\frac{15}{42}+\frac{4}{42}=\frac{19}{42}$

10.
$$\begin{aligned}2\frac{3}{4}+5\frac{2}{3}&=\frac{11}{4}+\frac{17}{3}\\&=\frac{11\cdot 3}{4\cdot 3}+\frac{17\cdot 4}{3\cdot 4}\\&=\frac{33}{12}+\frac{68}{12}\\&=\frac{101}{12}\\&=8\frac{5}{12}\end{aligned}$$

11. $\frac{17}{18}-\frac{5}{9}=\frac{17}{18}-\frac{5\cdot 2}{9\cdot 2}=\frac{17}{18}-\frac{10}{18}=\frac{7}{18}$

12. $\frac{6}{7}-\frac{2}{3}=\frac{6\cdot 3}{7\cdot 3}-\frac{2\cdot 7}{3\cdot 7}=\frac{18}{21}-\frac{14}{21}=\frac{4}{21}$

13.
$$\begin{aligned}3\frac{1}{5}-1\frac{3}{8}&=\frac{16}{5}-\frac{11}{8}\\&=\frac{16\cdot 8}{5\cdot 8}-\frac{11\cdot 5}{8\cdot 5}\\&=\frac{128}{40}-\frac{55}{40}\\&=\frac{73}{40}\\&=1\frac{33}{40}\end{aligned}$$

14. $\frac{25}{7}\times\frac{14}{45}=\frac{5\cdot 5}{7}\times\frac{2\cdot 7}{5\cdot 9}=\frac{10}{9}=1\frac{1}{9}$

15. $12\times 3\frac{1}{2}=\frac{12}{1}\times\frac{7}{2}=\frac{6\cdot 2}{1}\times\frac{7}{2}=42$

16. $4\div\frac{8}{7}=\frac{4}{1}\cdot\frac{7}{8}=\frac{4}{1}\times\frac{7}{2\cdot 4}=\frac{7}{2}=3\frac{1}{2}$

17. $2\frac{1}{3}\div 3\frac{1}{4}=\frac{7}{3}\div\frac{13}{4}=\frac{7}{3}\times\frac{4}{13}=\frac{28}{39}$

18. $\frac{3}{8}=8\overline{)3.000}=0.375$

$$\begin{array}{r}0.375\\8\overline{)3.000}\\\underline{2\ 4}\\60\\\underline{56}\\40\\\underline{40}\\0\end{array}$$

19. $\frac{5}{6}=6\overline{)5.000}=0.8\overline{3}$

$$\begin{array}{r}0.833\\6\overline{)5.000}\\\underline{4\ 8}\\20\\\underline{18}\\20\\\underline{18}\\2\end{array}$$

20. $\frac{3}{200}=200\overline{)3.000}=0.015$

$$\begin{array}{r}0.015\\200\overline{)3.000}\\\underline{2\ 00}\\1\ 000\\\underline{1\ 000}\\0\end{array}$$

21.
$$\begin{array}{r}15.230\\3.600\\+\ \ 0.821\\\hline 19.651\end{array}$$

22.
$$\begin{array}{r}3.28\\\times\ 0.63\\\hline 984\\1\ 968\ \\\hline 2.0664\end{array}$$

23. $\dfrac{3.015}{6.7} = 6.7_\wedge \overline{)3.0_\wedge 15} = 0.45$

$$\begin{array}{r} 0.45 \\ 6.7_\wedge\overline{)3.0_\wedge15} \\ \underline{2\,6\ \ 8} \\ 3\ \ 35 \\ \underline{3\ \ 35} \\ 0 \end{array}$$

24.
$$\begin{array}{r} 12.130 \\ -\ \ 9.884 \\ \hline 2.246 \end{array}$$

0.5 Exercises

1. Answers may vary. Sample answers follow: 19% means 19 out of 100 parts. Percent means per 100. 19% is really a fraction with a denominator of 100. In this case it would be $\frac{19}{100}$.

3. $0.28 = 28\%$

5. $0.568 = 56.8\%$

7. $0.076 = 7.6\%$

9. $2.39 = 239\%$

11. $3\% = 0.03$

13. $0.4\% = 0.004$

15. $250\% = 2.5$

17. $3.75\% = 0.0375$

19. 8% of 65
$0.08 \times 65 = 5.2$

21. 10% of 130
$0.10 \times 130 = 13$

23. 112% of 65
$1.12 \times 65 = 72.8$

25. 36 is what percent of 24?
$\dfrac{36}{24} = 1.50 = 150\%$

27. What percent of 340 is 17?
$\dfrac{17}{340} = \dfrac{1}{20} = 0.05 = 5\%$

29. 10 is what percent of 50?
$\dfrac{10}{50} = \dfrac{20}{100} = 20\%$

31. $\dfrac{60}{75} = 0.80 = 80\%$
Her grade was 80%.

33. $0.15 \times 32.80 = 4.92$
$32.80 + 4.92 = 37.72$
\$4.92 tip, \$37.72 new total

35. $\dfrac{380}{1850} \approx 0.21 = 21\%$
About 21% of the budget is for food.

37. $1.5\% \times 36{,}000 = 0.015 \times 36{,}000 = 540$
They can expect 540 gifts to be exchanged.

39. $0.08 \times 12.50 = 1$
$12.50 + 1 = 13.50$
\$1 per hour raise
\$13.50 per hour new rate

41. a. $4\% \text{ of } 920{,}000 = 0.04 \times 920{,}000 = 36{,}800$
His sales commission was \$36,800.

b. $600 \times 12 + 36{,}800 = 7200 + 36{,}800$
$= 44{,}000$
His annual salary was \$44,000.

Cumulative Review

43.
$$\begin{array}{r} 132.80 \\ 850.00 \\ +\ 850.00 \\ \hline 1832.80 \end{array} \qquad \begin{array}{r} 160.00 \\ 42.60 \\ 775.00 \\ 82.76 \\ +\ 159.45 \\ \hline 1219.81 \end{array}$$
Balance: $1832.80 - 1219.81 = 612.99$
The balance at the end of the month is \$612.99.

44. $33 \times 188.50 = 6220.50$
He will pay a total of \$6220.50.

45. Distance: $43{,}556.8 - 42{,}443.8 = 1113$
$\dfrac{1113}{42} = 26.5$
Her car achieved 26.5 miles per gallons.

46. Total $= 4.6 + 4.5 + 2.9 = 12$
Average $= \dfrac{12}{3} = 4.0$ inches per month.

Quick Quiz 0.5

1. 4.5% of $260 = 0.045 \times 260 = 11.7$

2. $\frac{63}{420} = 0.15 = 15\%$

3. 114% of $85 = 1.14 \times 85 = 96.9$

4. Answers may vary. Possible solution:
Move the decimal point two places to the right and add the % symbol.

0.6 Exercises

1. $693 \times 307 \approx 700 \times 300 = 210{,}000$

3. $2862 \times 5986 \approx 3000 \times 6000 \approx 18{,}000{,}000$

5. $14 + 73 + 80 + 21 + 56 \approx 10 + 70 + 80 + 20 + 60$
$= 240$

7. $41\overline{)829{,}346} \approx 40\overline{)800{,}000}$ with quotient $20{,}000$

9. $\frac{2714}{31{,}500} \approx \frac{3000}{30{,}000} = 0.1$

11. 17% of $21{,}365.85 \approx 0.20 \times 20{,}000 \approx \$4000$

13. $18.5 \times 22.50 \approx 20 \times 20 \approx 400$
The estimated cost is \$400.

15. $4 \times 22 \times 82 \approx 4 \times 20 \times 80 = 6400$
The store receives approximately \$6400 each hour.

17. $\frac{117.7}{3.8} \approx \frac{100}{4} = 25$
The car achieves about 25 miles per gallon.

19. $\frac{104{,}100{,}000}{411} \approx \frac{100{,}000{,}000}{400} = \$250{,}000$

21. $26\% \times 217{,}970{,}000{,}000$
$\approx 0.30 \times 200{,}000{,}000{,}000$
$\approx 60{,}000{,}000{,}000$
There is an estimated \$60,000,000,000 spent on education in California in 2000.

23. $34\% \times 45{,}068 \approx 0.30 \times 50{,}000 \approx 15{,}000$
The annual expenditure is about \$15,000.

25. $\frac{4176}{38{,}573} = \frac{4000}{40{,}000} = 0.10 = 10$
The estimated percent is 10%.

Cumulative Review

27. $0.6\% \times 350 = 0.006 \times 350 = 2.1$

28. $\frac{12}{20} = \frac{60}{100} = 0.60 = 60$
She scores a basket 60% of the time.

29. Tip = $15\% \times 5.85 = 0.88$
Total = $5.85 + 0.88 = \$6.73$
The total amount is \$6.73.

30. Defective Fords = $12\% \times 450 = 0.12 \times 450 = 54$
Replacements = $5 \times 54 = 270$ tires
They need 270 new tires.

Quick Quiz 0.6

1. $5687.98 + 8945.34 + 3429.57$
$\approx 6000 + 9000 + 3000$
$\approx 18{,}000$

2. $786 \times 543 \approx 800 \times 500 \approx 400{,}000$

3. $34{,}987\overline{)567{,}238} \approx 30{,}000\overline{)600{,}000}$ with quotient 20; $600\,00$; remainder 00

4. Answers may vary. Possible solution:
Look at the number directly to the right of the hundred thousands position (the ten thousands position). Since the digit is 9, which is greater than or equal to 5, round up. Since the hundred thousands position is a 9, change the 9 to a 0 and round the 7 in the millions position up to 8 (568,000,000).

0.7 Exercises

1. Area $= 3\frac{1}{3} \times 4\frac{1}{3} = \frac{10}{3} \times \frac{13}{3} = \frac{130}{9}$
Square feet $= \frac{130}{9} \times \frac{9}{1} = 130$
Cost = $130 \times 2.35 = \$305.50$
The total cost is \$305.50.

3. a. $\text{Length} = 2\times 15\frac{1}{2} + 2\times 25\frac{2}{3}$
$= 2\times\frac{31}{2} + 2\times\frac{77}{3}$
$= 31 + \frac{154}{3}$
$= \frac{93}{3} + \frac{154}{3}$
$= \frac{247}{3}$
$= 82\frac{1}{3}$

She needs $82\frac{1}{3}$ feet of fencing.

b. $\text{Cost} = 2.10\times\frac{247}{3} = \$172.90$

She should purchase the prepackaged 90 feet.
She would save \$172.90 – \$155.00 = \$17.90.

5.

Day	Jog	Walk	Rest	Walk
1	$1\frac{1}{2}$	$1\frac{3}{4}$	$2\frac{1}{2}$	1
3	$2\frac{2}{3}$	$3\frac{1}{9}$	$4\frac{4}{9}$	$1\frac{7}{9}$

Increase $1\frac{1}{3}\times 1\frac{1}{3} = \frac{4}{3}\times\frac{4}{3} = \frac{16}{9}$

$\frac{16}{9}\times 1\frac{1}{2} = \frac{16}{9}\times\frac{3}{2} = \frac{48}{18} = 2\frac{2}{3}$ mi

$\frac{16}{9}\times 1\frac{3}{4} = \frac{16}{9}\times\frac{7}{4} = \frac{28}{9} = 3\frac{1}{9}$ mi

$\frac{16}{9}\times 2\frac{1}{2} = \frac{16}{9}\times\frac{5}{2} = \frac{40}{9} = 4\frac{4}{9}$ mi

$\frac{16}{9}\times 1 = \frac{16}{9} = 1\frac{7}{9}$ miles

7. Betty will have a more demanding schedule on day 3 because Melinda increases each activity by $\frac{2}{3}$ by day 3 and Betty increases each activity by $\frac{7}{9}$ by day 3.

9. Increase: $1\frac{1}{2}\times\frac{1}{3} = \frac{3}{2}\times\frac{1}{3} = \frac{1}{2}$

Day 7: $1\frac{1}{2} + \left(6\times\frac{1}{2}\right) = \frac{3}{2} + \frac{6}{2} = \frac{9}{2} = 4\frac{1}{2}$ miles

She will jog $4\frac{1}{2}$ miles in day 7.

11. 1st increase: 30% × 350 = 66,200 = 19,860
1990 price: 66,200 + 19,860 = 86,060
2nd increase: 15% × 86,060 = 12,909
2000 price: 86,060 + 12,909 = \$98,969.00
The median house price in Atlanta in 2005 was \$98,696.00.

13. a. Percent available = 100 – 28 = 72%
0.72 × 62,500 = \$45,000
They have \$45,000 available.

b. 0.31 × 45,000 = \$13,950
\$13,950 is budgeted for food.

15. $\frac{138.97+67.76+5.18}{1150} = \frac{211.91}{1150} \approx 0.18$
About 18% of Fred's gross pay is deducted for federal, state, and local taxes.

17. $\frac{790.47}{1150} \approx 0.69 = 69\%$
He actually takes home 69% of his gross pay.

Quick Quiz 0.7

1. $\frac{525}{1.5} = 350$ stones
It will require 350 stones to pave the courtyard.

2. 4% of 345,000 = 0.04 × 345,000 = 13,800
$\frac{13{,}800}{16{,}200+13{,}800} = \frac{13{,}800}{30{,}000} = 0.46 = 46\%$
Her commission was 46% of her total income.

3. 2(40) + 2(95) = 80 + 190 = 270
4.50 × 270 = \$1215
The fence will cost \$1215.

4. Answers may vary. Possible solution: Multiply the number of kilometers by 0.62 to obtain miles.

Putting Your Skills to Work

1. \$100, \$300, \$2000, \$8000, \$8000, \$8000, \$12,000

2. \$25 + \$25 + \$25 + \$50 + \$200 + \$20 + \$20 = \$365
The minimum monthly payment is \$365.

3. \$2000 − \$400 = \$1600
\$1600 ÷ \$240 = 6.6 ≈ 7
It will take them 7 more months.

Chapter 0 Review Problems

1. $\frac{36}{48}=\frac{12\times 3}{12\times 4}=\frac{3}{4}$

2. $\frac{15}{30}=\frac{5\times 3}{5\times 10}=\frac{3}{10}$

3. $\frac{36}{82}=\frac{18\times 2}{41\times 2}=\frac{18}{41}$

4. $\frac{18}{30}=\frac{6\times 3}{6\times 5}=\frac{3}{5}$

5. $7\frac{1}{8}=\frac{7\times 8+1}{8}=\frac{56+1}{8}=\frac{57}{8}$

6. $\frac{34}{5}=5\overline{)34}=6\frac{4}{5}$ (quotient 6; $34-30=4$)

7. $\frac{80}{3}=3\overline{)80}=26\frac{2}{3}$ (quotient 26; $80-6=20$, $20-18=2$)

8. $\frac{5}{8}=\frac{?}{24}\Rightarrow\frac{5\cdot 3}{8\cdot 3}=\frac{15}{24}$

9. $\frac{1}{7}=\frac{?}{35}\Rightarrow\frac{1\cdot 5}{7\cdot 5}=\frac{5}{35}$

10. $\frac{3}{5}=\frac{?}{75}\Rightarrow\frac{3\times 15}{5\times 15}=\frac{45}{75}$

11. $\frac{2}{5}=\frac{?}{55}\Rightarrow\frac{2\times 11}{5\times 11}=\frac{22}{55}$

12. $\frac{3}{5}+\frac{1}{4}=\frac{3\times 4}{5\times 4}+\frac{1\times 5}{4\times 5}=\frac{12}{20}+\frac{5}{20}=\frac{17}{20}$

13. $\frac{7}{12}+\frac{5}{8}=\frac{7\times 2}{12\times 2}+\frac{5\times 3}{8\times 3}=\frac{14}{24}+\frac{15}{24}=\frac{29}{24}=1\frac{5}{24}$

14. $\frac{7}{20}-\frac{1}{12}=\frac{21}{60}-\frac{5}{60}=\frac{16}{60}=\frac{4}{15}$

15. $\frac{7}{10}-\frac{4}{15}=\frac{7\cdot 3}{10\cdot 3}-\frac{4\cdot 2}{5\cdot 2}=\frac{21}{30}-\frac{8}{30}=\frac{13}{30}$

16. $3\frac{1}{6}+2\frac{3}{5}=\frac{19}{6}+\frac{13}{5}=\frac{95}{30}+\frac{78}{30}=\frac{173}{30}=5\frac{23}{30}$

17. $2\frac{7}{10}+3\frac{3}{4}=\frac{27}{10}+\frac{15}{4}=\frac{54}{20}+\frac{75}{20}=\frac{129}{20}=6\frac{9}{20}$

18.
$$\begin{aligned}6\frac{2}{9}-3\frac{5}{12}&=\frac{56}{9}-\frac{41}{12}\\&=\frac{56\cdot 4}{9\cdot 4}-\frac{41\cdot 3}{12\cdot 3}\\&=\frac{224}{36}-\frac{123}{36}\\&=\frac{101}{36}\\&=2\frac{29}{36}\end{aligned}$$

19.
$$\begin{aligned}3\frac{1}{15}-1\frac{3}{20}&=\frac{46}{15}-\frac{23}{20}\\&=\frac{184}{60}-\frac{69}{60}\\&=\frac{115}{60}\\&=1\frac{55}{60}\\&=1\frac{11}{12}\end{aligned}$$

20. $6\times\frac{5}{11}=\frac{6}{1}\times\frac{5}{11}=\frac{30}{11}=2\frac{8}{11}$

21. $2\frac{1}{3}\times 4\frac{1}{2}=\frac{7}{3}\times\frac{9}{2}=\frac{7}{3}\times\frac{3\cdot 3}{2}=\frac{21}{2}=10\frac{1}{2}$

22. $16\times 3\frac{1}{8}=\frac{16}{1}\times\frac{25}{8}=\frac{8\cdot 2}{1}\times\frac{25}{8}=50$

23. $\frac{4}{7}\times 5=\frac{4}{7}\times\frac{5}{1}=\frac{20}{7}=2\frac{6}{7}$

24. $\frac{3}{8}\div 6=\frac{3}{8}\times\frac{1}{6}=\frac{3}{8}\times\frac{1}{2\cdot 3}=\frac{1}{16}$

25. $\frac{\frac{8}{3}}{\frac{5}{9}}=\frac{8}{3}\div\frac{5}{9}=\frac{8}{3}\times\frac{9}{5}=\frac{8}{3}\times\frac{3\cdot 3}{5}=\frac{24}{5}=4\frac{4}{5}$

26. $\frac{15}{16}\div 6\frac{1}{4}=\frac{15}{16}\div\frac{25}{4}=\frac{15}{16}\times\frac{4}{25}=\frac{3}{4}\times\frac{1}{5}=\frac{3}{20}$

27. $2\frac{6}{7}\div\frac{10}{21}=\frac{20}{7}\times\frac{21}{10}=\frac{2\cdot 10}{7}\times\frac{3\cdot 7}{10}=6$

28.
$$\begin{array}{r} 1.634 \\ 3.007 \\ +\ 2.560 \\ \hline 7.201 \end{array}$$

29.
$$\begin{array}{r} 24.831 \\ -\ 17.094 \\ \hline 7.737 \end{array}$$

30.
$$\begin{array}{r} 47.251 \\ -\ 17.690 \\ \hline 29.561 \end{array}$$

31.
$$\begin{array}{r} 1.900 \\ 2.530 \\ +\ 0.006 \\ \hline 4.436 \end{array}$$

32.
$$\begin{array}{r} 5.35 \\ \times\ 0.007 \\ \hline 0.03745 \end{array}$$

33. $362.341 \times 1000 = 362{,}341$

34. $2.6 \times 0.03 \times 1.02 = 0.07956$

35. $2.51 \times 100 \times 0.05 = 125.5$

36. $0.186 \div 100 = 0.00186$

37. $71.32 \div 1000 = 0.07132$

38. $0.523 \div 0.4 = 1.3075$

39.
$$\begin{array}{r} 90 \\ 0.015\wedge\overline{)1.350\wedge} \\ \underline{1\ 35} \\ 00 \end{array}$$

40.
$$\frac{4.186}{2.3} = \begin{array}{r} 1.82 \\ 2.3\wedge\overline{)4.1\wedge 86} \\ \underline{2\ 3} \\ 1\ 8\ \ 8 \\ \underline{1\ 8\ \ 4} \\ 46 \\ \underline{46} \end{array} = 1.82$$

41.
$$\begin{array}{r} 0.5 \\ 0.38\wedge\overline{)0.19\wedge 0} \\ \underline{19\ \ 0} \\ 0 \end{array}$$

42.
$$\frac{3}{8} = \begin{array}{r} 0.375 \\ 8\overline{)3.000} \\ \underline{2\ 4} \\ 60 \\ \underline{56} \\ 40 \\ \underline{40} \\ 0 \end{array} = 0.375$$

43. $0.36=\frac{36}{100}=\frac{9}{25}$

44. $1.4\% = 0.014$

45. $36.1\% = 0.361$

46. $0.02\% = 0.0002$

47. $125.3\% = 1.253$

48. 30% of 400
$0.30 \times 400 = 120$

49. 7.2% of 55
$0.072 \times 55 = 3.96$

50. 76 is what percent of 80?
$\frac{76}{80}=0.95=95\%$

51. What percent of 120 is 15?
$\frac{15}{120} = 0.125 = 12.5\%$

52. $\frac{750}{1250} = 0.6 = 60\%$

53. 80% of 22,299,041
$0.8 \times 22{,}299{,}041 \approx 17{,}839{,}233$
About 17,839,233 people reported having at least a 4-year high school education.

54. $\frac{720}{960} = 0.75 = 75\%$
75% of the class had a math deficiency.

55. $234{,}897 \times 1{,}936{,}112 \approx 200{,}000 \times 2{,}000{,}000$
$= 400{,}000{,}000{,}000$

56. $357 + 923 + 768 + 417 \approx 400 + 900 + 800 + 400$
$= 2500$

57. $634{,}318 - 284{,}000 \approx 600{,}000 - 300{,}000$
$\approx 300{,}000$

58. $21\frac{1}{5} - 8\frac{4}{5} - 1\frac{2}{3} \approx 21 - 9 - 2 = 10$

59. 18% of 56,297 $\approx 0.2 \times 60{,}000 = \$12{,}000$

60. $12{,}482 \div 389 \approx 10{,}000 \div 400 = 25$

61. $38.5 \times 8.35 \approx 40 \times 8 = \$320$
Her estimated salary is $320.

62. $\frac{3875}{5} \approx \frac{4000}{5} = \$800$
Each family owes $800.

63. Area $= 12\frac{1}{2} \times 9\frac{2}{3}$
$= \frac{25}{2} \times \frac{29}{3}$
$= 120\frac{5}{6}$ sq feet
$= \frac{120\frac{5}{6}}{9}$
$= 13\frac{23}{54}$ sq yards

Cost $= 26.00 \times 13\frac{23}{54} = 26 \times \frac{725}{54} = \$349.07$
The total cost is $349.07.

64. Increase = 4600 − 2300 = 2300
$\frac{2300}{2300} = 1$ or 100%
The percent increase is 100%.

65. Maximum distance $= 7\frac{2}{3} \times 240$
$= \frac{23}{3} \times 240$
$= 1840$ miles
The plane can fly 1840 miles.
Longest trip $= 80\% \times 1840$
$= 0.8 \times 1840$
$= 1472$ miles
The longest trip he would take is 1472 miles.

66. Maximum speed distance $= 6\frac{1}{4} \times 240$
$= \frac{25}{4} \times 240$
$= 1500$ miles
The plane can fly 1500 miles.
$0.70 \times 1500 = 1050$
The longest trip he would take is 1050 miles.

67. Volume $= 4 \times 2\frac{1}{2} \times \frac{3}{4}$
$= \frac{4}{1} \times \frac{5}{2} \times \frac{3}{4}$
$= \frac{15}{2}$
$= 7\frac{1}{2}$ cubic inches

The volume of the matchbox is $7\frac{1}{2}$ in.^{3}

68. Outer volume $= 16\times 12\frac{1}{2}\times 7\frac{1}{2}$
$= \frac{16}{1}\times\frac{25}{2}\times\frac{15}{2}$
$= 1500$ cubic inches

$\frac{1500}{\frac{15}{2}} = \frac{1500}{1}\times\frac{2}{15} = 200$ boxes

The crate can hold 200 boxes.

69. 8% of 5785
$0.08 \times 5785 = \$462.80$
His commission was $462.80.

70. $\frac{34.5}{0.75} = 46$ packages
He took 46 packages to the post office.

71. Amount paid = 225 × 4 × 12 = $10,800
Interest = 10,800 − 9214.50 = $1585.50
They paid $1585.50 more than the car loan.

72. Regular pay = 7.50 × 40 = 300
O/T hours = 49 − 40 = 9
O/T pay rate = 1.5 × 7.5 = 11.25
O/T pay = 11.25 × 9 = 101.25
Total pay = 300 + 101.25 = $401.25
He was paid a total of $401.25.

How Am I Doing? Chapter 0 Test

1. $\frac{16}{18} = \frac{2\times 8}{2\times 9} = \frac{8}{9}$

2. $\frac{48}{36} = \frac{4\times 12}{3\times 12} = \frac{4}{3}$

3. $6\frac{3}{7} = \frac{6\times 7+3}{7} = \frac{45}{7}$

4. $\frac{105}{9} = 9\overline{)105}$ with quotient 11: $\; 105 - 9 = 15,\; 15 - 9 = 6$; $= 11\frac{6}{9} = 11\frac{2}{3}$

5. $\frac{2}{3}+\frac{5}{6}+\frac{3}{8} = \frac{2\cdot 8}{3\cdot 8}+\frac{5\cdot 4}{6\cdot 4}+\frac{3\cdot 3}{8\cdot 3}$
$= \frac{16}{24}+\frac{20}{24}+\frac{9}{24}$
$= \frac{45}{24}$
$= \frac{15}{8}$
$= 1\frac{7}{8}$

6. $1\frac{1}{8}+3\frac{3}{4} = \frac{9}{8}+\frac{15}{4} = \frac{9}{8}+\frac{30}{8} = \frac{9+30}{8} = \frac{39}{8} = 4\frac{7}{8}$

7. $3\frac{2}{3}-2\frac{5}{6} = \frac{11}{3}-\frac{17}{6} = \frac{22}{6}-\frac{17}{6} = \frac{5}{6}$

8. $\frac{5}{7}\times\frac{28}{15} = \frac{1}{1}\times\frac{4}{3} = \frac{4}{3}$ or $1\frac{1}{3}$

9. $\frac{5}{18}\times\frac{3}{4} = \frac{5\times 3}{6\times 3\times 4} = \frac{5}{24}$

10. $\frac{7}{4}\div\frac{1}{2} = \frac{7}{4}\times\frac{2}{1} = \frac{7}{2}\times\frac{1}{1} = \frac{7}{2}$ or $3\frac{1}{2}$

11. $2\frac{1}{2}\times 3\frac{1}{4} = \frac{5}{2}\times\frac{13}{4} = \frac{65}{8} = 8\frac{1}{8}$

12. $5\frac{3}{8}\div 2\frac{3}{4} = \frac{43}{8}\div\frac{11}{4}$
$= \frac{43}{2\cdot 4}\times\frac{4}{11}$
$= \frac{43}{22}$
$= 1\frac{21}{22}$

13.
$$\begin{array}{r} 1.60 \\ 3.24 \\ +\ 9.80 \\ \hline 14.64 \end{array}$$

14.
$$\begin{array}{r} 7.0046 \\ -\ 3.0149 \\ \hline 3.9897 \end{array}$$

15.
$$\begin{array}{r} 32.8 \\ \times\ 0.04 \\ \hline 1.312 \end{array}$$

16. $0.07385 \times 1000 = 73.85$

17.
$$\begin{array}{r} 230 \\ 0.056_{\wedge})\overline{12.880_{\wedge}} \\ \underline{11\,2} \\ 1\,68 \\ \underline{1\,68} \\ 0 \end{array}$$

18. $26{,}325.9 \div 100 = 263.259$

19. $0.073 = 7.3\%$

20. $196.5\% = 1.965$

21. What is 3.5% of 180?
$0.035 \times 180 = 6.3$

22. What is 2% of 16.8?
$(0.02)(16.8) = 0.336$

23. 39 is what percent of 650?
$\frac{39}{650} = 0.06 = 6\%$

24. $\frac{138}{460} = 0.3 = 30\%$

25. $4 \div \frac{2}{9} = \frac{4}{1} \times \frac{9}{2} = \frac{2}{1} \times \frac{9}{1} = 18$
18 chips are in the stack.

26. $52{,}344\overline{)4{,}678{,}987} \approx 50000\overset{100}{\overline{)5000000}}$

27.
$$\begin{aligned} 285.36 + 311.85 + 113.6 &\approx 300 + 300 + 100 \\ &\approx 700 \end{aligned}$$

28. Commission: $(0.03)(870{,}000) = 26{,}100$
Total income: $26{,}100 + 14{,}000 = 40{,}100$
Percent: $\frac{26{,}100}{40{,}100} \approx 0.65 = 65\%$
Her commission was 65% of her total income.

29.
$$\begin{aligned} 210 \div 3\frac{1}{2} &= \frac{210}{1} \div \frac{7}{2} \\ &= \frac{210}{1} \times \frac{2}{7} \\ &= \frac{30 \times 7}{1} \times \frac{2}{7} \\ &= 60 \text{ tiles} \end{aligned}$$
They will need 60 tiles.

Chapter 1

1.1 Exercises

	Number	Int	RaN	IR	ReN
1.	23	X	X		X
3.	π			X	X
5.	–6.666...		X		X
7.	–2.3434...		X		X
9.	$\sqrt{2}$			X	X

11. 20,000 leagues under the sea $\Rightarrow -20{,}000$

13. Lost $37\frac{1}{2} \Rightarrow -37\frac{1}{2}$

15. Rise 7° $\Rightarrow +7$

17. Additive inverse of 8 is –8.

19. Opposite of –2.73 is 2.73.

21. $|-1.3| = 1.3$

23. $\left|\frac{5}{6}\right| = \frac{5}{6}$

25. $-6 + (-5) = -11$

27. $-20 + (-30) = -50$

29. $-\frac{7}{20} + \frac{13}{20} = \frac{6}{20} = \frac{3}{10}$

31. $\left(-\frac{2}{13}\right) + \left(-\frac{5}{13}\right) = -\frac{7}{13}$

33. $-\frac{2}{5} + \frac{3}{7} = -\frac{14}{35} + \frac{15}{35} = \frac{1}{35}$

35. $-10.3 + (-8.9) = -19.2$

37. $(+0.6) + (-0.2) = 0.4$

39. $-5.26 + (-8.9) = -14.16$

41. $-8 + 5 + (-3) = -3 + (-3) = -6$

43. $-2 + (-8) + 10 = -10 + 10 = 0$

45. $-\frac{3}{10} + \frac{3}{4} = -\frac{6}{20} + \frac{15}{20} = \frac{9}{20}$

47. $-14 + 9 + (-3) = -5 + (-3) = -8$

49. $8 + (-11) = -3$

51. $-83 + 142 = 59$

53. $-\frac{4}{9} + \frac{5}{6} = -\frac{8}{18} + \frac{15}{18} = \frac{7}{18}$

55. $-\frac{1}{10} + \frac{1}{2} = -\frac{1}{10} + \frac{5}{10} = \frac{4}{10} = \frac{2}{5}$

57. $5.18 + (-7.39) = -2.21$

59. $4 + (-8) + 16 = -4 + 16 = 12$

61.
$$\begin{aligned} 34 + (-18) + 11 + (-27) &= 16 + 11 + (-27) \\ &= 27 + (-27) \\ &= 0 \end{aligned}$$

63. $17.85 + (-2.06) + 0.15 = 15.79 + 0.15 = 15.94$

65. Profit = 214 – 47 = \$167
Her profit was \$167.

67. $-2300 + (-1500) = -3800$
He owed –\$3800.

69. $-15 + 3 + 21 = -12 + 21 = 9$
His total was a 9 yard gain.

71.
$$\begin{aligned} 8000 + (-3000) + (-1500) &= 5000 + (-1500) \\ &= 3500 \end{aligned}$$
The new population was 3500.

73. $30 + 14 + (-12) = 32$
Her total earnings were \$32,000,000.

75.
$$\begin{aligned} -13 + ? &= 5 \\ ? &= 18 \end{aligned}$$

Cumulative Review

77. $\frac{15}{16} + \frac{1}{4} = \frac{15}{16} + \frac{4}{16} = \frac{19}{16}$ or $1\frac{3}{16}$

78. $\left(\frac{3}{7}\right)\left(\frac{14}{9}\right) = \frac{3 \cdot 14}{7 \cdot 9} = \frac{3 \cdot 7 \cdot 2}{7 \cdot 3 \cdot 3} = \frac{2}{3}$

79. $\frac{2}{15}-\frac{1}{20}=\frac{8}{60}-\frac{3}{60}=\frac{5}{60}=\frac{1}{12}$

80. $2\frac{1}{2}\div 3\frac{2}{5}=\frac{5}{2}\div\frac{17}{5}=\frac{5}{2}\cdot\frac{5}{17}=\frac{25}{34}$

81. $0.72+0.8=1.52$

82. $1.63-0.98=0.65$

83. $\begin{array}{r} 1.63 \\ \times\ 0.7 \\ \hline 1.141 \end{array}$

84. $0.208\div 0.8=0.26$

Quick Quiz 1.1

1. $-18+(-16)=-34$

2. $-2.7+8.6+(-5.4)=5.9+(-5.4)=0.5$

3. $-\frac{5}{6}+\frac{7}{24}=-\frac{20}{24}+\frac{7}{24}=-\frac{13}{24}$

4. Answers may vary. Possible solution: Adding a negative number to a negative number will always result in a number further in the negative direction. However, adding numbers of opposite sign could result in a negative number if the absolute value of the negative number is larger than that of the positive number, or a positive number, if the absolute value of the positive number is greater than that of the negative number, or 0 if their absolute values are equal.

1.2 Exercises

1. First change subtracting −3 to adding a positive three. Then use the rules for addition of two real numbers with different signs. Thus, $-8-(-3)=-8+3=-5$

3. $18-35=18+(-35)=-17$

5. $19-23=19+(-23)=-4$

7. $-14-(-3)=-14+3=-11$

9. $-52-(-60)=-52+60=8$

11. $0-(-5)=0+5=5$

13. $-18-(-18)=-18+18=0$

15. $-17-(-20)=-17+20=3$

17. $\frac{2}{5}-\frac{4}{5}=\frac{2}{5}+\left(-\frac{4}{5}\right)=-\frac{2}{5}$

19. $\left(+\frac{3}{4}\right)-\left(-\frac{3}{5}\right)=\left(+\frac{15}{20}\right)+\left(+\frac{12}{20}\right)=\frac{27}{20}=1\frac{7}{20}$

21. $\left(-\frac{3}{4}\right)-\left(+\frac{5}{6}\right)=\left(-\frac{9}{12}\right)+\left(-\frac{10}{12}\right)=-\frac{19}{12}=-1\frac{7}{12}$

23. $(-0.6)-(+0.3)=(-0.6)+(-0.3)=-0.9$

25. $(+2.64)-(-1.83)=(+2.64)+(+1.83)=4.47$

27. $\frac{3}{5}-4=\frac{3}{5}+\left(-\frac{20}{5}\right)=-\frac{17}{5}=-3\frac{2}{5}$

29. $-\frac{2}{3}-4=-\frac{2}{3}+\left(-\frac{12}{3}\right)=-\frac{14}{3}$ or $-4\frac{2}{3}$

31. $34-87=34+(-87)=-53$

33. $-25-48=-25+(-48)=-73$

35. $2.3-(-4.8)=2.3+4.8=7.1$

37. $8-\left(-\frac{3}{4}\right)=8+\frac{3}{4}=8\frac{3}{4}$ or $\frac{35}{4}$

39. $\frac{5}{6}-7=\frac{5}{6}+\left(-\frac{42}{6}\right)=-\frac{37}{6}=-6\frac{1}{6}$

41. $-\frac{2}{7}-\frac{4}{5}=-\frac{10}{35}+\left(-\frac{28}{35}\right)=-\frac{38}{35}$ or $-1\frac{3}{35}$

43. $-135-(-126.5)=-135+126.5=-8.5$

45. $\frac{1}{5}-6=\frac{1}{5}+\left(-\frac{30}{5}\right)=-\frac{29}{5}$ or $-5\frac{4}{5}$

47. $4.5-(-1.56)=4.5+1.56=6.06$

49. $-3-2.047=-3+(-2.047)=-5.047$

51. $-2-(-9)=-2+9=7$

53. $-35-13=-35+(-13)=-48$

55. $7+(-6)-3=7+(-6)+(-3)=1+(-3)=-2$

57. $-13+12-(-1)=-13+12+1=-1+1=0$

59. $7+(-42)-27 = 7+(-42)+(-27)$
$= -35+(-27)$
$= -62$

61. $-6.4-(-2.7)+5.3 = -6.4+2.7+5.3$
$= -3.7+5.3$
$= 1.6$

63. $112-(-37) = 112+37 = 149$
The difference in the two amounts is $149.

Cumulative Review

65. $-37+16 = -21$

66. $-37+(-14) = -51$

67. $-3+(-6)+(-10) = -9+(-10) = -19$

68. $-5+20 = 15$
The afternoon temperature was 15°F.

69. $\left(\frac{4}{5}\right)\left(8\frac{1}{3}\right) = \left(\frac{4}{5}\right)\left(\frac{25}{3}\right) = \frac{20}{3} = 6\frac{2}{3}$

There were $6\frac{2}{3}$ miles covered in snow.

Quick Quiz 1.2

1. $-8-(-15) = -8+15 = 7$

2. $-1.3-0.6 = -1.3+(-0.6) = -1.9$

3. $\frac{5}{8}-\left(-\frac{2}{7}\right) = \frac{5}{8}+\frac{2}{7} = \frac{35}{56}+\frac{16}{56} = \frac{51}{56}$

4. Answers may vary. Possible solution:
The result could be positive, if the absolute value of the number subtracted is greater than the other number [−2 − (−3) = −2 + 3 = 1] or the result could be zero if the two numbers are the same [−2 − (−2) = −2 + 2 = 0], or the result could be negative if the absolute value of the number being subtracted is less than the other number [−2 − (−1) = −2 + 1 = −1].

1.3 Exercises

1. To multiply two real numbers, multiply the absolute values. The sign of the result is positive if both numbers have the same sign, but negative if the two numbers have opposite signs.

3. $5(-4) = -20$

5. $0(-12) = 0$

7. $14(3.5) = 49$

9. $(-1.32)(-0.2) = 0.264$

11. $1.8(-2.5) = -4.5$

13. $\frac{3}{8}(-4) = \frac{3}{8}\left(-\frac{4}{1}\right) = -\frac{3}{2} = -1\frac{1}{2}$

15. $\left(-\frac{3}{5}\right)\left(-\frac{15}{11}\right) = \frac{9}{11}$

17. $\left(\frac{12}{13}\right)\left(\frac{-5}{24}\right) = -\frac{5}{26}$

19. $(-36)\div(-9) = 4$

21. $-48\div(-8) = \frac{-48}{-8} = 6$

23. $-120\div(-8) = 15$

25. $156\div(-13) = -12$

27. $-9.1\div 0.07 = -\frac{9.1}{0.07} = -130$

29. $0.54\div(-0.9) = \frac{0.54}{-0.9} = -0.6$

31. $-6.3\div 7 = \frac{-6.3}{7} = -0.9$

33. $\left(-\frac{1}{5}\right)\div\left(\frac{2}{3}\right) = \left(-\frac{1}{5}\right)\left(\frac{3}{2}\right) = -\frac{3}{10}$

35. $\left(-\frac{5}{7}\right)\div\left(-\frac{3}{28}\right) = \frac{5}{7}\cdot\left(\frac{28}{3}\right) = \frac{20}{3} = 6\frac{2}{3}$

37. $-\frac{7}{12}\div\left(-\frac{5}{6}\right) = \left(\frac{7}{12}\right)\left(\frac{6}{5}\right) = \frac{7}{10}$

39. $\frac{-6}{-\frac{3}{7}} = \left(-\frac{6}{1}\right)\left(-\frac{7}{3}\right) = 14$

41. $\frac{\frac{-2}{3}}{\frac{8}{15}} = \left(-\frac{2}{3}\right)\left(\frac{15}{8}\right) = -\frac{5}{4} = -1\frac{1}{4}$

43. $\dfrac{\frac{8}{3}}{-4} = \dfrac{8}{3} \cdot \left(-\dfrac{1}{4}\right) = -\dfrac{8}{12} = -\dfrac{2}{3}$

45. $(-1)(-2)(-3)(4) = -(1)(2)(3)(4) = -24$

47. $-2(4)(3)(-1)(-3) = -72$

49. $(-3)(-2)\left(\dfrac{1}{3}\right)(-4)(2) = -(3)(2)\left(\dfrac{1}{3}\right)(4)(2) = -16$

51. $25(-0.04)(-0.3)(-1) = -1(-0.3)(-1)$
$= 0.3(-1)$
$= -0.3$

53. $\left(-\dfrac{4}{5}\right)\left(-\dfrac{6}{7}\right)\left(-\dfrac{1}{3}\right) = -\left(\dfrac{24}{35}\right)\left(\dfrac{1}{3}\right) = -\dfrac{8}{35}$

55. $\left(-\dfrac{3}{4}\right)\left(-\dfrac{7}{15}\right)\left(-\dfrac{8}{21}\right)\left(-\dfrac{5}{9}\right)$
$= +\left(\dfrac{3}{4}\right)\left(\dfrac{7}{3\cdot 5}\right)\left(\dfrac{4\cdot 2}{3\cdot 7}\right)\left(\dfrac{5}{9}\right)$
$= \dfrac{2}{27}$

57. $-36 \div (-4) = \dfrac{-36}{-4} = 9$

59. $12 + (-8) = 4$

61. $8 - (-9) = 8 + 9 = 17$

63. $6(-12) = -72$

65. $-37 \div 37 = \dfrac{-37}{37} = -1$

67. $17.60 \div 4 = \dfrac{17.60}{4} = 4.40$
He gave \$4.40 to each person and to himself.

69. $\dfrac{14{,}136}{60} = 235.60$
Her monthly bill is \$235.60.

71. $5(4) = +20$, gained 20 yards in small gains.

73. $-10(7) = -70$, lost 70 yards in medium losses.

75. Total $= -70 + 90 = 20$, gained 20 yards

Cumulative Review

76. $(-17.4) + (8.31) + (2.40) = -9.09 + 2.40 = -6.69$

77. $-\dfrac{3}{4} + \left(-\dfrac{2}{3}\right) + \left(-\dfrac{5}{12}\right) = -\dfrac{9}{12} + \left(-\dfrac{8}{12}\right) + \left(-\dfrac{5}{12}\right)$
$= -\dfrac{17}{12} + \left(-\dfrac{5}{12}\right)$
$= -\dfrac{22}{12}$
$= -\dfrac{11}{6}$
$= -1\dfrac{5}{6}$

78. $(-47) - (-32) = (-47) + 32 = -15$

79. $(-37) - (51) = (-37) + (-51) = -88$

Quick Quiz 1.3

1. $\left(-\dfrac{3}{8}\right)(5) = -\dfrac{3}{8} \cdot \dfrac{5}{1} = -\dfrac{15}{8}$ or $-1\dfrac{7}{8}$

2. $-4(3)(-5)(-2) = -12(-5)(-2) = 60(-2) = -120$

3. $-2.4 \div -0.7 = \dfrac{-2.4}{-0.7} = 4$

4. Answers may vary. Possible solution: Multiplying an even number of negative numbers results in a positive number, whereas multiplying an odd number of negative numbers results in a negative number.

1.4 Exercises

1. The base is 4 and the exponent is 4. Thus you multiply $(4)(4)(4)(4) = 256$.

3. The answer is negative. When you raise a negative number to an odd power the result is always negative.

5. If you have parentheses surrounding the -2, then the base is -2 and the exponent is 4. The result is 16. If you do not have parentheses, then the base is 2. You evaluate to obtain 16 and then take the opposite of 16, which is -16. Thus, $(-2)^4 = 16$, but $-2^4 = -16$.

7. $(5)(5)(5)(5)(5)(5)(5) = 5^7$

9. $(w)(w) = w^2$

11. $(p)(p)(p)(p) = p^4$

13. $(3q)(3q)(3q) = (3q)^3$ or 3^3q^3

15. $3^3 = 27$

17. $3^4 = 81$

19. $6^3 = 216$

21. $(-3)^3 = (-3)(-3)(-3) = -27$

23. $(-4)^2 = (-4)(-4) = 16$

25. $-5^2 = -(5)(5) = -25$

27. $\left(\frac{1}{4}\right)^2 = \left(\frac{1}{4}\right)\left(\frac{1}{4}\right) = \frac{1}{16}$

29. $\left(\frac{2}{5}\right)^3 = \left(\frac{2}{5}\right)\left(\frac{2}{5}\right)\left(\frac{2}{5}\right) = \frac{8}{125}$

31. $(2.1)^2 = (2.1)(2.1) = 4.41$

33. $(0.2)^4 = (0.2)(0.2)(0.2)(0.2) = 0.0016$

35. $(-16)^2 = (-16)(-16) = 256$

37. $-16^2 = -(16)(16) = -256$

39. $5^3 + 6^2 = 125 + 36 = 161$

41. $10^2 - 11^2 = 100 - 121 = -21$

43. $(-4)^2 - (12)^2 = 16 - 144 = -128$

45. $2^5 - (-3)^2 = 32 - 9 = 23$

47. $(-4)^3(-3)^2 = (-64)(9) = -576$

Cumulative Review

48. $$\begin{aligned}(-11)+(-13)+6+(-9)+8 &= -24+6+(-9)+8\\ &= -18+(-9)+8\\ &= -27+8\\ &= -19\end{aligned}$$

49. $$\begin{aligned}\frac{3}{4} \div \left(-\frac{9}{20}\right) &= \left(\frac{3}{4}\right)\left(-\frac{20}{9}\right)\\ &= \left(\frac{3}{4}\right)\left(-\frac{4 \cdot 5}{3 \cdot 3}\right)\\ &= -\frac{5}{3}\\ &= -1\frac{2}{3}\end{aligned}$$

50. $-17 - (-9) = -17 + 9 = -8$

51. $(-2.1)(-1.2) = 2.52$

Quick Quiz 1.4

1. $(-4)^4 = (-4)(-4)(-4)(-4) = 256$

2. $(1.8)^2 = (1.8)(1.8) = 3.24$

3. $\left(\frac{3}{4}\right)^3 = \left(\frac{3}{4}\right)\left(\frac{3}{4}\right)\left(\frac{3}{4}\right) = \frac{27}{64}$

4. Answers may vary. Possible solution:
If you have parentheses surrounding the –2, then the base is –2 and the exponent is 6. The result is 64. If you do not have parentheses, then the base is 2. You evaluate to obtain 64 and then take the opposite of 64, which is –64. Thus, $(-2)^6 = 64$ and $-2^6 = -64$.

1.5 Exercises

1. 3(4) + 6(5)

3. a. 3(4) + 6(5) = 12 + 6(5) = 18(5) = 90

b. 3(4) + 6(5) = 12 + 30 = 42

5. $$\begin{aligned}(2-5)^2 \div 3 \times 4 &= (-3)^2 \div 3 \times 4\\ &= 9 \div 3 \times 4\\ &= 3 \times 4\\ &= 12\end{aligned}$$

7. $9 + 4(5+2-8) = 9 + 4(-1) = 9 + (-4) = 5$

9. $8-2^3\cdot 5+3=8-8\cdot 5+3$
$=8-8\cdot 5+3$
$=8-40+3$
$=-29$

11. $4+42\div 3\cdot 2-8=4+14\cdot 2-8$
$=4+28-8$
$=32-8$
$=24$

13. $3\cdot 5+7\cdot 3-5\cdot 3=15+21-15=21$

15. $8-5(2)^3\div(-8)=8-5(8)\div(-8)$
$=8-40\div(-8)$
$=8+5$
$=13$

17. $3(5-7)^2-6(3)=3(-2)^2-6(3)$
$=3(4)-6(3)$
$=12-18$
$=-6$

19. $5\cdot 6-(3-5)^2+8\cdot 2=5\cdot 6-(-2)^2+8\cdot 2$
$=5\cdot 6-4+8\cdot 2$
$=30-4+16$
$=42$

21. $\frac{1}{2}\div\frac{2}{3}+6\cdot\frac{1}{4}=\frac{1}{2}\cdot\frac{3}{2}+\frac{6}{1}\cdot\frac{1}{4}$
$=\frac{3}{4}+\frac{6}{4}$
$=\frac{9}{4}$
$=2\frac{1}{4}$

23. $0.8+0.3(0.6-0.2)^2=0.8+0.3(0.4)^2$
$=0.8+0.3(0.16)$
$=0.8+0.048$
$=0.848$

25. $\frac{3}{8}\left(-\frac{1}{6}\right)-\frac{7}{8}+\frac{1}{2}=-\frac{3}{48}-\frac{7}{8}+\frac{1}{2}$
$=-\frac{1}{16}-\frac{14}{16}+\frac{8}{16}$
$=-\frac{15}{16}+\frac{8}{16}$
$=-\frac{7}{16}$

27. $(3-7)^2\div 8+3=(-4)^2\div 8+3$
$=16\div 8+3$
$=2+3$
$=5$

29. $\left(\frac{3}{4}\right)^2(-16)+\frac{4}{5}\div\frac{-8}{25}=\frac{9}{16}(-16)+\frac{4}{5}\div\frac{-8}{25}$
$=-9+\frac{4}{5}\div\frac{-8}{25}$
$=-9+\frac{4}{5}\left(\frac{25}{-8}\right)$
$=-9+\left(\frac{5}{-2}\right)$
$=-\frac{18}{2}+\left(-\frac{5}{2}\right)$
$=-\frac{23}{2}$
$=-11\frac{1}{2}$

31. $-6.3-(-2.7)(1.1)+(3.3)^2$
$=-6.3-(-2.7)(1.1)+10.89$
$=-6.3+2.97+10.89$
$=7.56$

33. $\left(\frac{1}{2}\right)^3+\left(\frac{1}{4}\right)-\left(\frac{1}{6}-\frac{1}{12}\right)-\frac{2}{3}\cdot\left(\frac{1}{4}\right)^2$
$=\left(\frac{1}{2}\right)^3+\frac{1}{4}-\left(\frac{1}{12}\right)-\frac{2}{3}\cdot\left(\frac{1}{4}\right)^2$
$=\frac{1}{8}+\frac{1}{4}-\frac{1}{12}-\frac{2}{3}\cdot\frac{1}{16}$
$=\frac{1}{8}+\frac{1}{4}-\frac{1}{12}-\frac{1}{24}$
$=\frac{6}{24}$
$=\frac{1}{4}$

35. $2(-2)+4(-1)+10(0)+2(+1)$

37. $2(-2)+4(-1)+10(0)+2(+1)$
$=-4+4(-1)+10(0)+2(+1)$
$=0(-1)+10(0)+2(+1)$
$=0+10(0)+2(+1)$
$=10(0)+2(+1)$
$=0+2(+1)$
$=2(+1)$
$=+2$ or 2 above par

43. $4x(3000-2y) = 4x(3000)+4x(-2y)$
$= 12{,}000x - 8xy$

The area is $(12{,}000x-8xy)\text{ ft}^2$.

Cumulative Review

44. $-18+(-20)+36+(-14) = -38+36+(-14)$
$= -2+(-14)$
$= -16$

45. $(-2)^6 = (-2)(-2)(-2)(-2)(-2)(-2) = 64$

46. $-27-(-41) = -27+41 = 14$

47. $25 \div 5(2)+(-6) = 5(2)+(-6) = 10+(-6) = 4$

48. $(12-10)^2+(-3)(-2) = (2)^2+(-3)(-2)$
$= 4+(-3)(-2)$
$= 4+6$
$= 10$

Quick Quiz 1.6

1. $5(-3a-7b) = 5(-3a)+5(-7b) = -15a-35b$

2. $-2x(x-4y+8)$
$= -2x(x)+(-2x)(-4y)+(-2x)(8)$
$= -2x^2+8xy-16x$

3. $-3ab(4a-5b-9)$
$= -3ab(4a)+(-3ab)(-5b)+(-3ab)(-9)$
$= -12a^2b+15ab^2+27ab$

4. Answers may vary. Possible solution:
Distribute $\left(-\frac{3}{7}\right)$ to each term within the parentheses.
$\left(-\frac{3}{7}\right)(21x^2-14x+3)$
$= \left(-\frac{3}{7}\right)(21x^2)+\left(-\frac{3}{7}\right)(-14x)+\left(-\frac{3}{7}\right)(3)$
$= -9x^2+6x-\frac{9}{7}$

1.7 Exercises

1. A term is a number, a variable, or a product of numbers and variables.

3. The two terms $5x$ and $-8x$ are like terms because they both have the variable x with the exponent of one.

5. The only like terms are $7xy$ and $-14xy$ because the other two have different exponents even though they have the same variables.

7. $-14b^2-11b^2 = (-14-11)b^2 = -25b^2$

9. $5a^3-7a^2+a^3 = 5a^3+1a^3-7a^2$
$= (5+1)a^3-7a^2$
$= 6a^3-7a^2$

11. $3x+2y-8x-7y = (3-8)x+(2-7)y = -5x-5y$

13. $1.3x-2.6y+5.8x-0.9y$
$= (1.3+5.8)x+(-2.6-0.9)y$
$= 7.1x-3.5y$

15. $1.6x-2.8y-3.6x-5.9y$
$= (1.6-3.6)x+(-2.8-5.9)y$
$= -2x-8.7y$

17. $3p-4q+2p+3+5q-21$
$= (3+2)p+(-4+5)q+3-21$
$= 5p+q-18$

19. $2ab+5bc-6ac-2ab = (2-2)ab+5bc-6ac$
$= 5bc-6ac$

21. $2x^2-3x-5-7x+8-x^2$
$= (2-1)x^2+(-3-7)x-5+8$
$= x^2-10x+3$

23. $2y^2-8y+9-12y^2-8y+3$
$= (2-12)y^2+(-8-8)y+9+3$
$= -10y^2-16y+12$

25. $\frac{1}{3}x-\frac{2}{3}y-\frac{2}{5}x+\frac{4}{7}y$
$= \left(\frac{1}{3}-\frac{2}{5}\right)x+\left(-\frac{2}{3}+\frac{4}{7}\right)y$
$= \left(\frac{5}{15}-\frac{6}{15}\right)x+\left(-\frac{14}{21}+\frac{12}{21}\right)y$
$= -\frac{1}{15}x-\frac{2}{21}y$

27. $\frac{3}{4}a^2 - \frac{1}{3}b - \frac{1}{5}a^2 - \frac{1}{2}b$
$= \left(\frac{3}{4} - \frac{1}{5}\right)a^2 + \left(-\frac{1}{3} - \frac{1}{2}\right)b$
$= \left(\frac{25}{20} - \frac{4}{20}\right)a^2 + \left(-\frac{2}{6} - \frac{3}{6}\right)b$
$= -\frac{11}{20}a^2 - \frac{5}{6}b$

29. $3rs - 8r + s - 5rs + 10r - s$
$= (3-5)rs + (-8+10)r + (1-1)s$
$= -2rs + 2r$

31. $5(2a-b) - 3(5b-6a) = 10a - 5b - 15b + 18a$
$= 28a - 20b$

33. $-3b(5a-3b) + 4(-3ab - 5b^2)$
$= -15ab + 9b^2 - 12ab - 20b^2$
$= -27ab - 11b^2$

35. $6(c - 2d^2) - 2(4c - d^2) = 6c - 12d^2 - 8c + 2d^2$
$= (6-8)c + (-12+2)d^2$
$= -2c - 10d^2$

37. $4(2-x) - 3(-5-12x) = 8 - 4x + 15 + 36x$
$= 32x + 23$

39. $3a + 2b + 4a + 7b = 7a + 9b$
He needs $(7a + 9b)$ to enclose the pool.

41. $2(5x-10) + 2(2x+6) = 10x - 20 + 4x + 12$
$= 14x - 8$
The perimeter is $(14x - 8)$ feet.

Cumulative Review

43. $-\frac{3}{4} - \frac{1}{3} = -\frac{9}{12} - \frac{4}{12} = -\frac{13}{12}$ or $-1\frac{1}{12}$

44. $\left(\frac{2}{3}\right)\left(-\frac{9}{16}\right) = -\frac{18}{48} = -\frac{3 \cdot 6}{8 \cdot 6} = -\frac{3}{8}$

45. $\frac{4}{5} + \left(-\frac{1}{25}\right) + \left(-\frac{3}{10}\right) = \frac{40}{50} + \left(-\frac{2}{50}\right) + \left(-\frac{15}{50}\right)$
$= \frac{23}{50}$

46. $\left(\frac{5}{7}\right) \div \left(-\frac{14}{3}\right) = \left(\frac{5}{7}\right)\left(-\frac{3}{14}\right) = -\frac{15}{98}$

Quick Quiz 1.7

1. $3xy - \frac{2}{3}x^2y - \frac{5}{6}xy + \frac{7}{3}x^2y$
$= \left(3 - \frac{5}{6}\right)xy + \left(-\frac{2}{3} + \frac{7}{3}\right)x^2y$
$= \frac{13}{6}xy + \frac{5}{3}x^2y$

2. $8.2a^2b + 5.5ab^2 - 7.6a^2b - 9.9ab^2$
$= (8.2 - 7.6)a^2b + (5.5 - 9.9)ab^2$
$= 0.6a^2b - 4.4ab^2$

3. $2(3x - 5y) - 2(-7x - 4y) = 6x - 10y + 14x + 8y$
$= (6+14)x + (-10+8)y$
$= 20x - 2y$

4. Answers may vary. Possible solution:
Use the distributive property to remove the parentheses. Then simplify by combining like terms.
$1.2(3.5x - 2.2y) - 4.5(2.0x + 1.5y)$
$= 4x - 2.64y - 9x - 6.75y$
$= -4.8x - 9.39y$

1.8 Exercises

1. If $x = 3$, then $-2x + 1 = -2(3) + 1 = -6 + 1 = -5$.

3. If $y = -10$, then
$\frac{2}{5}y - 8 = \frac{2}{5}(-10) - 8 = -4 - 8 = -12.$

5. If $x = \frac{1}{2}$, then
$5x + 10 = 5\left(\frac{1}{2}\right) + 10 = \frac{5}{2} + 10 = \frac{25}{2} = 12\frac{1}{2}.$

7. If $x = 7$, then $2 - 4x = 2 - 4(7) = 2 - 28 = -26$.

9. If $x = 2.4$, then
$3.5 - 2x = 3.5 - 2(2.4) = 3.5 - 4.8 = -1.3$.

11. If $x = -\frac{3}{4}$, then
$9x + 13 = 9\left(-\frac{3}{4}\right) + 13 = -\frac{27}{4} + \frac{52}{4} = \frac{25}{4} = 6\frac{1}{4}.$

13. If $x = -2$, then
$x^2 - 3x = (-2)^2 - 3(-2) = 4 + 6 = 10.$

15. If $y = -1$, then $5y^2 = 5(-1)^2 = 5(1) = 5$.

17. If $x = 2$, then $-3x^3 = -3(2)^3 = -3(8) = -24$.

19. If $x = -2$, then $-5x^2 = -5(-2)^2 = -5(4) = -20$.

21. If $x = -3$, then
$$\begin{aligned} 2x^2 + 3x &= 2(-3)^2 + 3(-3) \\ &= 2(9) - 9 \\ &= 18 - 9 \\ &= 9. \end{aligned}$$

23. If $x = 3$, then
$(2x)^2 + x = [2(3)]^2 + 3 = [6]^2 + 3 = 36 + 3 = 39.$

25. If $x = -2$, then
$2 - (-x)^2 = 2 - [-(-2)]^2 = 2 - (2)^2 = 2 - 4 = -2.$

27. If $a = -2$, then
$$\begin{aligned} 10a + (4a)^2 &= 10(-2) + [4(-2)]^2 \\ &= 10(-2) + (-8)^2 \\ &= -10 + 64 \\ &= 44. \end{aligned}$$

29. If $x = \frac{1}{2}$, then
$$\begin{aligned} 4x^2 - 6x &= 4\left(\frac{1}{2}\right)^2 - 6\left(\frac{1}{2}\right) \\ &= 4\left(\frac{1}{4}\right) - 6\left(\frac{1}{2}\right) \\ &= 1 - 3 \\ &= -2. \end{aligned}$$

31. If $x = 3$, then
$x^2 - 7x + 3 = (3)^2 - 7(3) + 3 = 9 - 21 + 3 = -9.$

33. If $x = -4$, then
$$\begin{aligned} \frac{1}{2}x^2 - 3x + 9 &= \frac{1}{2}(-4)^2 - 3(-4) + 9 \\ &= \frac{1}{2}(16) + 12 + 9 \\ &= 8 + 12 + 9 \\ &= 29. \end{aligned}$$

35. If $r = -1$ and $s = 3$, then
$$\begin{aligned} 2r^2 + 3s^2 - rs &= 2(-1)^2 + 3(3)^2 - (-1)(3) \\ &= 2(1) + 3(9) - (-3) \\ &= 2 + 27 + 3 \\ &= 32. \end{aligned}$$

37. If $a = 5$, $b = 9$, and $c = -1$, then
$$\begin{aligned} a^3 + 2abc - 3c^2 &= 5^3 + 2(5)(9)(-1) - 3(-1)^2 \\ &= 125 - 90 - 3(1) \\ &= 125 - 90 - 3 \\ &= 32. \end{aligned}$$

39. If $a = -1$ and $b = -2$, then
$$\frac{a^2 + ab}{3b} = \frac{(-1)^2 + (-1)(-2)}{3(-2)} = \frac{1+2}{-6} = -\frac{1}{2}.$$

41. $A = ab$, $b = 22$, $a = 16$
$A = (22)(16) = 352$
The area is 352 square feet.

43. $A = s^2$
$$\begin{aligned} \text{Increase} &= A_{\text{new}} - A_{\text{old}} \\ &= (3.2)^2 - (3)^2 \\ &= 10.24 - 9 \\ &= 1.24 \end{aligned}$$
The area is increased by 1.24 square centimeters.

45. $A = \frac{1}{2}a(b_1 + b_2)$, $a = 4$, $b_1 = 9$, $b_2 = 7$
$$\begin{aligned} A &= \frac{1}{2}(4)(9+7) \\ &= \frac{4(16)}{2} \\ &= 32 \end{aligned}$$
The area is 32 square inches.

47. $A = \frac{1}{2}ab$, $a = 400$, $b = 280$
$A = \frac{1}{2}(400)(280) = 56{,}000$
The area is 56,000 square feet.

49. $r = \frac{d}{2} = \frac{6}{2} = 3$
$$\begin{aligned} A &= \pi r^2 \\ &= (3.14)(3)^2 \\ &= (3.14)(9) \\ &= 28.26 \end{aligned}$$
The area is 28.26 square feet.

51. $C = \frac{5}{9}(F - 32)$
$C = \frac{5}{9}(-109.3 - 32)$
$C = \frac{5}{9}(-141.3)$
$C = 5(-15.7)$
$C = -78.5$
The temperature is −78.5°C.

53. $A = \frac{1}{2}ab,\ a = 20,\ b = 12$
$A = \frac{1}{2}(20)(12) = 120$
Cost = 19.50(120) = 2340
The cost is $2340.

55. $F = \frac{9}{5}C + 32$
$F = \frac{9}{5}(50) + 32 = 90 + 32 = 122$
$F = \frac{9}{5}(-55) + 32 = -99 + 32 = -67$
The temperatures range from 122°F to −67°F.

Cumulative Review

57. $(-2)^4 - 4 \div 2 - (-2) = 16 - 4 \div 2 - (-2)$
$= 16 - 2 + 2$
$= 16$

58. $3(x - 2y) - (x^2 - y) - (x - y)$
$= 3x - 6y - x^2 + y - x + y$
$= -x^2 + 2x - 4y$

Quick Quiz 1.8

1. If $x = -2$, then
$2x^2 - 4x - 14 = 2(-2)^2 - 4(-2) - 14$
$= 2(4) - 4(-2) - 14$
$= 8 + 8 - 14$
$= 2.$

2. If $a = \frac{1}{2}$ and $b = -\frac{1}{3}$, then
$5a - 6b = 5\left(\frac{1}{2}\right) - 6\left(-\frac{1}{3}\right)$
$= \frac{5}{2} + 2$
$= \frac{5}{2} + \frac{4}{2}$
$= \frac{9}{2}$ or $4\frac{1}{2}$.

3. If $x = -2$ and $y = 3$, then
$x^3 + 2x^2y + 5y + 2$
$= (-2)^3 + 2(-2)^2(3) + 5(3) + 2$
$= -8 + 24 + 15 + 2$
$= 33.$

4. Answers may vary. Possible solution:
The formula for the area of a circle is $A = \pi r^2$. Therefore, first find the radius from the diameter $\left(r = \frac{d}{2}\right)$. Use $\pi \approx 3.14$.
$A = 3.14(6\text{ m})^2 = 3.14(36\text{ m}^2) = 113.04\text{ m}^2$

1.9 Exercises

1. $-3x - 2y = -(3x + 2y)$

3. To simplify expressions with grouping symbols, we use the *distributive* property.

5. $6x - 3(x - 2y) = 6x - 3x + 6y = 3x + 6y$

7. $5(c - 3d) - (3c + d) = 5c - 15d - 3c - d$
$= (5 - 3)c + (-15 - 1)d$
$= 2c - 16d$

9. $-3(x + 3y) + 2(2x + y) = -3x - 9y + 4x + 2y$
$= x - 7y$

11. $2[4x^2 - 2(x - 3)] = 2[4x^2 - 2x + 6]$
$= 8x^2 - 4x + 12$

13. $2[5(x + y) - 2(3x - 4y)] = 2[5x + 5y - 6x + 8y]$
$= 2(-x + 13y)$
$= -2x + 26y$

15. $[10-4(x-2y)]+3(2x+y)$
$=10-4x+8y+6x+3y$
$=(-4+6)x+(8+3)y+10$
$=2x+11y+10$

17. $5[3a-2a(3a+6b)+6a^2]$
$=5[3a-6a^2-12ab+6a^2]$
$=5[3a-12ab]$
$=15a-60b$

19. $6(2a^2-3a-4)-3(a-2)$
$=12a^2-18a-24-3a+6$
$=12a^2-21a-18$

21. $3a^2-4[2b-3b(b+2)]=3a^2-4(2b-3b^2-6b)$
$=3a^2-4(-4b-3b^2)$
$=3a^2+16b+12b^2$

23. $5b+\{-[3a+2(5a-2b)]-1\}$
$=5b+[-3a-2(5a-2b)]-1$
$=5b+(-3a-10a+4b)-1$
$=5b-13a+4b-1$
$=-13a+9b-1$

25. $3\{3b^2+2[5b-(2-b)]\}=3\{3b^2+2[5b-2+b]\}$
$=3\{3b^2+2[6b-2]\}$
$=3\{3b^2+12b-4\}$
$=9b^2+36b-12$

27. $-4\{3a^2-2[4a^2-(b+a^2)]\}$
$=-4[3a^2-2(4a^2-b-a^2)]$
$=-4[3a^2-2(3a^2-b)]$
$=-4(3a^2-6a^2+2b)$
$=-4(-3a^2+2b)$
$=12a^2-8b$

Cumulative Review

29. If $C = 1064.18$, then
$F=1.8C+32$
$=1.8(1064.18)+32$
$=1915.524+32$
$=1947.52$
The melting point is 1947.52°F.

30. $A=\pi r^2$
$=3.14(380)^2$
$=3.14(144,400)$
$=453,416$
The area is 453,416 sq ft.

31. If $k = 2.205\,p$
If $p = 120$, $k = 2.205(120) = 264.6$.
If $p = 150$, $k = 2.205(150) = 330.75$.
Great Danes weigh on average from 264.6 to 330.75 kg.

32. $k = 2.205p$
$k = 2.205(9) = 19.845$
$k = 2.205(14) = 30.87$
Miniature Pinschers weigh on average 19.845 to 30.87 kg.

Quick Quiz 1.9

1. $2[3x-2(5x+y)]=2(3x-10x-2y)$
$=2(-7x-2y)$
$=-14x-4y$

2. $3[x-3(x+4)+5y]=3(x-3x-12+5y)$
$=3(-2x+5y-12)$
$=-6x+15y-36$

3. $-4\{2a+2[2ab-b(1-a)]\}$
$=-4\{2a+2[2ab-b+ab]\}$
$=-4\{2a+2[3ab-b]\}$
$=-4\{2a+6ab-2b\}$
$=-8a-24ab+8b$

4. Answers may vary. Possible solution:
First combine like terms within the square brackets. Then use the distributive property to remove graphing symbols and combine like terms after each step.
$3\{2-3[4x-2(x+3)+5x]\}$
$=3\{2-3[9x-2(x+3)]\}$
$=3\{2-3[9x-2x-6]\}$
$=3\{2-3[7x-6]\}$
$=3\{2-21x+18\}$
$=3\{20-21x\}$
$=60-63x$

Putting Your Skills to Work

1. SHELL: \$3.55
ARCO: \$3.43 + \$0.45 = \$3.88
Shelby should choose SHELL.

2. SHELL: 3($3.55) = $10.65
ARCO: 3($3.43) + $0.45 = $10.74
Shelby should choose SHELL.

3. SHELL: 4($3.55) = $14.20
ARCO: 4($3.43) + $0.45 = $14.17
Shelby should choose ARCO.

4. SHELL: 10($3.55) = $35.50
ARCO: 10($3.43) + $0.45 = $34.75
Shelby should choose ARCO.

5. Let x = gallons of gas.
$$3.55x = 3.43x + 0.45$$
$$3.55x - 3.43x = 0.45$$
$$0.12x = 0.45$$
$$x = 3.75$$
They cost the same for 3.75 gallons.

6. Answers may vary.

7. Answers may vary.

Chapter 1 Review Problems

1. $(-6) + (-2) = -8$

2. $-12 + 7.8 = -4.2$

3. $(+5) + (-2) + (-12) = (+3) + (-12) = -9$

4. $(+3.7) + (-1.8) = 1.9$

5. $\left(+\frac{1}{2}\right)+\left(-\frac{5}{6}\right)=\left(+\frac{3}{6}\right)+\left(-\frac{5}{6}\right)=-\frac{2}{6}=-\frac{1}{3}$

6. $\left(-\frac{3}{11}\right)+\left(-\frac{1}{22}\right)=\left(-\frac{6}{22}\right)+\left(-\frac{1}{22}\right)=-\frac{7}{22}$

7. $\left(+\frac{3}{4}\right)+\left(-\frac{1}{12}\right)+\left(-\frac{1}{2}\right)$
$=\left(+\frac{9}{12}\right)+\left(-\frac{1}{12}\right)+\left(-\frac{6}{12}\right)$
$=\frac{2}{12}$
$=\frac{1}{6}$

8. $\frac{2}{15}+\frac{1}{6}+\left(-\frac{4}{5}\right)=\frac{4}{30}+\frac{5}{30}+\left(-\frac{24}{30}\right)$
$=-\frac{15}{30}$
$=-\frac{1}{2}$

9. $(+5) - (-3) = (+5) + (+3) = 8$

10. $(-2) - (-15) = (-2) + (+15) = 13$

11. $(-30) - (+3) = (-30) + (-3) = -33$

12. $(+8) - (-1.2) = (+8) + (+1.2) = 9.2$

13. $\left(-\frac{7}{8}\right)+\left(-\frac{3}{4}\right)=\left(-\frac{7}{8}\right)+\left(-\frac{6}{8}\right)=-\frac{13}{8}=-1\frac{5}{8}$

14. $-\frac{3}{8}+\frac{5}{6}=-\frac{9}{24}+\frac{20}{24}=\frac{11}{24}$

15. $-20.8 - 1.9 = -20.8 + (-1.9) = -22.7$

16. $-151 - (-63) = -151 + 63 = -88$

17. $87 \div (-29) = -3$

18. $-10.4 \div (-0.8) = 13$

19. $\frac{-24}{-\frac{3}{4}}=-24\div\left(-\frac{3}{4}\right)=\left(\frac{-24}{1}\right)\left(-\frac{4}{3}\right)=32$

20. $-\frac{2}{3}\div\left(-\frac{4}{5}\right)=-\frac{2}{3}\cdot\left(-\frac{5}{4}\right)=\frac{10}{12}=\frac{5}{6}$

21. $\left(\frac{5}{7}\right)\div\left(-\frac{5}{25}\right)=\frac{5}{7}\cdot\left(-\frac{25}{5}\right)=-\frac{25}{7}$ or $-3\frac{4}{7}$

22. $(-6)(3)(4) = (-18)(4) = -72$

23. $-1(-4)(-3)(-5) = 4(-3)(-5) = -12(-5) = 60$

24. $(-5)\left(-\frac{1}{2}\right)(4)(-3)=\left(\frac{5}{2}\right)(4)(-3)=10(-3)=-30$

25. $-5+(-2)-(-3)=-5+(-2)+(+3)$
$=(-7)+(+3)$
$=-4$

26. $6-(-4)+(-2)+(8)=6+4+(-2)+(8)$
$=10+(-2)+(8)$
$=8+8$
$=16$

27. $(-16)+(-13)=-29$

28. $(-11)-(-12)=-11+12=1$

29. $-\frac{4}{3}+\frac{2}{3}+\frac{1}{6}=-\frac{2}{3}+\frac{1}{6}$
$=-\frac{4}{6}+\frac{1}{6}$
$=-\frac{3}{6}$
$=-\frac{1}{2}$

30. $-\frac{6}{7}+\frac{1}{2}+\left(-\frac{3}{14}\right)=-\frac{12}{14}+\frac{7}{14}+\left(-\frac{3}{14}\right)$
$=-\frac{5}{14}+\left(-\frac{3}{14}\right)$
$=-\frac{8}{14}$
$=-\frac{4}{7}$

31. $(-3)(-2)(-5)=-(3)(2)(5)=-30$

32. $-6+(-2)-(-3)=-6+(-2)+3=-8+3=-5$

33. $3.5(-2.6)=-9.1$

34. $(-5.4)\div(-6)=0.9$

35. $5-(-3.5)+1.6=5+3.5+1.6=10.1$

36. $-8+2-(-4.8)=-6+4.8=-1.2$

37. $17+3.4+(-16)+(-2.5)=20.4+(-16)+(-2.5)$
$=4.4+(-2.5)$
$=1.9$

38. $37+(-44)+12.5+(-6.8)=-7+12.5+(-6.8)$
$=5.5+(-6.8)$
$=-1.3$

39. $(-8)(3)=-24$; there were 24 yards lost.

40. $-34+12=-22$
The temperature was −22°F.

41. $14{,}494-(-282)=14{,}494+282=14{,}776$
The difference in height is 14,776 feet.

42. A: $2.28+2.45+(-6.58)=-1.85$
B: $-3.35+2.75+(-1.84)=-2.44$
Stock B had the greatest loss.

43. $(3)^5=(-3)(-3)(-3)(-3)(-3)=-243$

44. $(-2)^6=64$

45. $(-5)^4=625$

46. $\left(-\frac{2}{3}\right)^3=\left(-\frac{2}{3}\right)\left(-\frac{2}{3}\right)\left(-\frac{2}{3}\right)=-\frac{8}{27}$

47. $-9^2=-81$

48. $(0.6)^2=0.36$

49. $\left(\frac{5}{6}\right)^2=\left(\frac{5}{6}\right)\left(\frac{5}{6}\right)=\frac{25}{36}$

50. $\left(\frac{3}{4}\right)^3=\left(\frac{3}{4}\right)\left(\frac{3}{4}\right)\left(\frac{3}{4}\right)=\frac{27}{64}$

51. $(5)(-4)+(3)(-2)^3=(5)(-4)+(3)(-8)$
$=-20+(-24)$
$=-44$

52. $20-(-10)-(-6)+(-5)-1$
$=20+10+6+(-5)-1$
$=30+6+(-5)+(-1)$
$=36+(-5)+(-1)$
$=31+(-1)$
$=30$

53. $(3-6)^2+(-12)+-3(-2)=(-3)^2+(-12)+6$
$=9+(-12)+6$
$=-3+6$
$=3$

54. $7(-3x+y)=7(-3x)+7(y)=-21x+7y$

55. $3x(6-x+3y)=3x(6)+3x(-x)+3x(3y)$
$=18x-3x^2+9xy$

56. $-(7x^2-3x+11)$
$=-1(7x^2)+(-1)(-3x)+(-1)(11)$
$=-7x^2+3x-11$

57. $(2xy+x-y)(-3y^2)$
$=(2xy)(-3y^2)+x(-3y^2)-y(-3y^2)$
$=-6xy^3-3xy^2+3y^3$

58. $3a^2b-2bc+6bc^2-8a^2b-6bc^2+5bc$
$=(3-8)a^2b+(-2+5)bc+(6-6)bc^2$
$=-5a^2b+3bc$

59. $9x+11y-12x-15y=-3x-4y$

60. $4x^2-13x+7-9x^2-22x-16$
$=(4-9)x^2+(-13-22)x+7-16$
$=-5x^2-35x-9$

61. $-x+\frac{1}{2}+14x^2-7x-1-4x^2$
$=(14-4)x^2+(-7-1)x+\frac{1}{2}-1$
$=10x^2-8x-\frac{1}{2}$

62. If $x=-7$, then
$7x-6=7(-7)-6=-49-6=-55$.

63. If $x=8$, then $7-\frac{3}{4}x=7-\frac{3}{4}(8)=7-6=1$.

64. $x^2+3x-4=(-3)^2+3(-3)-4$
$=9-9-4$
$=-4$

65. If $x=3$, then
$-x^2+5x-9=-3^2+5(3)-9=-9+15-9=-3$.

66. If $x=-1$, then
$2x^3-x^2+6x+9=2(-1)^3-(-1)^2+6(-1)+9$
$=-2-1-6+9$
$=0$.

67. If $a=-1$, $b=5$, and $c=-2$, then
$b^2-4ac=(5)^2-4(-1)(-2)=25-8=17$.

68. If $m=-4$, $M=15$, $G=-1$, and $r=-2$, then
$\frac{mMG}{r^2}=\frac{-4(15)(-1)}{(-2)^2}=\frac{60}{4}=15$.

69. If $p=6000$, $r=18\%$, and $t=\frac{3}{4}$, then
$I=prt=6000(0.18)\left(\frac{3}{4}\right)=810$.
The interest is \$810.

70. $F=\frac{9C+160}{5}$
$=\frac{9(20)+160}{5}$
$=\frac{180+160}{5}$
$=\frac{340}{5}$
$=68$

$F=\frac{9C+160}{5}$
$=\frac{9(25)+160}{5}$
$=\frac{225+160}{5}$
$=\frac{385}{5}$
$=77$
The range is 68°F to 77°F.

71. If $r=4$, $A=\pi r^2$
$=3.14(4)^2$
$=3.14(16)$
$=50.24$ square feet.
Cost = (50.24 sq ft)(\$1.50/sq ft) = \$75.36
The total cost is \$75.36.

72. $P=180S-R-C$
$=180(56)-300-1200$
$=10,080-300-1200$
$=8580$
The daily profit is \$8580.

73. $A = \frac{1}{2}a(b_1 + b_2),\ a = 200,\ b_1 = 700,\ b_2 = 300$

$A = \frac{1}{2}(200)(700 + 300)$
$= 100(1000)$
$= 100,000$ square feet

Cost = 2(100,000) = \$200,000

The area is $100,000\text{ ft}^2$ and the cost is \$200,000.

74. $A = \frac{1}{2}ab,\ a = 3.8,\ b = 5.5$

$A = \frac{1}{2}(3.8)(5.5) = 10.45$ square feet

Cost = 66(10.45) = \$689.70

The area is 10.45 ft^2 and the cost is \$689.70.

75. $5x - 7(x - 6) = 5x - 7x + 42 = -2x + 42$

76. $3(x-2) - 4(5x+3) = 3x - 6 - 20x - 12$
$= -17x - 18$

77. $2[3 - (4 - 5x)] = 2(3 - 4 + 5x)$
$= 2(-1 + 5x)$
$= -2 + 10x$

78. $-3x[x + 3(x - 7)] = -3x(x + 3x - 21)$
$= -3x(4x - 21)$
$= -12x^2 + 63x$

79. $2xy^3 - 6x^3y - 4x^2y^2 + 3(xy^3 - 2x^2y - 3x^2y^2)$
$= 2xy^3 - 6x^3y - 4x^2y^2 + 3xy^3 - 6x^2y - 9x^2y^2$
$= (2+3)xy^3 - 6x^3y + (-4-9)x^2y^2 - 6x^2y$
$= 5xy^3 - 6x^3y - 13x^2y^2 - 6x^2y$

80. $-5(x + 2y - 7) + 3x(2 - 5y)$
$= -5x - 10y + 35 + 6x - 15xy$
$= x - 10y + 35 - 15xy$

81. $-(a + 3b) + 5[2a - b - 2(4a - b)]$
$= -(a + 3b) + 5(2a - b - 8a + 2b)$
$= -(a + 3b) + 5(-6a + b)$
$= -a - 3b - 30a + 5b$
$= -31a + 2b$

82. $-5\{2a - [5a - b(3 + 2a)]\}$
$= -5\{2a - [5a - 3b - 2ab]\}$
$= -5\{2a - 5a + 3b + 2ab\}$
$= -5\{-3a + 3b + 2ab\}$
$= 15a - 15b - 10ab$

83. $-3\{2x - [x - 3y(x - 2y)]\}$
$= -3[2x - (x - 3xy + 6y^2)]$
$= -3(2x - x + 3xy - 6y^2)$
$= -3(x + 3xy - 6y^2)$
$= -3x - 9xy + 18y^2$

84. $2\{3x + 2[x + 2y(x - 4)]\}$
$= 2[3x + 2(x + 2xy - 8y)]$
$= 2(3x + 2x + 4xy - 16y)$
$= 2(5x + 4xy - 16y)$
$= 10x + 8xy - 32y$

85. $-6.3 + 4 = -2.3$

86. $4 + (-8) + 12 = -4 + 12 = 8$

87. $-\frac{2}{3} - \frac{4}{5} = -\frac{10}{15} + \left(-\frac{12}{15}\right) = -\frac{22}{15} = -1\frac{7}{15}$

88. $-\frac{7}{8} - \left(-\frac{3}{4}\right) = -\frac{7}{8} + \frac{6}{8} = -\frac{1}{8}$

89. $3 - (-4) + (-8) = 3 + 4 + (-8) = 7 + (-8) = -1$

90. $-1.1 - (-0.2) + 0.4 = -1.1 + 0.2 + 0.4$
$= -0.9 + 0.4$
$= -0.5$

91. $\left(-\frac{9}{10}\right)\left(-2\frac{1}{4}\right) = \left(-\frac{9}{10}\right)\left(-\frac{9}{4}\right) = \frac{81}{40}$ or $2\frac{1}{40}$

92. $3.6 \div (-0.45) = -8$

93. $-14.4 \div (-0.06) = 240$

94. $(-8.2)(3.1) = -25.42$

95. $400 + 1000 - 800 = 1400 - 800 = 600$
Her score was \$600.

96. $(-0.3)^4 = (-0.3)(-0.3)(-0.3)(-0.3) = 0.0081$

97. $-0.5^4 = -(0.5)(0.5)(0.5)(0.5) = -0.0625$

98. $9(5) - 5(2)^3 + 5 = 9(5) - 5(8) + 5$
$= 45 - 5(8) + 5$
$= 45 - 40 + 5$
$= 5 + 5$
$= 10$

99. $3.8x-0.2y-8.7x+4.3y$
$=(3.8-8.7)x+(-0.2+4.3)y$
$=-4.9x+4.1y$

100. If $p=-2$ and $q=3$, then
$\frac{2p+q}{3q}=\frac{2(-2)+3}{3(3)}=\frac{-4+3}{9}=-\frac{1}{9}.$

101. If $s=-3$ and $t=-2$, then
$\frac{4s-7t}{s}=\frac{4(-3)-7(-2)}{-3}=\frac{-12+14}{-3}=-\frac{2}{3}.$

102. $F=\frac{9}{5}C+32,\ C=38.6$
$F=\frac{9}{5}(38.6)+32$
$F=69.48+32$
$F=101.48°$
Your dog does not have a fever; in fact, its temperature is below normal.

103. $-7(x-3y^2+4)+3y(4-6y)$
$=-7x+21y^2-28+12y-18y^2$
$=-7x+3y^2+12y-28$

104. $-2\{6x-3[7y-2y(3-x)]\}$
$=-2\{6x-3[7y-6y+2xy]\}$
$=-2\{6x-3[y+2xy]\}$
$=-2\{6x-3y-6xy\}$
$=-12x+6y+12xy$

How Am I Doing? Chapter 1 Test

1. $-2.5+6.3+(-4.1)=3.8+(-4.1)=-0.3$

2. $-5-(-7)=-5+7=2$

3. $\left(-\frac{2}{3}\right)(7)=-\frac{14}{3}=-4\frac{2}{3}$

4. $(-5)(-2)(7)(-1)=-(10)(7)(1)=-(70)(1)=-70$

5. $(-12)\div(-3)=4$

6. $(-1.8)\div(0.6)=-3$

7. $(-4)^3=(-4)(-4)(-4)=-64$

8. $(1.6)^2=(1.6)(1.6)=2.56$

9. $\left(\frac{2}{3}\right)^4=\left(\frac{2}{3}\right)\left(\frac{2}{3}\right)\left(\frac{2}{3}\right)\left(\frac{2}{3}\right)=\frac{16}{81}$

10. $(0.2)^2-(2.1)(-3)+0.46$
$=0.04-(2.1)(-3)+0.46$
$=0.04-(-6.3)+0.46$
$=0.04+6.3+0.46$
$=6.34+0.46$
$=6.8$

11. $3(4-6)^3+12\div(-4)+2$
$=3(-2)^3+12\div(-4)+2$
$=3(-8)+12\div(-4)+2$
$=-24+(-3)+2$
$=-25$

12. $-5x(x+2y-7)=-5x(x)-5x(2y)-5x(-7)$
$=-5x^2-10xy+35x$

13. $-2ab^2(-3a-2b+7ab)$
$=-2ab^2(-3a)-2ab^2(-2b)-2ab^2(7ab)$
$=6a^2b^2+4ab^3-14a^2b^3$

14. $6ab-\frac{1}{2}a^2b+\frac{3}{2}ab+\frac{5}{2}a^2b$
$=\left(6+\frac{3}{2}\right)ab+\left(-\frac{1}{2}+\frac{5}{2}\right)a^2b$
$=\left(\frac{12}{2}+\frac{3}{2}\right)ab+\frac{4}{2}a^2b$
$=\frac{15}{2}ab+2a^2b$

15. $2.3x^2y-8.1xy^2+3.4xy^2-4.1x^2y$
$=(2.3-4.1)x^2y+(-8.1+3.4)xy^2$
$=-1.8x^2y-4.7xy^2$

16. $3(2-a)-4(-6-2a)=6-3a+24+8a$
$=5a+30$

17. $5(3x-2y)-(x+6y)=15x-10y-x-6y$
$=(15-1)x+(-10-6)y$
$=14x-16y$

18. $x^3-3x^2y+2y-5=3^3-3(3)^2(-4)+2(-4)-5$
$=27-3(9)(-4)-8-5$
$=27+108-8-5$
$=122$

19. If $x = -3$, then

$$\begin{aligned}3x^2 - 7x - 11 &= 3(-3)^2 - 7(-3) - 11\\ &= 3(9) - 7(-3) - 11\\ &= 27 + 21 - 11\\ &= 37.\end{aligned}$$

20. $$\begin{aligned}2a - 3b &= 2\left(\frac{1}{3}\right) - 3\left(-\frac{1}{2}\right)\\ &= \frac{2}{3} + \frac{3}{2}\\ &= \frac{4}{6} + \frac{9}{6}\\ &= \frac{13}{6} \text{ or } 2\frac{1}{6}\end{aligned}$$

21. $k = 1.61r = 1.61(60) = 96.6$
You are traveling at 96.6 kilometers per hour.

22. $A = \frac{1}{2}a(b_1 + b_2),\ a = 120,\ b_1 = 200,\ b_2 = 180$

$$\begin{aligned}A &= \frac{1}{2}(120)(200 + 180)\\ &= 60(380)\\ &= 22{,}800\end{aligned}$$

The area is 22,800 square feet.

23. $A = \frac{1}{2}ab,\ a = 6.8,\ b = 8.5$

$A = \frac{1}{2}(6.8)(8.5) = 28.9$ square feet

Cost = 0.80(28.9) = \$23.12
The cost is \$23.12.

24. $A = 60 \times 10 = 600$ sq ft

$600 \text{ sq ft} \times \frac{1 \text{ can}}{200 \text{ sq ft}} = 3 \text{ cans}$

You should buy 3 cans.

25. $$\begin{aligned}3[x - 2y(x + 2y) - 3y^2] &= 3[x - 2xy - 4y^2 - 3y^2]\\ &= 3[x - 2xy - 7y^2]\\ &= 3x - 6xy - 21y^2\end{aligned}$$

26. $$\begin{aligned}-3\{a + b[3a - b(1 - a)]\} &= -3[a + b(3a - b + ab)]\\ &= -3(a + 3ab - b^2 + ab^2)\\ &= -3a - 9ab + 3b^2 - 3ab^2\end{aligned}$$

Chapter 2

2.1 Exercises

1. When we use the equals sign, we indicate two expressions are equal in value.

3. The solution of an equation is a value of the variable that makes the equation true.

5. Answers may vary. A sample answer is to isolate the variable x.

7.
$$x+11=15$$
$$x+11+(-11)=15+(-11)$$
$$x=4$$
Check: $4+11 \stackrel{?}{=} 15$
$15=15$ ✓

9.
$$20=9+x$$
$$20+(-9)=9+(-9)+x$$
$$11=x$$
Check: $20 \stackrel{?}{=} 9+11$
$20=20$ ✓

11.
$$x-3=14$$
$$x-3+3=14+3$$
$$x=17$$
Check: $17-3 \stackrel{?}{=} 14$
$14=14$ ✓

13.
$$0=x+5$$
$$0+(-5)=x+5+(-5)$$
$$-5=x$$
Check: $0 \stackrel{?}{=} -5+5$
$0=0$ ✓

15.
$$x-6=-19$$
$$x-6+6=-19+6$$
$$x=-13$$
Check: $-13-6 \stackrel{?}{=} -19$
$-19=-19$ ✓

17.
$$-12+x=50$$
$$-12+12+x=50+12$$
$$x=62$$
Check: $-12+62 \stackrel{?}{=} 50$
$50=50$ ✓

19.
$$3+5=x-7$$
$$8=x-7$$
$$8+7=x-7+7$$
$$15=x$$
Check: $3+5 \stackrel{?}{=} 15-7$
$8=8$ ✓

21.
$$32-17=x-6$$
$$15=x-6$$
$$15+6=x-6+6$$
$$21=x$$
Check: $32-17 \stackrel{?}{=} 21-6$
$15=15$ ✓

23.
$$4+8+x=6+6$$
$$12+x=12$$
$$12+(-12)+x=12+(-12)$$
$$x=0$$
Check: $4+8+0 \stackrel{?}{=} 6+6$
$12=12$ ✓

25.
$$18-7+x=7+9-5$$
$$11+x=11$$
$$11+(-11)+x=11+(-11)$$
$$x=0$$
Check: $18-7+0 \stackrel{?}{=} 7+9-5$
$11=11$ ✓

27.
$$-12+x-3=15-18+9$$
$$-15+x=6$$
$$-15+15+x=6+15$$
$$x=21$$
Check: $-12+21-3 \stackrel{?}{=} 15-18+9$
$6=6$ ✓

29. $-7+x=2,\ x \stackrel{?}{=} 5$
$$-7+5 \stackrel{?}{=} 2$$
$$-2 \neq 2$$
$x = 5$ is not a solution.
$$-7+x=2$$
$$-7+7+x=2+7$$
$$x=9$$

31. $-11+5=x+8,\ x \stackrel{?}{=} 6$
$-11+5 \stackrel{?}{=} 6+8$
$-6 \neq 14$
$x=6$ is not the solution.

$$\begin{aligned} -11+5&=x+8\\ -6&=x+8\\ -6+(-8)&=x+8+(-8)\\ -14&=x \end{aligned}$$

33. $x-23=-56,\ x \stackrel{?}{=} -33$
$-33-23 \stackrel{?}{=} -56$
$-56=-56$
$x=-33$ is a solution.

35. $15-3+20=x-3,\ x \stackrel{?}{=} 35$
$15-3+20 \stackrel{?}{=} 35-3$
$32=32$
$x=35$ is the solution.

37.
$$\begin{aligned} 2.5+x&=0.7\\ 2.5+(-2.5)+x&=0.7+(-2.5)\\ x&=-1.8 \end{aligned}$$

39.
$$\begin{aligned} 12.5+x-8.2&=4.9\\ x+4.3&=4.9\\ x+4.3+(-4.3)&=4.9+(-4.3)\\ x&=0.6 \end{aligned}$$

41.
$$\begin{aligned} x-\frac{1}{4}&=\frac{3}{4}\\ x-\frac{1}{4}+\frac{1}{4}&=\frac{3}{4}+\frac{1}{4}\\ x&=\frac{4}{4}\\ x&=1 \end{aligned}$$

43.
$$\begin{aligned} \frac{2}{3}+x&=\frac{1}{6}+\frac{1}{4}\\ \frac{8}{12}+x&=\frac{2}{12}+\frac{3}{12}\\ \frac{8}{12}+x&=\frac{5}{12}\\ \frac{8}{12}+\left(-\frac{8}{12}\right)+x&=\frac{5}{12}+\left(-\frac{8}{12}\right)\\ x&=-\frac{3}{12}\\ x&=-\frac{1}{4} \end{aligned}$$

45.
$$\begin{aligned} 3+x&=-12+8\\ 3+x&=-4\\ 3+(-3)+x&=-4+(-3)\\ x&=-7 \end{aligned}$$

47.
$$\begin{aligned} 5\frac{1}{6}+x&=8\\ \frac{31}{6}+x&=\frac{48}{6}\\ \frac{31}{6}+\left(-\frac{31}{6}\right)+x&=\frac{48}{6}+\left(-\frac{31}{6}\right)\\ x&=\frac{17}{6} \text{ or } 2\frac{5}{6} \end{aligned}$$

49.
$$\begin{aligned} \frac{5}{12}-\frac{5}{6}&=x-\frac{3}{2}\\ \frac{5}{12}-\frac{10}{12}&=x-\frac{18}{12}\\ -\frac{5}{12}&=x-\frac{18}{12}\\ -\frac{5}{12}+\frac{18}{12}&=x-\frac{18}{12}+\frac{18}{12}\\ \frac{13}{12}&=x\\ 1\frac{11}{12}&=x \end{aligned}$$

51.
$$\begin{aligned} 1.6+x-3.2&=-2+5.6\\ x-1.6&=3.6\\ x-1.6+1.6&=3.6+1.6\\ x&=5.2 \end{aligned}$$

53.
$$\begin{aligned} x-18.225&=1.975\\ x-18.225+18.225&=1.975+18.225\\ x&=20.2 \end{aligned}$$

Cumulative Review

55.
$$\begin{aligned} x+3y-5x-7y+2x&=(1-5+2)x+(3-7)y\\ &=-2x-4y \end{aligned}$$

56.
$$\begin{aligned} &y^2+y-12-3y^2-5y+16\\ &=(1-3)y^2+(1-5)y-12+16\\ &=-2y^2-4y+4 \end{aligned}$$

Quick Quiz 2.1

1.
$$\begin{aligned} x-4.7&=9.6\\ x-4.7+4.7&=9.6+4.7\\ x&=14.3 \end{aligned}$$

2.
$$\begin{aligned}-8.6+x&=-12.1\\ -8.6+8.6+x&=-12.1+8.6\\ x&=-3.5\end{aligned}$$

3.
$$\begin{aligned}3-12+7&=8+x-2\\ -2&=6+x\\ -2-6&=6-6+x\\ -8&=x\end{aligned}$$

4. Answers may vary. Possible solution: Substitute 3.8 for x in the equation. Simplify. If the resultant equation is true, $x = 3.8$ is the solution. If the resultant equation is not true, $x = 3.8$ is not the solution.

2.2 Exercises

1. To solve the equation $6x = -24$, divide each side of the equation by $\underline{6}$.

3. To solve the equation $\frac{1}{7}x = -2$, multiply each side of the equation by $\underline{7}$.

5.
$$\begin{aligned}\frac{1}{9}x&=4\\ 9\left(\frac{1}{9}x\right)&=9(4)\\ x&=36\end{aligned}$$
Check: $\frac{1}{9}(36) \stackrel{?}{=} 4$
$4 = 4$ ✓

7.
$$\begin{aligned}\frac{1}{2}x&=-15\\ 2\left(\frac{1}{2}x\right)&=2(-15)\\ x&=-30\end{aligned}$$
Check: $\frac{1}{2}(-30) \stackrel{?}{=} -15$
$-15 = -15$ ✓

9.
$$\begin{aligned}\frac{x}{5}&=16\\ 5\left(\frac{x}{5}\right)&=5(16)\\ x&=80\end{aligned}$$
Check: $\frac{80}{5} \stackrel{?}{=} 16$
$16 = 16$ ✓

11.
$$\begin{aligned}-3&=\frac{x}{5}\\ 5(-3)&=5\left(\frac{x}{5}\right)\\ -15&=x\end{aligned}$$
Check: $-3 \stackrel{?}{=} \frac{-15}{5}$
$-3 = -3$ ✓

13.
$$\begin{aligned}13x&=52\\ \frac{13x}{13}&=\frac{52}{13}\\ x&=4\end{aligned}$$
Check: $13(4) \stackrel{?}{=} 52$
$52 = 52$ ✓

15.
$$\begin{aligned}56&=7x\\ \frac{56}{7}&=\frac{7x}{7}\\ 8&=x\end{aligned}$$
Check: $56 \stackrel{?}{=} 7(8)$
$56 = 56$ ✓

17.
$$\begin{aligned}-16&=6x\\ \frac{-16}{6}&=\frac{6x}{6}\\ -\frac{8}{3}&=x\end{aligned}$$
Check: $-16 \stackrel{?}{=} 6\left(-\frac{8}{3}\right)$
$-16 = -16$ ✓

19.
$$\begin{aligned}1.5x&=75\\ \frac{1.5x}{1.5}&=\frac{75}{1.5}\\ x&=50\end{aligned}$$
Check: $1.5(50) \stackrel{?}{=} 75$
$75 = 75$ ✓

21.
$$\begin{aligned}-15&=-x\\ \frac{-15}{-1}&=\frac{-x}{-1}\\ 15&=x\end{aligned}$$
Check: $-15 \stackrel{?}{=} (-1)(15)$
$-15 = -15$ ✓

23. $-112 = 16x$
$\frac{-112}{16} = \frac{16x}{16}$
$-7 = x$
Check: $-112 \stackrel{?}{=} 16(-7)$
$-112 = -112$ ✓

25. $0.4x = 0.08$
$\frac{0.4x}{0.4} = \frac{0.08}{0.4}$
$x = 0.2$ or $\frac{1}{5}$
Check: $(0.4)(0.2) \stackrel{?}{=} 0.08$
$0.08 = 0.08$ ✓

27. $-3.9x = -15.6$
$\frac{-3.9x}{-3.9} = \frac{-15.6}{-3.9}$
$x = 4$
Check: $-3.9(4) \stackrel{?}{=} -15.6$
$-15.6 = -15.6$ ✓

29. $-3x = 21,\ x \stackrel{?}{=} 7$
$-3(7) \stackrel{?}{=} 21$
$-21 \neq 21$
$x = 7$ is not the solution.
$-3x = 21$
$\frac{-3x}{-3} = \frac{21}{-3}$
$x = -7$

31. $-x = 15,\ x \stackrel{?}{=} -15$
$-(-15) \stackrel{?}{=} 15$
$15 = 15$
$x = -15$ is the solution.

33. $7y = -0.21$
$\frac{7y}{7} = \frac{-0.21}{7}$
$y = -0.03$

35. $-56 = -21t$
$\frac{-56}{-21} = \frac{-21t}{-21}$
$\frac{8}{3} = t$

37. $4.6y = -3.22$
$\frac{4.6y}{4.6} = \frac{-3.22}{4.6}$
$y = -0.7$

39. $4x + 3x = 21$
$7x = 21$
$\frac{7x}{7} = \frac{21}{7}$
$x = 3$

41. $2x - 7x = 20$
$-5x = 20$
$\frac{-5x}{-5} = \frac{20}{-5}$
$x = -4$

43. $-8x - 4x = -5$
$-12x = -5$
$\frac{-12x}{-12} = \frac{-5}{-12}$
$x = \frac{5}{12}$

45. $12 - 19 = -7x$
$-7 = -7x$
$\frac{-7}{-7} = \frac{-7x}{-7}$
$1 = x$

47. $8m = -14 + 30$
$8m = 16$
$\frac{8m}{8} = \frac{16}{8}$
$m = 2$

49. $\frac{2}{3}x = 18$
$\frac{3}{2}\left(\frac{2}{3}x\right) = \frac{3}{2}(18)$
$x = 27$

51. $-2.5133x = 26.38965$
$\frac{-2.5133x}{-2.5133} = \frac{26.38965}{-2.5133}$
$x = -10.5$

Cumulative Review

53. $(-6)(-8) + (-3)(2) = 48 - 6 = 42$

54. $(-3)^3 + (-20) \div 2 = -27 + (-20) \div 2$
$= -27 + (-10)$
$= -37$

55. $5 + (2-6)^2 = 5 + (-4)^2 = 5 + 16 = 21$

56. Find 25% of 30.
25% of 30 = 0.25 × 30 = 7.5
The whale will lose 7.5 tons.
30 − 7.5 = 22.5
The whale will weigh 22.5 tons.

57. Find 35% of 20.
35% of 20 = 0.35 × 20 = 7
The number of earthquakes is expected to increase by 7.
20 + 7 = 27
A total of 27 earthquakes can be expected.

Quick Quiz 2.2

1. $2.5x = -95$
$\frac{2.5x}{2.5} = \frac{-95}{2.5}$
$x = -38$

2. $-3.9x = -54.6$
$\frac{-3.9x}{-3.9x} = \frac{-54.6}{-3.9}$
$x = 14$

3. $7x - 12x = 60$
$-5x = 60$
$\frac{-5x}{-5} = \frac{60}{-5}$
$x = -12$

4. Answers may vary. Possible solution:
Change $36\frac{2}{3}$ to an improper fraction. Substitute that value for x in the equation. Simplify. If the resultant equation is true, $x = 36\frac{2}{3}$ is the solution.

2.3 Exercises

1. $4x + 13 = 21$
$4x + 13 + (-13) = 21 + (-13)$
$4x = 8$
$\frac{4x}{4} = \frac{8}{4}$
$x = 2$
Check: $4(2) + 13 \stackrel{?}{=} 21$
$21 = 21$ ✓

3. $4x - 11 = 13$
$4x - 11 + 11 = 13 + 11$
$4x = 24$
$\frac{4x}{4} = -\frac{24}{4}$
$x = 6$
Check: $4(6) - 11 \stackrel{?}{=} 13$
$13 = 13$ ✓

5. $7x - 18 = -46$
$7x - 18 + 18 = -46 + 18$
$7x = -28$
$\frac{7x}{7} = \frac{-28}{7}$
$x = -4$
Check: $7(-4) - 18 \stackrel{?}{=} -46$
$-28 - 18 \stackrel{?}{=} -46$
$-46 = -46$ ✓

7. $-4x + 17 = -35$
$-4x + 17 + (-17) = -35 + (-17)$
$-4x = -52$
$\frac{-4x}{-4} = \frac{-52}{-4}$
$x = 13$
Check: $-4(13) + 17 \stackrel{?}{=} -35$
$-52 + 17 \stackrel{?}{=} -35$
$-35 = -35$ ✓

9. $2x + 3.2 = 9.4$
$2x + 3.2 + (-3.2) = 9.4 + (-3.2)$
$2x = 6.2$
$\frac{2x}{2} = \frac{6.2}{2}$
$x = 3.1$
Check: $2(3.1) + 3.2 \stackrel{?}{=} 9.4$
$6.2 + 3.2 \stackrel{?}{=} 9.4$
$9.4 = 9.4$ ✓

11. $\frac{1}{4}x + 6 = 13$
$\frac{1}{4}x + 6 - 6 = 13 - 6$
$\frac{1}{4}x = 7$
$4\left(\frac{1}{4}x\right) = 4(7)$
$x = 28$

Check: $\frac{1}{4}(28)+6 \stackrel{?}{=} 13$
$7+6 \stackrel{?}{=} 13$
$13 = 13$ ✓

13. $\frac{1}{3}x+5=-4$
$\frac{1}{3}x+5+(-5)=-4+(-5)$
$\frac{1}{3}x=-9$
$3\left(\frac{1}{3}x\right)=3(-9)$
$x=-27$
Check: $\frac{1}{3}(-27)+5 \stackrel{?}{=} -4$
$-9+5 \stackrel{?}{=} -4$
$-4=-4$ ✓

15. $8x=48+2x$
$8x+(-2x)=48+2x+(-2x)$
$6x=48$
$\frac{6x}{6}=\frac{48}{6}$
$x=8$
Check: $8(8) \stackrel{?}{=} 48+2(8)$
$64 \stackrel{?}{=} 48+16$
$64=64$ ✓

17. $-6x=-27+3x$
$-6x+(-3x)=-27+3x+(-3x)$
$-9x=-27$
$\frac{-9x}{-9}=\frac{-27}{-9}$
$x=3$
Check: $-6(3) \stackrel{?}{=} -27+3(3)$
$-18 \stackrel{?}{=} -27+9$
$-18=-18$ ✓

19. $44-2x=6x$
$44-2x+2x=6x+2x$
$44=8x$
$\frac{44}{8}=\frac{8x}{8}$
$\frac{11}{2}=x$ or $x=5.5$
Check: $44-2\left(\frac{11}{2}\right) \stackrel{?}{=} 6\left(\frac{11}{2}\right)$
$44-11 \stackrel{?}{=} 33$
$33=33$ ✓

21. $54-2x=-8x$
$54-2x+2x=-8x+2x$
$54=-6x$
$\frac{54}{-6}=\frac{-6x}{-6}$
$-9=x$
Check: $54-2(-9) \stackrel{?}{=} -8(-9)$
$54+18 \stackrel{?}{=} 72$
$72=72$ ✓

23. $2y+3y=12-y,\ y \stackrel{?}{=} 2$
$2(2)+3(2) \stackrel{?}{=} 12-2$
$4+6 \stackrel{?}{=} 10$
$10=10$
$y=2$ is the solution.

25. $7x+6-3x=2x-5+x,\ x \stackrel{?}{=} 11$
$7(11)+6-3(11) \stackrel{?}{=} 2(11)-5+11$
$77+6-33 \stackrel{?}{=} 22-5+11$
$50 \neq 28$
$x=11$ is not a solution.
$7x+6-3x=2x-5+x$
$4x+6=3x-5$
$4x+(-3x)+6=3x+(-3x)-5$
$x+6=-5$
$x+6+(-6)=-5+(-6)$
$x=-11$

27. $14-2x=-5x+11$
$14-2x+5x=-5x+5x+11$
$14+3x=11$
$3x+14+(-14)=11+(-14)$
$3x=-3$
$\frac{3x}{3}=\frac{-3}{3}$
$x=-1$

29. $x-6=8-x$
$x+x-6=8-x+x$
$2x-6=8$
$2x-6+6=8+6$
$2x=14$
$\frac{2x}{2}=\frac{14}{2}$
$x=7$

31.
$$0.6y+0.8=0.1-0.1y$$
$$0.6y+0.1y+0.8=0.1-0.1y+0.1y$$
$$0.7y+0.8=0.1$$
$$0.7y+0.8-0.8=0.1-0.8$$
$$0.7y=-0.7$$
$$\frac{0.7y}{0.7}=\frac{-0.7}{0.7}$$
$$y=-1$$

33.
$$5x-9=3x+23$$
$$5x+(-3x)-9=3x+(-3x)+23$$
$$2x-9=23$$
$$2x-9+9=23+9$$
$$2x=32$$
$$\frac{2x}{2}=\frac{32}{2}$$
$$x=16$$

35. $-3 + 10y + 6 = 15 + 12y - 18$

Left
$$10y+3=12y-3$$
$$10y+(-12y)+3=12y+(-12y)-3$$
$$-2y+3=-3$$
$$-2y+3+(-3)=-3+(-3)$$
$$-2y=-6$$
$$\frac{-2y}{-2}=\frac{-6}{-2}$$
$$y=3$$

Right
$$10y+3=12y-3$$
$$10y+(-10y)+3=12y+(-10y)-3$$
$$3=2y-3$$
$$3+3=2y-3+3$$
$$6=2y$$
$$\frac{6}{2}=\frac{2y}{2}$$
$$3=y$$

Neither approach is better.

37.
$$5(x+3)=35$$
$$5x+15=35$$
$$5x+15-15=35-15$$
$$5x=20$$
$$\frac{5x}{5}=\frac{20}{5}$$
$$x=4$$

Check: $5(4+3) \stackrel{?}{=} 35$
$$5(7) \stackrel{?}{=} 35$$
$$35=35 \checkmark$$

39.
$$5(4x-3)+8=-2$$
$$20x-15+8=-2$$
$$20x-7=-2$$
$$20x-7+7=-2+7$$
$$20x=5$$
$$\frac{20x}{20}=\frac{5}{20}$$
$$x=0.25 \text{ or } \frac{1}{4}$$

Check: $5[4(0.25)-3]+8 \stackrel{?}{=} -2$
$$5(1-3)+8 \stackrel{?}{=} -2$$
$$5(-2)+8 \stackrel{?}{=} -2$$
$$-10+8 \stackrel{?}{=} -2$$
$$-2=-2 \checkmark$$

41.
$$7x-3(5-x)=10$$
$$7x-15+3x=10$$
$$10x-15=10$$
$$10x=25$$
$$\frac{10x}{10}=\frac{25}{10}$$
$$x=2.5 \text{ or } \frac{5}{2}$$

Check: $7(2.5)-3[5-(2.5)] \stackrel{?}{=} 10$
$$17.5-3(2.5) \stackrel{?}{=} 10$$
$$17.5-7.5 \stackrel{?}{=} 10$$
$$10=10 \checkmark$$

43.
$$0.5x-0.3(2-x)=4.6$$
$$0.5x-0.6-0.3x=4.6$$
$$0.8x-0.6=4.6$$
$$0.8x-0.6+0.6=4.6+0.6$$
$$0.8x=5.2$$
$$\frac{0.8x}{0.8}=\frac{5.2}{0.8}$$
$$x=6.5 \text{ or } 6\frac{1}{2} \text{ or } \frac{13}{2}$$

Check: $0.5(6.5)-0.3(2-6.5) \stackrel{?}{=} 4.6$
$$0.5(6.5)-0.3(-4.5) \stackrel{?}{=} 4.6$$
$$3.25+1.35 \stackrel{?}{=} 4.6$$
$$4.6=4.6 \checkmark$$

45.
$$\begin{aligned}4(a-3)+2&=2(a-5)\\4a-12+2&=2a-10\\4a-10&=2a-10\\4a-2a-10&=2a-2a-10\\2a-10&=-10\\2a-10+10&=-10+10\\2a&=0\\\frac{2a}{2}&=\frac{0}{2}\\a&=0\end{aligned}$$
Check: $4(0-3)+2 \stackrel{?}{=} 2(0-5)$
$$\begin{aligned}4(-3)+2&\stackrel{?}{=}2(-5)\\-12+2&\stackrel{?}{=}-10\\-10&=-10\ \checkmark\end{aligned}$$

47.
$$\begin{aligned}-2(x+3)+4&=3(x+4)+2\\-2x-6+4&=3x+12+2\\-2x-2&=3x+14\\-2x+(-3x)-2&=3x+(-3x)+14\\-5x-2&=14\\-5x-2+2&=14+2\\-5x&=16\\\frac{-5x}{-5}&=\frac{16}{-5}\\x&=-3.2\text{ or }-\frac{16}{5}\end{aligned}$$
Check: $-2[(-3.2)+3]+4 \stackrel{?}{=} 3[(-3.2)+4]+2$
$$\begin{aligned}-2(-0.2)+4&\stackrel{?}{=}3(0.8)+2\\0.4+4&\stackrel{?}{=}2.4+2\\4.4&=4.4\ \checkmark\end{aligned}$$

49.
$$\begin{aligned}-3(y-3y)+4&=-4(3y-y)+6+13y\\-3(-2y)+4&=-4(2y)+6+13y\\6y+4&=-8y+6+13y\\6y+4&=5y+6\\6y-5y+4&=5y-5y+6\\y+4&=6\\y+4-4&=6-4\\y&=2\end{aligned}$$
Check:
$$\begin{aligned}-3[2-3(2)]+4&\stackrel{?}{=}-4[3(2)-2]+6+13(2)\\-3(2-6)+4&\stackrel{?}{=}-4(6-2)+6+26\\-3(-4)+4&\stackrel{?}{=}-4(4)+32\\12+4&\stackrel{?}{=}-16+32\\16&=16\ \checkmark\end{aligned}$$

51.
$$\begin{aligned}5.7x+3&=4.2x-3\\5.7x-4.2x+3&=4.2x-4.2x-3\\1.5x+3&=-3\\1.5x+3-3&=-3-3\\1.5x&=-6\\\frac{1.5x}{1.5}&=\frac{-6}{1.5}\\x&=-4\end{aligned}$$

53.
$$\begin{aligned}5z+7-2z&=32-2z\\3z+7&=32-2z\\3z+2z+7&=32-2z+2z\\5z+7&=32\\5z+7-7&=32-7\\5z&=25\\\frac{5z}{5}&=\frac{25}{5}\\z&=5\end{aligned}$$

55.
$$\begin{aligned}-0.3a+1.4&=-1.2-0.7a\\-0.3a+0.7a+1.4&=-1.2-0.7a+0.7a\\0.4a+1.4&=-1.2\\0.4a+1.4+(-1.4)&=-1.2+(-1.4)\\0.4a&=-2.6\\\frac{0.4a}{0.4}&=\frac{-2.6}{0.4}\\a&=-6.5\end{aligned}$$

57.
$$\begin{aligned}6x+8-3x&=11-12x-13\\3x+8&=-12x-2\\3x+12x+8&=-12x+12x-2\\15x+8&=-2\\15x+8-8&=-2-8\\15x&=-10\\\frac{15x}{15}&=\frac{-10}{15}\\x&=-\frac{2}{3}\end{aligned}$$

59.
$$\begin{aligned}-3.5x+1.3&=-2.7x+1.5\\-3.5x+3.5x+1.3&=-2.7x+3.5x+1.5\\1.3&=0.8x+1.5\\1.3-1.5&=0.8x+1.5-1.5\\-0.2&=0.8x\\\frac{-0.2}{0.8}&=\frac{0.8x}{0.8}\\-0.25&=x\text{ or }x=-\frac{1}{4}\end{aligned}$$

61.
$$\begin{aligned} 5(4+x) &= 3(3x-1)-9 \\ 20+5x &= 9x-3-9 \\ 20+5x &= 9x-12 \\ 20+5x-5x &= 9x-5x-12 \\ 20 &= 4x-12 \\ 20+12 &= 4x-12+12 \\ 32 &= 4x \\ \frac{32}{4} &= \frac{4x}{4} \\ 8 &= x \end{aligned}$$

63.
$$\begin{aligned} 4x+3.2-1.9x &= 0.3x-4.9 \\ 2.1x+3.2 &= 0.3x-4.9 \\ 2.1x+(-0.3x)+3.2 &= 0.3x+(-0.3x)-4.9 \\ 1.8x+3.2 &= -4.9 \\ 1.8x+3.2+(-3.2) &= -4.9+(-3.2) \\ 1.8x &= -8.1 \\ \frac{1.8x}{1.8} &= \frac{-8.1}{1.8} \\ x &= -4.5 \text{ or } -\frac{9}{2} \end{aligned}$$

Cumulative Review

65. $-3y(2x+y)+5(3xy-y^2)$
$= -6xy-3y^2+15xy-5y^2$
$= (-6+15)xy+(-3-5)y^2$
$= 9xy-8y^2$

66. $-\{2(x-3)+3[x-(2x-5)]\}$
$= -\{2x-6+3[x-2x+5]\}$
$= -\{2x-6+3[-x+5]\}$
$= -\{2x-6-3x+15\}$
$= -\{-x+9\}$
$= x-9$

67. We multiply and then add.
$18.5\times\$18.66 = \$345.21$
$9.25\times\$57.56 = 532.43$
$5.0\times\$29.26 = 146.30$
$\$1023.94$
The value was \$1023.94 on October 5, 2007.

68. a. 30% of \$899 = 0.30 × \$899 = \$269.70
\$899 − \$269.70 = \$629.30
With a total discount of 30%, the sale price is \$629.30.

b. 20% of \$899 = 0.20 × \$899 = \$179.80
\$899 − \$179.80 = \$719.20
The price after the 20% discount is \$719.20.
10% of \$719.20 = 0.10 × \$719.20 = \$71.92
\$719.20 − \$71.92 = \$647.28
The sale price after both discounts is \$647.28.

Quick Quiz 2.3

1.
$$\begin{aligned} 7x-6 &= -4x-10 \\ 7x+4x-6 &= -4x+4x-10 \\ 11x-6 &= -10 \\ 11x-6+6 &= -10+6 \\ 11x &= -4 \\ \frac{11x}{11} &= \frac{-4}{11} \\ x &= -\frac{4}{11} \end{aligned}$$

2.
$$\begin{aligned} -3x+6.2 &= -5.8 \\ -3x+6.2-6.2 &= -5.8-6.2 \\ -3x &= -12 \\ \frac{-3x}{-3} &= \frac{-12}{-3} \\ x &= 4 \end{aligned}$$

3.
$$\begin{aligned} 2(3x-2) &= 4(5x+3) \\ 6x-4 &= 20x+12 \\ 6x-6x-4 &= 20x-6x+12 \\ -4 &= 14x+12 \\ -4-12 &= 14x+12-12 \\ -16 &= 14x \\ \frac{-16}{14} &= \frac{14x}{14} \\ -\frac{8}{7} &= x \end{aligned}$$

4. Answers may vary. Possible solution: Use the distributive property to remove parentheses. Combine like terms on the left side of the equation. Move the variable terms to the left side of the equation and the constants to the right side of the equation. Simplify.

2.4 Exercises

1.
$$\frac{1}{2}x+\frac{2}{3}=\frac{1}{6}$$
$$6\left(\frac{1}{2}x\right)+6\left(\frac{2}{3}\right)=6\left(\frac{1}{6}\right)$$
$$3x+4=1$$
$$3x+4+(-4)=1+(-4)$$
$$3x=-3$$
$$\frac{3x}{3}=\frac{-3}{3}$$
$$x=-1$$

Check: $\frac{1}{2}(-1)+\frac{2}{3}\stackrel{?}{=}\frac{1}{6}$
$$-\frac{1}{2}+\frac{2}{3}\stackrel{?}{=}\frac{1}{6}$$
$$-\frac{3}{6}+\frac{4}{6}\stackrel{?}{=}\frac{1}{6}$$
$$\frac{1}{6}=\frac{1}{6}\ \checkmark$$

3.
$$\frac{2}{3}x=\frac{1}{15}x+\frac{3}{5}$$
$$15\left(\frac{2}{3}x\right)=15\left(\frac{1}{15}x\right)+15\left(\frac{3}{5}\right)$$
$$10x=x+9$$
$$10x+(-x)=x+(-x)+9$$
$$9x=9$$
$$\frac{9x}{9}=\frac{9}{9}$$
$$x=1$$

Check: $\frac{2}{3}(1)\stackrel{?}{=}\frac{1}{15}(1)+\frac{3}{5}$
$$\frac{2}{3}\stackrel{?}{=}\frac{1}{15}+\frac{9}{15}$$
$$\frac{2}{3}\stackrel{?}{=}\frac{10}{15}$$
$$\frac{2}{3}=\frac{2}{3}\ \checkmark$$

5.
$$\frac{x}{2}+\frac{x}{5}=\frac{7}{10}$$
$$10\left(\frac{x}{2}\right)+10\left(\frac{x}{5}\right)=10\left(\frac{7}{10}\right)$$
$$5x+2x=7$$
$$7x=7$$
$$\frac{7x}{7}=\frac{7}{7}$$
$$x=1$$

Check: $\frac{1}{2}+\frac{1}{5}\stackrel{?}{=}\frac{7}{10}$
$$\frac{5}{10}+\frac{2}{10}\stackrel{?}{=}\frac{7}{10}$$
$$\frac{7}{10}=\frac{7}{10}\ \checkmark$$

7.
$$5-\frac{1}{3}x=\frac{1}{12}x$$
$$12(5)-12\left(\frac{1}{3}x\right)=12\left(\frac{1}{12}x\right)$$
$$60-4x=x$$
$$60-4x+4x=x+4x$$
$$60=5x$$
$$\frac{60}{5}=\frac{5x}{5}$$
$$12=x$$

Check: $5-\frac{1}{3}(12)\stackrel{?}{=}\frac{1}{12}(12)$
$$5-4\stackrel{?}{=}1$$
$$1=1\ \checkmark$$

9.
$$2+\frac{y}{2}=\frac{3y}{4}-3$$
$$4(2)+4\left(\frac{y}{2}\right)=4\left(\frac{3y}{4}\right)-4(3)$$
$$8+2y=3y-12$$
$$8=y-12$$
$$20=y$$

Check: $2+\left(\frac{20}{2}\right)\stackrel{?}{=}\frac{3(20)}{4}-3$
$$2+10\stackrel{?}{=}15-3$$
$$12=12\ \checkmark$$

11. $\frac{x-3}{5} = 1 - \frac{x}{3}$

$15\left(\frac{x-3}{5}\right) = 15(1) - 15\left(\frac{x}{3}\right)$

$3(x-3) = 15 - 5x$

$3x - 9 = 15 - 5x$

$3x + 5x - 9 = 15 - 5x + 5x$

$8x - 9 = 15$

$8x - 9 + 9 = 15 + 9$

$8x = 24$

$\frac{8x}{8} = \frac{24}{8}$

$x = 3$

Check: $\frac{3-3}{5} \stackrel{?}{=} 1 - \frac{3}{3}$

$\frac{0}{5} \stackrel{?}{=} 1 - 1$

$0 = 0$ ✓

13. $\frac{x+3}{4} = \frac{x}{2} + \frac{1}{6}$

$12\left(\frac{x+3}{4}\right) = 12\left(\frac{x}{2}\right) + 12\left(\frac{1}{6}\right)$

$3(x+3) = 6x + 2$

$3x + 9 = 6x + 2$

$3x + (-6x) + 9 = 6x + (-6x) + 2$

$-3x + 9 = 2$

$-3x + 9 + (-9) = 2 + (-9)$

$-3x = -7$

$\frac{-3x}{-3} = \frac{-7}{-3}$

$x = \frac{7}{3}$

Check: $\frac{\frac{7}{3}+3}{4} \stackrel{?}{=} \frac{\frac{7}{3}}{2} + \frac{1}{6}$

$\frac{\frac{7}{3}+\frac{9}{3}}{4} \stackrel{?}{=} \frac{7}{3} \cdot \frac{1}{2} + \frac{1}{6}$

$\frac{16}{3} \cdot \frac{1}{4} \stackrel{?}{=} \frac{7}{6} + \frac{1}{6}$

$\frac{4}{3} \stackrel{?}{=} \frac{8}{6}$

$\frac{4}{3} = \frac{4}{3}$ ✓

15. $0.6x + 5.9 = 3.8$

$10(0.6x) + 10(5.9) = 10(3.8)$

$6x + 59 = 38$

$6x + 59 - 59 = 38 - 59$

$6x = -21$

$\frac{6x}{6} = \frac{-21}{6}$

$x = -\frac{7}{2} = -3.5$

Check: $0.6(-3.5) + 5.9 \stackrel{?}{=} 3.8$

$-2.1 + 5.9 \stackrel{?}{=} 3.8$

$3.8 = 3.8$

17. $\frac{1}{2}(y-2) + 2 = \frac{3}{8}(3y-4),\ y \stackrel{?}{=} 4$

$\frac{1}{2}(4-2) + 2 \stackrel{?}{=} \frac{3}{8}(3 \cdot 4 - 4)$

$1 + 2 \stackrel{?}{=} \frac{3}{8}(8)$

$3 = 3$

$y = 4$ is a solution.

19. $\frac{1}{2}\left(y - \frac{1}{5}\right) = \frac{1}{5}(y+2),\ y \stackrel{?}{=} \frac{5}{8}$

$\frac{1}{2}\left(\frac{5}{8} - \frac{1}{5}\right) \stackrel{?}{=} \frac{1}{5}\left(\frac{5}{8} + 2\right)$

$\frac{1}{2}\left(\frac{17}{40}\right) \stackrel{?}{=} \frac{1}{5}\left(\frac{21}{8}\right)$

$\frac{17}{80} \neq \frac{42}{80}$

$y = \frac{5}{8}$ is not a solution.

21. $\frac{3}{4}(3x+1) = 2(3-2x) + 1$

$\frac{9}{4}x + \frac{3}{4} = 6 - 4x + 1$

$\frac{9}{4}x + \frac{3}{4} = -4x + 7$

$4\left(\frac{9}{4}x\right) + 4\left(\frac{3}{4}\right) = 4(-4x) + 4(7)$

$9x + 3 = -16x + 28$

$9x + 16x + 3 = -16x + 16x + 28$

$25x + 3 = 28$

$25x + 3 - 3 = 28 - 3$

$25x = 25$

$\frac{25x}{25} = \frac{25}{25}$

$x = 1$

23.
$$
\begin{aligned}
2(x-2) &= \frac{2}{5}(3x+1)+2 \\
2x-4 &= \frac{6}{5}x+\frac{2}{5}+2 \\
2x-4 &= \frac{6}{5}x+\frac{12}{5} \\
5(2x)-5(4) &= 5\left(\frac{6}{5}x\right)+5\left(\frac{12}{5}\right) \\
10x-20 &= 6x+12 \\
10x-6x-20 &= 6x-6x+12 \\
4x-20 &= 12 \\
4x-20+20 &= 12+20 \\
4x &= 32 \\
\frac{4x}{4} &= \frac{32}{4} \\
x &= 8
\end{aligned}
$$

25.
$$
\begin{aligned}
0.3x-0.2(3-5x) &= -0.5(x-6) \\
0.3x-0.6+x &= -0.5x+3 \\
1.3x-0.6 &= -0.5x+3 \\
10(1.3x)-10(0.6) &= 10(-0.5x)+10(3) \\
13x-6 &= -5x+30 \\
13x+5x-6 &= -5x+5x+30 \\
18x-6 &= 30 \\
18x-6+6 &= 30+6 \\
18x &= 36 \\
\frac{18x}{18} &= \frac{36}{18} \\
x &= 2
\end{aligned}
$$

27.
$$
\begin{aligned}
-8(0.1x+0.4)-0.9 &= -0.1 \\
-0.8x-3.2-0.9 &= -0.1 \\
-0.8x-4.1 &= -0.1 \\
10(-0.8x)-10(4.1) &= 10(-0.1) \\
-8x-41 &= -1 \\
-8x-41+41 &= -1+41 \\
-8x &= 40 \\
\frac{-8x}{-8} &= \frac{40}{-8} \\
x &= -5
\end{aligned}
$$

29.
$$
\begin{aligned}
\frac{1}{3}(y+2) &= 3y-5(y-2) \\
\frac{1}{3}y+\frac{2}{3} &= 3y-5y+10 \\
\frac{1}{3}y+\frac{2}{3} &= -2y+10 \\
3\left(\frac{1}{3}y\right)+3\left(\frac{2}{3}\right) &= 3(-2y)+3(10) \\
y+2 &= -6y+30 \\
y+6y+2 &= -6y+6y+30 \\
7y+2 &= 30 \\
7y+2-2 &= 30-2 \\
7y &= 28 \\
y &= 4
\end{aligned}
$$

31.
$$
\begin{aligned}
\frac{1+2x}{5}+\frac{4-x}{3} &= \frac{1}{15} \\
15\left(\frac{1+2x}{5}\right)+15\left(\frac{4-x}{3}\right) &= 15\left(\frac{1}{15}\right) \\
3(1+2x)+5(4-x) &= 1 \\
3+6x+20-5x &= 1 \\
x+23 &= 1 \\
x+23-23 &= 1-23 \\
x &= -22
\end{aligned}
$$

33.
$$
\begin{aligned}
\frac{1}{5}(x+3) &= 2x-3(2-x)-3 \\
\frac{1}{5}x+\frac{3}{5} &= 2x-6+3x-3 \\
\frac{1}{5}x+\frac{3}{5} &= 5x-9 \\
5\left(\frac{1}{5}x\right)+5\left(\frac{3}{5}\right) &= 5(5x)-5(9) \\
x+3 &= 25x-45 \\
x-x+3 &= 25x-x-45 \\
3 &= 24x-45 \\
3+45 &= 24x-45+45 \\
48 &= 24x \\
\frac{48}{24} &= \frac{24x}{24} \\
2 &= x
\end{aligned}
$$

35.
$$\begin{aligned}
\frac{1}{3}(x-2) &= 3x-2(x-1)+\frac{16}{3} \\
\frac{1}{3}x-\frac{2}{3} &= 3x-2x+2+\frac{16}{3} \\
\frac{1}{3}x-\frac{2}{3} &= x+\frac{22}{3} \\
3\left(\frac{1}{3}x\right)-3\left(\frac{2}{3}\right) &= 3(x)+3\left(\frac{22}{3}\right) \\
x-2 &= 3x+22 \\
x-3x-2 &= 3x-3x+22 \\
-2x-2 &= 22 \\
-2x-2+2 &= 22+2 \\
-2x &= 24 \\
\frac{-2x}{-2} &= \frac{24}{-2} \\
x &= -12
\end{aligned}$$

37.
$$\begin{aligned}
\frac{4}{5}x-\frac{2}{3} &= \frac{3x+1}{2} \\
30\left(\frac{4}{5}x\right)-30\left(\frac{2}{3}\right) &= 30\left(\frac{3x+1}{2}\right) \\
24x-20 &= 15(3x+1) \\
24x-20 &= 45x+15 \\
24x+(-45x)-20 &= 45x+(-45x)+15 \\
-21x-20 &= 15 \\
-21x-20+20 &= 15+20 \\
-21x &= 35 \\
\frac{-21x}{-21} &= \frac{35}{-21} \\
x &= -\frac{5}{3}
\end{aligned}$$

39.
$$\begin{aligned}
0.4x-0.5(2x+3) &= -0.7(x+3) \\
0.4x-x-1.5 &= -0.7x-2.1 \\
-0.6x-1.5 &= -0.7x-2.1 \\
10(-0.6x)-10(1.5) &= 10(-0.7x)-10(2.1) \\
-6x-15 &= -7x-21 \\
-6x+7x-15 &= -7x+7x-21 \\
x-15 &= -21 \\
x-15+15 &= -21+15 \\
x &= -6
\end{aligned}$$

41.
$$\begin{aligned}
-1+5(x-2) &= 12x+3-7x \\
-1+5x-10 &= 5x+3 \\
5x-11 &= 5x+3 \\
5x-5x-11 &= 5x-5x+3 \\
-11 &= 3, \text{ no solution}
\end{aligned}$$

43.
$$\begin{aligned}
9(x+3)-6 &= 24-2x-3+11x \\
9x+27-6 &= 9x+21 \\
9x+21 &= 9x+21 \\
9x-9x+21 &= 9x-9x+21 \\
21 &= 21
\end{aligned}$$
Infinite number of solutions

45.
$$\begin{aligned}
-3(4x-1) &= 5(2x-1)+8 \\
-12x+3 &= 10x-5+8 \\
-12x+3 &= 10x+3 \\
-12x-10x+3 &= 10x-10x+3 \\
-22x+3 &= 3 \\
-22x+3-3 &= 3-3 \\
-22x &= 0 \\
\frac{-22x}{-22} &= \frac{0}{-22} \\
x &= 0
\end{aligned}$$

47.
$$\begin{aligned}
3(4x+1)-2x &= 2(5x-3) \\
12x+3-2x &= 10x-6 \\
10x+3 &= 10x-6 \\
10x+(-10x)+3 &= 10x+(-10x)-6 \\
3 &= -6
\end{aligned}$$
No solution

Cumulative Review

49.
$$\begin{aligned}
\left(-3\frac{1}{4}\right)\left(5\frac{1}{3}\right) &= \left(-\frac{13}{4}\right)\left(\frac{16}{3}\right) \\
&= -\frac{13\cdot \cancel{4}\cdot 4}{\cancel{4}\cdot 3} \\
&= -\frac{52}{3} \text{ or } -17\frac{1}{3}
\end{aligned}$$

50.
$$\begin{aligned}
5\frac{1}{2}\div 1\frac{1}{4} &= \frac{11}{2}\div\frac{5}{4} \\
&= \frac{11}{2}\cdot\frac{4}{5} \\
&= \frac{11\cdot \cancel{2}\cdot 2}{\cancel{2}\cdot 5} \\
&= \frac{22}{5} \text{ or } 4\frac{2}{5}
\end{aligned}$$

51. 30% of 440 = 0.30 × 440 = 132
440 + 132 = 572
30% of 750 = 0.3 × 750 = 225
750 + 225 = 975
The weight range for females is 572 – 975 grams.

52. Find the area of the seating area.

$$\begin{aligned}\text{Area} &= \frac{B+b}{2}h\\ &= \frac{150+88}{2}\cdot 200\\ &= \frac{238}{2}\cdot 200\\ &= 119\cdot 200\\ &= 23{,}800\ \text{ft}^2\end{aligned}$$

Find the area required for each seat.

$\text{Area} = L\cdot W = 2.5\cdot 3 = 7.5\ \text{ft}^2$

Now divide.

$23{,}800 \div 7.5 \approx 3173$

The auditorium will hold approximately 3173 seats.

Quick Quiz 2.4

1.
$$\begin{aligned}\frac{3}{4}x+\frac{5}{12} &= \frac{1}{3}x-\frac{1}{6}\\ 12\left(\frac{3}{4}x\right)+12\left(\frac{5}{12}\right) &= 12\left(\frac{1}{3}x\right)-12\left(\frac{1}{6}\right)\\ 9x+5 &= 4x-2\\ 9x-4x+5 &= 4x-4x-2\\ 5x+5 &= -2\\ 5x+5-5 &= -2-5\\ 5x &= -7\\ \frac{5x}{5} &= \frac{-7}{5}\\ x &= -\frac{7}{5}\end{aligned}$$

2.
$$\begin{aligned}\frac{2}{3}x-\frac{3}{5}+\frac{7}{5}x+\frac{1}{3} &= 1\\ 15\left(\frac{2}{3}x\right)-15\left(\frac{3}{5}\right)+15\left(\frac{7}{5}x\right)+15\left(\frac{1}{3}\right) &= 15(1)\\ 10x-9+21x+5 &= 15\\ 31x-4 &= 15\\ 31x-4+4 &= 15+4\\ 31x &= 19\\ \frac{31x}{31} &= \frac{19}{31}\\ x &= \frac{19}{31}\end{aligned}$$

3.
$$\begin{aligned}\frac{2}{3}(x+2)+\frac{1}{4} &= \frac{1}{2}(5-3x)\\ \frac{2}{3}x+\frac{4}{3}+\frac{1}{4} &= \frac{5}{2}-\frac{3}{2}x\\ \frac{2}{3}x+\frac{19}{12} &= \frac{5}{2}-\frac{3}{2}x\\ 12\left(\frac{2}{3}x\right)+12\left(\frac{19}{12}\right) &= 12\left(\frac{5}{2}\right)-12\left(\frac{3}{2}x\right)\\ 8x+19 &= 30-18x\\ 8x+18x+19 &= 30-18x+18x\\ 26x+19 &= 30\\ 26x+19-19 &= 30-19\\ 26x &= 11\\ \frac{26x}{26} &= \frac{11}{26}\\ x &= \frac{11}{26}\end{aligned}$$

How Am I Doing? Sections 2.1–2.4

1.
$$\begin{aligned}5-8+x &= -12\\ -3+x &= -12\\ -3+3+x &= -12+3\\ x &= -9\end{aligned}$$

2.
$$\begin{aligned}-2.8+x &= 4.7\\ -2.8+2.8+x &= 4.7+2.8\\ x &= 7.5\end{aligned}$$

3.
$$\begin{aligned}-45 &= -5x\\ \frac{-45}{-5} &= \frac{-5x}{-5}\\ 9 &= x\end{aligned}$$

4.
$$\begin{aligned}12x-6x &= -48\\ 6x &= -48\\ \frac{6x}{6} &= \frac{-48}{6}\\ x &= -8\end{aligned}$$

5.
$$\begin{aligned}-1.2x+3.5 &= 2.7\\ -1.2x+3.5+(-3.5) &= 2.7+(-3.5)\\ -1.2x &= -0.8\\ \frac{-1.2x}{-1.2} &= \frac{-0.8}{-1.2}\\ x &= \frac{2}{3}\end{aligned}$$

6.
$$\begin{aligned}-14x+9&=2x+7\\-14x-2x+9&=2x-2x+7\\-16x+9&=7\\-16x+9-9&=7-9\\-16x&=-2\\\frac{-16x}{-16}&=\frac{-2}{-16}\\x&=\frac{1}{8}\end{aligned}$$

7.
$$\begin{aligned}14x+2(7-2x)&=20\\14x+14-4x&=20\\10x+14&=20\\10x+14+(-14)&=20+(-14)\\10x&=6\\\frac{10x}{10}&=\frac{6}{10}\\x&=\frac{3}{5}\end{aligned}$$

8.
$$\begin{aligned}0.5(1.2x-3.4)&=-1.4x+5.8\\0.6x-1.7&=-1.4x+5.8\\0.6x+1.4x-1.7&=-1.4x+1.4x+5.8\\2x-1.7&=5.8\\2x-1.7+1.7&=5.8+1.7\\2x&=7.5\\\frac{2x}{2}&=\frac{7.5}{2}\\x&=3.75\text{ or }\frac{15}{4}\end{aligned}$$

9.
$$\begin{aligned}3(x+6)&=-2(4x-1)+x\\3x+18&=-8x+2+x\\3x+18&=-7x+2\\3x+7x+18&=-7x+7x+2\\10x+18&=2\\10x+18+(-18)&=2+(-18)\\10x&=-16\\\frac{10x}{10}&=-\frac{16}{10}\\x&=-\frac{8}{5}\text{ or }-1.6\end{aligned}$$

10.
$$\begin{aligned}\frac{x}{3}+\frac{x}{4}&=\frac{5}{6}\\12\left(\frac{x}{3}\right)+12\left(\frac{x}{4}\right)&=12\left(\frac{5}{6}\right)\\4x+3x&=10\\7x&=10\\\frac{7x}{7}&=\frac{10}{7}\\x&=\frac{10}{7}\end{aligned}$$

11.
$$\begin{aligned}\frac{1}{4}(x+3)&=4x-2(x-3)\\\frac{1}{4}x+\frac{3}{4}&=4x-2x+6\\\frac{1}{4}x+\frac{3}{4}&=2x+6\\4\left(\frac{1}{4}x\right)+4\left(\frac{3}{4}\right)&=4(2x)+4(6)\\x+3&=8x+24\\x+(-x)+3&=8x+(-x)+24\\3&=7x+24\\3+(-24)&=7x+24+(-24)\\-21&=7x\\\frac{-21}{7}&=\frac{7x}{7}\\-3&=x\end{aligned}$$

12.
$$\begin{aligned}\frac{1}{2}(x-1)+2&=3(2x-1)\\\frac{1}{2}x-\frac{1}{2}+2&=6x-3\\\frac{1}{2}x+\frac{3}{2}&=6x-3\\2\left(\frac{1}{2}x\right)+2\left(\frac{3}{2}\right)&=2(6x)-2(3)\\x+3&=12x-6\\x+(-x)+3&=12x+(-x)-6\\3&=11x-6\\3+6&=-11x-6+6\\9&=11x\\\frac{9}{11}&=\frac{11x}{11}\\\frac{9}{11}&=x\end{aligned}$$

13. $\frac{1}{7}(7x-14)-2=\frac{1}{3}(x-2)$

$$x-2-2=\frac{1}{3}x-\frac{2}{3}$$
$$x-4=\frac{1}{3}x-\frac{2}{3}$$
$$3(x)-3(4)=3\left(\frac{1}{3}x\right)-3\left(\frac{2}{3}\right)$$
$$3x-12=x-2$$
$$3x-x-12=x-x-2$$
$$2x-12=-2$$
$$2x-12+12=-2+12$$
$$2x=10$$
$$\frac{2x}{2}=\frac{10}{2}$$
$$x=5$$

14. $0.2(x-3)=4(0.2x-0.1)$

$$0.2x-0.6=0.8x-0.4$$
$$10(0.2x)-10(0.6)=10(0.8x)-10(0.4)$$
$$2x-6=8x-4$$
$$2x-2x-6=8x-2x-4$$
$$-6=6x-4$$
$$-6+4=6x-4+4$$
$$-2=6x$$
$$\frac{-2}{6}=\frac{6x}{6}$$
$$-\frac{1}{3}=x$$

2.5 Exercises

1. Multiply each term by 5. Then subtract 160 from each side. Then divide each side by 9. We would obtain $\frac{5F-160}{9}=C$.

3. a. $A=60\text{ m}^2$, $a=12\text{ m}$

$$A=\frac{1}{2}ab$$
$$60=\frac{1}{2}(12)(b)$$
$$60=6b$$
$$\frac{60}{6}=\frac{6b}{6}$$
$$10=b$$

base = 10 m

b. $A=88\text{ m}^2$, $b=11\text{ m}$

$$A=\frac{1}{2}ab$$
$$88=\frac{1}{2}a(11)$$
$$88=\frac{11}{2}a$$
$$\frac{2}{11}(88)=\frac{2}{11}\left(\frac{11}{2}a\right)$$
$$16=a$$

altitude = 16 m

5. a. $4x+3y=18$

$$4x+(-4x)+3y=18+(-4x)$$
$$3y=18-4x$$
$$\frac{3y}{3}=\frac{18-4x}{3}$$
$$y=\frac{18-4x}{3}\text{ or }y=-\frac{4}{3}x+6$$

b. $y=-\frac{4}{3}(-3)+6$

$$y=4+6$$
$$y=10$$

7. $A=\frac{1}{2}bh$

$$2(A)=2\left(\frac{1}{2}bh\right)$$
$$2A=bh$$
$$\frac{2A}{h}=\frac{bh}{h}$$
$$\frac{2A}{h}=b$$

9. $I=Prt$

$$\frac{I}{rt}=\frac{Prt}{rt}$$
$$\frac{I}{rt}=P$$

11. $y=mx+b$

$$y-b=mx$$
$$\frac{y-b}{x}=\frac{mx}{x}$$
$$\frac{y-b}{x}=m$$

13.
$$\begin{aligned}8x-12y&=24\\8x+(-8x)-12y&=24+(-8x)\\-12y&=24-8x\\\frac{-12y}{-12}&=\frac{24-8x}{-12}\\y&=\frac{8}{12}x-\frac{24}{12}\\y&=\frac{2}{3}x-2\end{aligned}$$

15.
$$\begin{aligned}y&=-\frac{2}{3}x+4\\3y&=3\left(-\frac{2}{3}x\right)+3(4)\\3y&=-2x+12\\3y+(-12)&=-2x+12+(-12)\\3y-12&=-2x\\\frac{3y-12}{-2}&=\frac{-2x}{-2}\\\frac{3y}{-2}-\frac{12}{-2}&=x\\-\frac{3}{2}y+6&=x\end{aligned}$$

17.
$$\begin{aligned}ax+by&=c\\by&=c-ax\\\frac{by}{b}&=\frac{c-ax}{b}\\y&=\frac{c-ax}{b}\end{aligned}$$

19.
$$\begin{aligned}A&=\pi r^2\\\frac{A}{\pi}&=\frac{\pi r^2}{\pi}\\\frac{A}{\pi}&=r^2\end{aligned}$$

21.
$$\begin{aligned}g&=\frac{GM}{r^2}\\r^2(g)&=r^2\left(\frac{GM}{r^2}\right)\\gr^2&=GM\\\frac{gr^2}{g}&=\frac{GM}{g}\\r^2&=\frac{GM}{g}\end{aligned}$$

23.
$$\begin{aligned}A&=P(1+rt)\\A&=P+Prt\\A-P&=Prt\\\frac{A-P}{Pr}&=\frac{Prt}{Pr}\\\frac{A-P}{Pr}&=t\end{aligned}$$

25.
$$\begin{aligned}S&=2\pi rh+2\pi r^2\\S-2\pi r^2&=2\pi rh+2\pi r^2-2\pi r^2\\S-2\pi r^2&=2\pi rh\\\frac{S-2\pi r^2}{2\pi r}&=\frac{2\pi rh}{2\pi r}\\\frac{S-2\pi r^2}{2\pi r}&=h\end{aligned}$$

27.
$$\begin{aligned}K&=\frac{1}{2}mv^2\\2(K)&=2\left(\frac{1}{2}mv^2\right)\\2K&=mv^2\\\frac{2K}{v^2}&=\frac{mv^2}{v^2}\\\frac{2K}{v^2}&=m\end{aligned}$$

29.
$$\begin{aligned}V&=LWH\\\frac{V}{WH}&=\frac{LWH}{WH}\\\frac{V}{WH}&=L\end{aligned}$$

31.
$$\begin{aligned}V&=\frac{1}{3}\pi r^2h\\3(V)&=3\left(\frac{1}{3}\pi r^2h\right)\\3V&=\pi r^2h\\\frac{3V}{\pi h}&=\frac{\pi r^2h}{\pi h}\\\frac{3V}{\pi h}&=r^2\end{aligned}$$

33.
$$\begin{aligned}P&=2L+2W\\P-2L&=2W\\\frac{P-2L}{2}&=\frac{2W}{2}\\\frac{P-2L}{2}&=W\end{aligned}$$

35.
$$c^2 = a^2 + b^2$$
$$c^2 - b^2 = a^2 + b^2 - b^2$$
$$c^2 - b^2 = a^2$$

37.
$$F = \frac{9}{5}C + 32$$
$$F - 32 = \frac{9}{5}C$$
$$\frac{5}{9}(F - 32) = \frac{5}{9}\left(\frac{9}{5}C\right)$$
$$\frac{5}{9}(F - 32) = C$$
$$C = \frac{5(F-32)}{9} \text{ or } C = \frac{5F - 160}{9}$$

39.
$$P = \frac{E^2}{R}$$
$$RP = R \cdot \frac{E^2}{R}$$
$$RP = E^2$$
$$\frac{RP}{P} = \frac{E^2}{P}$$
$$R = \frac{E^2}{P}$$

41.
$$A = \frac{\pi r^2 S}{360}$$
$$360A = 360 \cdot \frac{\pi r^2 S}{360}$$
$$360A = \pi r^2 S$$
$$\frac{360A}{\pi r^2} = \frac{\pi r^2 S}{\pi r^2}$$
$$\frac{360A}{\pi r^2} = S$$

43. $W = \frac{P - 2L}{2}$, $P = 5.8$, $L = 2.1$

$$W = \frac{5.8 - 2(2.1)}{2} = 0.8$$

The width is 0.8 mile.

45. $L = \frac{V}{WH}$, $V = 5940$, $W = 22$, $H = 9$

$$L = \frac{5940}{(22)(9)} = 30$$

The length is 30 feet.

47. a.
$$V = 1100x + 7050$$
$$V - 7050 = 1100x$$
$$\frac{V - 7050}{1100} = \frac{1100x}{1100}$$
$$\frac{V - 7050}{1100} = x$$

b. $V = 36,750$

$$x = \frac{36,750 - 7050}{1100} = 27 \text{ years}$$

Year = 1985 + 27 = 2012

The year is 2012.

49. $A = \frac{1}{2}ab$

If $b \to 2b$, $A = \frac{1}{2}a(2b) = ab$

A doubles.

51. $A = \pi r^2$

If $r \Rightarrow 2r$, $A = \pi(2r)^2 = 4\pi r^2$

A is quadrupled if r doubles.

Cumulative Review

53. 20% of \$80 = $0.20 \times \$80 = \$16$

54. 0.5% of 200 = $0.005 \times 200 = 1$

55.
$$\left(3\frac{1}{4}\right)(12,000) = \frac{13}{4}(12,000)$$
$$= 39,000 \text{ square feet}$$

The company needs 39,000 square feet of plastic.

56.
$$\text{Total} = 4\frac{1}{3} + 2\frac{3}{4} + 3\frac{1}{2}$$
$$= \frac{13}{3} + \frac{11}{4} + \frac{7}{2}$$
$$= \frac{52}{12} + \frac{33}{12} + \frac{42}{12}$$
$$= \frac{127}{12}$$
$$= 10\frac{7}{12}$$

The spotlight was used $10\frac{7}{12}$ hours.

Quick Quiz 2.5

1.
$$A = 3x + 2w$$
$$A - 2w = 3x + 2w - 2w$$
$$A - 2w = 3x$$
$$\frac{A - 2w}{3} = \frac{3x}{3}$$
$$\frac{A - 2w}{3} = x$$

2.
$$A = \frac{1}{3}h(a + b)$$
$$A = \frac{1}{3}ah + \frac{1}{3}bh$$
$$3(A) = 3\left(\frac{1}{3}ah\right) + 3\left(\frac{1}{3}bh\right)$$
$$3A = ah + bh$$
$$3A - bh = ah$$
$$\frac{3A - bh}{h} = \frac{ah}{h}$$
$$\frac{3A - bh}{h} = a$$

3.
$$2ax(3 - y) = 2axy - 5$$
$$6ax - 2axy = 2axy - 5$$
$$6ax = 2axy - 5 + 2axy$$
$$6ax = 4axy - 5$$
$$6ax + 5 = 4axy$$
$$\frac{6ax + 5}{4ax} = \frac{4axy}{4ax}$$
$$\frac{6ax + 5}{4ax} = y$$

4. Answers may vary. Possible solution: Add 9 to both sides. Then multiply both sides by $\frac{8}{3}$. We would obtain:
$$y = \frac{3}{8}x - 9$$
$$y + 9 = \frac{3}{8}x - 9 + 9$$
$$y + 9 = \frac{3}{8}x$$
$$\frac{8}{3}(y + 9) = \frac{8}{3} \cdot \frac{3}{8}x$$
$$\frac{8(y + 9)}{3} = x$$

2.6 Exercises

1. $5 > -6$ is equivalent to $-6 < 5$. Both statements imply that 5 is to the right of –6 on a number line.

3. $9 \ ? -3$
Use >, since 9 is to the right of –3 on the number line.
$9 > -3$

5. $-4 \ ? -2$
Use <, since –4 is to the left of –2 on the number line.
$-4 < -2$

7. $\frac{3}{5} \ ? \ \frac{4}{7}$
$\frac{21}{35} \ ? \ \frac{20}{35}$
Use >, since 21 > 20.
$\frac{3}{5} > \frac{4}{7}$

9. $-4.2 \ ? \ 3.5$
Use <, since –4.2 is to the left of 3.5 on the number line.
$-4.2 < 3.5$

11. $-\frac{10}{3} \ ? -3$
$-3\frac{1}{3} \ ? -3$
Use <, since $-3\frac{1}{3}$ is to the left of –3 on the number line.
$-\frac{10}{3} < -3$

13. $-\frac{5}{8} \ ? -\frac{3}{5}$
$-0.625 \ ? -0.6$
Use <, since –0.625 is to the left of –0.6 on the number line.
$-\frac{5}{8} < -\frac{3}{5}$

15. $\frac{123}{4986} \approx 0.024669$
0.0247 is greater.

17. $x > 7$
x is greater than 7. All of the points to the right of 7 are shaded.

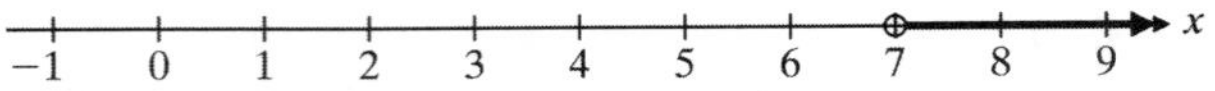

19. x is greater than or equal to −6. All of the points to the right of −6 are shaded. The closed circle indicates that we do include the point for −6.

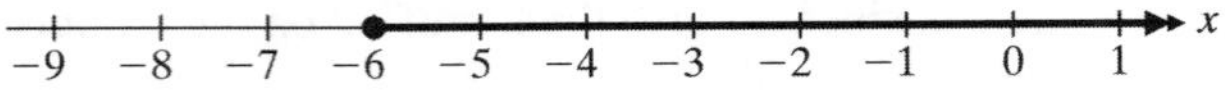

21. $x < -\frac{1}{4}$

x is less than $-\frac{1}{4}$. All of the points to the left of $-\frac{1}{4}$ are shaded.

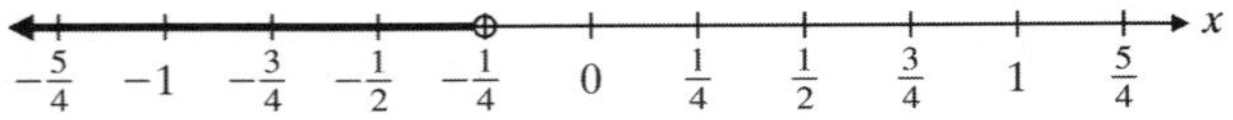

23. $x \le -5.3$
x is less than or equal to −5.3. All of the points to the left of −5.3 are shaded. The closed circle indicates that we do include the point for −5.3.

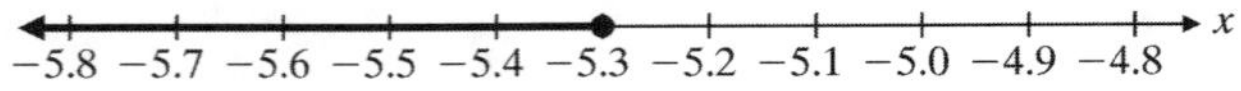

25. $25 < x$
25 is less than x is equivalent to x is greater than 25. All of the points to the right of 25 are shaded.

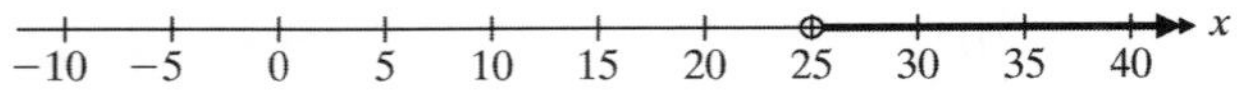

27. x is greater than or equal to $-\frac{2}{3}$.

$x \ge -\frac{2}{3}$

29. x is less than −20.
$x < -20$

31. x is less than or equal to 3.7.
$x \le 3.7$

33. Since the number of hours must not be less than 12, then the number of hours must be greater than or equal to 12. Thus we write $c \ge 12$.

35. Since the height must be at least 48 inches, the height must be greater than or equal to 48 inches. Thus we write $h \ge 48$.

Cumulative Review

37. 16 is what percent of 800?
$\frac{16}{800} = 0.02 = 2\%$
2% are accepted.

38. $\frac{3}{8} = 0.375 = 37.5\%$

39. Increase = 10 − 8 = 2

Percent increase $= \frac{2}{8} = 0.25 = 25\%$

100% + 25% = 125%
New price = 125% of 12 = 1.25(12) = $15

40. We subtract.
$414 - (-81) = 414 + 81 = 495$ ft
The top of the cliff is 495 feet from the tree trunk.

Quick Quiz 2.6

1. $x > 5$
x is greater than 5. All of the points to the right of 5 are shaded.

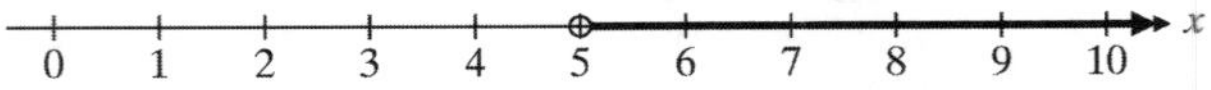

2. $x \le -3.5$
x is less than or equal to −3.5. All of the points to the left of −3.5 are shaded. The closed circle indicates that we do include the point for −3.5.

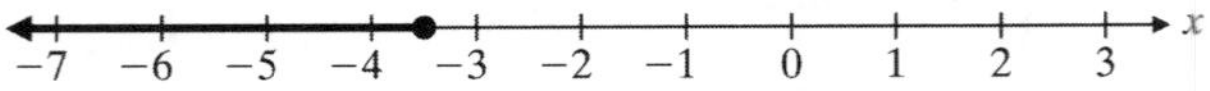

3. x is greater than or equal to −2.
$x \ge -2$

4. Answers may vary. Possible solution: $12 < x$ is the same as $x > 12$, but written in different ways. The graphs will be exactly the same.

2.7 Exercises

1.
$$x + 7 \le 4$$
$$x + 7 - 7 \le 4 - 7$$
$$x \le -3$$

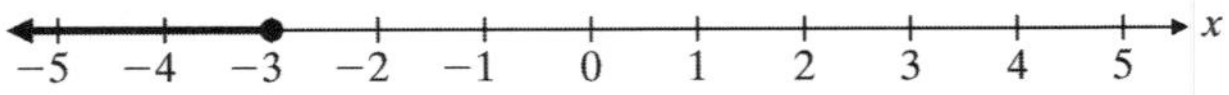

3. $$\begin{aligned} 5x &\le 25 \\ \frac{5x}{5} &\le \frac{25}{5} \\ x &\le 5 \end{aligned}$$

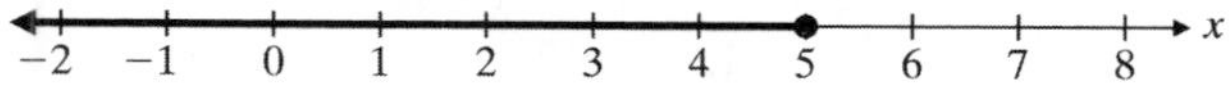

5. $$\begin{aligned} -2x &< 18 \\ \frac{-2x}{-2} &> \frac{18}{-2} \\ x &> -9 \end{aligned}$$

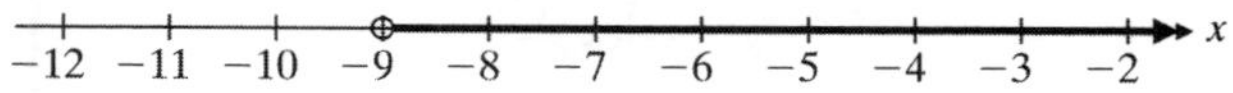

7. $$\begin{aligned} \frac{1}{2}x &\ge 4 \\ 2\left(\frac{1}{2}x\right) &\ge 2(4) \\ x &\ge 8 \end{aligned}$$

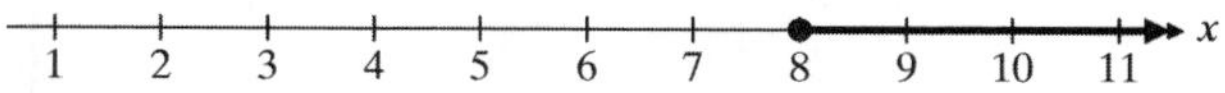

9. $$\begin{aligned} -\frac{1}{4}x &> 3 \\ -4\left(-\frac{1}{4}x\right) &< -4(3) \\ x &< -12 \end{aligned}$$

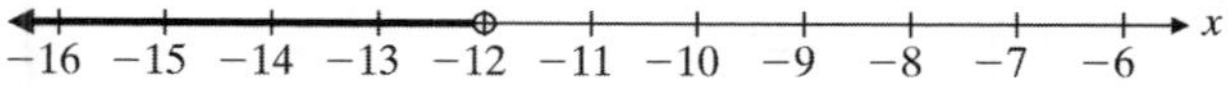

11. $$\begin{aligned} 2x-3 &< 4 \\ 2x-3+3 &< 4+3 \\ 2x &< 7 \\ \frac{2x}{2} &< \frac{7}{2} \\ x &< \frac{7}{2} \end{aligned}$$

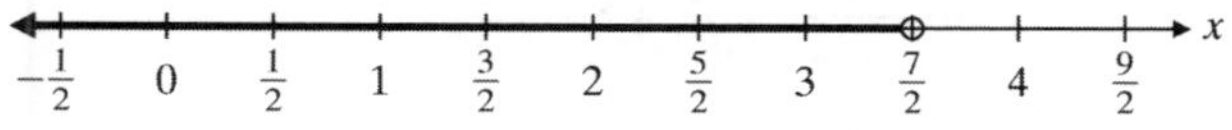

13. $$\begin{aligned} 8-5x &> 13 \\ 8-5x-8 &> 13-8 \\ -5x &> 5 \\ \frac{-5x}{-5} &< \frac{5}{-5} \\ x &< -1 \end{aligned}$$

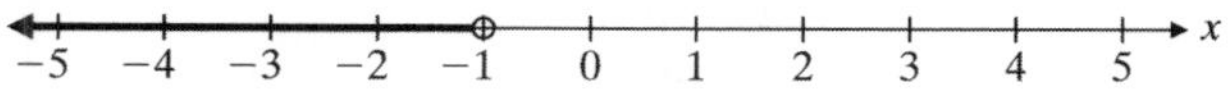

15. $$\begin{aligned} -4+5x &< -3x+8 \\ -4x+5x+3x &< -3x+3x+8 \\ -4+8x &< 8 \\ -4+4+8x &< 8+4 \\ 8x &< 12 \\ \frac{8x}{8} &< \frac{12}{8} \\ x &< \frac{3}{2} \end{aligned}$$

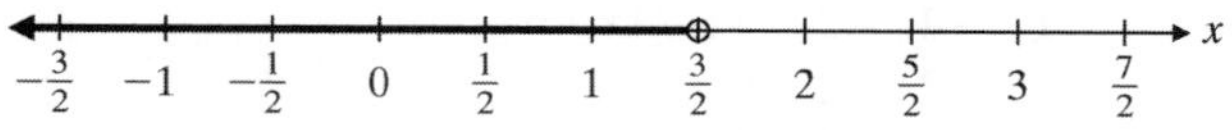

17. $$\begin{aligned} \frac{5x}{6}-5 &> \frac{x}{6}-9 \\ 6\left(\frac{5x}{6}\right)-6(5) &> 6\left(\frac{x}{6}\right)-6(9) \\ 5x-30 &> x-54 \\ 5x-30-x &> x-54-x \\ 4x-30 &> -54 \\ 4x-30+30 &> -54+30 \\ 4x &> -24 \\ \frac{4x}{4} &> \frac{-24}{4} \\ x &> -6 \end{aligned}$$

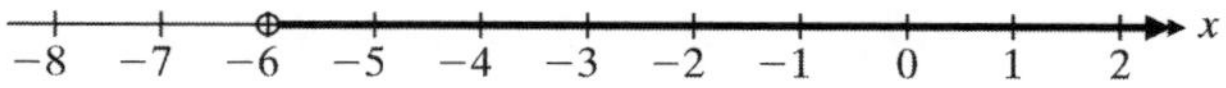

19. $$\begin{aligned} 2(3x+4) &> 3(x+3) \\ 6x+8 &> 3x+9 \\ 6x+8-3x &> 3x+9-3x \\ 3x+8 &> 9 \\ 3x+8-8 &> 9-8 \\ 3x &> 1 \\ \frac{3x}{3} &> \frac{1}{3} \\ x &> \frac{1}{3} \end{aligned}$$

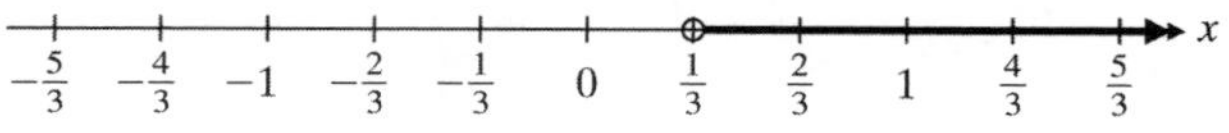

21. $$\begin{aligned} 5 &> 3 \\ 5+(-2) &> 3+(-2) \\ 3 &> 1 \end{aligned}$$

Adding any number to both sides of an inequality does not reverse the direction.

23.
$$\begin{aligned}3x+8&<7x-4\\3x+(-7x)+8&<7x+(-7x)-4\\-4x+8&<-4\\-4x+8+(-8)&<-4+(-8)\\-4x&<-12\\\frac{-4x}{-4}&>\frac{-12}{-4}\\x&>3\end{aligned}$$

25.
$$\begin{aligned}6x-2&\ge 4x+6\\6x-4x-2&\ge 4x-4x+6\\2x-2&\ge 6\\2x-2+2&\ge 6+2\\2x&\ge 8\\\frac{2x}{2}&\ge\frac{8}{2}\\x&\ge 4\end{aligned}$$

27.
$$\begin{aligned}0.3(x-1)&<0.1x-0.5\\0.3x-0.3&<0.1x-0.5\\10(0.3x)-10(0.3)&<10(0.1x)-10(0.5)\\3x-3&<x-5\\3x-3-x&<x-5-x\\2x-3&<-5\\2x-3+3&<-5+3\\2x&<-2\\\frac{2x}{2}&<\frac{-2}{2}\\x&<-1\end{aligned}$$

29.
$$\begin{aligned}3+5(2-x)&\ge -3(x+5)\\3+10-5x&\ge -3x-15\\13-5x&\ge -3x-15\\13-5x+3x&\ge -3x-15+3x\\-2x+13&\ge -15\\-2x+13-13&\ge -15-13\\-2x&\ge -28\\\frac{-2x}{-2}&\le\frac{-28}{-2}\\x&\le 14\end{aligned}$$

31.
$$\begin{aligned}\frac{x+6}{7}-\frac{6}{14}&>\frac{x+3}{2}\\14\left(\frac{x+6}{7}\right)-14\left(\frac{6}{14}\right)&>14\left(\frac{x+3}{2}\right)\\2(x+6)-6&>7(x+3)\\2x+12-6&>7x+21\\2x+6&>7x+21\\2x+6-7x&>7x+21-7x\\-5x+6&>21\\-5x+6-6&>21-6\\-5x&>15\\\frac{-5x}{-5}&<\frac{15}{-5}\\x&<-3\end{aligned}$$

33.
$$\begin{aligned}\frac{1}{6}-\frac{1}{2}(3x+2)&<\frac{1}{3}\left(x-\frac{1}{2}\right)\\\frac{1}{6}-\frac{3}{2}x-1&<\frac{1}{3}x-\frac{1}{6}\\-\frac{3}{2}x-\frac{5}{6}&<\frac{1}{3}x-\frac{1}{6}\\6\left(-\frac{3}{2}x\right)-6\left(\frac{5}{6}\right)&<6\left(\frac{1}{3}x\right)-6\left(\frac{1}{6}\right)\\-9x-5&<2x-1\\-9x-2x-5&<2x-2x-1\\-11x-5&<-1\\-11x-5+5&<-1+5\\-11x&<4\\\frac{-11x}{-11}&>\frac{4}{-11}\\x&>-\frac{4}{11}\end{aligned}$$

35.
$$\begin{aligned}\frac{75+83+86+x}{4}&\ge 80\\\frac{244+x}{4}&\ge 80\\4\left(\frac{244+x}{4}\right)&\ge 4(80)\\244+x&\ge 320\\x&\ge 76\end{aligned}$$
The student must get a 76 or higher.

37.
$$\begin{aligned}268+4x&\ge 300\\268+4x-268&\ge 300-268\\4x&\ge 32\\\frac{4x}{4}&\ge\frac{32}{4}\\x&\ge 8\end{aligned}$$
It will take 8 days or more.

Cumulative Review

39. $A = \pi r^2,\ r = \dfrac{9}{2} = 4.5$

$A \approx 3.14(4.5)^2 \approx 63.6$

The area of the basketball is 63.6 in.^2.

40. $B = 150,\ b = 100,\ h = 60$

$$\begin{aligned} A &= \frac{B+b}{2}h \\ &= \frac{150+100}{2}\cdot 60 \\ &= \frac{250}{2}\cdot 60 \\ &= 125\cdot 60 \\ &= 7500\text{ ft}^2 \end{aligned}$$

The area of the seating area is 7500 square feet.

Quick Quiz 2.7

1.
$$\begin{aligned} 4(x-3) &\le 2x \\ 4x-12 &\le 2x \\ 4x-12-2x &\le 2x-2x \\ 2x-12 &\le 0 \\ 2x-12+12 &\le 0+12 \\ 2x &\le 12 \\ \frac{2x}{2} &\le \frac{12}{2} \\ x &\le 6 \end{aligned}$$

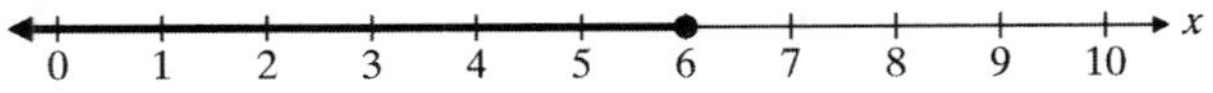

2.
$$\begin{aligned} 4-3(x+2)+5(x-4) &> 0 \\ 4-3x-6+5x-20 &> 0 \\ 2x-22 &> 0 \\ 2x-22+22 &> 0+22 \\ 2x &> 22 \\ \frac{2x}{2} &> \frac{22}{2} \\ x &> 11 \end{aligned}$$

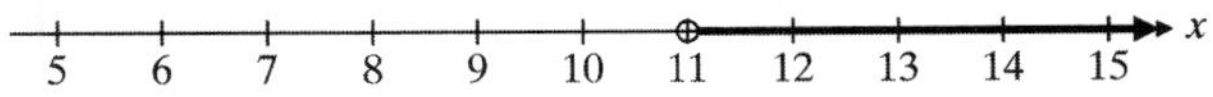

3.
$$\begin{aligned} \frac{1}{2}x+\frac{1}{4} &\le \frac{1}{8}(5x-3) \\ \frac{1}{2}x+\frac{1}{4} &\le \frac{5}{8}x-\frac{3}{8} \\ 8\left(\frac{1}{2}x\right)+8\left(\frac{1}{4}\right) &\le 8\left(\frac{5}{8}x\right)-8\left(\frac{3}{8}\right) \\ 4x+2 &\le 5x-3 \\ 4x+2-5x &\le 5x-3-5x \\ -x+2 &\le -3 \\ -x+2-2 &\le -3-2 \\ -x &\le -5 \\ \frac{-1x}{-1} &\ge \frac{-5}{-1} \\ x &\ge 5 \end{aligned}$$

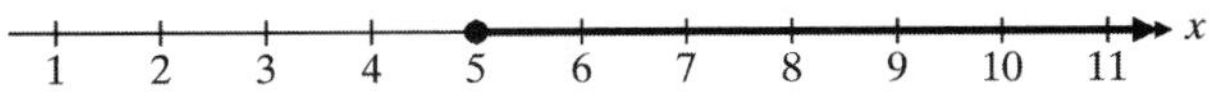

4. Answers may vary. Possible solution:
Use the distributive property to remove parentheses. Combine like terms on the left. Collect x-terms on the left and integers on the right. Simplify to obtain $-9x < -36$. Divide both sides by -9 and reverse the equality sign to obtain $x > 4$.

Putting Your Skills to Work

1. Find 5% of \$2500.
$0.05 \times 2500 = 125$
He saves \$125 for 2 years or 24 months.
$125 \times 24 = 3000$
Michael will have saved \$3000.

2. $\dfrac{5000}{24} = 208\dfrac{1}{3}$
Michael will need to save about \$209 per month.

3. Answers may vary.

4. Answers may vary.

5. Answers may vary.

Chapter 2 Review Problems

1.
$$\begin{aligned} 5x &= -35 \\ \frac{5x}{5} &= \frac{-35}{5} \\ x &= -7 \end{aligned}$$

2.
$$\begin{aligned} x-19 &= -22 \\ x-19+19 &= -22+19 \\ x &= -3 \end{aligned}$$

3.
$$\begin{aligned} 6-18x &= 4-17x \\ 6-18x+18x &= 4-17x+18x \\ 6 &= 4+x \\ 6-4 &= 4-4+x \\ 2 &= x \end{aligned}$$

4.
$$\begin{aligned} 18-10x &= 63+5x \\ 18-10x+10x &= 63+5x+10x \\ 18 &= 63+15x \\ 18-63 &= 63-63+15x \\ -45 &= 15x \\ \frac{-45}{15} &= \frac{15x}{15} \\ -3 &= x \end{aligned}$$

5.
$$\begin{aligned} 6x-2(x+3) &= 5 \\ 6x-2x-6 &= 5 \\ 4x-6 &= 5 \\ 4x-6+6 &= 5+6 \\ 4x &= 11 \\ \frac{4x}{4} &= \frac{11}{4} \\ x &= \frac{11}{4} = 2.75 \end{aligned}$$

6.
$$\begin{aligned} 1-2(6-x) &= 3x+2 \\ 1-12+2x &= 3x+2 \\ 2x-11 &= 3x+2 \\ 2x-11+11 &= 3x+2+11 \\ 2x &= 3x+13 \\ -3x+2x &= -3x+3x+13 \\ -x &= 13 \\ \frac{-1x}{-1} &= \frac{13}{-1} \\ x &= -13 \end{aligned}$$

7.
$$\begin{aligned} x-(0.5x+2.6) &= 17.6 \\ x-0.5x-2.6 &= 17.6 \\ 0.5x-2.6 &= 17.6 \\ 10(0.5x)-10(2.6) &= 10(17.6) \\ 5x-26 &= 176 \\ 5x-26+26 &= 176+26 \\ 5x &= 202 \\ \frac{5x}{5} &= \frac{202}{5} \\ x &= 40.4 \end{aligned}$$

8.
$$\begin{aligned} -0.2(x+1) &= 0.3(x+11) \\ 10[-0.2(x+1)] &= 10[0.3(x+11)] \\ -2(x+1) &= 3(x+11) \\ -2x-2 &= 3x+33 \\ -2x-2-33 &= 3x+33-33 \\ -2x-35 &= 3x \\ 2x-2x-35 &= 2x+3x \\ -35 &= 5x \\ \frac{-35}{5} &= \frac{5x}{5} \\ -7 &= x \end{aligned}$$

9.
$$\begin{aligned} 3(x-2) &= -4(5+x) \\ 3x-6 &= -20-4x \\ 3x+4x-6 &= -20-4x+4x \\ 7x-6 &= -20 \\ 7x-6+6 &= -20+6 \\ 7x &= -14 \\ \frac{7x}{7} &= \frac{-14}{7} \\ x &= -2 \end{aligned}$$

10.
$$\begin{aligned} -\frac{2}{5}x &= 16 \\ -\frac{5}{2}\left(-\frac{2}{5}x\right) &= -\frac{5}{2}(16) \\ x &= -40 \end{aligned}$$

11.
$$\begin{aligned} \frac{3}{4}x &= 15 \\ \frac{4}{3}\left(\frac{3}{4}x\right) &= \frac{4}{3}(15) \\ x &= 20 \end{aligned}$$

12.
$$\begin{aligned} 4(2x+3) &= 5(x-3) \\ 8x+12 &= 5x-15 \\ 8x+(-5x)+12 &= 5x+(-5x)-15 \\ 3x+12 &= -15 \\ 3x+12+(-12) &= -15+(-12) \\ 3x &= -27 \\ \frac{3x}{3} &= \frac{-27}{3} \\ x &= -9 \end{aligned}$$

13.
$$\begin{aligned} 3(x-3) &= 13x+21 \\ 3x-9 &= 13x+21 \\ 3x+(-3x)-9 &= 13x+(-3x)+21 \\ -9 &= 10x+21 \\ -21-9 &= 10x+21+(-21) \\ -30 &= 10x \\ \frac{-30}{10} &= \frac{10x}{10} \\ -3 &= x \end{aligned}$$

14.
$$\begin{aligned} 1.2x-0.8 &= 0.8x+0.4 \\ 1.2x-0.8-0.8x &= 0.8x+0.4-0.8x \\ 0.4x-0.8 &= 0.4 \\ 0.4x-0.8+0.8 &= 0.4+0.8 \\ 0.4x &= 1.2 \\ \frac{0.4x}{0.4} &= \frac{1.2}{0.4} \\ x &= 3 \end{aligned}$$

15.
$$\begin{aligned} 3.2-0.6x &= 0.4(x-2) \\ 3.2-0.6x &= 0.4x-0.8 \\ 3.2-0.6x+0.6x &= 0.4x-0.8+0.6x \\ 3.2 &= x-0.8 \\ 3.2+0.8 &= x-0.8+0.8 \\ 4 &= x \end{aligned}$$

16.
$$\begin{aligned} 12-x+2 &= 3x-10+4x \\ -x+14 &= 7x-10 \\ -x+14+10 &= 7x-10+10 \\ -x+24 &= 7x \\ x-x+24 &= x+7x \\ 24 &= 8x \\ \frac{24}{8} &= \frac{8x}{8} \\ 3 &= x \end{aligned}$$

17.
$$\begin{aligned} 28 &= 5x-(3x+12) \\ 28 &= 5x-3x-12 \\ 28 &= 2x-12 \\ 28+12 &= 2x-12+12 \\ 40 &= 2x \\ \frac{40}{2} &= \frac{2x}{2} \\ 20 &= x \end{aligned}$$

18.
$$\begin{aligned} 12-5x &= -7x-2 \\ 12-5x+7x &= -7x+7x-2 \\ 12+2x &= -2 \\ 12-12+2x &= -2-12 \\ 2x &= -14 \\ \frac{2x}{2} &= \frac{-14}{2} \\ x &= -7 \end{aligned}$$

19.
$$\begin{aligned} 2(3-x) &= 1-(x-2) \\ 6-2x &= 1-x+2 \\ 6-2x+x &= 3-x+x \\ 6-x &= 3 \\ 6+(-6)-x &= 3+(-6) \\ -x &= -3 \\ \frac{-x}{-1} &= \frac{-3}{-1} \\ x &= 3 \end{aligned}$$

20.
$$\begin{aligned} 4(x+5)-7 &= 2(x+3) \\ 4x+20-7 &= 2x+6 \\ 4x+13 &= 2x+6 \\ 4x+13-13 &= 2x+6-13 \\ 4x &= 2x-7 \\ -2x+4x &= -2x+2x-7 \\ 2x &= -7 \\ \frac{2x}{2} &= \frac{-7}{2} \\ x &= -\frac{7}{2} \text{ or } -3.5 \end{aligned}$$

21.
$$\begin{aligned} 0.9y+3 &= 0.4y+1.5 \\ 0.9y+3-0.4y &= 0.4y+1.5-0.4y \\ 0.5y+3 &= 1.5 \\ 0.5y+3-3 &= 1.5-3 \\ 0.5y &= -1.5 \\ \frac{0.5y}{0.5} &= \frac{-1.5}{0.5} \\ y &= -3 \end{aligned}$$

22.
$$\begin{aligned} 7y-3.4 &= 11.3 \\ 7y-3.4+3.4 &= 11.3+3.4 \\ 7y &= 14.7 \\ \frac{7y}{7} &= \frac{14.7}{7} \\ y &= 2.1 \end{aligned}$$

23.
$$\begin{aligned}3&=2x+5-3(x-1)\\3&=2x+5-3x+3\\3&=-x+8\\3+(-8)&=-x+8+(-8)\\-5&=-x\\\frac{-5}{-1}&=\frac{-x}{-1}\\5&=x\end{aligned}$$

24.
$$\begin{aligned}2(5x-1)-7&=3(x-1)+5-4x\\10x-2-7&=3x-3+5-4x\\10x-9&=-x+2\\10x+x-9&=-x+x+2\\11x-9&=2\\11x-9+9&=2+9\\11x&=11\\\frac{11x}{11}&=\frac{11}{11}\\x&=1\end{aligned}$$

25.
$$\begin{aligned}\frac{3}{4}x-3&=\frac{1}{2}x+2\\4\left(\frac{3}{4}x\right)-4(3)&=4\left(\frac{1}{2}x\right)+4(2)\\3x-12&=2x+8\\3x-12+12&=2x+8+12\\3x&=2x+20\\-2x+3x&=-2x+2x+20\\x&=20\end{aligned}$$

26.
$$\begin{aligned}1&=\frac{5x}{6}+\frac{2x}{3}\\6(1)&=6\left(\frac{5x}{6}\right)+6\left(\frac{2x}{3}\right)\\6&=5x+4x\\6&=9x\\\frac{6}{9}&=\frac{9x}{9}\\\frac{2}{3}&=x\end{aligned}$$

27.
$$\begin{aligned}\frac{7x}{5}&=5+\frac{2x}{5}\\5\left(\frac{7x}{5}\right)&=5(5)+5\left(+\frac{2x}{5}\right)\\7x&=25+2x\\7x-2x&=25+2x-2x\\5x&=25\\\frac{5x}{5}&=\frac{25}{5}\\x&=5\end{aligned}$$

28.
$$\begin{aligned}\frac{7x-3}{2}-4&=\frac{5x+1}{3}\\6\left(\frac{7x-3}{2}\right)-6(4)&=6\left(\frac{5x+1}{3}\right)\\3(7x-3)-24&=2(5x+1)\\21x-9-24&=10x+2\\21x-33&=10x+2\\21x+(-10x)-33&=10x+(-10x)+2\\11x-33&=2\\11x-33+33&=2+33\\11x&=35\\\frac{11x}{11}&=\frac{35}{11}\\x&=\frac{35}{11}\end{aligned}$$

29.
$$\begin{aligned}\frac{3x-2}{2}+\frac{x}{4}&=2+x\\4\left(\frac{3x-2}{2}\right)+4\left(\frac{x}{4}\right)&=4(2)+4(x)\\2(3x-2)+x&=8+4x\\6x-4+x&=8+4x\\7x-4&=4x+8\\7x-4+4&=4x+8+4\\7x&=4x+12\\-4x+7x&=-4x+4x+12\\3x&=12\\\frac{3x}{3}&=\frac{12}{3}\\x&=4\end{aligned}$$

30.
$$\frac{-3}{2}(x+5)=1-x$$
$$-\frac{3}{2}x-\frac{15}{2}=1-x$$
$$2\left(-\frac{3}{2}x\right)-2\left(\frac{15}{2}\right)=2(1)-2(x)$$
$$-3x-15=2-2x$$
$$-3x+3x-15=2-2x+3x$$
$$-15=2+x$$
$$-15+(-2)=2+(-2)+x$$
$$-17=x$$

31.
$$\frac{-4}{3}(2x+1)=-x-2$$
$$-\frac{8}{3}x-\frac{4}{3}=-x-2$$
$$3\left(-\frac{8}{3}x\right)-3\left(\frac{4}{3}\right)=3(-x)-3(2)$$
$$-8x-4=-3x-6$$
$$-8x-4+4=-3x-6+4$$
$$-8x=-3x-2$$
$$3x-8x=3x-3x-2$$
$$-5x=-2$$
$$\frac{-5x}{-5}=\frac{-2}{-5}$$
$$x=\frac{2}{5}\text{ or }0.4$$

32.
$$\frac{1}{3}(x-2)=\frac{x}{4}+2$$
$$\frac{1}{3}x-\frac{2}{3}=\frac{x}{4}+2$$
$$12\left(\frac{1}{3}x\right)-12\left(\frac{2}{3}\right)=12\left(\frac{x}{4}\right)+12(2)$$
$$4x-8=3x+24$$
$$4x+(-3x)-8=3x+(-3x)+24$$
$$x-8=24$$
$$x-8+8=24+8$$
$$x=32$$

33.
$$\frac{1}{5}(x-3)=2-\frac{x}{2}$$
$$\frac{1}{5}x-\frac{3}{5}=2-\frac{x}{2}$$
$$10\left(\frac{1}{5}x\right)-10\left(\frac{3}{5}\right)=10(2)-10\left(\frac{x}{2}\right)$$
$$2x-6=20-5x$$
$$2x+5x-6=20-5x+5x$$
$$7x-6=20$$
$$7x-6+6=20+6$$
$$7x=26$$
$$\frac{7x}{7}=\frac{26}{7}$$
$$x=\frac{26}{7}$$

34.
$$\frac{3}{4}-\frac{2}{3}x=\frac{1}{3}x+\frac{3}{4}$$
$$12\left(\frac{3}{4}\right)-12\left(\frac{2}{3}x\right)=12\left(\frac{1}{3}x\right)+12\left(\frac{3}{4}\right)$$
$$9-8x=4x+9$$
$$9-8x+8x=4x+9+8x$$
$$9=12x+9$$
$$9-9=12x+9-9$$
$$0=12x$$
$$\frac{0}{12}=\frac{12x}{12}$$
$$0=x$$

35.
$$x+\frac{1}{8}+\frac{1}{2}x=\frac{1}{4}\left(-3x+\frac{1}{2}\right)$$
$$\frac{3}{2}x+\frac{1}{8}=-\frac{3}{4}x+\frac{1}{8}$$
$$8\left(\frac{3}{2}x\right)+8\left(\frac{1}{8}\right)=8\left(-\frac{3}{4}x\right)+8\left(\frac{1}{8}\right)$$
$$12x+1=-6x+1$$
$$12x+1+6x=-6x+1+6x$$
$$18x+1=1$$
$$18x+1-1=1-1$$
$$18x=0$$
$$\frac{18x}{18}=\frac{0}{18}$$
$$x=0$$

36.
$$\begin{aligned}
\frac{3}{2}x-\frac{5}{6}+x&=\frac{1}{2}x+\frac{2}{3}\\
6\left(\frac{3}{2}x\right)-6\left(\frac{5}{6}\right)+6x&=6\left(\frac{1}{2}x\right)+6\left(\frac{2}{3}\right)\\
9x-5+6x&=3x+4\\
15x-5&=3x+4\\
15x+(-3x)-5&=3x+(-3x)+4\\
12x-5&=4\\
12x-5+5&=4+5\\
12x&=9\\
\frac{12x}{12}&=\frac{9}{12}\\
x&=\frac{3}{4}
\end{aligned}$$

37.
$$\begin{aligned}
-\frac{8}{3}x-8+2x-5&=-\frac{5}{3}\\
-\frac{8}{3}x-13+2x&=-\frac{5}{3}\\
3\left(-\frac{8}{3}x\right)-3(13)+3(2x)&=3\left(-\frac{5}{3}\right)\\
-8x-39+6x&=-5\\
-2x-39&=-5\\
-2x-39+39&=-5+39\\
-2x&=34\\
\frac{-2x}{-2}&=\frac{34}{-2}\\
x&=-17
\end{aligned}$$

38.
$$\begin{aligned}
3.5(2x+3)&=2.4x-1\\
7x+10.5&=2.4x-1\\
10(7x)+10(10.5)&=10(2.4x)-10(1)\\
70x+105&=24x-10\\
70x+105-24x&=24x-10-24x\\
46x+105&=-10\\
46x+105-105&=-10-105\\
46x&=-115\\
\frac{46x}{46}&=\frac{-115}{46}\\
x&=-2.5
\end{aligned}$$

39.
$$\begin{aligned}
\frac{7}{12}(x-3)&=\frac{1}{3}x+4\\
12\left[\frac{7}{12}(x-3)\right]&=12\left[\frac{1}{3}x+4\right]\\
7(x-3)&=4x+48\\
7x-21&=4x+48\\
7x-4x-21&=4x-4x+48\\
3x-21&=48\\
3x-21+21&=48+21\\
3x&=69\\
\frac{3x}{3}&=\frac{69}{3}\\
x&=23
\end{aligned}$$

40.
$$\begin{aligned}
\frac{1}{6}+\frac{1}{3}(x-3)&=\frac{1}{2}(x+9)\\
\frac{1}{6}+\frac{1}{3}x-1&=\frac{1}{2}x+\frac{9}{2}\\
\frac{1}{3}x-\frac{5}{6}&=\frac{1}{2}x+\frac{9}{2}\\
6\left(\frac{1}{3}x\right)-6\left(\frac{5}{6}\right)&=6\left(\frac{1}{2}x\right)+6\left(\frac{9}{2}\right)\\
2x-5&=3x+27\\
2x-2x-5&=3x-2x+27\\
-5&=x+27\\
-5-27&=x+27-27\\
-32&=x
\end{aligned}$$

41.
$$\begin{aligned}
\frac{1}{7}(x+5)-\frac{6}{14}&=\frac{1}{2}(x+3)\\
\frac{1}{7}x+\frac{5}{7}-\frac{6}{14}&=\frac{1}{2}x+\frac{3}{2}\\
\frac{1}{7}x+\frac{2}{7}&=\frac{1}{2}x+\frac{3}{2}\\
14\left(\frac{1}{7}x\right)+14\left(\frac{2}{7}\right)&=14\left(\frac{1}{2}x\right)+14\left(\frac{3}{2}\right)\\
2x+4&=7x+21\\
2x+(-2x)+4&=7x+(-2x)+21\\
4&=5x+21\\
4+(-21)&=5x+21+(-21)\\
-17&=5x\\
\frac{-17}{5}&=\frac{5x}{5}\\
-\frac{17}{5}&=x \text{ or } x=-3.4
\end{aligned}$$

42.
$$\frac{1}{6}(8x+3)=\frac{1}{2}(2x+7)$$
$$\frac{4}{3}x+\frac{1}{2}=x+\frac{7}{2}$$
$$6\left(\frac{4}{3}x\right)+6\left(\frac{1}{2}\right)=6(x)+6\left(\frac{7}{2}\right)$$
$$8x+3=6x+21$$
$$8x+3-3=6x+21-3$$
$$8x=6x+18$$
$$-6x+8x=-6x+6x+18$$
$$2x=18$$
$$\frac{2x}{2}=\frac{18}{2}$$
$$x=9$$

43.
$$-\frac{2}{5}(3x+1)=\frac{1}{3}(2-x)$$
$$-\frac{6}{5}x-\frac{2}{5}=\frac{2}{3}-\frac{1}{3}x$$
$$15\left(-\frac{6}{5}x\right)-15\left(\frac{2}{5}\right)=15\left(\frac{2}{3}\right)-15\left(\frac{1}{3}x\right)$$
$$-18x-6=10-5x$$
$$-18x+5x-6=10-5x+5x$$
$$-13x-6=10$$
$$-13x-6+6=10+6$$
$$-13x=16$$
$$\frac{-13x}{-13}=\frac{16}{-13}$$
$$x=-\frac{16}{13}$$

44.
$$\frac{7}{9}x+\frac{2}{3}=5+\frac{1}{3}x$$
$$9\left(\frac{7}{9}x+\frac{2}{3}\right)=9\left(5+\frac{1}{3}x\right)$$
$$7x+6=45+3x$$
$$7x-3x+6=45+3x-3x$$
$$4x+6=45$$
$$4x+6-6=45-6$$
$$4x=39$$
$$\frac{4x}{4}=\frac{39}{4}$$
$$x=\frac{39}{4}\text{ or }9.75$$

45.
$$3x-y=10$$
$$3x+(-3x)-y=-3x+10$$
$$-y=-3x+10$$
$$\frac{-y}{-1}=\frac{-3x+10}{-1}$$
$$y=3x-10$$

46.
$$5x+2y+7=0$$
$$5x+2y=-7$$
$$2y=-5x-7$$
$$y=\frac{-5x-7}{2}$$

47.
$$A=P(1+rt)$$
$$A=P+Prt$$
$$A+(-P)=P+(-P)+Prt$$
$$A-P=Prt$$
$$\frac{A-P}{Pt}=\frac{Prt}{Pt}$$
$$\frac{A-P}{Pt}=r$$

48.
$$A=4\pi r^2+2\pi rh$$
$$A-4\pi r^2=2\pi rh$$
$$\frac{A-4\pi r^2}{2\pi r}=\frac{2\pi rh}{2\pi r}$$
$$\frac{A-4\pi r^2}{2\pi r}=h$$

49.
$$H=\frac{1}{3}(a+2p+3)$$
$$H=\frac{1}{3}a+\frac{2}{3}p+1$$
$$3(H)=3\left(\frac{1}{3}a\right)+3\left(\frac{2}{3}p\right)+3(1)$$
$$3H=a+2p+3$$
$$3H+(-a)+(-3)=a+(-a)+2p+3+(-3)$$
$$3H-a-3=2p$$
$$\frac{3H-a-3}{2}=\frac{2p}{2}$$
$$\frac{3H-a-3}{2}=p$$

50.
$$\begin{aligned} ax+by&=c \\ ax-ax+by&=c-ax \\ by&=c-ax \\ \frac{by}{b}&=\frac{c-ax}{b} \\ y&=\frac{c-ax}{b} \end{aligned}$$

51. a.
$$\begin{aligned} x&=\frac{ABC}{10} \\ 10(x)&=10\left(\frac{ABC}{10}\right) \\ 10x&=ABC \\ \frac{10x}{BC}&=\frac{ABC}{BC} \\ \frac{10x}{BC}&=A \end{aligned}$$

b. $A=\frac{10x}{BC}=\frac{10(6)}{(-1)(1.5)}=\frac{60}{-1.5}=-40$

52. a.
$$\begin{aligned} 2l+2w&=P \\ 2w&=P-2l \\ \frac{2w}{2}&=\frac{P-2l}{2} \\ w&=\frac{P-2l}{2} \end{aligned}$$

b.
$$\begin{aligned} w&=\frac{P-2l}{2} \\ &=\frac{34-2(10.5)}{2} \\ &=\frac{34-21}{2} \\ &=\frac{13}{2} \\ &=6.5 \end{aligned}$$

53. a.
$$\begin{aligned} V&=lwh \\ \frac{V}{lw}&=\frac{lwh}{lw} \\ \frac{V}{lw}&=h \end{aligned}$$

b. $V=48,\ l=2,\ w=4$

$h=\frac{48}{2(4)}=6$

54.
$$\begin{aligned} 9+2x&\le 6-x \\ 9+2x+x&\le 6-x+x \\ 9+3x&\le 6 \\ 9+3x-9&\le 6-9 \\ 3x&\le -3 \\ \frac{3x}{3}&\le\frac{-3}{3} \\ x&\le -1 \end{aligned}$$

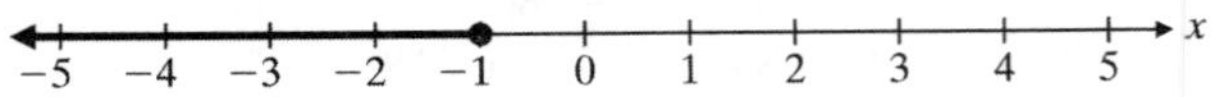

55.
$$\begin{aligned} 4x-7&\ge -x+8 \\ 4x-7+x&\ge -x+8+x \\ 5x-7&\ge 8 \\ 5x-7+7&\ge 8+7 \\ 5x&\ge 15 \\ \frac{5x}{5}&\ge\frac{15}{5} \\ x&\ge 3 \end{aligned}$$

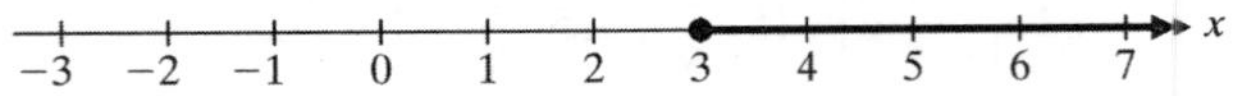

56.
$$\begin{aligned} 2x-3+x&>5(x+1) \\ 3x-3&>5x+5 \\ 3x-3-5x&>5x+5-5x \\ -2x-3&>5 \\ -2x-3+3&>5+3 \\ -2x&>8 \\ \frac{-2x}{-2}&<\frac{8}{-2} \\ x&<-4 \end{aligned}$$

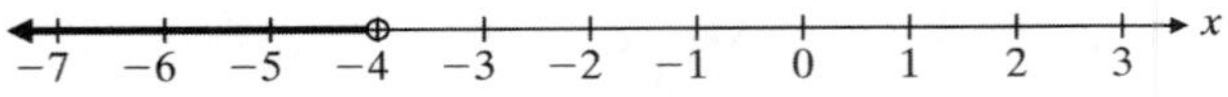

57.
$$\begin{aligned} -x+4&<3x+16 \\ -x+4-4&<3x+16-4 \\ -x&<3x+12 \\ -3x-x&<-3x+3x+12 \\ -4x&<12 \\ \frac{-4x}{-4}&>\frac{12}{-4} \\ x&>-3 \end{aligned}$$

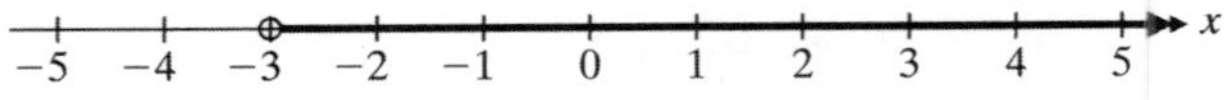

58.
$$8-\frac{1}{3}x \le x$$
$$3(8)-3\left(\frac{1}{3}x\right) \le 3x$$
$$24-x \le 3x$$
$$24-x+(-3x) \le 3x+(-3x)$$
$$24-4x \le 0$$
$$24+(-24)-4x \le 0+(-24)$$
$$-4x \le -24$$
$$\frac{-4x}{-4} \ge \frac{-24}{-4}$$
$$x \ge 6$$

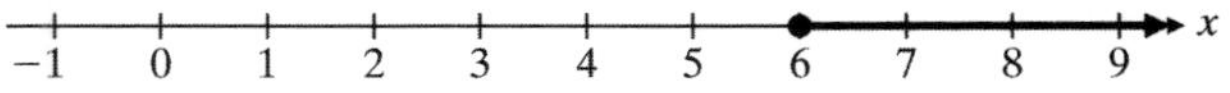

59.
$$7-\frac{3}{5}x > 4$$
$$5(7)-5\left(\frac{3}{5}x\right) > 5(4)$$
$$35-3x > 20$$
$$35+(-35)-3x > 20+(-35)$$
$$-3x > -15$$
$$\frac{-3x}{-3} < \frac{-15}{-3}$$
$$x < 5$$

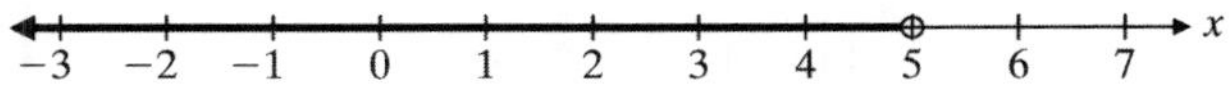

60.
$$-4x-14 < 4-2(3x-1)$$
$$-4x-14 < 4-6x+2$$
$$-4x-14 < 6-6x$$
$$-4x-14+6x < 6-6x+6x$$
$$2x-14 < 6$$
$$2x-14+14 < 6+14$$
$$2x < 20$$
$$\frac{2x}{2} < \frac{20}{2}$$
$$x < 10$$

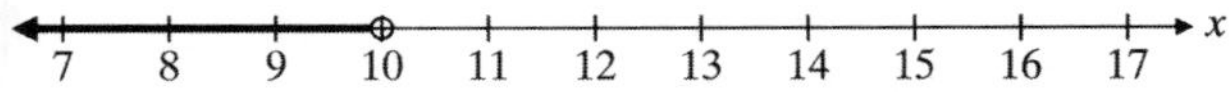

61.
$$3(x-2)+8 < 7x+14$$
$$3x-6+8 < 7x+14$$
$$3x+2 < 7x+14$$
$$3x-2+2 < 7x+14-2$$
$$3x < 7x+12$$
$$-7x+3x < -7x+7x+12$$
$$-4x < 12$$
$$\frac{-4x}{-4} > \frac{12}{-4}$$
$$x > -3$$

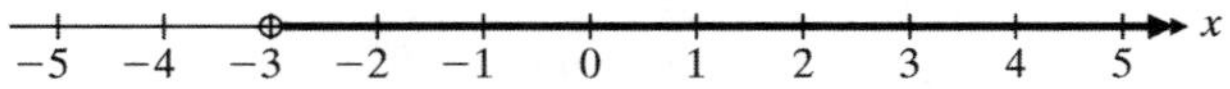

62.
$$\frac{1}{2}(2x+3) > 10$$
$$x+\frac{3}{2} > 10$$
$$2(x)+2\left(\frac{3}{2}\right) > 2(10)$$
$$2x+3 > 20$$
$$2x+3-3 > 20-3$$
$$2x > 17$$
$$\frac{2x}{2} > \frac{17}{2}$$
$$x > \frac{17}{2}$$

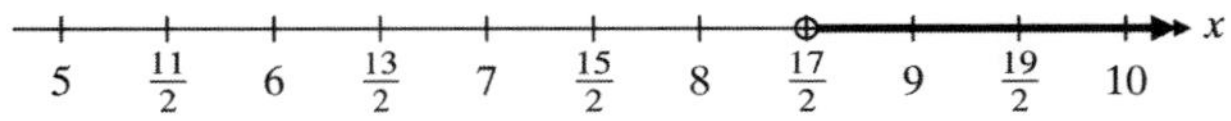

63.
$$5(1-x) < 3(x-1)-2(3-x)$$
$$5-5x < 3x-3-6+2x$$
$$5-5x < 5x-9$$
$$5+9-5x < 5x-9+9$$
$$14-5x < 5x$$
$$14-5x+5x < 5x+5x$$
$$14 < 10x$$
$$\frac{14}{10} < \frac{10x}{10}$$
$$\frac{7}{5} < x \text{ or } x > \frac{7}{5}$$

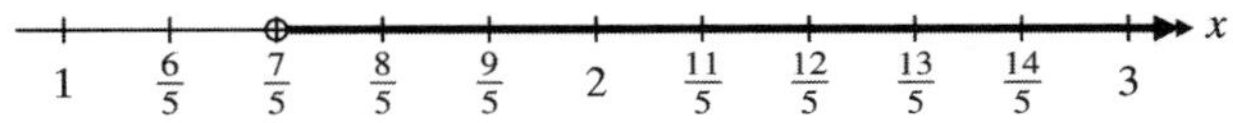

64. $15h \le 480$
$$\frac{15h}{15} \le \frac{480}{15}$$
$h \le 32$ hours
Julian can work a maximum of 32 hours.

65. $110n \le 2420$
$$\frac{110n}{110} \le \frac{2420}{110}$$
$$n \le 22 \text{ times}$$
A substitute teacher can be hired a maximum of 22 times.

66.
$$\begin{aligned}10(2x+4)-13 &= 8(x+7)-3\\ 20x+40-13 &= 8x+56-3\\ 20x+27 &= 8x+53\\ 20x+27-8x &= 8x+53-8x\\ 12x+27 &= 53\\ 12x+27-27 &= 53-27\\ 12x &= 26\\ \frac{12x}{12} &= \frac{26}{12}\\ x &= \frac{13}{6}\end{aligned}$$

67.
$$\begin{aligned}-9x+15-2x &= 4-3x\\ -11x+15 &= 4-3x\\ -11x+15+3x &= 4-3x+3x\\ -8x+15 &= 4\\ -8x+15-15 &= 4-15\\ -8x &= -11\\ \frac{-8x}{-8} &= \frac{-11}{-8}\\ x &= \frac{11}{8}\end{aligned}$$

68.
$$\begin{aligned}-2(x-3) &= -4x+3(3x+2)\\ -2x+6 &= -4x+9x+6\\ -2x+6 &= 5x+6\\ -2x+6-6 &= 5x+6-6\\ -2x &= 5x\\ 2x-2x &= 2x+5x\\ 0 &= 7x\\ \frac{0}{7} &= \frac{7x}{7}\\ 0 &= x\end{aligned}$$

69.
$$\begin{aligned}\frac{1}{2}+\frac{5}{4}x &= \frac{2}{5}x-\frac{1}{10}+4\\ 20\left(\frac{1}{2}\right)+20\left(\frac{5}{4}x\right) &= 20\left(\frac{2}{5}x\right)-20\left(\frac{1}{10}\right)+20(4)\\ 10+25x &= 8x-2+80\\ 10+25x &= 8x+78\\ 10+25x+(-8x) &= 8x+(-8x)+78\\ 10+17x &= 78\\ 10+(-10)+17x &= 78+(-10)\\ 17x &= 68\\ \frac{17x}{17} &= \frac{68}{17}\\ x &= 4\end{aligned}$$

70.
$$\begin{aligned}\frac{1}{6}x-\frac{2}{3} &= \frac{1}{3}(x-4)\\ \frac{1}{6}x-\frac{2}{3} &= \frac{1}{3}x-\frac{4}{3}\\ 6\left(\frac{1}{6}x\right)-6\left(\frac{2}{3}\right) &= 6\left(\frac{1}{3}x\right)-6\left(\frac{4}{3}\right)\\ x-4 &= 2x-8\\ x+(-x)-4 &= 2x+(-x)-8\\ -4 &= x-8\\ -4+8 &= x-8+8\\ 4 &= x\end{aligned}$$

71.
$$\begin{aligned}\frac{1}{2}(x-3) &= \frac{1}{4}(3x-1)\\ \frac{1}{2}x-\frac{3}{2} &= \frac{3}{4}x-\frac{1}{4}\\ 4\left(\frac{1}{2}x\right)-4\left(\frac{3}{2}\right) &= 4\left(\frac{3}{4}x\right)-4\left(\frac{1}{4}\right)\\ 2x-6 &= 3x-1\\ 2x-6+1 &= 3x-1+1\\ 2x-5 &= 3x\\ -2x+2x-5 &= -2x+3x\\ -5 &= x\end{aligned}$$

72.
$$\begin{aligned}10+3f &= 5e-2f\\ 10+3f+2f &= 5e-2f+2f\\ 10+5f &= 5e\\ 10+5f-10 &= 5e-10\\ 5f &= 5e-10\\ \frac{5f}{5} &= \frac{5e-10}{5}\\ f &= \frac{5e}{5}-\frac{10}{5}\\ f &= e-2\end{aligned}$$

73.
$$W=\frac{2a+4b}{3}$$
$$3(W)=3\left(\frac{2a+4b}{3}\right)$$
$$3W=2a+4b$$
$$3W-4b=2a+4b-4b$$
$$3W-4b=2a$$
$$\frac{3W-4b}{2}=\frac{2a}{2}$$
$$\frac{3W-4b}{2}=a$$

74.
$$5-\frac{1}{2}x>4$$
$$2(5)-2\left(\frac{1}{2}x\right)>2(4)$$
$$10-x>8$$
$$-10+10-x>-10+8$$
$$-x>-2$$
$$\frac{-x}{-1}<\frac{-2}{-1}$$
$$x<2$$

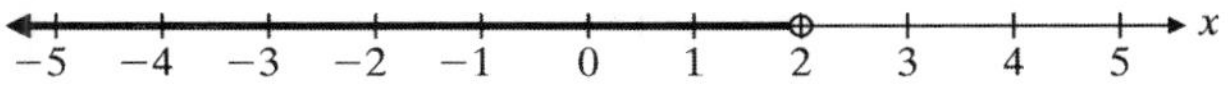

75.
$$2(x-1)\ge 3(2+x)$$
$$2x-2\ge 6+3x$$
$$2x-2-3x>6+3x-3x$$
$$-x-2\ge 6$$
$$-x-2+2\ge 6+2$$
$$-x\ge 8$$
$$x\le -8$$

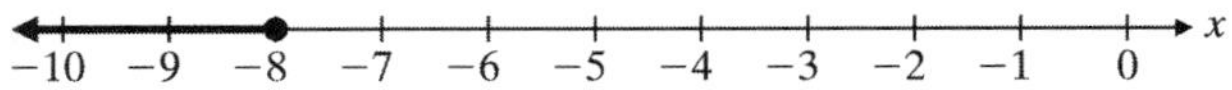

76.
$$\frac{1}{3}(x+2)\le\frac{1}{2}(3x-5)$$
$$\frac{1}{3}x+\frac{2}{3}\le\frac{3}{2}x-\frac{5}{2}$$
$$6\left(\frac{1}{3}x\right)+6\left(\frac{2}{3}\right)\le 6\left(\frac{3}{2}x\right)-6\left(\frac{5}{2}\right)$$
$$2x+4\le 9x-15$$
$$2x+4+15\le 9x-15+15$$
$$2x+19\le 9x$$
$$-2x+2x+19\le -2x+9x$$
$$19\le 7x$$
$$\frac{19}{7}\le\frac{7x}{7}$$
$$\frac{19}{7}\le x \text{ or } x\ge\frac{19}{7}$$

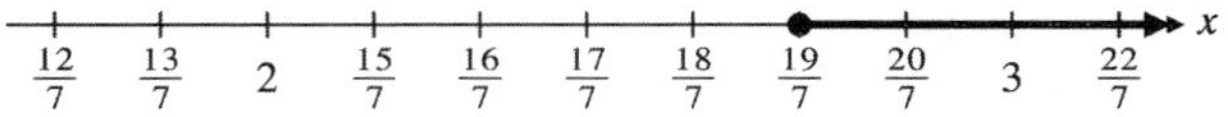

77. $4(2-x)-(-5x+1)\ge -8$
$$8-4x+5x-1\ge -8$$
$$x+7\ge -8$$
$$x+7-7\ge -8-7$$
$$x\ge -15$$

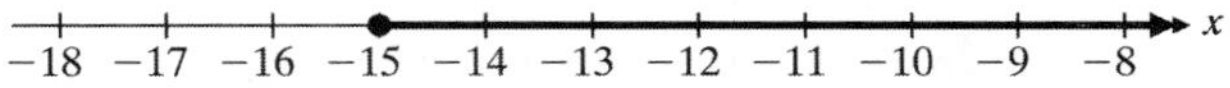

How Am I Doing? Chapter 2 Test

1.
$$3x+5.6=11.6$$
$$3x+5.6-5.6=11.6-5.6$$
$$3x=6$$
$$\frac{3x}{3}=\frac{6}{3}$$
$$x=2$$

2.
$$9x-8=-6x-3$$
$$9x+6x-8=-6x+6x-3$$
$$15x-8=-3$$
$$15x-8+8=-3+8$$
$$15x=5$$
$$\frac{15x}{15}=\frac{5}{15}$$
$$x=\frac{1}{3}$$

3.
$$\begin{aligned} 2(2y-3) &= 4(2y+2) \\ 4y-6 &= 8y+8 \\ 4y-6+6 &= 8y+8+6 \\ 4y &= 8y+14 \\ -8y+4y &= -8y+8y+14 \\ -4y &= 14 \\ \frac{-4y}{-4} &= \frac{14}{-4} \\ y &= -\frac{7}{2} \text{ or } -3.5 \end{aligned}$$

4.
$$\begin{aligned} \frac{1}{7}y+3 &= \frac{1}{2}y \\ 14\left(\frac{1}{7}y\right)+14(3) &= 14\left(\frac{1}{2}y\right) \\ 2y+42 &= 7y \\ 2y-2y+42 &= 7y-2y \\ 42 &= 5y \\ \frac{42}{5} &= \frac{5y}{5} \\ 8\frac{2}{5} &= y \text{ or } y = 8.4 \end{aligned}$$

5.
$$\begin{aligned} 4(7-4x) &= 3(6-2x) \\ 28-16x &= 18-6x \\ 28-16x+6x &= 18-6x+6x \\ 28-10x &= 18 \\ 28+(-28)-10x &= 18+(-28) \\ -10x &= -10 \\ \frac{-10x}{-10} &= \frac{-10}{-10} \\ x &= 1 \end{aligned}$$

6.
$$\begin{aligned} 0.8x+0.18-0.4x &= 0.3(x+0.2) \\ 0.4x+0.18 &= 0.3x+0.06 \\ 100(0.4x)+100(0.18) &= 100(0.3x)+100(0.06) \\ 40x+18 &= 30x+6 \\ 40x+18-18 &= 30x+6-18 \\ 40x &= 30x-12 \\ -30x+40x &= -30x+30x-12 \\ 10x &= -12 \\ \frac{10x}{10} &= \frac{-12}{10} \\ x &= -\frac{6}{5} \text{ or } -1.2 \end{aligned}$$

7.
$$\begin{aligned} \frac{2y}{3}+\frac{1}{5}-\frac{3y}{5}+\frac{1}{3} &= 1 \\ 15\left(\frac{2y}{3}\right)+15\left(\frac{1}{5}\right)-15\left(\frac{3y}{5}\right)+15\left(\frac{1}{3}\right) &= 15(1) \\ 10y+3-9y+5 &= 15 \\ y+8 &= 15 \\ y+8-8 &= 15-8 \\ y &= 7 \end{aligned}$$

8.
$$\begin{aligned} 3-2y &= 2(3y-2)-5y \\ 3-2y &= 6y-4-5y \\ 3-2y &= y-4 \\ 3-2y+2y &= y+2y-4 \\ 3 &= 3y-4 \\ 3+4 &= 3y-4+4 \\ 7 &= 3y \\ \frac{7}{3} &= \frac{3y}{3} \\ \frac{7}{3} &= y \end{aligned}$$

9.
$$\begin{aligned} 5(20-x)+10x &= 165 \\ 100-5x+10x &= 165 \\ 100+5x &= 165 \\ -100+100+5x &= -100+165 \\ 5x &= 65 \\ \frac{5x}{5} &= \frac{65}{5} \\ x &= 13 \end{aligned}$$

10.
$$\begin{aligned} 5(x+40)-6x &= 9x \\ 5x+200-6x &= 9x \\ 200-x &= 9x \\ 200-x+x &= 9x+x \\ 200 &= 10x \\ \frac{200}{10} &= \frac{10x}{10} \\ 20 &= x \end{aligned}$$

11.
$$\begin{aligned} -2(2-3x) &= 76-2x \\ -4+6x &= 76-2x \\ -76-4+6x &= -76+76-2x \\ -80+6x &= -2x \\ -80+6x-6x &= -2x-6x \\ -80 &= -8x \\ \frac{-80}{-8} &= \frac{-8x}{-8} \\ 10 &= x \end{aligned}$$

12.
$$\begin{aligned}
20-(2x+6)&=5(2-x)+2x\\
20-2x-6&=10-5x+2x\\
-2x+14&=-3x+10\\
3x-2x+14&=3x-3x+10\\
x+14&=10\\
x+14-14&=10-14\\
x&=-4
\end{aligned}$$

13.
$$\begin{aligned}
2x-3&=12-6x+3(2x+3)\\
2x-3&=12-6x+6x+9\\
2x-3&=21\\
2x-3+3&=21+3\\
2x&=24\\
\frac{2x}{2}&=\frac{24}{2}\\
x&=12
\end{aligned}$$

14.
$$\begin{aligned}
\frac{1}{3}x-\frac{3}{4}x&=\frac{1}{12}\\
12\left(\frac{1}{3}x\right)-12\left(\frac{3}{4}x\right)&=12\left(\frac{1}{12}\right)\\
4x-9x&=1\\
-5x&=1\\
\frac{-5x}{-5}&=\frac{1}{-5}\\
x&=-\frac{1}{5}\text{ or }-0.2
\end{aligned}$$

15.
$$\begin{aligned}
\frac{3}{5}x+\frac{7}{10}&=\frac{1}{3}x+\frac{3}{2}\\
30\left(\frac{3}{5}x\right)+30\left(\frac{7}{10}\right)&=30\left(\frac{1}{3}x\right)+30\left(\frac{3}{2}\right)\\
18x+21&=10x+45\\
18x+21-21&=10x+45-21\\
18x&=10x+24\\
-10x+18x&=-10x+10x+24\\
8x&=24\\
\frac{8x}{8}&=\frac{24}{8}\\
x&=3
\end{aligned}$$

16.
$$\begin{aligned}
\frac{15x-2}{28}&=\frac{5x-3}{7}\\
28\left(\frac{15x-2}{28}\right)&=28\left(\frac{5x-3}{7}\right)\\
15x-2&=4(5x-3)\\
15x-2&=20x-12\\
15x-2+12&=20x-12+12\\
15x+10&=20x\\
-15x+15x+10&=-15x+20x\\
10&=5x\\
\frac{10}{5}&=\frac{5x}{5}\\
2&=x
\end{aligned}$$

17.
$$\begin{aligned}
\frac{2}{3}(x+8)+\frac{3}{5}&=\frac{1}{5}(11-6x)\\
\frac{2}{3}x+\frac{16}{3}+\frac{3}{5}&=\frac{11}{5}-\frac{6}{5}x\\
\frac{2}{3}x+\frac{89}{15}&=\frac{11}{5}-\frac{6}{5}x\\
15\left(\frac{2}{3}x\right)+15\left(\frac{89}{15}\right)&=15\left(\frac{11}{5}\right)-15\left(\frac{6}{5}x\right)\\
10x+89&=33-18x\\
10x+18x+89&=33-18x+18x\\
28x+89&=33\\
28x+89+(-89)&=33+(-89)\\
28x&=-56\\
\frac{28x}{28}&=\frac{-56}{28}\\
x&=-2
\end{aligned}$$

18.
$$\begin{aligned}
A&=3w+2P\\
A-2P&=3w+2P-2P\\
A-2P&=3w\\
\frac{A-2P}{3}&=\frac{3w}{3}\\
\frac{A-2P}{3}&=w
\end{aligned}$$

19. $$\frac{2w}{3} = 4 - \frac{1}{2}(x+6)$$
$$\frac{2w}{3} = 4 - \frac{1}{2}x - 3$$
$$\frac{2w}{3} = 1 - \frac{1}{2}x$$
$$6\left(\frac{2w}{3}\right) = 6(1) - 6\left(\frac{1}{2}x\right)$$
$$4w = 6 - 3x$$
$$\frac{4w}{4} = \frac{6-3x}{4}$$
$$w = \frac{6-3x}{4}$$

20. $$A = \frac{1}{2}h(a+b)$$
$$A = \frac{1}{2}ha + \frac{1}{2}hb$$
$$2(A) = 2\left(\frac{1}{2}ha\right) + 2\left(\frac{1}{2}hb\right)$$
$$2A = ha + hb$$
$$2A - hb = ha$$
$$\frac{2A-hb}{h} = \frac{ha}{h}$$
$$\frac{2A-hb}{h} = a$$

21. $$5ax(2-y) = 3axy + 5$$
$$10ax - 5axy = 3axy + 5$$
$$10ax - 5axy + 5axy = 3axy + 5 + 5axy$$
$$10ax - 5 = 8axy$$
$$\frac{10ax-5}{8ax} = \frac{8axy}{8ax}$$
$$\frac{10ax-5}{8ax} = y$$

22. $$V = \frac{1}{3}Bh$$
$$3V = 3\left(\frac{1}{3}Bh\right)$$
$$3V = Bh$$
$$\frac{3V}{h} = \frac{Bh}{h}$$
$$\frac{3V}{h} = B$$

23. $B = \frac{3V}{h}$, $V = 140$, $h = 14$

$$B = \frac{3(140)}{14} = 30 \text{ square inches}$$

24. $$3(x-2) \geq 5x$$
$$3x - 6 \geq 5x$$
$$3x + (-5x) - 6 \geq 5x + (-5x)$$
$$-2x - 6 \geq 0$$
$$-2x - 6 + 6 \geq 0 + 6$$
$$-2x \geq 6$$
$$\frac{-2x}{-2} \leq \frac{6}{-2}$$
$$x \leq -3$$

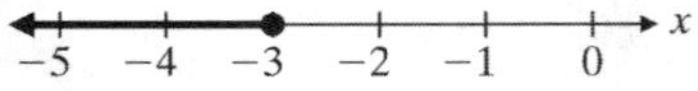

25. $$2 - 7(x+1) - 5(x+2) < 0$$
$$2 - 7x - 7 - 5x - 10 < 0$$
$$-12x - 15 < 0$$
$$-12x - 15 + 15 < 0 + 15$$
$$-12x < 15$$
$$\frac{-12x}{-12} > \frac{15}{-12}$$
$$x > -\frac{5}{4}$$

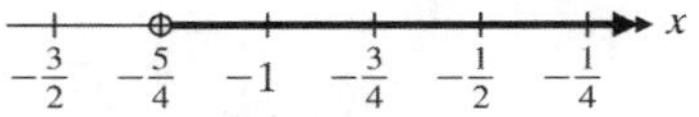

26. $$5 + 8x - 4 < 2x + 13$$
$$8x + 1 < 2x + 13$$
$$8x + 1 - 1 < 2x + 13 - 1$$
$$8x < 2x + 12$$
$$-2x + 8x < -2x + 2x + 12$$
$$6x < 12$$
$$\frac{6x}{6} = \frac{12}{6}$$
$$x < 2$$

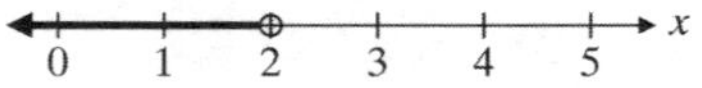

27.
$$\frac{1}{4}x+\frac{1}{16}\le\frac{1}{8}(7x-2)$$
$$\frac{1}{4}x+\frac{1}{16}\le\frac{7}{8}x-\frac{1}{4}$$
$$16\left(\frac{1}{4}x\right)+16\left(\frac{1}{16}\right)\le16\left(\frac{7}{8}x\right)-16\left(\frac{1}{4}\right)$$
$$4x+1\le14x-4$$
$$4x+1+4\le14x-4+4$$
$$4x+5\le14x$$
$$-4x+4x+5\le-4x+14x$$
$$5\le10x$$
$$\frac{5}{10}\le\frac{10x}{10}$$
$$\frac{1}{2}\le x$$

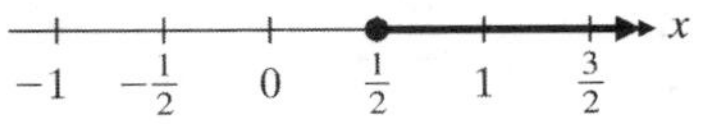

Cumulative Test Chapters 0–2

1. $\frac{6}{7}-\frac{2}{3}=\frac{18}{21}-\frac{14}{21}=\frac{4}{21}$

2. $1\frac{3}{4}+2\frac{1}{5}=1\frac{15}{20}+2\frac{4}{20}=\frac{35}{20}+\frac{44}{20}=\frac{79}{20}$ or $3\frac{19}{20}$

3. $4\frac{3}{8}\div\frac{1}{2}=\frac{35}{8}\div\frac{1}{2}=\frac{35}{8}\cdot\frac{2}{1}=\frac{35\cdot\cancel{2}}{4\cdot\cancel{2}\cdot1}=\frac{35}{4}$ or $8\frac{3}{4}$

4.
$$\begin{array}{r} 1.23 \\ \times\ 0.56 \\ \hline 738 \\ 615 \\ \hline 0.6888 \end{array}$$

5.
$$\begin{array}{r} 0.12 \\ 1.2_\wedge\overline{)0.1_\wedge44} \\ 1\ 2 \\ \hline 24 \\ 24 \\ \hline \end{array}$$

6. What is 4.3% of 830?
$0.043(830)=35.69$

7. $(-3)(-5)(-1)(2)(-1)=(3)(5)(1)(2)(1)=30$

8. $3st-8s^2t+12st^2+s^2t-5st+2st^2$
$=(3-5)st+(-8+1)s^2t+(12+2)st^2$
$=-2st-7s^2t+14st^2$

9. $(5x)^2=5^2(x)^2=25x^2$

10. $2\{3x-4[5-3y(2-x)]\}$
$=2[3x-4(5-6y+3xy)]$
$=2(3x-20+24y-12xy)$
$=6x-40+48y-24xy$

11.
$$-2(x+5)=4x-15$$
$$-2x-10=4x-15$$
$$-2x-10+2x=4x-15+2x$$
$$-10=6x-15$$
$$-10+15=6x-15+15$$
$$5=6x$$
$$\frac{5}{6}=\frac{6x}{6}$$
$$\frac{5}{6}=x$$

12.
$$\frac{1}{3}(x+5)=2x-5$$
$$\frac{1}{3}x+\frac{5}{3}=2x-5$$
$$3\left(\frac{1}{3}x\right)+3\left(\frac{5}{3}\right)=3(2x)-3(5)$$
$$x+5=6x-15$$
$$x+5+15=6x-15+15$$
$$x+20=6x$$
$$-x+x+20=-x+6x$$
$$20=5x$$
$$\frac{20}{5}=\frac{5x}{5}$$
$$4=x$$

13.
$$\frac{2y}{3}-\frac{1}{4}=\frac{1}{6}+\frac{y}{4}$$
$$12\left(\frac{2y}{3}\right)-12\left(\frac{1}{4}\right)=12\left(\frac{1}{6}\right)+12\left(\frac{y}{4}\right)$$
$$8y-3=2+3y$$
$$8y-3-3y=2+3y-3y$$
$$5y-3=2$$
$$5y-3+3=2+3$$
$$5y=5$$
$$\frac{5y}{5}=\frac{5}{5}$$
$$y=1$$

14.
$$\begin{aligned} 3b-4c &= 9-b \\ 3b-4c+b &= 9-b+b \\ 4b-4c &= 9 \\ 4b-4c+4c &= 9+4c \\ 4b &= 4c+9 \\ \frac{4b}{4} &= \frac{4c+9}{4} \\ b &= \frac{4c+9}{4} \end{aligned}$$

15.
$$\begin{aligned} H &= \frac{2}{3}(b+4a) \\ H &= \frac{2}{3}b+\frac{8}{3}a \\ 3(H) &= 3\left(\frac{2}{3}b\right)+3\left(\frac{8}{3}a\right) \\ 3H &= 2b+8a \\ 3H-8a &= 2b+8a-8a \\ 3H-8a &= 2b \\ \frac{3H-8a}{2} &= \frac{2b}{2} \\ \frac{3H-8a}{2} &= b \end{aligned}$$

16.
$$\begin{aligned} I &= Prt \\ \frac{I}{Pr} &= \frac{Prt}{Pr} \\ \frac{I}{Pr} &= t \end{aligned}$$

17.
$$\begin{aligned} A &= \frac{ha}{2}+\frac{hb}{2} \\ 2(A) &= 2\left(\frac{ha}{2}\right)+2\left(\frac{hb}{2}\right) \\ 2A &= ha+hb \\ 2A-hb &= ha+hb-hb \\ 2A-hb &= ha \\ \frac{2A-hb}{h} &= \frac{ha}{h} \\ \frac{2A-hb}{h} &= a \text{ or } a=\frac{2A}{h}-b \end{aligned}$$

18.
$$\begin{aligned} 7x-2 &\le 3x+10-2x \\ 7x-2 &\le x+10 \\ 7x-2-x &\le x+10-x \\ 6x-2 &\le 10 \\ 6x-2+2 &\le 10+2 \\ 6x &\le 12 \\ \frac{6x}{6} &\le \frac{12}{6} \\ x &\le 2 \end{aligned}$$

19.
$$\begin{aligned} \frac{1}{2}(x-5) &\ge x-4 \\ \frac{1}{2}x-\frac{5}{2} &\ge x-4 \\ 2\left(\frac{1}{2}x\right)-2\left(\frac{5}{2}\right) &\ge 2(x)-2(4) \\ x-5 &\ge 2x-8 \\ x-5-2x &\ge 2x-8-2x \\ -x-5 &\ge -8 \\ -x-5+5 &\ge -8+5 \\ -x &\ge -3 \\ \frac{-x}{-1} &\le \frac{-3}{-1} \\ x &\le 3 \end{aligned}$$

20.
$$\begin{aligned} 4(2-x) &> 1-5x-8 \\ 8-4x &> -7-5x \\ -8+8-4x &> -8-7-5x \\ -4x &> -15-5x \\ -4x+5x &> -15-5x+5x \\ x &> -15 \end{aligned}$$

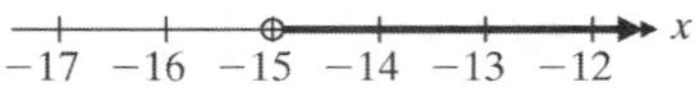

21.
$$\begin{aligned} x+\frac{5}{9} &\le \frac{1}{3}+\frac{7}{9}x \\ 9(x)+9\left(\frac{5}{9}\right) &\le 9\left(\frac{1}{3}\right)+9\left(\frac{7}{9}x\right) \\ 9x+5 &\le 3+7x \\ 9x+5-7x &\le 3+7x-7x \\ 2x+5 &\le 3 \\ 2x+5-5 &\le 3-5 \\ 2x &\le -2 \\ \frac{2x}{2} &\le \frac{-2}{2} \\ x &\le -1 \end{aligned}$$

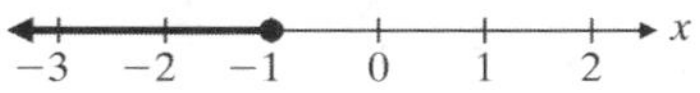

22.

$$\begin{aligned}
7x-13 &\le 3(4-5x)-3\\
7x-13 &\le 12-15x-3\\
7x-13 &\le 9-15x\\
7x+15x-13 &\le 9-15x+15x\\
22x-13 &\le 9\\
22x-13+13 &\le 9+13\\
22x &\le 22\\
\frac{22x}{22} &\le \frac{22}{22}\\
x &\le 1
\end{aligned}$$

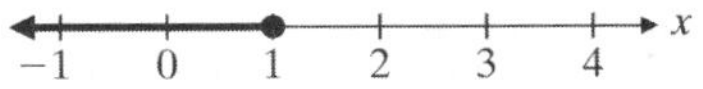

23.

$$\begin{aligned}
\frac{0+82+89+87+x}{5} &\ge 70\\
\frac{258+x}{5} &\ge 70\\
5\left(\frac{258+x}{5}\right) &\ge 5(70)\\
258+x &\ge 350\\
258+x-258 &\ge 350-258\\
x &\ge 92
\end{aligned}$$

He must receive a 92 or better.

Chapter 3

3.1 Exercises

1. a quantity increased by 5: $x + 5$

3. twelve less than a number: $x - 12$

5. one-eight of a quantity: $\frac{1}{8}x$ or $\frac{x}{8}$

7. twice a quantity: $2x$

9. three more than half of a number: $3+\frac{1}{2}x$

11. double a quantity increased by nine: $2x + 9$

13. one-third of the sum of a number and seven: $\frac{1}{3}(x+7)$

15. one-third of a number reduced by twice the same number: $\frac{1}{3}x-2x$

17. five times a quantity decreased by eleven: $5x - 11$

19. Since the value of the IBM stock is being compared to the value of the AT&T stock, we let the variable represent the value of the AT&T stock.
x = value of a share of AT&T stock
The value of a share of IBM stock is \$74.50 more than the value of a share of AT&T stock.
$x + 74.50$ = value of a share of IBM stock

21. Since the length of the rectangle is being compared to the width, we let the variable represent the width of a rectangle.
w = width of a rectangle
The length is 7 inches more than twice the width.
$2w + 7$ = length of rectangle

23. Since the numbers of boxes of cookies sold by Sarah and Imelda are being compared to the number sold by Keiko, we let the variable represent the number of boxes sold by Keiko.
x = number of boxes of cookies sold by Keiko
The number of boxes sold by Sarah was 43 fewer than the number sold by Keiko.
$x - 43$ = number of boxes of cookies sold by Sarah
The number of boxes sold by Imelda was 53 more than the number sold by Keiko.
$x + 53$ = number of boxes of cookies sold by Imelda

25. Since the first and third angles are being compared to the second angle, we let the variable represent the second angle.
x = second angle
The first angle is 16 degrees less than the second angle.
$s - 16$ = first angle
The third angle is double the second.
$2s$ = third angle

27. Since the exports of Japan are being compared to the exports of Canada, we let the variable represent the value of the exports of Canada.
v = value of exports of Canada
The value of the exports of Japan was twice the value of the exports of Canada.
$2v$ = value of exports of Japan

29. Since the first and third angles are being compared to the second angle, we let the variable represent the second angle.
x = second angle
The first angle is triple the second angle.
$3x$ = first angle
The third angle is 14 degrees less than the second angle.
$x - 14$ = third angle

31. x = number of men aged 16 to 24
Since $302 - 220 = 82$, 82 more men aged 25 to 34 rented kayaks than men aged 16 to 24.
$x + 82$ = number of men aged 25 to 34
Since $220 - 195 = 25$, 25 fewer men aged 35 to 44 rented kayaks than men aged 16 to 24.
$x - 25$ = number of men aged 35 to 44
Since $220 - 110 = 110$, 110 fewer men aged 45+ rented kayaks than men aged 16 to 24.
$x - 110$ = number of men aged 45+

Cumulative Review

33.
$$x+\frac{1}{2}(x-3)=9$$
$$x+\frac{1}{2}x-\frac{3}{2}=9$$
$$\frac{3}{2}x-\frac{3}{2}=9$$
$$2\left(\frac{3}{2}x\right)-2\left(\frac{3}{2}\right)=2(9)$$
$$3x-3=18$$
$$3x-3+3=18+3$$
$$3x=21$$
$$\frac{3x}{3}=\frac{21}{3}$$
$$x=7$$

34.
$$\frac{3}{5}x-3(x-1)=9$$
$$\frac{3}{5}x-3x+3=9$$
$$-\frac{12}{5}x+3=9$$
$$5\left(-\frac{12}{5}x\right)+5(3)=5(9)$$
$$-12x+15=45$$
$$-12x+15-15=45-15$$
$$-12x=30$$
$$\frac{-12x}{-12}=\frac{30}{-12}$$
$$x=-\frac{5}{2} \text{ or } -2\frac{1}{2}$$

Quick Quiz 3.1

1. Ten greater than a number: $x + 10$ or $10 + x$

2. Five less than double a number: $2x - 5$

3. Since the first and third angles are being compared to the second, we let the variable represent the second angle.
   x = second angle
   The first angle is 15 degrees more than the second angle.
   $x + 15$ = first angle
   The third angle is double the second.
   $2x$ = third angle

4. Answers may vary. Possible solution: "one-third of the sum" means you multiply $\frac{1}{3}$ times the sum. The sum will be a quantity in parentheses because the $\frac{1}{3}$ is multiplied by the whole sum; $\frac{1}{3}(x+7)$.

3.2 Exercises

1. x = the number
$$x-543=718$$
$$x-543+543=718+543$$
$$x=1261$$
The number is 1261.
Check:
Does 1261 minus 543 give 718?
$$1261-543 \stackrel{?}{=} 718$$
$$718=718 \checkmark$$

3. x = the number
$$\frac{x}{8}=296$$
$$8\left(\frac{x}{8}\right)=8(296)$$
$$x=2368$$
The number is 2368.
Check:
Is 2368 divided by 8 equal to 296?
$$\frac{2368}{8} \stackrel{?}{=} 296$$
$$296=296 \checkmark$$

5. x = the number
$$x+17=199$$
$$x+17-17=199-17$$
$$x=182$$
The number is 182.
Check:
Is 17 greater than 182 equal to 199?
$$182+17 \stackrel{?}{=} 199$$
$$199=199 \checkmark$$

7. x = the number

$$2x+7=93$$
$$2x+7-7=93-7$$
$$2x=86$$
$$\frac{2x}{2}=\frac{86}{2}$$
$$x=43$$

The number is 43.
Check:
When 43 is doubled and then increased by 7, is the result 93?

$$2(43)+7 \stackrel{?}{=} 93$$
$$86+7 \stackrel{?}{=} 93$$
$$93=93 \checkmark$$

9. x = the number

$$18-\frac{2}{3}x=12$$
$$3(18)-3\left(\frac{2}{3}x\right)=3(12)$$
$$54-2x=36$$
$$54-2x-54=36-54$$
$$-2x=-18$$
$$\frac{-2x}{-2}=\frac{-18}{-2}$$
$$x=9$$

The number is 9.
Check:
When 18 is reduced by two-thirds of 9, is the result 12?

$$18-\frac{2}{3}(9) \stackrel{?}{=} 12$$
$$18-6 \stackrel{?}{=} 12$$
$$12=12 \checkmark$$

11. x = the number

$$3x-8=5x$$
$$3x-8-3x=5x-3x$$
$$-8=2x$$
$$\frac{-8}{2}=\frac{2x}{2}$$
$$-4=x$$

The number is –4.
Check:
Is 8 less than triple –4 the same as 5 times –4?

$$3(-4)-8 \stackrel{?}{=} 5(-4)$$
$$-12-8 \stackrel{?}{=} -20$$
$$-20=-20 \checkmark$$

13. x = the number

$$x+\frac{1}{2}x+\frac{1}{3}x=22$$
$$6x+6\left(\frac{1}{2}x\right)+6\left(\frac{1}{3}x\right)=6(22)$$
$$6x+3x+2x=132$$
$$11x=132$$
$$\frac{11x}{11}=\frac{132}{11}$$
$$x=12$$

The number is 12.
Check:
When 12, half of 12, and one-third of 12 are added, is the result 22?

$$12+\frac{1}{2}(12)+\frac{1}{3}(12) \stackrel{?}{=} 22$$
$$12+6+4 \stackrel{?}{=} 22$$
$$22=22 \checkmark$$

15. x = number of used bikes
$4x$ = number of new bikes

$$4x=60$$
$$\frac{4x}{4}=\frac{60}{4}$$
$$x=15$$

Used bikes in stock: 15

17. x = number of wildfires between 1/1/03 and 9/25/03.

$$2x-14,725=82,955$$
$$2x=97,680$$
$$\frac{2x}{2}=\frac{97,680}{2}$$
$$x=48,840$$

There were 48,840 wildfires between 1/1/03 and 9/25/03.

19. x = number of CDs

$$218+11x=284$$
$$218+11x-218=284-218$$
$$11x=66$$
$$\frac{11x}{11}=\frac{66}{11}$$
$$x=6$$

Kyle purchased 6 CDs.

21. x = number of items of jewelry

$$\begin{aligned} 2(38)+49+11.50x &= 171 \\ 76+49+11.50x &= 171 \\ 125+11.50x &= 171 \\ 125+11.50x-125 &= 171-125 \\ 11.50x &= 46 \\ \frac{11.50x}{11.50} &= \frac{46}{11.50} \\ x &= 4 \end{aligned}$$

She could buy 4 items of jewelry.

23. $r=\frac{d}{t}$, $d = 6404$, $t = 20$

$$r=\frac{6404}{20}=320.2$$

It traveled at 320.2 mi/hr

25. $J = 2.5E$, $E = 220$
$J = 2.5(220) = 550$
The astronaut would weigh 550 lb on Jupiter.

27. Nell: $r = 12$, $t = 2.5$
$d = rt = 12(2.5) = 30$
Kristin: $r = 14$, $t = 2.5$
$d = rt = 14(2.5) = 35$
Distance apart = 35 – 30 = 5
They will be 5 miles apart.

29. r = rate
Highway route: $5r = 320$
$r = 64$
Highway route at 64 mph.

Mountain route: $6r = 312$
$r = 52$
Mountain route at 52 mph.
Difference: 64 – 52 = 12 mph
The highway route was 12 mph faster than the mountain route.

31. x = score on the final label

$$\begin{aligned} \frac{84+87+93+89+89+94+94}{10} &= 90 \\ \frac{630+3x}{10} &= 90 \\ 630+3x &= 900 \\ 3x &= 270 \\ x &= 90 \end{aligned}$$

She needs a 90 on the final lab.

33. x = number of cricket chirps
F = Fahrenheit temperature

a. $F-40=\frac{x}{4}$

b.
$$\begin{aligned} 90-40 &= \frac{x}{4} \\ 50 &= \frac{x}{4} \\ x &= 200 \end{aligned}$$

200 chirps per minute should be recorded.

c.
$$\begin{aligned} F-40 &= \frac{148}{4} \\ F-40 &= 37 \\ F &= 77 \end{aligned}$$

The temperature would be 77°F.

Cumulative Review

34.
$$\begin{aligned} 5x(2x^2-6x-3) &= 5x(2x^2)-5x(6x)-5x(3) \\ &= 10x^3-30x^2-15x \end{aligned}$$

35.
$$\begin{aligned} &-2a(ab-3b+5a) \\ &= -2a(ab)-2a(-3b)-2a(5a) \\ &= -2a^2b+6ab-10a^2 \end{aligned}$$

36.
$$\begin{aligned} 7x-3y-12x-8y+5y &= (7-12)x+(-3-8+5)y \\ &= -5x-6y \end{aligned}$$

37.
$$\begin{aligned} &5x^2y-7xy^2-8xy-9x^2y \\ &= (5-9)x^2y-7xy^2-8xy \\ &= -4x^2y-7xy^2-8xy \end{aligned}$$

Quick Quiz 3.2

1. x = the number

$$\begin{aligned} 3x-15 &= 36 \\ 3x-15+15 &= 36+15 \\ 3x &= 51 \\ \frac{3x}{3} &= \frac{51}{3} \\ x &= 17 \end{aligned}$$

The number is 17.

2. x = the number

$$\frac{1}{2}x+\frac{1}{6}x+\frac{1}{8}x=38$$
$$24\left(\frac{1}{2}x\right)+24\left(\frac{1}{6}x\right)+24\left(\frac{1}{8}x\right)=24(38)$$
$$12x+4x+3x=912$$
$$19x=912$$
$$\frac{19x}{19}=\frac{912}{19}$$
$$x=48$$

The number is 48.

3. x = score on next test

$$\frac{84+89+73+80+x}{5}=80$$
$$\frac{326+x}{5}=80$$
$$5\left(\frac{326+x}{5}\right)=5(80)$$
$$326+x=400$$
$$326+x-326=400-326$$
$$x=74$$

James must score 74.

4. Answers may vary. Possible solution: Since we want to know how many pairs of socks, let x = the number of pairs of socks. Set up the equation to represent the total amount spent.
2($23) + 0.75x$ = $60.25

3.3 Exercises

1. x = long length
$x - 17$ = short length

$$x+x-17=47$$
$$2x-17=47$$
$$2x=64$$
$$x=32$$

$x - 17 = 15$
The long piece is 32 meters long.
The short piece is 15 meters long.

3. x = James' salary
$x + 3400$ = Lin's salary

$$x+(x+3400)=82,300$$
$$2x+3400=82,300$$
$$2x+3400-3400=82,300-3400$$
$$2x=78,900$$
$$\frac{2x}{2}=\frac{78,900}{2}$$
$$x=39,450$$

$x + 3400 = 39{,}450 + 3400 = 42{,}850$
James earns $39,450 per year.
Lin earns $42,850 per year.

5. x = hours David worked
$x + 15$ = hours Sarah worked
$x - 5$ = hours Kate worked

$$x+x+15+x-5=100$$
$$3x+10=100$$
$$3x=90$$
$$x=30$$

$x + 15 = 45$
$x - 5 = 25$
David worked 30 hours.
Sarah worked 45 hours.
Kate worked 25 hours.

7. x = number born in Oklahoma
$x + 8$ = number born in Texas
$x - 9$ = number born in Arizona

$$x+x+8+x-9=32$$
$$3x-1=32$$
$$3x=33$$
$$x=11$$

$x + 8 = 11 + 8 = 19$
$x - 9 = 11 - 9 = 2$
11 were born in Oklahoma.
19 were born in Texas.
2 were born in Arizona.

9. x = width
$2x - 3$ = length

$$P=2L+2W$$
$$42=2(2x-3)+2x$$
$$42=4x-6+2x$$
$$42=6x-6$$
$$48=6x$$
$$8=x$$

$2x - 3 = 2(8) - 3 = 16 - 3 = 13$
The width is 8 meters. The length is 13 meters.

11. x = width
$3x + 20$ = length

$$P=2L+2W$$
$$96=2(3x+20)+2x$$
$$96=6x+40+2x$$
$$96=8x+40$$
$$56=8x$$
$$7=x$$

$3x + 20 = 3(7) + 20 = 41$
The width is 7 centimeters.
The length is 41 centimeters.

13. w = width
$3w - 5$ = length
$2w + 2(3w - 35) = 190$
$2w + 6w - 70 = 190$
$8w = 260$
$w = 32.5$
$3w - 35 = 62.5$
The width is 32.5 centimeters.
The length is 62.5 centimeters.

15. x = speed of jackal
$2x$ = speed of cheetah
$x + 10$ = speed of elk
$x + 2x + x + 10 = 150$
$4x + 10 = 150$
$4x = 140$
$x = 35$
$2x = 70$
$x + 10 = 45$
35 miles per hour is the speed of the jackal.
70 miles per hour is the speed of the cheetah.
45 miles per hour is the speed of the elk.

17. x = length C
$2x$ = length A
$1.5x$ = length B
$x + 3$ = length D
$x + 2x + 1.5x + x + 3 = 58$
$5.5x + 3 = 58$
$5.5x = 55$
$x = 10$
x = length C = 10"
$2x$ = length A = 20"
$1.5x$ = length B = 15"
$x + 3$ = length D = 13"

19. x = length of side
$x + 2$ = length of new side
$4(x + 2) = 5x - 3$
$4x + 8 = 5x - 3$
$11 = x$
The original square was 11 meters by 11 meters.

Cumulative Review

21. $-4x(2x^2 - 3x + 8)$
$= -4x(2x^2) - 4x(-3x) - 4x(8)$
$= -8x^3 + 12x^2 - 32x$

22. $5a(ab + 6b - 2a) = 5a(ab) + 5a(6b) - 5a(2a)$
$= 5a^2b + 30ab - 10a^2$

23. $-7x + 10y - 12x - 8y - 2$
$= (-7 - 12)x + (10 - 8)y - 2$
$= -19x + 2y - 2$

24. $3x^2y - 6xy^2 + 7xy + 6x^2y$
$= (3 + 6)x^2y - 6xy^2 + 7xy$
$= 9x^2y - 6xy^2 + 7xy$

Quick Quiz 3.3

1. x = width
$2x + 80$ = length
$2(x) + 2(2x + 80) = 610$
$2x + 4x + 160 = 610$
$6x + 160 = 610$
$6x + 160 - 160 = 610 - 160$
$6x = 450$
$\frac{6x}{6} = \frac{450}{6}$
$x = 75$
$2x + 180 = 2(75) + 80 = 230$
width: 75 yd; length: 230 yd

2. x = number of students in art history
$3x$ = number of students in psychology
$x + 14$ = number of students in algebra
$x + 3x + (x + 14) = 79$
$5x + 14 = 79$
$5x + 14 - 14 = 79 - 14$
$5x = 65$
$\frac{5x}{5} = \frac{65}{5}$
$x = 13$
$3x = 3(13) = 39$
$x + 14 = 13 + 14 = 27$
There are 13 students in art history, 39 in psychology, and 27 in algebra.

3. x = length of second side

$\frac{2}{3}x$ = length of first side

$x - 15$ = length of third side

$$x+\frac{2}{3}x+(x-15)=73$$
$$\frac{8}{3}x-15=73$$
$$3\left(\frac{8}{3}x\right)-3(15)=3(73)$$
$$8x-45=219$$
$$8x-45+45=219+45$$
$$8x=264$$
$$\frac{8x}{8}=\frac{264}{8}$$
$$x=33$$

$$\frac{2}{3}x=\frac{2}{3}(33)=22$$

$x - 15 = 33 - 15 = 18$
first side: 22 meters
second side: 33 meters
third side: 18 meters

4. Answers may vary. Possible solution:
Since Laurie's salary is given in terms of Don's, let Don's salary = x.
Then Laurie's salary = x + 2600. The sum of their salaries is 71,200, since together they earned \$71,200. Solve: $x + x + 2600 = 71{,}200$.

How Am I Doing? Sections 3.1–3.3

1. three times a quantity is then decreased by 40:
$3x - 40$

2. one-half of a number is increased by 12:
$\frac{1}{2}x+12$

3. a number multiplied by 5 and then divided by 12: $\frac{5x}{12}$ or $\frac{5}{12}x$

4. one-fourth of the sum of a number and 10:
$\frac{1}{4}(x+10)$ or $\frac{x+10}{4}$

5. x = the number
$$2x+5=-47$$
$$2x+5-5=-47-5$$
$$2x=-52$$
$$\frac{2x}{2}=\frac{-52}{2}$$
$$x=-26$$
The number is –26.

6. x = the number
$$\frac{x}{4}+8=25$$
$$4\left(\frac{x}{4}\right)+4(8)=4(25)$$
$$x+32=100$$
$$x+32-32=100-32$$
$$x=68$$
The number is 68.

7. x = number of months
$$310x=12{,}400$$
$$\frac{310x}{310}=\frac{12{,}400}{310}$$
$$x=40$$
They will be equal in 40 months.

8. x = hours of overtime
$$24(40)+\frac{3}{2}(24)x=1140$$
$$960+36x=1140$$
$$36x=180$$
$$\frac{36x}{36}=\frac{180}{36}$$
$$x=5$$
Bill worked 5 hours overtime.

9. x = 4th person's salary
$x + 2500$ = 5th person's salary
$40,000 + 13,000$ = 3rd person's salary

$$40,000 + 35,000 + 53,500 + x + x + 2500 = 2x + 131,000$$
$$\frac{2x + 131,000}{5} = 50,200$$
$$5\left(\frac{2x + 131,000}{5}\right) = 5(50,200)$$
$$2x + 131,000 = 251,000$$
$$2x = 120,000$$
$$\frac{2x}{2} = \frac{120,000}{2}$$
$$x = 60,000$$

$x + 2500 = 60,000 + 2500 = 62,500$
The 4th person's salary was \$60,000. The 5th person's salary was \$62,500.

10. x = length of short piece
$1.5x$ = length of long piece

$$x + 1.5x = 10$$
$$2.5x = 10$$
$$\frac{2.5x}{2.5} = \frac{10}{2.5}$$
$$x = 4$$

$1.5x = 1.5(4) = 6$
The long piece is 6 feet. The short piece is 4 feet.

11. x = number of pages on Sunday
$2x - 25$ = number of pages on Monday
$3x - 15$ = number of pages on Tuesday

$$x + 2x - 25 + 3x - 15 = 260$$
$$6x - 40 = 260$$
$$6x = 300$$
$$\frac{6x}{6} = \frac{300}{6}$$
$$x = 50$$

$2x - 25 = 2(50) - 25 = 75$
$3x - 15 = 3(50) - 15 = 135$
He read 50 pages on Sunday, 75 pages on Monday, and 135 pages on Tuesday.

12. x = price of SSC
$x + 12,000$ = price of Pagani
$2x - 118,000$ = price of Bugatti

$$x + (x + 12,000) + (2x - 118,000) = 2,514,000$$
$$4x - 106,000 = 2,514,000$$
$$4x = 2,620,000$$
$$\frac{4x}{4} = \frac{2,620,000}{4}$$
$$x = 655,000$$

$x + 12,000 = 655,000 + 12,000 = 667,000$

$$2x-118{,}000 = 2(655{,}000)-118{,}000$$
$$= 1{,}192{,}000$$

SSC: $655,000
Pagani: $667,000
Bugatti: $1,192,000

3.4 Exercises

1. x = number of bags
$$0.75x+3(6.50)=27.75$$
$$0.75x+19.50=27.75$$
$$0.75x=8.25$$
$$\frac{0.75x}{0.75}=\frac{8.25}{0.75}$$
$$x=11$$
He filled 11 bags.

3. x = overtime hours
$$8(40)+12x=380$$
$$320+12x=380$$
$$12x=60$$
$$\frac{12x}{12}=\frac{60}{12}$$
$$x=5$$
Kate needs 5 hours of overtime per week.

5. x = time to pay off the uniforms
$$600+105x=1817.75$$
$$105x=1217.75$$
$$\frac{105x}{105}=\frac{1217.75}{105}$$
$$x\approx 11.6$$
It will take about 12 weeks.

7. x = original price
$$0.28x=100.80$$
$$\frac{0.28x}{0.28}=\frac{100.80}{0.28}$$
$$x=360$$
The original price was $360.

9. x = last year's salary
$0.03x$ = raise
$$x+0.03x=22{,}660$$
$$1.03x=22{,}660$$
$$\frac{1.03x}{1.03}=\frac{22{,}660}{1.03}$$
$$x=22{,}000$$
Last year's salary was $22,000.

11. x = investment
$0.06x$ = interest
$$x+0.06x=12{,}720$$
$$1.06x=12{,}720$$
$$\frac{1.06x}{1.06}=\frac{12{,}720}{1.06}$$
$$x=12{,}000$$
She invested $12,000.

13. x = amount earning 7%
$5000-x$ = amount earning 5%
$$0.07x+0.05(5000-x)=310$$
$$0.07x+250-0.05x=310$$
$$0.02x+250=310$$
$$0.02x=60$$
$$\frac{0.02x}{0.02}=\frac{60}{0.02}$$
$$x=3000$$
$5000-x=2000$
They invested $3000 at 7% and $2000 at 5%.

15. x = amount invested at 12%
$400{,}000-x$ = amount invested at 8%
$$0.12x+0.08(400{,}000-x)=38{,}000$$
$$0.12x+32{,}000-0.08x=38{,}000$$
$$0.04x+32{,}000=38{,}000$$
$$0.04x=6000$$
$$\frac{0.04x}{0.04}=\frac{6000}{0.04}$$
$$x=150{,}000$$
$400{,}000-x=250{,}000$
They invested $150,000 at 12% and $250,000 at 8%.

17. x = amount invested

$\frac{x}{2}$ = amount invested at 5%

$\frac{x}{3}$ = amount invested at 4%

$$x-\frac{x}{2}-\frac{x}{3}=\frac{6x-3x-2x}{6}$$
$$=\frac{x}{6}$$
= amount invested at 3.5%

$$0.05\left(\frac{x}{2}\right)+0.04\left(\frac{x}{3}\right)+(0.035)\left(\frac{x}{6}\right)=530$$
$$\left(\frac{0.05}{2}+\frac{0.04}{3}+\frac{0.035}{6}\right)x=530$$
$$\frac{53}{1200}x=530$$
$$x=12{,}000$$

He invested \$12,000.

19. x = number of quarters
$x - 4$ = number of nickels
$$0.25x+0.05(x-4)=3.70$$
$$25x+5(x-4)=370$$
$$25x+5x-20=370$$
$$30x=390$$
$$\frac{30x}{30}=\frac{390}{30}$$
$$x=13$$
$x-4=9$
She has 13 quarters and 9 nickels.

21. x = number of dimes
$x + 3$ = number of quarters
$2(x + 3) = 2x + 6$ = number of nickels
$$0.25(x+3)+0.10x+0.05(2x+6)=3.75$$
$$25(x+3)+10x+5(2x+6)=375$$
$$25x+75+10x+10x+30=375$$
$$45x+105=375$$
$$45x=270$$
$$\frac{45x}{45}=\frac{270}{45}$$
$$x=6$$
$x+3=6+3=9$
$2x+6=12+6=18$
He has 9 quarters, 6 dimes, and 18 nickels.

23. x = number of regular stamps
$x - 8$ = number of breast cancer stamps
$$0.37x+0.4(x-8)=19.90$$
$$0.37x+0.4x-3.2=19.90$$
$$0.77x=23.1$$
$$\frac{0.77x}{0.77}=\frac{23.1}{0.77}$$
$$x=30$$
$x-8=16$
$x+3=30-8=22$
He bought 30 regular stamps and 22 breast cancer stamps.

25. x = number of \$50 bills
$x + 16$ = number of \$10 bills
$3x + 1$ = number of \$20 bills
$$50x+10(x+16)+20(3x+1)=1380$$
$$50x+10x+160+60x+20=1380$$
$$120x+180=1380$$
$$120x=1200$$
$$\frac{120x}{120}=\frac{1200}{120}$$
$$x=10$$
$x + 16 = 10 + 16 = 26$
$3x + 1 = 3(10) + 1 = 31$
They had ten \$50 bills, twenty-six \$10 bills, and thirty-one \$20 bills.

27. x = amount of sales
$$18{,}000+0.04x=55{,}000$$
$$0.04x=37{,}000$$
$$\frac{0.04x}{0.04}=\frac{37{,}000}{0.04}$$
$$x=925{,}000$$
She must sell \$925,000 worth of furniture.

Cumulative Review

29. $5(3) + 6 \div (-2) = 15 + (-3) = 12$

30.
$$\begin{aligned}5(-3)-2(12-15)^2\div 9&=5(-3)-2(-3)^2\div 9\\&=5(-3)-2(9)\div 9\\&=-15-18\div 9\\&=-15-2\\&=-17\end{aligned}$$

31. If $a = -1$ and $b = 4$, then
$$\begin{aligned}a^2-2ab+b^2&=(-1)^2-2(-1)(4)+(4)^2\\&=1+8+16\\&=25\end{aligned}$$

32. If $a = -1$ and $b = 4$, then
$$\begin{aligned}a^3+ab^2-b-5&=(-1)^3+(-1)(4)^2-4-5\\&=-1+(-1)(16)-4-5\\&=-1-16-4-5\\&=-26\end{aligned}$$

Quick Quiz 3.4

1. x = number of months
$$224+114x=1250$$
$$114x=1026$$
$$\frac{114x}{114}=\frac{1026}{114}$$
$$x=9$$
He can rent the machine for 9 months.

2. x = last year's cost
$0.7x$ = increase
$$x + 0.7x = 12,412$$
$$1.07x = 12,412$$
$$\frac{1.07x}{1.07} = \frac{12,412}{1.07}$$
$$x = 11,600$$
It cost \$11,600 before the increase.

3. x = amount invested at 4%
$5000 - x$ = amount invested at 5%
$$0.04x + 0.05(5000 - x) = 228$$
$$0.04x + 250 - 0.05x = 228$$
$$250 - 0.01x = 228$$
$$-0.01x = -22$$
$$\frac{-0.01x}{-0.01} = \frac{-22}{-0.01}$$
$$x = 2200$$
$5000 - x = 5000 - 2200 = 2800$
They invested \$2200 at 4% and \$2800 at 5%.

4. Answers may vary. Possible solution:
Since each amount is given in terms of quarters, let x = the number of quarters. Then
$2x$ = number of dimes, and
$x + 1$ = number of nickels. Now set up an equation, and solve, given the value of each coin and the total.
$0.25x + 0.10(2x) + 0.05(x + 1) = 2.75$

3.5 Exercises

1. Perimeter is the distance around a plane figure.

3. Area is a measure of the amount of surface in a region.

5. The sum of the interior angles of any triangle is 180°.

7. $a = 24$, $b = 13$
$$A = \frac{1}{2}ab = \frac{1}{2}(24)(13) = 156 \text{ in.}^2$$

9. $a = 14$, $b = 7$
$$A = ab = (14)(7) = 98 \text{ in.}^2$$

11. $A = 90$, $b = 30$
$$A = \frac{1}{2}ab$$
$$90 = \frac{1}{2}a(30)$$
$$90 = 15a$$
$$\frac{90}{15} = \frac{15a}{15}$$
$$6 = a$$
$$a = 6$$
The altitude is 6 feet.

13. x = length of adjacent side
$$2(14) + 2x = 46$$
$$28 + 2x = 46$$
$$28 + 2x - 28 = 46 - 28$$
$$2x = 18$$
$$\frac{2x}{2} = \frac{18}{2}$$
$$x = 9$$
The length is 9 inches.

15. $A = \pi r^2$, $r = 7.00$
$$A \approx (3.14)(7.00)^2 = 153.86 \text{ square feet}$$

17. $C = \pi d$, $d = 3032$
$C = (3.14)(3032)$
$C = 9520.48$ miles

19. $A = 600$, $b_1 = 20$, $b_2 = 30$
$$A = \frac{1}{2}a(b_1 + b_2)$$
$$600 = \frac{1}{2}a(20 + 30)$$
$$600 = \frac{1}{2}a(50)$$
$$600 = 25a$$
$$\frac{600}{25} = \frac{25a}{25}$$
$$24 = a$$
The attitude is 24 inches.

21. length = 3(80) = 240
width = 2(280) = 560
$$A = LW = 240(560) = 134,400 \text{ m}^2$$

23. $C = 2\pi r$, $C = 31.4$

$$31.4 = 2(3.14)r$$
$$31.4 = 6.28r$$
$$\frac{31.4}{6.28} = \frac{6.28r}{6.28}$$
$$5 = r$$

The radius is 5 centimeters.

25. $r = \frac{d}{2}$, $d = 64$

$$r = \frac{64}{2} = 32$$

$C = 2\pi r$, $r = 32$

$C = 2(3.14)(32) = 200.96$ centimeters

27. x = measure of 3rd angle

Sum of the angles is 180°.

$$x + 47 + 59 = 180$$
$$x + 106 = 180$$
$$x = 74$$

The measure of the 3rd angle is 74°.

29. x = measure of 3rd angle

$2x$ = measure of the equal angles

Sum of the angles is 180°.

$$x + 2x + 2x = 180$$
$$5x = 180$$
$$x = 36$$

$2x = 2(36) = 72$

The measure of the 3rd angle is 36°.

The measure of the equal angles is 72°.

31. x = measure of the equal angles

Sum of the angles is 180°.

$$x + x + 120 = 180$$
$$2x + 120 = 180$$
$$2x = 60$$
$$x = 30$$

The measure of the equal angles is 30°.

33. x = measure of 2nd angle

$2x$ = measure of 1st angle

$2(2x) + 19 = 4 + 19$ = measure of 3rd angle

$$x + 2x + 4x + 19 = 180$$
$$7x + 19 = 180$$
$$7x = 161$$
$$x = 23$$

$2x = 46$

$4x + 19 = 111$

1st angle measures 46°.

2nd angle measures 23°.

3rd angle measures 111°.

35. Find the circumference of the circle.

$C = 2\pi r$, $r = 3$ ft

$C = 2\pi(3) \approx 2(3.14)(3) = 18.84$ ft

She will need to buy 19 feet of fencing.

37. Number of cans of paint $= \frac{40}{20} = 2$

Area painted $= 2(102) = 204$

Area of one side $= \frac{204}{4} = 51$

$$A = \frac{1}{2}a(b_1 + b_2)$$
$$51 = \frac{1}{2}a(8 + 9)$$
$$102 = a(17)$$
$$6 = a$$

Altitude is 6 feet.

39. $V = 235.5$, $r = 15$

$V = \pi r^2 h$, $r = 10$, $h = 8$

$$V = (3.14)(10)^2(8)$$
$$= (3.14)(100)(8)$$
$$= 2512$$

The volume is 2512 cubic inches.

41. $V = \frac{4}{3}\pi r^3$, $r = \frac{d}{2} = \frac{8}{2} = 4$

$$V = \frac{4}{3}(3.14)(4)^3 = 267.95$$

$267.95 > 175$

Yes, it will stay aloft.

43. $r = 6$, $h = 4$

a. $V = 4\pi r^2 h$

$$= 4(3.14)(6)^2(4)$$
$$= 452.16$$

The volume is $452.16\ \text{cm}^3$.

b. $S = 2\pi rh + 2\pi r^2$

$$= 2(3.14)(6)(4) + 2(3.14)(6)^2$$
$$= 150.72 + 226.08$$
$$= 376.8$$

The total surface area is $376.8\ \text{cm}^2$.

45. Area of the $\frac{3}{4}$ circle:

$A = \frac{3}{4}\pi r^2,\ r = 8$

$A = \frac{3}{4}(3.14)(8)^2 = 150.72$

Area of the $\frac{1}{4}$ circle:

$A = \frac{1}{4}\pi r^2,\ r = 4$

$A = \frac{1}{4}(3.14)(4)^2 = 12.56$

Total area = 150.72 + 12.56 = 163.28
Goat eats 163.28 square feet.

47. a. $A = l_1 w_1 + \frac{1}{4}\pi r^2 + l_2 w_2$

$l_1 = 1.5,\ w_1 = 9.5,\ r = 1.5,$
$l_2 = 1.5,\ w_2 = 4.5$

$A = (1.5)(9.5) + \frac{1}{4}(3.14)(1.5)^2 + (1.5)(4.5)$
$= 14.25 + 1.77 + 6.75$
$= 22.77 \text{ yards}^2$

b. Cost = (2.50)(22.77) = \$56.93

49. x = length of straight edge
$P = 2\pi r + 2x,\ r = 7,\ x = 15$
$P = 2(3.14)(7) + 2(15) = 43.96 + 30 = 73.96$
The perimeter is 73.96 inches.

51. Perimeter $= \frac{120.93}{1.39} = 87$ feet
w = width, $4w - 3$ = length
$2w + 2(4w - 3) = 87$
$2w + 8w - 6 = 87$
$10w = 93$
$w = 9.3$
$4w - 3 = 34.2$
The width is 9.3 feet, and the length is 34.2 feet.

53. Moon's radius = 1080(5280) = 5,702,400
4-ft poles: $C = 2\pi r,\ r = 5{,}702{,}404$
$C = 2(3.14)(5{,}702{,}404) = 35{,}811{,}097.12$
3-ft poles: $C = 2\pi r,\ r = 5{,}702{,}404$
$C = 2(3.14)(5{,}702{,}403) = 35{,}811{,}090.84$
$C_4 - C_3 = 35{,}811{,}097.12 - 35{,}811{,}090.84$
$= 6.28$ feet

Cumulative Review

54. $2(x-6) - 3[4 - x(x+2)]$
$= 2x - 12 - 3[4 - x^2 - 2x]$
$= 2x - 12 - 12 + 3x^2 + 6x$
$= 3x^2 + 8x - 24$

55. $-5(x + 3) - 2[4 + 3(x - 1)]$
$= -5x - 15 - 2(4 + 3x - 3)$
$= -5x - 15 - 2(3x + 1)$
$= -5 - 15 - 6x - 2$
$= -11x - 17$

56. The increase is 5% of 203 billion.
5% of 203 = 0.05 × 203 = 10.15 ≈ 10
The number of pieces of mail increased by about 10 billion.
203 billion + 10 billion = 213 billion
Approximately 213 billion pieces of mail were delivered in 2006.

57. Subtract to find the increase.
92,558 − 36,276 = 56,282
Now find the percent increase.
$\frac{56{,}282}{36{,}276} \approx 1.55 = 155\%$

Quick Quiz 3.5

1. $a = 8$ ft, $b_1 = 12$ ft, $b_2 = 17$ ft

$A = \frac{1}{2}a(b_1 + b_2)$
$= \frac{1}{2}(8)(12 + 17)$
$= \frac{1}{2}(8)(29)$
$= 4(29)$
$= 116 \text{ ft}^2$

2. $r = 3$ in.

$V = \frac{4}{3}\pi r^3$
$= \frac{4}{3}(3.14)(3)^3$
$= \frac{4}{3}(3.14)(27)$
$\approx 113 \text{ in.}^3$

3. Area of square: $A = s^2 = (5)^2 = 25 \text{ yd}^2$

Area of rectangle: $A = lw = (5)(14) = 70 \text{ yd}^2$

Total area: $25 \text{ yd}^2 + 70 \text{ yd}^2 = 95 \text{ yd}^2$

Cost: $95 \times \$24 = \$2280$

4. Answers may vary. Possible solution:
Since the sum of the interior angles of a triangle is 180°, let x = measure of the third angle and set up, and solve, the equation:
$135° + 11° + x = 180°$.

3.6 Exercises

1. The cost is greater than \$67,000.
$x > 67{,}000$

3. The number of people must not be more than 90.
$x \le 90$

5. The height is not more than 1500 feet.
$h \le 1500$

7. Ramon's grade cannot be less than 93.
$x \ge 93$

9. 3rd length: x

$$87 + 64 + x \le 291$$
$$151 + x \le 291$$
$$x \le 140$$

It must be 140 centimeters or less.

11. x = number of deliveries in week 6

$$\frac{18 + 40 + 21 + 7 + 36 + x}{6} \ge 24$$
$$\frac{122 + x}{6} \ge 24$$
$$122 + x \ge 144$$
$$x \ge 22$$

She must make 22 or more deliveries.

13. $A = LW,\ W = 8$

$$8L \le 60$$
$$L \le 7.5$$

Depth can be no more than 7.5 feet.

15. x = number of miles

$$2(29.95) + 0.49x \le 100$$
$$59.90 + 0.49x \le 100$$
$$0.49x \le 40.10$$
$$x \le 81.8$$

They could drive at most 81.8 miles.

17.

$$\frac{5}{9}(F - 32) < 110$$
$$F(5 - 32) < 990$$
$$F - 32 < 198$$
$$F < 230$$

Temperatures less than 230°F.

19.

$$19.2x + 124.8 > 528$$
$$19.2x > 403.2$$
$$x > 21$$

$1990 + 21 = 2011$
The amount of federal aid will exceed \$528 billion after 2011.

21. x = sales, in dollars

$$16{,}000 + 0.06(x - 30{,}000) > 22{,}000$$
$$16{,}000 + 0.06x - 1800 > 22{,}000$$
$$0.06x + 14{,}200 > 22{,}000$$
$$0.06x > 7800$$
$$x > 130{,}000$$

Frank must have more than \$130,000 in sales.

23.

$$18n > 5000 + 7n$$
$$11n > 5000$$
$$n > 454.5$$

Need to manufacture and sell at least 455 discs.

Cumulative Review

25.

$$10 - 3x > 14 - 2x$$
$$-x > 4$$
$$x < -4$$

26.

$$2x - 3 \ge 7x - 18$$
$$-5x \ge -15$$
$$x \le 3$$

27.

$$30 - 2(x + 1) \le 4x$$
$$30 - 2x - 2 \le 4x$$
$$28 - 2x \le 4x$$
$$-6x \le -28$$
$$x \ge \frac{14}{3}$$
$$x \ge 4\frac{2}{3}$$

28.

$$2(x + 3) - 22 < 4(x - 2)$$
$$2x + 6 - 22 < 4x - 8$$
$$2x - 16 < 4x - 8$$
$$-2x < 8$$
$$x > -4$$

Quick Quiz 3.6

1. x = length of garden
$8x \le 79.2$
$x \le 9.9$
The length must be no more than 9.9 feet.

2. x = number of hours fourth student can work
$\frac{15+17+12+x}{4} < 15.5$
$\frac{44+x}{4} < 15.5$
$44+x < 62$
$x < 18$
The fourth student must work fewer than 18 hours per week.

3. x = sales, in dollars
$17,000+0.04(x-10,000) > 28,000$
$17,000+0.04x-400 > 28,000$
$0.04x+16,600 > 28,000$
$0.04x > 11,400$
$x > 285,000$
Chris must have more than \$285,000 in sales.

4. Answers may vary. Possible solution:
Since we are solving for the number of miles, let x = miles. The charge is \$37.50 each day for two days, so that total cost is 2(37.50). It costs 19.5¢ per mile, so that total cost is $0.19x$. Now set up the inequality and solve. If he doesn't want to spend more than \$150, the equation is:
$2(37.50) + 0.19x \le 150$.

Putting Your Skills to Work

1. All plans include 100 minutes.
Plan A: \$39.99
Plan B: \$39.99
Plan C: \$49.99
Plan D: \$59.99
He should choose either plan A or B.

2. Plan A: 300 min = \$39.99
50 min = \$0.20×50 = \$10
Total = \$39.99 + \$10 = \$49.99
Plans B, C, and D include 350 minutes.
Plan B: \$39.99
Plan C: \$49.99
Plan D: \$59.99
He should choose plan B.

3. Plan A: 300 min = \$39.99
170 min = \$0.20×170 = \$34
Total = \$39.99 + \$34 = \$73.99
Plan B: 450 min = \$39.99
20 min = \$0.45×20 = \$9
Total = \$39.99 + \$9 = \$48.99
Plans C and D include 470 minutes.
Plan C: \$49.99
Plan D: \$59.99
He should choose plan B.

4. Plan A: 300 min = \$39.99
250 min = \$0.20×250 = \$50
Total = \$39.99 + \$50 = \$89.99
Plan B: 450 min = \$39.99
100 min = \$0.45×100 = \$45
Total = \$39.99 + \$45 = \$84.99
Plans C and D include 550 minutes.
Plan C: \$49.99
Plan D: \$59.99
He should choose plan C.

5. 11 hr = 11 × 60 = 660 min
Plan A: 300 min = \$39.99
360 min = \$0.20×360 = \$72
Total = \$39.99 + \$72 = \$111.99
Plan B: 450 min = \$39.99
210 min = \$0.45×210 = \$94.50
Total = \$39.99 + \$94.50 = \$134.49
Plan C: 600 min = \$49.99
60 min = \$0.20×60 = \$12
Total = \$49.99 + \$12 = \$61.99
Plan D: 900 min = \$59.99
He should choose plan D.

6. 16 hours + 40 min = 16 × 60 + 40 = 1000 min
Plan A: 300 min = \$39.99
700 min = \$0.20×700 = \$140
Total = \$179.99
Plan B: 450 min = \$39.99
550 min = \$0.45×550 = \$247.50
Total = \$251.49
Plan C: 600 min = \$49.99
400 min = \$0.20×400 = \$80
Total = \$129.99
Plan D: 900 min = \$59.99
100 min = \$0.40×100 = \$40
Total = \$99.99
He should choose plan D.

7. Calling Plan C + Bundle D = \$49.99 + \$35.00
= \$84.99

The total cost is \$84.99.

8. Calling Plan D + Bundle D = \$59.99 + \$35.00
= \$94.99

The total cost is \$94.99.

9. Calling Plan C: 600 min = \$49.99
150 min = \$0.20×150 = \$30
Total = \$49.99 + 30 = \$79.99
Calling Plan C + Bundle D = \$79.99 + \$35.00
= \$114.99

10. Answers may vary.

Chapter 3 Review Problems

1. 19 more than a number: $x + 19$

2. two-thirds of a number: $\frac{2}{3}x$

3. half a number: $\frac{1}{2}x$ or $\frac{x}{2}$

4. 18 less than a number: $x - 18$

5. triple the sum of a number and 4: $3(x + 4)$

6. twice a number decreased by 3: $2x - 3$

7. Since the numbers of working people and unemployed people are being compared to the number of retired people, we let the variable represent the number of retired people.
r = number of retired people
The number of working people is four times the number of retired people.
$4r$ = number of working people
The number of unemployed people is one-half the number of retired people.
$\frac{1}{2}r = 0.5r$ = number of unemployed people

8. Since the length of the rectangle is being compared to the width, we let the variable represent the width of a rectangle.
w = width of a rectangle
The length is 5 meters more than triple the width.
$3w + 5$ = length of rectangle

9. Since angles A and C are being compared to angle B, we let the variable represent the number of degrees in angle B.
b = the number of degrees in angle B
The number of degrees in angle A is double the number of degrees in angle B.
$2b$ = number of degrees in angle A
The number of degrees in angle C is 17 less than the number of degrees in angle B.
$b - 17$ = number of degrees in angle C

10. Since the numbers of students in biology and geology are being compared to the number of students in algebra, we let the variable represent the number of students in algebra.
a = number of students in algebra
There are 29 more students in biology than in algebra.
$a + 29$ = number of students in biology
There are one-half as many students in geology as in algebra.
$\frac{1}{2}a = 0.5a$ = number of students in geology

11. x = the number

$$3x - 14 = -5$$
$$3x - 14 + 14 = -5 + 14$$
$$3x = 9$$
$$\frac{3x}{3} = \frac{9}{3}$$
$$x = 3$$

The number is 3.

12. x = the number

$$2x - 7 = -21$$
$$2x - 7 + 7 = -21 + 7$$
$$2x = -14$$
$$\frac{2x}{2} = \frac{-14}{2}$$
$$x = -7$$

The number is −7.

13. x = cost of chair

$$210 + 6x = 450$$
$$210 + 6x - 210 = 450 - 210$$
$$6x = 240$$
$$\frac{6x}{6} = \frac{240}{6}$$
$$x = 40$$

One chair costs \$40.

14. x = David's age
$2x$ = Jon's age
$2x = 32$
$\frac{2x}{2} = \frac{32}{2}$
$x = 16$
David is 16 years old.

15. t_1 = time for first car
t_2 = time for other car
$800 = 60t_1$
$\frac{800}{60} = \frac{60t_1}{60}$
$13.3 \approx t_1$
$800 = 65t_2$
$\frac{800}{65} = \frac{65t_2}{65}$
$12.3 \approx t_2$
The first car took 13.3 hours. The other car took 12.3 hours.

16. x = score on last test
$\frac{83+86+91+77+x}{5} = 85$
$\frac{337+x}{5} = 85$
$337 + x = 425$
$x = 88$
Zach needs a grade of 88.

17. x = the measure of the third side
$x + 2(31) = 85$
$x + 62 = 85$
$x = 23$
The third side measures 23 feet.

18. 1st angle: x
2nd angle: $3x$
3rd angle: $2x - 12$
$x + 3x + 2x - 12 = 180$
$6x - 12 = 180$
$6x = 192$
$x = 32$
$3x = 3(32) = 96$
$2x - 12 = 2(32) - 12 = 52$
The angles measure 32°, 96°, and 52°.

19. x = length of one piece
$\frac{3}{5}x$ = length of other
$x + \frac{3}{5}x = 50$
$5x + 3x = 250$
$8x = 250$
$x = 31.25$
$\frac{3}{5}x = 18.75$
The lengths are 31.25 yd and 18.75 yd.

20. x = George's salary
$2000 + \frac{1}{2}x$ = Heather's salary
$x + 2000 + \frac{1}{2}x = 65,000$
$\frac{3}{2}x + 2000 = 65,000$
$\frac{3}{2}x = 63,000$
$x = 42,000$
$2000 + \frac{1}{2}x = 2000 + \frac{1}{2}(42,000) = 23,000$
George earns $42,000 and Heather earns $23,000.

21. x = number of kilowatt-hours
$25 + 0.15x = 71.50$
$0.15x = 46.50$
$x = 310$
310 kilowatt-hours were used.

22. x = number of miles driven
$0.25x + 39(3) = 187$
$0.25x + 117 = 187$
$0.25x = 70$
$x = 280$
He drove 280 miles.

23. x = amount withdrawn
$0.055(7400 - x) = 242$
$407 - 0.055x = 242$
$-0.055x = -165$
$x = 3000$
They withdrew $3000.

24. x = original price
$0.18x = 36$
$x = 200$
The original price was $200.

25. x = amount invested at 12%
$9000 - x$ = amount at 8%
$$0.12x + 0.08(9000 - x) = 1000$$
$$12x + 8(9000 - x) = 100{,}000$$
$$12x + 72{,}000 - 8x = 100{,}000$$
$$4x = 28{,}000$$
$$x = 7000$$
$9000 - x = 2000$
They invested \$7000 at 12% and \$2000 at 8%.

26. x = amount at 4.5%
$5000 - x$ = amount at 6%
$$0.045x + 0.06(5000 - x) = 270$$
$$45x + 60(5000 - x) = 270{,}000$$
$$45x + 300{,}000 - 60x = 270{,}000$$
$$-15x = -30{,}000$$
$$x = 2000$$
$5000 - x = 5000 - 2000 = 3000$
He invested \$2000 at 4.5% and \$3000 at 6%.

27. x = number of dimes
$x + 3$ = number of quarters
$2(x + 3) = 2x + 6$ = number of nickels
$$0.05(2x + 6) + 0.10x + 0.25(x + 3) = 3.75$$
$$5(2x + 6) + 10x + 25(x + 3) = 375$$
$$10x + 30 + 10x + 25x + 75 = 375$$
$$45x = 270$$
$$x = 6$$
$x + 3 = 6 + 3 = 9$
$2x + 6 = 2(6) + 6 = 18$
She has 18 nickels, 6 dimes, and 9 quarters.

28. n = number of nickels
$n + 2$ = number of quarters
$n - 3$ = number of dimes
$$0.05n + 0.25(n + 2) + 0.10(n - 3) = 9.80$$
$$0.05n + 0.25n + 0.50 + 0.10n - 0.30 = 9.80$$
$$0.4n + 0.20 = 9.80$$
$$0.4n = 9.60$$
$$n = 24$$
$n + 2 = 24 + 2 = 26$
$n - 3 = 24 - 3 = 21$
There were 24 nickels, 21 dimes, and 26 quarters.

29. $\text{S.A.} = 4\pi r^2,\ r = \frac{d}{2} = \frac{2.6}{2} = 1.3$
$\text{S.A.} = 4(3.14)(1.3)^2$
$\text{S.A.} = 21.2264 \text{ in}^2$
The surface area is about 21.23 in.^2

30. P = (number of sides)(length of side)
$P = 6(4)$
$P = 24$ feet
The trench would be 24 feet.

31. x = measure of third angle
$$62 + 47 + x = 180$$
$$109 + x = 180$$
$$x = 71$$
The angle is 71°.

32. $A = \frac{1}{2}ab = \frac{1}{2}(8)(10.5) = 42$
The area is 42 mi^2.

33. $l = 11,\ w = 6,\ h = 7$
$V = lwh = 11(6)(7) = 462$
The volume is 462 ft^3.

34. $V = \frac{4}{3}\pi r^3 = \frac{4}{3}(3.14)(4.5)^3 = 381.51$
The volume is 381.51 in.^3

35.
$$\begin{aligned} A &= \pi r^2 \\ &= 3.14\left(\frac{23}{2}\right)^2 \\ &= 3.14(11.5)^2 \\ &= 415.265 \end{aligned}$$
The area is 415.27 ft^2.

36. x = length of 3rd side
P = sum of sides
$$17 + 17 + x = 46$$
$$34 + x = 46$$
$$x = 12$$
The 3rd side is 12 in.

37.
$$\begin{aligned} C &= 2.50[2(18)(22) + 2(18)(19) - 100] \\ &= 2.50(1376) \\ &= 3440 \end{aligned}$$
It will cost \$3440.

38. $r = 5,\ h = 14$
$$\begin{aligned} S &= 2\pi rh + 2\pi r^2 \\ &= 2(3.14)(5)(14) + 2(3.14)(5)^2 \\ &= 439.60 + 157.00 \\ &= 596.60 \text{ m}^2 \end{aligned}$$
Cost = 40(596.60) = \$23,864.00
The cost is \$23,864.

39. x = miles driven
$$0.25x + 3(20) \le 100$$
$$0.25x + 60 \le 100$$
$$0.25x \le 40$$
$$x \le 160$$
Distance must be no more than 160 miles.

40. x = sales
$$25{,}000 + 0.05x > 38{,}000$$
$$0.05x > 13{,}000$$
$$x > 260{,}000$$
She had more than $260,000 in sales.

41. $d = rt$
$$1590 = 8.75t + 20(t - 24)$$
$$1590 = 28.75t - 480$$
$$1590 = 28.75t - 480$$
$$2070 = 28.75t$$
$$72 = t$$
It took 72 months.
CP distance = rt = 8.75(72) = 630 miles
UP distance = rt = 20(72 − 24) = 960 miles

42. x = amount spent on ties
$$3(17.95) + x < 70$$
$$53.85 + x < 70$$
$$x < 16.15$$
He must spend less than $16.15 on ties.

43. x = length of commercials
$4x + 4$ = length of entertainment
$$x + 4x + 4 = 30$$
$$5x = 26$$
$$x = 5.2$$
$4x + 4 = 4(5.2) + 4 = 24.8$

The entertainment was 24.8 min or $24\frac{4}{5}$ min.

44. x = weight of each ball
$$3x = x + 8$$
$$2x = 8$$
$$x = 4$$
Each ball weighs 4 oz.

45. x = number of years to even the cost
$$83x + 744 > 41x + 870$$
$$42x + 744 > 870$$
$$42x > 126$$
$$x > 3$$
It will take more than 3 years.

46. x = number of months
$$1100x + 42{,}000 < 1800x$$
$$42{,}000 < 700x$$
$$60 < x$$
It will take more than 60 months.

47. x = height
$2.5x$ = width
$$3(2.5x) + 2x = 38$$
$$7.5x + 2x = 38$$
$$9.5x = 38$$
$$x = 4$$
$2.5x = 10$
Width = 10 feet, height = 4 feet

48.

Year	Amount	Total
1	1200	1200
2	2400	3600
3	3600	7200
4	4800	12,000
5	6000	18,000
6	7200	25,200
7	8400	33,600
8	9600	43,200
9	10,800	54,000
10	12,000	66,000

It will take 10 years before he has $66,000 in his retirement plan.

49. x = number of free throws
$3x$ = number of field goals
$$1x + 2(3x) + 3(12) = 99$$
$$7x + 36 = 99$$
$$7x = 63$$
$$x = 9$$
$3x = 27$
They made 9 free throws and 27 field goals.

50. x = width
$2x - 10$ = length
$P = 2L + 2W$
$$76 = 2(2x - 10) + 2x$$
$$76 = 4x - 20 + 2x$$
$$76 = 6x - 20$$
$$96 = 6x$$
$$16 = x$$
$2x - 10 = 22$
Length = 22 feet
Width = 16 feet

51. x = hours Carl worked
$2x - 8$ = hours Ryan worked
$$x+2x-8=28$$
$$3x-8=28$$
$$3x=36$$
$$x=12$$
$2x - 8 = 2(12) - 8 = 16$
Carl worked 12 hours.
Ryan worked 16 hours.

52. x = amount invested at 9.75%
$x + 2000$ = amount invested at 8.5%
$$0.0975x=0.085(x+2000)$$
$$0.0975x=0.085x+170$$
$$0.0125x=170$$
$$x=13,600$$
$x + 2000 = 13,600 + 2000 = 15,600$
\$13,600 was invested at 9.75%.
\$15,600 was invested at 8.5%.

53. x = measure of 2nd angle
$x + 20$ = measure of 1st angle
$2x$ = measure of 3rd angle
$$x+x+20+2x=180$$
$$4x+20=180$$
$$4x=160$$
$$x=40$$
$x + 20 = 60$
$2x = 80$
1st angle = 60°
2nd angle = 40°
3rd angle = 80°

54. $A=\frac{1}{2}a(b_1+b_2),\ a=12,\ b_1=15,\ b_2=9$
$$A=\frac{1}{2}(12)(15+9)=6(34)=204$$
Cost = 6(204) = \$1224
The cedar cost \$1224.

55. x = score on the final exam
$$\frac{88+77+95+92+2x}{6}=90$$
$$\frac{352+2x}{6}=90$$
$$352+2x=540$$
$$2x=188$$
$$x=94$$
She must make 94 on the final exam.

56. x = business miles driven
$2x$ = pleasure miles driven
$$x+2x=27,000$$
$$3x=27,000$$
$$x=9000$$
Deduction = (0.33)(9000) = \$2970
No, her correct deduction was \$2970.

57. $d=rt,\ r=\frac{32}{3},\ t=60$
$$d=\frac{32}{3}(60)=640 \text{ miles}$$
Boston to Denver: $d = 1800$
$$1800=\frac{32}{3}t$$
$$169\approx t$$
It will take about 169 minutes or 2 hr 49 min.

58. $d=rt, d=80,\ r=0.8\left(\frac{50}{6}\right)$
$$80=0.8\left(\frac{50}{6}\right)t$$
$$12=t$$
It will take him 12 seconds.

59. x = number of Explorers
$2x$ = number of Caravans
$$5x+7(2x)=95$$
$$5x+14x=95$$
$$19x=95$$
$$x=5$$
$2x = 10$
They need 5 Explorers and 10 Caravans.

60. x = smallest angle
$3x$ = largest angle
$2x - 12$ = third angle
$$x+3x+2x-12=180$$
$$6x-12=180$$
$$6x=192$$
$$x=32$$
$3x = 3(32) = 96$
$2x - 12 = 2(32) - 12 = 52$
Smallest angle = 32°
Largest angle = 96°
Third angle = 52°

61. $V = \frac{4}{3}\pi r^3,\ r = \frac{3}{2} = 1.5$

$V = \frac{4}{3}(3.14)(1.5)^3$

$V = 14.13$ cubic meters

$80 + 75 + 2(14.13) = 183.26$ kg

The total weight is 183.26 kg.

How Am I Doing? Chapter 3 Test

1. x = number

$2x - 11 = 59$

$2x = 70$

$x = 35$

The number is 35.

2. x = number

$\frac{1}{2}x + \frac{1}{9}x + \frac{1}{12}x = 25$

$36\left(\frac{1}{2}x + \frac{1}{9}x + \frac{1}{12}x\right) = 36(25)$

$18x + 4x + 3x = 900$

$25x = 900$

$x = 36$

The number is 36.

3. x = number

$2(x+5) = 3x + 14$

$2x + 10 = 3x + 14$

$10 = x + 14$

$-4 = x$

The number is −4.

4. $\frac{2}{3}x$ = 1st side

x = 2nd side

$x - 14$ = 3rd side

$\frac{2}{3}x + x + x - 14 = 66$

$\frac{8}{3}x - 14 = 66$

$\frac{8}{3}x = 80$

$x = 30$

$\frac{2}{3}x = \frac{2}{3}(30) = 20$

$x - 14 = 30 - 14 = 16$

The three sides are 20 m, 30 m, and 16 m, respectively.

5. w = width

$2w + 7$ = length

$2w + 2(2w+7) = 134$

$2w + 4w + 14 = 134$

$6w + 14 = 134$

$6w = 120$

$w = 20$

$2w + 7 = 2(20) + 7 = 47$

width = 20 m

length = 47 m

6. $2x$ = 1st pollutant

x = 2nd pollutant

$0.75x$ = 3rd pollutant

$2x + x + 0.75x = 15$

$3.75x = 15$

$x = 4$

$2x = 2(4) = 8$

$0.75x = 0.75(4) = 3$

1st pollutant: 8 ppm;

2nd pollutant: 4 ppm;

3rd pollutant: 3 ppm

7. x = number of months

$116x + 200 = 1940$

$116x = 1740$

$x = 15$

Raymond will be able to rent the computer for 15 months.

8. x = last year's tuition

$x + 0.08x = 22,680$

$1.08x = 22,680$

$x = 21,000$

Last year's tuition was $21,000.

9. x = amount at 14%

$4000 - x$ = amount at 11%

$0.14x + 0.11(4000 - x) = 482$

$100[0.14x + 0.11(4000 - x)] = 100(482)$

$14x + 11(4000 - x) = 48,200$

$14x + 44,000 - 11x = 48,200$

$3x + 44,000 = 48,200$

$3x = 4200$

$x = 1400$

$4000 - x = 4000 - 1400 = 2600$

He invested $1400 at 14% and $2600 at 11%.

10. $2x$ = number of nickels
$x - 1$ = number of dimes
x = number of quarters

$$\begin{aligned} 0.05(2x)+0.10(x-1)+0.25(x) &= 3.50 \\ 5(2x)+10(x-1)+25x &= 350 \\ 10x+10x-10+25x &= 350 \\ 45x-10 &= 350 \\ 45x &= 360 \\ x &= 8 \end{aligned}$$

$2x = 2(8) = 16$
$x - 1 = 8 - 1 = 7$
She has: 16 nickels; 7 dimes; 8 quarters.

11. $C = 2\pi r$, $r = 34$
$C = 2(3.14)(34) = 213.52$
The circumference is 213.52 inches.

12. $A = \frac{1}{2}a(b_1 + b_2) = \frac{1}{2}(16)(10+14) = 192$

The area is 192 in.^2

13. $r = 10$

$V = \frac{4}{3}\pi r^3 = \frac{4}{3}(3.14)(10)^3 = 418$

The volume is 4187 in.^3

14. $A = ab$, $a = 8$, $b = 12$
$A = 8(12) = 96$
The area is 96 square centimeters.

15. $C = 12(1.5) + 9(2) + \frac{1}{2}(2)(1.5)$

$= 18 + 18 + 1.5$

$= 37.5 \text{ yd}^2$

Cost = 12(37.5) = \$450
The total cost is \$450.

16. x = score

$$\begin{aligned} \frac{76+84+78+x}{4} &\geq 80 \\ \frac{238}{4} &\geq 80 \\ 238 + x &\geq 320 \\ x &\geq 82 \end{aligned}$$

She must score at least an 82.

17. x = sales

$$\begin{aligned} 15{,}000 + 0.05(x - 10{,}000) &> 20{,}000 \\ 15{,}000 + 0.05x - 500 &> 20{,}000 \\ 0.05x + 14{,}500 &> 20{,}000 \\ 0.05x &> 5500 \\ x &> 110{,}000 \end{aligned}$$

She must have sales of more than \$110,000.

18. $A = LW$, $L = 150$, $W = 100$
$A = 150(100) = 15{,}000$ square feet

$\text{Time} = \frac{15{,}000}{25} = 600$ hours

It will take Andrew 600 hours.

Cumulative Test for Chapters 0–3

1.

$$\begin{array}{r} 3.69 \\ 2.4_\wedge \overline{)\,8.8_\wedge 56} \\ \underline{7\,2} \\ 1\,6\,5 \\ \underline{1\,4\,4} \\ 2\,16 \\ \underline{2\,16} \end{array}$$

2. $\frac{3}{8} + \frac{5}{12} + \frac{1}{2} = \frac{9}{24} + \frac{10}{24} + \frac{12}{24} = \frac{9+10+12}{24} = \frac{31}{24}$ or $1\frac{7}{24}$

3. $4(4x - y + 5) - 3(6x - 2y)$
$= 16x - 4y + 20 - 18x + 6y$
$= -2x + 2y + 20$

4. $12 - 3(2-4) + 12 \div 4 = 12 - 3(-2) + 12 \div 4$
$= 12 + 6 + 3$
$= 21$

5.

$$\begin{aligned} H &= \frac{1}{2}(3a + 5b) \\ 2H &= 2\left(\frac{1}{2}\right)(3a + 5b) \\ 2H &= 3a + 5b \\ 2H + (-3a) &= 3a + (-3a) + 5b \\ 2H - 3a &= 5b \\ \frac{2H - 3a}{5} &= \frac{5b}{5} \\ \frac{2H - 3a}{5} &= b \end{aligned}$$

6. $5x-3\le 2(4x+1)+4$
$5x-3\le 8x+2+4$
$5x-3\le 8x+6$
$5x\le 8x+9$
$-3x\le 9$
$x\ge -3$

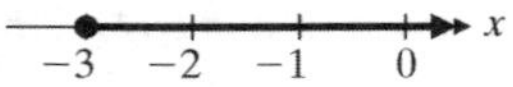

7. x = the number
$2x+15=1$
$2x=-14$
$x=-7$
The number is −7.

8. x = number of students in sociology
$x - 12$ = number of students in literature
$x+(x-12)=96$
$2x-12=96$
$2x=108$
$x=54$
$x - 12 = 54 - 12 = 42$
There are 42 students in literature and 54 in sociology.

9. w = width
$3w + 11$ = length
$P = 2L + 2W$ or $2L + 2W = P$
$2(3w+11)+2w=78$
$6w+22+2w=78$
$8w=56$
$w=7$
$3w + 11 = 32$
The dimensions are: length = 32 cm, width = 7 cm.

10. x = last year's sales
$x+0.15x=265,000$
$1.15x=265,000$
$x\approx 230,435$
Last year's sales were approximately $230,435.

11. x = amount invested at 15%
$7000 - x$ = amount invested at 7%
$0.15x+0.07(7000-x)=730$
$15x+7(7000-x)=73,000$
$15x+49,000-7x=73,000$
$8x=24,000$
$x=3000$
$7000 - x = 4000$
He invested $3000 at 15% and $4000 at 7%.

12. $x + 3$ = number of nickels
$3x$ = number of dimes
x = number of quarters
$0.05(x+3)+0.10(3x)+0.25(x)=2.55$
$5(x+3)+10(3x)+25x=255$
$5x+15+30x+25x=255$
$60x+15=255$
$60x=240$
$x=4$
$x + 3 = 4 + 3 = 7$
$3x = 3(4) = 12$
She has: 7 nickels; 12 dimes; 4 quarters.

13. $b = 25$, $a = 13$
$$A=\frac{1}{2}ab=\frac{1}{2}(13)(25)=162.5\text{ m}^2$$
Cost = 4.50(162.5) = $731.25
The total cost is $731.25.

14. $$V=\frac{4}{3}\pi r^3=\frac{4}{3}(3.14)(3.00)^3=113.04\text{ in.}^3$$
weight = (113.04)(1.50) = 169.56
The sphere is 169.56 lb.

15. $d = rt$, $d = 1625$, $r = 65$
$1625=65t$
$25=t$
It will take 25 days.

Chapter 4

4.1 Exercises

1. When you multiply exponential expressions with the same base, keep the base the same and add the exponents.

3. A sample example is: $\frac{2^2}{2^3} \stackrel{?}{=} \frac{1}{2^{3-2}}$

$$\frac{4}{8} \stackrel{?}{=} \frac{1}{2}$$
$$\frac{1}{2} = \frac{1}{2}$$

5. $6x^{11}y$: Coefficient is 6, bases are x, y and exponents are 11, 1.

7. $2 \cdot 2 \cdot a \cdot a \cdot a \cdot b = 2^2 a^3 b$

9. $(-5)(x)(y)(z)(y)(x)(x)(z)$
$= -5(x \cdot x \cdot x)(y \cdot y)(z \cdot z)$
$= -5x^3y^2z^2$

11. $(7^4)(7^6) = 7^{4+6} = 7^{10}$

13. $(5^{10})(5^{16}) = 5^{10+16} = 5^{26}$

15. $(x^4)(x^8) = x^{4+8} = x^{12}$

17. $t^{15} \cdot t = t^{15} \cdot t^1 = t^{15+1} = t^{16}$

19. $-5x^4(4x^2) = (-5 \cdot 4)(x^4 \cdot x^2)$
$= -20x^{4+2}$
$= -20x^6$

21. $(5x)(10x^2) = (5 \cdot 10)(x \cdot x^2) = 50x^{1+2} = 50x^3$

23. $(2xy^3)(9x^2y^5) = (2 \cdot 9)(x \cdot x^2)(y^3 \cdot y^5)$
$= 18x^{1+2}y^{3+5}$
$= 18x^3y^8$

25. $\left(\frac{2}{5}xy^3\right)\left(\frac{1}{3}x^2y^2\right) = \left(\frac{2}{5} \cdot \frac{1}{3}\right)(x \cdot x^2)(y^3 \cdot y^2)$
$= \frac{2}{15}x^{1+2}y^{3+2}$
$= \frac{2}{15}x^3y^5$

27. $(1.1x^2z)(-2.5xy) = (1.1)(-2.5)(x^2 \cdot x)yz$
$= -2.75x^{2+1}yz$
$= -2.75x^3yz$

29. $(8a)(2a^3b)(0) = 0$

31. $(-16x^2y^4)(-5xy^3) = (-16)(-5)(x^2 \cdot x)(y^4 \cdot y^3)$
$= 80x^{2+1}y^{4+3}$
$= 80x^3y^7$

33. $(-8x^3y^2)(3xy^5) = (-8)(3)(x^3 \cdot x)(y^2 \cdot y^5)$
$= -24x^{3+1}y^{2+5}$
$= -24x^4y^7$

35. $(-2x^3y^2)(0)(-3x^4y) = 0$

37. $(8a^4b^3)(-3x^2y^5) = (8)(-3)a^4b^3x^2y^5$
$= -24a^4b^3x^2y^5$

39. $(2x^2y)(-3y^3z^2)(5xz^4)$
$= (2)(-3)(5)(x^2 \cdot x)(y \cdot y^3)(z^2 \cdot z^4)$
$= -30x^{2+1}y^{1+3}z^{2+4}$
$= -30x^3y^4z^6$

41. $\frac{y^{12}}{y^5} = y^{12-5} = y^7$

43. $\frac{y^5}{y^8} = \frac{1}{y^{8-5}} = \frac{1}{y^3}$

45. $\frac{11^{18}}{11^{30}} = \frac{1}{11^{30-18}} = \frac{1}{11^{12}}$

47. $\frac{2^{17}}{2^{10}} = 2^{17-10} = 2^7$

49. $\frac{a^{13}}{4a^5} = \frac{a^{13-5}}{4} = \frac{a^8}{4}$

51. $\frac{x^7}{y^9} = \frac{x^7}{y^9}$

53. $\frac{48x^5y^3}{24xy^3} = 2x^{5-1}y^{3-3} = 2x^4y^0 = 2x^4$

55. $\frac{16x^5y}{-32x^2y^3} = \frac{x^{5-2}}{-2y^{3-1}} = -\frac{x^3}{2y^2}$

57. $\frac{1.8f^4g^3}{54f^2g^8} = \frac{f^{4-2}}{30g^{8-3}} = \frac{f^2}{30g^5}$

59. $\frac{-85x^5y^{10}}{-5xy^7} = 17x^{5-1}y^{10-7} = 17x^4y^3$

61. $\frac{8^0x^2y^3}{16x^5y} = \frac{y^{3-1}}{16x^{5-2}} = \frac{y^2}{16x^3}$

63. $\frac{18a^6b^3c^0}{24a^5b^3} = \frac{3}{4}a^{6-5}b^{3-3}c^0 = \frac{3}{4}ab^0 = \frac{3}{4}a$

65. $\frac{12c^2}{42ab} = \frac{2c^2}{7ab}$

67. $(x^2)^6 = x^{2\cdot 6} = x^{12}$

69. $(x^3y)^5 = (x^3)^5 \cdot y^5 = x^{3\cdot 5}y^5 = x^{15}y^5$

71. $(rs^2)^6 = r^6(s^2)^6 = r^6s^{2\cdot 6} = r^6s^{12}$

73. $(3a^3b^2c)^3 = 3^3(a^3)^3(b^2)^3c^3$
$= 3^3a^{3\cdot 3}b^{2\cdot 3}c^3$
$= 27a^9b^6c^3$

75. $(-3a^4)^2 = (-3)^2(a^4)^2 = 9a^{4\cdot 2} = 9a^8$

77. $\left(\frac{x}{2m^4}\right)^7 = \frac{x^7}{2^7m^{4\cdot 7}} = \frac{x^7}{128m^{28}}$

79. $\left(\frac{5x}{7y^2}\right)^2 = \frac{5^2x^2}{7^2y^{2\cdot 2}} = \frac{25x^2}{49y^4}$

81. $(-3a^2b^3c^0)^4 = (-3a^2b^3)^4$
$= (-3)^4(a^2)^4(b^3)^4$
$= 81a^8b^{12}$

83. $(-2x^3y^0z)^3 = (-2x^3z)^3$
$= (-2)^3(x^3)^3z^3$
$= -8x^9z^3$

85. $\frac{(3x)^5}{(3x^2)^3} = \frac{3^5x^5}{3^3x^{2\cdot 3}} = \frac{3^5x^5}{3^3x^6} = \frac{3^{5-3}}{x^{6-5}} = \frac{9}{x}$

87. $(-5a^2b^3)^2(ab) = (-5)^2(a^2)^2(b^3)^2(ab)$
$= 25a^4b^6(ab)$
$= 25a^{4+1}b^{6+1}$
$= 25a^5b^7$

89. $\left(\frac{8}{y^5}\right)^2 = \frac{8^2}{y^{5\cdot 2}} = \frac{64}{y^{10}}$

91. $\left(\frac{2x}{y^3}\right)^4 = \frac{2^4x^4}{y^{3\cdot 4}} = \frac{16x^4}{y^{12}}$

Cumulative Review

92. $-3 - 8 = -3 + (-8) = -11$

93. $-17 + (-32) + (-24) + 27 = -49 + (-24) + 27$
$= -73 + 27$
$= -46$

94. $\left(-\frac{3}{5}\right)\left(-\frac{2}{15}\right) = \frac{3}{5} \cdot \frac{2}{15} = \frac{\cancel{3} \cdot 2}{5 \cdot \cancel{3} \cdot 5} = \frac{2}{25}$

95. $-\frac{5}{4} \div \frac{5}{16} = -\frac{5}{4} \cdot \frac{16}{5} = -\frac{\cancel{5} \cdot \cancel{4} \cdot 4}{\cancel{4} \cdot \cancel{5}} = -4$

96. Subtract to find the decrease.
4,100,000 − 3,400,254 = 699,764
The rainforest decreased by 699,764 km^2 between 1970 and 2006.
What percent of 4,100,000 is 699,764?

$$\frac{699,746}{4,100,000} \approx 0.1707, \text{ or approximately } 17.1\%$$

Approximately 17.1% of the rainforest was lost between 1970 and 2006.

97. Find 2.3% of the amount of rainforest in 2006.

$$2.3\% \text{ or } 3,400,254 = (0.023)(3,400,254) \approx 78,206 \text{ km}^2$$

Approximately 78,206 km^2 of rainforest were lost during 2007.

Quick Quiz 4.1

1. $(2x^2y^3)(-5xy^4) = (2)(-5)(x^2 \cdot x)(y^3 \cdot y^4) = -10x^{2+1}y^{3+4} = -10x^3y^7$

2. $\dfrac{-28x^6y^6}{35x^3y^8} = -\dfrac{4x^{6-3}}{5y^{8-6}} = -\dfrac{4x^3}{5y^2}$

3. $(-3x^3y^5)^4 = (-3)^4(x^3)^4(y^5)^4 = 81x^{3\cdot 4}y^{5\cdot 4} = 81x^{12}y^{20}$

4. Answers may vary. Possible solution:

$\dfrac{(4x^3)^2}{(2x^4)^3}$

In the numerator and the denominator, raise each factor inside the parentheses to the power.

$\dfrac{4^2(x^3)^2}{2^3(x^4)^3}$

Evaluate the constants to their respective powers, and multiply exponents on the variable expressions.

$\dfrac{16x^6}{8x^{12}}$

Divide the numbers and subtract exponents on the variable expressions. Because the larger exponent is in the denominator, the variable expression will be in the denominator.

$\dfrac{2}{x^6}$

4.2 Exercises

1. $x^{-4} = \dfrac{1}{x^4}$

3. $3^{-4} = \dfrac{1}{3^4} = \dfrac{1}{81}$

5. $\dfrac{1}{y^{-8}} = y^8$

7. $\dfrac{x^{-4}y^{-5}}{z^{-6}} = \dfrac{z^6}{x^4y^5}$

9. $a^3b^{-2} = \dfrac{a^3}{b^2}$

11. $(2x^{-3})^{-3} = 2^{-3}x^9 = \dfrac{x^9}{2^3} = \dfrac{x^9}{8}$

13. $3x^{-2} = \dfrac{3}{x^2}$

15. $(3xy^2)^{-2} = 3^{-2}x^{-2}y^{-4} = \dfrac{1}{3^2x^2y^4} = \dfrac{1}{9x^2y^4}$

17. $\dfrac{3xy^{-2}}{z^{-3}} = \dfrac{3xz^3}{y^2}$

19. $\dfrac{(3x)^{-2}}{(3x)^{-3}} = \dfrac{(3x)^3}{(3x)^2} = (3x)^{3-2} = 3x$

21. $a^{-1}b^3c^{-4}d = \dfrac{b^3d}{ac^4}$

23. $(8^{-2})(2^3) = \dfrac{1}{8^2} \cdot 2^3 = \dfrac{1}{64} \cdot 8 = \dfrac{1}{8}$

25. $\left(\dfrac{3x^0y^2}{z^4}\right)^{-2} = \left(\dfrac{3y^2}{z^4}\right)^{-2} = \dfrac{3^{-2}y^{-4}}{z^{-8}} = \dfrac{z^8}{3^2y^4} = \dfrac{z^8}{9y^4}$

27. $\frac{x^{-2}y^{-3}}{x^4y^{-2}} = \frac{y^2}{x^4x^2y^3} = \frac{1}{x^{4+2}y^{3-2}} = \frac{1}{x^6y}$

29. $123,780 = 1.2378 \cdot 10,000 = 1.2378 \times 10^5$

31. $0.063 = 6.3 \times 0.01 = 6.3 \times 10^{-2}$

33. Move the decimal point 11 places to the left.
$889,610,000,000 = 8.8961 \times 10^{11}$

35. Move the decimal point 6 places to the right.
$0.00000342 = 3.42 \times 10^{-6}$

37. $3.02 \times 10^5 = 3.02 \times 100,000 = 302,000$

39. $4.7 \times 10^{-4} = 4.7 \times \frac{1}{10,000} = 0.00047$

41. $9.83 \times 10^5 = 9.83 \times 100,000 = 983,000$

43. $0.0000237 = 2.37 \times 10^{-5}$ miles per hour

45. $1.496 \times 10^8 = 149,600,000$ km

47. $(42,000,000)(150,000,000)$
$= (4.2 \times 10^7)(1.5 \times 10^8)$
$= 6.2 \times 10^{15}$

49. $\frac{(5,000,000)(16,000)}{8,000,000,000} = \frac{(5 \times 10^6)(1.6 \times 10^4)}{8 \times 10^9}$
$= \frac{8 \times 10^{10}}{8 \times 10^9}$
$= 1.0 \times 10^1$

51. $(0.003)^4 = (3 \times 10^{-3})^4$
$= 3^4 \times 10^{-3(4)}$
$= 81 \times 10^{-12}$
$= 8.1 \times 10^{-11}$

53. $(150,000,000)(0.00005)(0.002)(30,000)$
$= (1.5 \times 10^8)(5 \times 10^{-5})(2 \times 10^{-3})(3 \times 10^4)$
$= 45 \times 10^4$
$= 4.5 \times 10^5$

55. $\frac{8.42 \times 10^{12}}{2.99 \times 10^8} = \frac{8.42}{2.99} \times 10^{12-8} \approx 2.82 \times 10^4$
Each individual would be assigned approximately $\$2.82 \times 10^4$, or \$28,200.

57. $d = rt$
$d = (0.00000275)(24)$
$= (2.75 \times 10^{-6})(24)$
$= 66 \times 10^{-6}$
$= 6.6 \times 10^{-5}$
It traveled 6.6×10^{-5} mile in a day.

59. $r = \frac{d}{t} = \frac{3.5 \times 10^9}{9.5}$
$= \frac{3.5}{9.5} \times 10^9$
$\approx 0.368 \times 10^9$
$= 3.68 \times 10^8$
New Horizons will travel approximately 3.68×10^8 miles per year.

61. Percent increase $= \frac{9.36 \times 10^{11} - 3.61 \times 10^{11}}{3.61 \times 10^{11}}$
$= 1.593$
$= 159.3\%$

Cumulative Review

63. $-2.7 - (-1.9) = -2.7 + 1.9 = -0.8$

64. $(-1)^{33} = -1$

65. $-\frac{3}{4} + \frac{5}{7} = \frac{-21}{28} + \frac{20}{28} = \frac{-21+20}{28} = -\frac{1}{28}$

Quick Quiz 4.2

1. $3x^{-3}y^2z^{-4} = \frac{3y^2}{x^3z^4}$

2. $\frac{4a^3b^{-4}}{8a^{-5}b^{-3}} = \frac{a^3a^5b^3}{2b^4} = \frac{a^{3+5}}{2b^{4-3}} = \frac{a^8}{2b}$

3. Move the decimal point 3 places to the right.
$0.00876 = 8.76 \times 10^{-3}$

4. Answers may vary. Possible solution:
$(4x^{-3}y^4)^{-3}$
Raise each factor inside the parentheses to the power.
$4^{-3}(x^{-3})^{-3}(y^4)^{-3}$
Multiply exponents on the variable expressions.
$4^{-3}x^9y^{-12}$
Rewrite as a fraction.
$\frac{x^9}{4^3y^{12}}$
Evaluate 4^3.
$\frac{x^9}{64y^{12}}$

4.3 Exercises

1. A polynomial in x is the sum of a finite number of terms of the form ax^n, where a is any real number and n is a whole number. An example is $3x^2-5x-9$.

3. The degree of a polynomial in x is the largest exponent of x in any of the terms of the polynomial.

5. $6x^3y$
The sum of the exponents is 3 + 1 = 4. Therefore this is a polynomial of degree 4. It has one term, so it is a monomial.

7. $20x^5+6x^3-7x$
The greatest degree of any term is 5, so this polynomial is of degree 5. It has three terms, so it is a trinomial.

9. $4x^2y^3-7x^3y^3$
The sum of the exponents on the first term is 2 + 3 = 5, and the sum of the exponents on the second term is 3 + 3 = 6. The greater of these is 6, so the degree of the polynomial is 6. The polynomial has two terms, so it is a binomial.

11.
$$\begin{aligned}&(-3x+15)+(8x-43)\\&=[(-3x)+8x]+[15+(-43)]\\&=[(-3+8)x]+[15-43]\\&=5x-28\end{aligned}$$

13.
$$\begin{aligned}&(6x^2+5x-6)+(-8x^2-3x+5)\\&=[6x^2+(-8x^2)]+[5x+(-3x)]+[-6+5]\\&=[(6-8)x^2]+[(5-3)x]+[-6+5]\\&=(-2x^2)+2x+(-1)\\&=-2x^2+2x-1\end{aligned}$$

15.
$$\begin{aligned}&\left(\frac{1}{2}x^2+\frac{1}{3}x-4\right)+\left(\frac{1}{3}x^2+\frac{1}{6}x-5\right)\\&=\left[\frac{1}{2}x^2+\frac{1}{3}x^2\right]+\left[\frac{1}{3}x+\frac{1}{6}x\right]+[(-4)+(-5)]\\&=\left[\left(\frac{1}{2}+\frac{1}{3}\right)x^2\right]+\left[\left(\frac{1}{3}+\frac{1}{6}\right)x\right]+[(-4)+(-5)]\\&=\left[\left(\frac{3}{6}+\frac{2}{6}\right)x^2\right]+\left[\left(\frac{2}{6}+\frac{1}{6}\right)x\right]+(-9)\\&=\frac{5}{6}x^2+\frac{3}{6}x+(-9)\\&=\frac{5}{6}x^2+\frac{1}{2}x-9\end{aligned}$$

17.
$$\begin{aligned}&(3.4x^3-7.1x+3.4)+(2.2x^2-6.1x-8.8)\\&=3.4x^3+2.2x^2+(-7.1-6.1)x+(3.4-8.8)\\&=3.4x^3+2.2x^2-13.2x-5.4\end{aligned}$$

19.
$$\begin{aligned}(2x-19)-(-3x+5)&=(2x-19)+(3x-5)\\&=(2+3)x+(-19-5)\\&=5x-24\end{aligned}$$

21.
$$\begin{aligned}&\left(\frac{2}{5}x^2-\frac{1}{2}x+5\right)-\left(\frac{1}{3}x^2-\frac{3}{7}x-6\right)\\&=\left(\frac{2}{5}x^2-\frac{1}{2}x+5\right)+\left(-\frac{1}{3}x^2+\frac{3}{7}x+6\right)\\&=\left(\frac{2}{5}-\frac{1}{3}\right)x^2+\left(-\frac{1}{2}+\frac{3}{7}\right)x+(5+6)\\&=\left(\frac{6}{15}-\frac{5}{15}\right)x^2+\left(-\frac{7}{14}+\frac{6}{14}\right)x+(5+6)\\&=\frac{1}{15}x^2-\frac{1}{14}x+11\end{aligned}$$

23.
$$\begin{aligned}&(4x^3+3x)-(x^3+x^2-5x)\\&=(4x^3+3x)+(-x^3-x^2+5x)\\&=(5-1)x^3-x^2+(3+5)x\\&=3x^3-x^2+8x\end{aligned}$$

25. $(0.5x^4 - 0.7x^2 + 8.3) - (5.2x^4 + 1.6x + 7.9)$
$= (0.5x^4 - 0.7x^2 + 8.3) + (-5.2x^4 - 1.6x - 7.9)$
$= (0.5 - 5.2)x^4 - 0.7x^2 - 1.6x + (8.3 - 7.9)$
$= -4.7x^4 - 0.7x^2 - 1.6x + 0.4$

27. $(8x+2)+(x-7)-(3x+1)$
$= (8x+2)+(x-7)+(-3x-1)$
$= (8+1-3)x+(2-7-1)$
$= 6x-6$

29. $(5x^2y - 6xy^2 + 2) + (-8x^2y + 12xy^2 - 6)$
$= (5-8)x^2y + (-6+12)xy^2 + (2-6)$
$= -3x^2y + 6xy^2 - 4$

31. $(3x^4 - 4x^2 - 18) - (2x^4 + 3x^3 + 6)$
$= (3x^4 - 4x^2 - 18) + (-2x^4 - 3x^3 - 6)$
$= (3-2)x^4 - 3x^3 - 4x^2 + (-18-6)$
$= x^4 - 3x^3 - 4x^2 - 24$

33. $x = 1990 - 1990 = 0$
$-2.06(0)^2 + 77.82(0) + 743 = 0 + 0 + 743$
$= 743$
There were 743 thousand, or 743,000, prisoners in 1990.

35. Find the population for each year.
2002: $x = 2002 - 1990 = 12$
$-2.06(12)^2 + 77.82(12) + 743$
$= -2.06(144) + 77.82(12) + 743$
$= -296.64 + 933.84 + 743$
$= 1380.2$ thousand
2007: $x = 2007 - 1990 = 17$
$-2.06(17)^2 + 77.82(17) + 743$
$= -2.06(289) + 77.82(17) + 743$
$= -595.34 + 1322.94 + 743$
$= 1470.6$ thousand
Subtract to find the increase.
$1470.6 - 1380.2 = 90.4$
The prison population increased by 90.4 thousand, or 90,400.

37. $(x)^2 + (12)(x) + (2x)(x) = x^2 + 12x + 2x^2$
$= (1+2)x^2 + 12x$
$= 3x^2 + 12x$

Cumulative Review

39. $3y - 8x = 2$
$-8x = -3y + 2$
$x = \dfrac{-3y+2}{-8}$
$x = \dfrac{3y-2}{8}$

40. $3A = 2BCD$
$\dfrac{3A}{2CD} = \dfrac{2BCD}{2CD}$
$\dfrac{3A}{2CD} = B$

41. $A = \dfrac{1}{2}h(b+c)$
$2A = h(b+c)$
$\dfrac{2A}{h} = \dfrac{h(b+c)}{h}$
$\dfrac{2A}{h} - c = b$ or $b = \dfrac{2A}{h} - c$

42. $B = \dfrac{5xy}{d}$
$d \cdot B = d \cdot \dfrac{5xy}{d}$
$Bd = 5xy$
$\dfrac{Bd}{B} = \dfrac{5xy}{B}$
$d = \dfrac{5xy}{B}$

Quick Quiz 4.3

1. $(3x^2 - 5x + 8) + (-7x^2 - 6x - 3)$
$= (3-7)x^2 + (-5-6)x + (8-3)$
$= -4x^2 - 11x + 5$

2. $(2x^2 - 3x - 7) - (-4x^2 + 6x + 9)$
$= (2x^2 - 3x - 7) + (4x^2 - 6x - 9)$
$= (2+4)x^2 + (-3-6)x + (-7-9)$
$= 6x^2 - 9x - 16$

3. $(5x-3) - (2x-4) + (-6x+7)$
$= (5x-3) + (-2x+4) + (-6x+7)$
$= (5-2-6)x + (-3+4+7)$
$= -3x + 8$

4. Answers may vary. Possible solution:
$2xy^2 - 5x^3y^4$
To determine the degree of the polynomial, first determine the degree of each term by finding the sum of the exponents on the variables in each term. The degree of the first term, $2xy^2$, is 1 + 2 = 3, and the degree of the second term, $-5x^3y^4$, is 3 + 4 = 7. The degree of the polynomial is the greater of these, which is 7. To determine whether the polynomial is a monomial, a binomial, or a trinomial, we must count the number of terms in the polynomial. The polynomial has two terms, $2xy^2$ and $-5x^3y^4$, so it is a binomial.

How Am I Doing? Sections 4.1–4.3

1. $(8x^2y^3)(-3xy^2) = (8)(-3)(x^2 \cdot x)(y^3 \cdot y^2)$
$= -24x^{2+1}y^{3+2}$
$= -24x^3y^5$

2. $-\dfrac{35xy^6}{25x^8y^3} = -\dfrac{7y^{6-3}}{5x^{8-1}} = -\dfrac{7y^3}{5x^7}$

3. $\dfrac{60x^7y^0}{15x^2y^9} = \dfrac{60x^7}{15x^2y^9} = \dfrac{4x^{7-2}}{y^9} = \dfrac{4x^5}{y^9}$

4. $(-3x^5y)^4 = (-3)^4(x^5)^4y^4 = 81x^{5\cdot 4}y^4 = 81x^{20}y^4$

5. $(4x^{-3}y^4)^{-2} = (4)^{-2}(x^{-3})^{-2}(y^4)^{-2}$
$= (4)^{-2}x^6y^{-8}$
$= \dfrac{x^6}{(4)^2y^8}$
$= \dfrac{x^6}{16y^8}$

6. $\dfrac{4x^4y^{-3}}{12x^{-1}y^2} = \dfrac{x^4 \cdot x}{3y^2 \cdot y^3} = \dfrac{x^{4+1}}{3y^{2+3}} = \dfrac{x^5}{3y^5}$

7. Move the decimal point four places to the left.
$58,740 = 5.874 \times 10^4$

8. Move the decimal point five places to the right.
$0.00009362 = 9.362 \times 10^{-5}$

9. $(2x^2 + 0.5x - 2) + (0.3x^2 - 0.9x - 3.4)$
$= (2 + 0.3)x^2 + (0.5 - 0.9)x + (-2 - 3.4)$
$= 2.3x^2 - 0.4x - 5.4$

10. $(3x^2 + 7x - 10) - (-x^2 + 5x - 2)$
$= (3x^2 + 7x - 10) + (x^2 - 5x + 2)$
$= (3 + 1)x^2 + (7 - 5)x + (-10 + 2)$
$= 4x^2 + 2x - 8$

11. $\left(\dfrac{1}{2}x^3 + \dfrac{1}{4}x^2 - 2x\right) - \left(\dfrac{1}{3}x^3 - \dfrac{1}{8}x^2 - 5x\right)$
$= \left(\dfrac{1}{2}x^3 + \dfrac{1}{4}x^2 - 2x\right) + \left(-\dfrac{1}{3}x^3 + \dfrac{1}{8}x^2 + 5x\right)$
$= \left(\dfrac{1}{2} - \dfrac{1}{3}\right)x^3 + \left(\dfrac{1}{4} + \dfrac{1}{8}\right)x^2 + (-2 + 5)x$
$= \left(\dfrac{3}{6} - \dfrac{2}{6}\right)x^3 + \left(\dfrac{2}{8} + \dfrac{1}{8}\right)x^2 + 3x$
$= \dfrac{1}{6}x^3 + \dfrac{3}{8}x^2 + 3x$

12. $\left(\dfrac{1}{16}x^2 + \dfrac{1}{8}\right) + \left(\dfrac{1}{4}x^2 - \dfrac{3}{10}x - \dfrac{1}{2}\right)$
$= \left(\dfrac{1}{16} + \dfrac{1}{4}\right)x^2 - \dfrac{3}{10}x + \left(\dfrac{1}{8} - \dfrac{1}{2}\right)$
$= \left(\dfrac{1}{16} + \dfrac{4}{16}\right)x^2 - \dfrac{3}{10}x + \left(\dfrac{1}{8} - \dfrac{4}{8}\right)$
$= \dfrac{5}{16}x^2 - \dfrac{3}{10}x - \dfrac{3}{8}$

4.4 Exercises

1. $-2x(6x^3 - x) = -2x(6x^3) - 2x(-x)$
$= -12x^4 + 2x^2$

3. $3x^2(7x - 3) = 3x^2(7x) + 3x^2(-3) = 21x^3 - 9x^2$

5. $2x^3(-2x^3 + 5x - 1)$
$= 2x^3(-2x^3) + 2x^3(5x) + 2x^3(-1)$
$= -4x^6 + 10x^4 - 2x^3$

7. $\dfrac{1}{2}(2x + 3x^2 + 5x^3) = \dfrac{1}{2}(2x) + \dfrac{1}{2}(3x^2) + \dfrac{1}{2}(5x^3)$
$= x + \dfrac{3}{2}x^2 + \dfrac{5}{2}x^3$

9. $(2x^3 - 4x^2 + 5x)(-x^2y) = -2x^5y + 4x^4y - 5x^3y$

11. $(3x^3+x^2-8x)(3xy)=9x^4y+3x^3y-24x^2y$

13. $(x^3-3x^2+5x-2)(3x)=3x^4-9x^3+15x^2-6x$

15. $(x^2y^2-6xy+8)(-2xy)$
$=-2x^3y^3+12x^2y^2-16xy$

17. $(-7x^3+3x^2+2x-1)(4x^2y)$
$=-28x^5y+12x^4y+8x^3y-4x^2y$

19. $(3d^4-4d^2+6)(-2c^2d)$
$=-6c^2d^5+8c^2d^3-12c^2d$

21. $6x^3(2x^4-x^2+3x+9)$
$=12x^7-6x^5+18x^4+54x^3$

23. $-2x^3(8x^3-5x^2+6x)=-16x^6+10x^5-12x^4$

25. $(x+5)(x+7)=x^2+7x+5x+35=x^2+12x+35$

27. $(x+6)(x+2)=x^2+2x+6x+12=x^2+8x+12$

29. $(x-8)(x+2)=x^2+2x-8x-16=x^2-6x-16$

31. $(x-5)(x-4)=x^2-4x-5x+20=x^2-9x+20$

33. $(5x-2)(-4x-3)=-20x^2-15x+8x+6$
$=-20x^2-7x+6$

35. $(2x-5)(x+3y)=2x^2+6xy-5x-15y$

37. $(5x+2)(3x-y)=15x^2-5xy+6x-2y$

39. $(4y+1)(5y-3)=20y^2-12y+5y-3$
$=20y^2-7y-3$

41. The signs are incorrect.
The result should be: $(x-2)(-3)=-3x+6$

43. $(5x+2)(5x+2)=25x^2+\underline{20x}+4$

45. $(4x-3y)(5x-2y)=20x^2-8xy-15xy+6y^2$
$=20x^2-23xy+6y^2$

47. $(7x-2)^2=(7x-2)(7x-2)$
$=49x^2-14x-14x+4$
$=49x^2-28x+4$

49. $(4a+2b)^2=(4a+2b)(4a+2b)$
$=16a^2+8ab+8ab+4b^2$
$=16a^2+16ab+4b^2$

51. $(0.2x+3)(4x-0.3)=0.8x^2-0.06x+12x-0.9$
$=0.8x^2+11.94x-0.9$

53. $\left(\frac{1}{2}x+\frac{1}{3}\right)\left(\frac{1}{2}x-\frac{1}{4}\right)=\frac{1}{4}x^2-\frac{1}{8}x+\frac{1}{6}x-\frac{1}{12}$
$=\frac{1}{4}x^2-\frac{3}{24}x+\frac{4}{24}x-\frac{1}{12}$
$=\frac{1}{4}x^2+\frac{1}{24}x-\frac{1}{12}$

55. $(2a-3b)(x-5b)=2ax-10ab-3bx+15b^2$

57. $(2x-3)(5x+2)=10x^2+4x-15x-6$
$=10x^2-11x-6$
The area is $(10x^2-11x-6)$ square units.

Cumulative Review

59. $3(x-6)=-2(x+4)+6x$
$3x-18=-2x-8+6x$
$3x-18=4x-8$
$-18=x-8$
$-10=x$

60. $3(w-7)-(4-w)=11w$
$3w-21-4+w=11w$
$4w-25=11w$
$-25=7w$
$-\frac{25}{7}=w$

61. x = number of \$10 bills
$x+1$ = number of \$20 bills
$3x-1$ = number of \$5 bills
$10x+20(x+1)+5(3x-1)=375$
$10x+20x+20+15x-5=375$
$45x+15=375$
$45x=360$
$x=8$
$x+1=8+1=9$
$3x-1=3(8)-1=23$
She had eight \$10's, nine \$20's, and twenty-three \$5's.

62. $17.7x + 240$; $x = 1990 - 1990 = 0$
$17.7(0) + 240 = 0 + 240 = 240$
Approximately \$240 billion in social security checks were sent in 1990.

63. $17.7x + 240$; $x = 2005 - 1990 = 15$
$17.7(15) + 240 = 265.5 + 240 = 505.5$
Approximately \$505.5 billion in social security checks were sent in 2005.

64. $17.7x + 240$; $x = 2010 - 1990 = 20$
$17.7(20) + 240 = 354 + 240 = 594$
Approximately \$594 billion in social security checks will be sent in 2010.

65. $17.7x + 240$; $x = 2014 - 1990 = 24$
$17.7(24) + 240 = 424.8 + 240 = 664.8$
Approximately \$664.8 billion in social security checks will be sent in 2014.

Quick Quiz 4.4

1. $(2x^2y^2 - 3xy + 4)(4xy^2)$
$= 8x^3y^4 - 12x^2y^3 + 16xy^2$

2. $(2x+3)(3x-5) = 6x^2 - 10x + 9x - 15$
$= 6x^2 - x - 15$

3. $(6a-4b)(2a-3b) = 12a^2 - 18ab - 8ab + 12b^2$
$= 12a^2 - 26ab + 12b^2$

4. Answers may vary. Possible solution:
$(7x-3)^2$
First write the square of the binomial as the product of the binomial and itself.
$(7x - 3)(7x - 3)$
Then use FOIL and collect like terms.
$49x^2 - 21x - 21x + 9 = 49x^2 - 42x + 9$

4.5 Exercises

1. In the special case of $(a + b)(a - b)$, a binomial times a binomial is a <u>binomial</u>.

3. $(4x-7)^2 = 16x^2 - 56x + 49$
The student left out the middle term which comes from the product of the two outer terms and the product of the two inner terms.

5. $(y-7)(y+7) = y^2 - 7^2 = y^2 - 49$

7. $(x-8)(x+8) = x^2 - (8)^2 = x^2 - 64$

9. $(6x-5)(6x+5) = (6x)^2 - 5^2 = 36x^2 - 25$

11. $(2x-7)(2x+7) = (2x)^2 - (7)^2 = 4x^2 - 49$

13. $(5x-3y)(5x+3y) = (5x)^2 - (3y)^2 = 25x^2 - 9y^2$

15. $(0.6x+3)(0.6x-3) = (0.6x)^2 - 3^2 = 0.36x^2 - 9$

17. $(2y+5)^2 = (2y)^2 + (2)(2y)(5) + (5)^2$
$= 4y^2 + 20y + 25$

19. $(5x-4)^2 = (5x)^2 - 2(5x)(4) + 4^2$
$= 25x^2 - 40x + 16$

21. $(7x+3)^2 = (7x)^2 + 2(7x)(3) + 3^2$
$= 49x^2 + 42x + 9$

23. $(3x-7)^2 = (3x)^2 - 2(3x)(7) + 7^2$
$= 9x^2 - 42x + 49$

25. $\left(\frac{2}{3}x + \frac{1}{4}\right)^2 = \left(\frac{2}{3}x\right)^2 + 2\left(\frac{2}{3}x\right)\left(\frac{1}{4}\right) + \left(\frac{1}{4}\right)^2$
$= \frac{4}{9}x^2 + \frac{1}{3}x + \frac{1}{16}$

27. $(9xy+4z)^2 = (9xy)^2 + 2(9xy)(4z) + (4z)^2$
$= 81x^2y^2 + 72xyz + 16z^2$

29. $(7x+3y)(7x-3y) = (7x)^2 - (3y)^2 = 49x^2 - 9y^2$

31. $(3c-5d)^2 = (3c)^2 - (2)(3c)(5d) + (5d)^2$
$= 9c^2 - 30cd + 25d^2$

33. $(9a-10b)(9a+10b) = (9a)^2 - (10b)^2$
$= 81a^2 - 100b^2$

35. $(5x+9y)^2 = (5x)^2 + 2(5x)(9y) + (9y)^2$
$= 25x^2 + 90xy + 81y^2$

37. $(x^2 - x + 5)(x - 3)$
$= (x^2 - x + 5)x + (x^2 - x + 5)(-3)$
$= x^3 - x^2 + 5x - 3x^2 + 3x - 15$
$= x^3 - 4x^2 + 8x - 15$

39. $(4x+1)(x^3-2x^2+x-1)$
$=4x(x^3-2x^2+x-1)+1(x^3-2x^2+x-1)$
$=4x^4-8x^3+4x^2-4x+x^3-2x^2+x-1$
$=4x^4-7x^3+2x^2-3x-1$

41. $(a^2-3a+2)(a^2+4a-3)=a^2(a^2+4a-3)-3a(a^2+4a-3)+2(a^2+4a-3)$
$=a^4+4a^3-3a^2-3a^3-12a^2+9a+2a^2+8a-6$
$=a^4+a^3-13a^2+17a-6$

43. $(x+3)(x-1)(3x-8)=(x^2-x+3x-3)(3x-8)$
$=(x^2+2x-3)(3x)+(x^2+2x-3)(-8)$
$=3x^3+6x^2-9x-8x^2-16x+24$
$=3x^3-2x^2-25x+24$

45. $(2x-5)(x-1)(x+3)=(2x^2-2x-5x+5)(x+3)$
$=(2x^2-7x+5)(x+3)$
$=(2x^2-7x+5)x+(2x^2-7x+5)3$
$=2x^3-7x^2+5x+6x^2-21x+15$
$=2x^3-x^2-16x+15$

47. $(a-5)(2a+3)(a+5)=(2a^2+3a-10a-15)(a+5)$
$=(2a^2-7a-15)(a+5)$
$=(2a^2-7a-15)a+(2a^2-7a-15)5$
$=2a^3-7a^2-15a+10a^2-35a-75$
$=2a^3+3a^2-50a-75$

49. $V=(2x+1)(3x-2)(4x+3)$
$=(6x^2-4x+3x-2)(4x+3)$
$=(6x^2-x-2)(4x+3)$
$=(6x^2-x-2)(4x)+(6x^2-x-2)(3)$
$=24x^3-4x^2-8x+18x^2-3x-6$
$=24x^3+14x^2-11x-6$

Cumulative Review

51. Let x = amount invested at 7%
$18{,}000-x$ = amount invested at 11%
$0.07x+0.11(18{,}000-x)=1540$
$7x+11(18{,}000-x)=154{,}000$
$7x+198{,}000-11x=154{,}000$
$-4x=-44{,}000$
$x=11{,}000$
$18{,}000-x=7000$
She invested \$11,000 at 7% and \$7000 at 11%.

52. Let x = length.

Then width $= 2 + \frac{1}{2}x$.

$$\begin{aligned} 2x + 2\left(2 + \frac{1}{2}x\right) &= 34 \\ 2x + 4 + x &= 34 \\ 3x + 4 &= 34 \\ 3x &= 30 \\ x &= 10 \end{aligned}$$

$$2 + \frac{1}{2}x = 2 + \frac{1}{2}(10) = 7$$

The dimensions of the room are: width = 7 m, length = 10 m.

Quick Quiz 4.5

1. $$\begin{aligned}(7x - 12y)(7x + 12y) &= (7x)^2 - (12y)^2 \\ &= 49x^2 - 144y^2\end{aligned}$$

2. $$\begin{aligned}(2x + 3)(x - 2)(3x + 1) &= (2x^2 - 4x + 3x - 6)(3x + 1) \\ &= (2x^2 - x - 6)(3x + 1) \\ &= (2x^2 - x - 6)(3x) + (2x^2 - x - 6)(1) \\ &= 6x^3 - 3x^2 - 18x + 2x^2 - x - 6 \\ &= 6x^3 - x^2 - 19x - 6\end{aligned}$$

3. $$\begin{aligned}(3x - 2)(5x^3 - 2x^2 - 4x + 3) &= (3x)(5x^3 - 2x^2 - 4x + 3) + (-2)(5x^3 - 2x^2 - 4x + 3) \\ &= 15x^4 - 6x^3 - 12x^2 + 9x - 10x^3 + 4x^2 + 8x - 6 \\ &= 15x^4 - 16x^3 - 8x^2 + 17x - 6\end{aligned}$$

4. Answers may vary. Possible solution:

To use the formula $(a + b)^2 = a^2 + 2ab + b^2$ to multiply $(6x - 9y)^2$, first identify a and b: $a = 6x$ and $b = -9y$. Then substitute these values for a and b in the formula and simplify.

$$\begin{aligned}(6x - 9y)^2 &= (6x)^2 + 2(6x)(-9y) + (-9y)^2 \\ &= 36x^2 - 108xy + 81y^2\end{aligned}$$

4.6 Exercises

1. $$\begin{aligned}\frac{25x^4 - 15x^2 + 20x}{5x} &= \frac{25x^4}{5x} - \frac{15x^2}{5x} + \frac{20x}{5x} \\ &= 5x^3 - 3x + 4\end{aligned}$$

3. $$\begin{aligned}\frac{8y^4 - 12y^3 - 4y^2}{4y^2} &= \frac{8y^4}{4y^2} - \frac{12y^3}{4y^2} - \frac{4y^2}{4y^2} \\ &= 2y^2 - 3y - 1\end{aligned}$$

5. $\dfrac{81x^7 - 36x^5 - 63x^3}{9x^3} = \dfrac{81x^7}{9x^3} - \dfrac{36x^5}{9x^3} - \dfrac{63x^3}{9x^3}$
$= 9x^4 - 4x^2 - 7$

7. $(48x^7 - 54x^4 + 36x^3) \div 6x^3$
$= \dfrac{48x^7}{6x^3} - \dfrac{54x^4}{6x^3} + \dfrac{36x^3}{6x^3}$
$= 8x^4 - 9x + 6$

9.
$$\begin{array}{r}
3x+5 \\
2x+1\overline{)\,6x^2+13x+5} \\
\underline{6x^2+\ \ 3x} \\
10x+5 \\
\underline{10x+5} \\
0
\end{array}$$

$\dfrac{6x^2+13x+5}{2x+1} = 3x+5$

Check: $(3x+5)(2x+1) = 6x^2 + 3x + 10x + 5$
$= 6x^2 + 13x + 5$

11.
$$\begin{array}{r}
x-3 \\
x-5\overline{)\,x^2-8x-17} \\
\underline{x^2-5x} \\
-3x-17 \\
\underline{-3x+15} \\
-32
\end{array}$$

$\dfrac{x^2-8x-17}{x-5} = x - 3 - \dfrac{32}{x-5}$

Check:
$(x-5)\left(x - 3 - \dfrac{32}{x-5}\right) = x^2 - 3x - 5x + 15 - 32$
$= x^2 - 8x - 17$

13.
$$\begin{array}{r}
3x^2-4x+8 \\
x+1\overline{)\,3x^3-\ \ x^2+4x-2} \\
\underline{3x^3+3x^2} \\
-4x^2+4x \\
\underline{-4x^2-4x} \\
8x-2 \\
\underline{8x+8} \\
-10
\end{array}$$

$\dfrac{3x^3-x^2+4x-2}{x+1} = 3x^2 - 4x + 8 - \dfrac{10}{x+1}$

Check: $(x+1)\left(3x^2 - 4x + 8 - \dfrac{10}{x+1}\right)$
$= 3x^3 - x^2 + 4x + 8 - 10$
$= 3x^3 - x^2 + 4x - 2$

15.
$$\begin{array}{r}
2x^2-3x-2 \\
2x+5\overline{)\,4x^3+\ \ 4x^2-19x-15} \\
\underline{4x^3+10x^2} \\
-6x^2-19x \\
\underline{-6x^2-15x} \\
-4x-15 \\
\underline{-4x-10} \\
-5
\end{array}$$

$\dfrac{4x^3+4x^2-19x-15}{2x+5} = 2x^2 - 3x - 2 - \dfrac{5}{2x+5}$

Check: $(2x+5)\left(2x^2 - 3x - 2 - \dfrac{5}{2x+5}\right)$
$= 4x^3 + 4x^2 - 19x - 10 - 5$
$= 4x^3 + 4x^2 - 19x - 15$

17.
$$\begin{array}{r}
2x^2+3x-1 \\
5x-2\overline{)\,10x^3+11x^2-11x+2} \\
\underline{10x^3\ \ -4x^2} \\
15x^2-11x \\
\underline{15x^2\ \ -6x} \\
-5x+2 \\
\underline{-5x+2} \\
0
\end{array}$$

$\dfrac{10x^3+11x^2-11x+2}{5x-2} = 2x^2 + 3x - 1$

19.
$$\begin{array}{r}
4y^2-2y+5 \\
3y-4\overline{)\,12y^3-22y^2+23y-20} \\
\underline{12y^3-16y^2} \\
-6y^2+23y \\
\underline{-6y^2+\ \ 8y} \\
15y-20 \\
\underline{15y-20} \\
0
\end{array}$$

$\dfrac{12y^3-22y^2+23y-20}{3y-4} = 4y^2 - 2y + 5$

21.
$$
\begin{array}{r}
y^2-4y-1 \\
y+3\overline{\big)\,y^3-y^2-13y-12} \\
\underline{y^3+3y^2} \\
-4y^2-13y \\
\underline{-4y^2-12y} \\
-y-12 \\
\underline{-y-3} \\
-9
\end{array}
$$

$$(y^3-y^2-13y-12)\div(y+3)=y^2-4y-1-\frac{9}{y+3}$$

23.
$$
\begin{array}{r}
y^3+2y^2-5y-10 \\
y-2\overline{\big)\,y^4+0y^3-9y^2+0y-5} \\
\underline{y^4-2y^3} \\
2y^3-9y^2 \\
\underline{2y^3-4y^2} \\
-5y^2+0y \\
\underline{-5y^2+10y} \\
-10y-5 \\
\underline{-10y+20} \\
-25
\end{array}
$$

$$(y^4-9y^2-5)\div(y-2)$$
$$=y^3+2y^2-5y-10-\frac{25}{y-2}$$

Cumulative Review

25. In 2000, the price was 120% of the 1995 price.
120% of \$2.36 = 1.20 × \$2.36 ≈ \$2.83
In 2005, the price was 113% of the 2000 price.
113% of \$2.83 = 1.13 × \$2.83 ≈ \$3.20

26. x = first page number
$x + 1$ = second page number
$$x+(x+1)=341$$
$$2x+1=341$$
$$2x=340$$
$$x=170$$
$x + 1 = 170 + 1 = 171$
The page numbers are 170 and 171.

27. a. $\frac{7+8+7+4}{4}=\frac{26}{4}=6.5$ hurricanes per year

b. $\frac{5+9+11+5}{4}=\frac{30}{4}=7.5$ hurricanes per year

c. $\frac{7.5-6.5}{6.5}=\frac{1}{6.5}\approx 0.154=15.4\%$

d. 15.4% of 7.5 = 0.154(7.5) ≈ 1.2
7.5 + 1.2 = 8.7 hurricanes per year

Quick Quiz 4.6

1.
$$\frac{20x^5-64x^4-8x^3}{4x^2}=\frac{20x^5}{4x^2}-\frac{64x^4}{4x^2}-\frac{8x^3}{4x^2}$$
$$=5x^3-16x^2-2x$$

2.
$$
\begin{array}{r}
4x^2-5x-2 \\
2x+3\overline{\big)\,8x^3+2x^2-19x-6} \\
\underline{8x^3+12x^2} \\
-10x^2-19x \\
\underline{-10x^2-15x} \\
-4x-6 \\
\underline{-4x-6} \\
0
\end{array}
$$

$$(8x^3+2x^2-19x-6)\div(2x+3)=4x^2-5x-2$$

3.
$$
\begin{array}{r}
x^2+2x+8 \\
x-2\overline{\big)\,x^3+0x^2+4x-3} \\
\underline{x^3-2x^2} \\
2x^2+4x \\
\underline{2x^2-4x} \\
8x-3 \\
\underline{8x-16} \\
13
\end{array}
$$

$$(x^3+4x-3)\div(x-2)=x^2+2x+8+\frac{13}{x-2}$$

4. Answers may vary. Possible solution:
Multiply the quotient and the divisor. Then add the remainder. You should get the original dividend.
$$(x-2)(x^2+2x+8)+13$$
$$=x^3+2x^2+8x-2x^2-4x-16+13$$
$$=x^3+4x-3$$
Yes, the answer checks.

Putting Your Skills to Work

1. $\frac{450+425+460}{3}=\frac{1335}{3}=445$
The average grocery expenses is \$445.

2. $445 \times 0.06 + 30 = 26.7 + 30 = 56.7$
Fourth month = \$445 – \$56.70 = \$388.30
Their grocery bill is \$388.30 in the fourth month.

3. $388.30 \times 0.20 = 77.66$
\$388.30 – \$77.66 = \$310.64
Their grocery bill is \$310.64 in the fifth month.

4. $\frac{445 - 310.64}{445} = 0.302$
Their savings is 30.2%.

5. Answers may vary.

Chapter 4 Review Problems

1. $(-6a^2)(3a^5) = (-6)(3)(a^2 \cdot a^5)$
$= -18a^{2+5}$
$= -18a^7$

2. $(5^{10})(5^{13}) = 5^{10+13} = 5^{23}$

3. $(3xy^2)(2x^3y^4) = (3 \cdot 2)(x \cdot x^3)(y^2 \cdot y^4)$
$= 6x^{1+3}y^{2+4}$
$= 6x^4y^6$

4. $(2x^3y^4)(-7xy^5) = (2)(-7)(x^3 \cdot x)(y^4 \cdot y^5)$
$= -14x^{3+1}y^{4+5}$
$= -14x^4y^9$

5. $\frac{7^{15}}{7^{27}} = \frac{1}{7^{27-15}} = \frac{1}{7^{12}}$

6. $\frac{x^{12}}{x^{17}} = \frac{1}{x^{17-12}} = \frac{1}{x^5}$

7. $\frac{y^{30}}{y^{16}} = y^{30-16} = y^{14}$

8. $\frac{9^{13}}{9^{24}} = \frac{1}{9^{24-13}} = \frac{1}{9^{11}}$

9. $\frac{-15xy^2}{25x^6y^6} = -\frac{3}{5x^{6-1}y^{6-2}} = -\frac{3}{5x^5y^4}$

10. $\frac{-12a^3b^6}{18a^2b^{12}} = -\frac{2a^{3-2}}{3b^{12-6}} = -\frac{2a}{3b^6}$

11. $(x^3)^8 = x^{3 \cdot 8} = x^{24}$

12. $(b^5)^6 = b^{5 \cdot 6} = b^{30}$

13. $(-3a^3b^2)^2 = (-3)^2(a^3)^2(b^2)^2$
$= 9a^{3 \cdot 2}b^{2 \cdot 2}$
$= 9a^6b^4$

14. $(3x^3y)^4 = 3^4(x^3)^4y^4 = 81x^{3 \cdot 4}y^4 = 81x^{12}y^4$

15. $\left(\frac{5ab^2}{c^3}\right)^2 = \frac{5^2a^2(b^2)^2}{(c^3)^2} = \frac{25a^2b^4}{c^6}$

16. $\left(\frac{x^0y^3}{4w^5z^2}\right)^3 = \frac{(y^3)^3}{4^3(w^5)^3(z^2)^3} = \frac{y^9}{64w^{15}z^6}$

17. $a^{-3}b^5 = \frac{b^5}{a^3}$

18. $m^8p^{-5} = \frac{m^8}{p^5}$

19. $\frac{2x^{-6}}{y^{-3}} = \frac{2y^3}{x^6}$

20. $(2x^{-5}y)^{-3} = 2^{-3}(x^{-5})^{-3}y^{-3}$
$= 2^{-3}x^{15}y^{-3}$
$= \frac{x^{15}}{2^3y^3}$
$= \frac{x^{15}}{8y^3}$

21. $(6a^4b^5)^{-2} = 6^{-2}(a^4)^{-2}(b^5)^{-2}$
$= 6^{-2}a^{-8}b^{-10}$
$= \frac{1}{6^2a^8b^{10}}$
$= \frac{1}{36a^8b^{10}}$

22. $\dfrac{3x^{-3}}{y^{-2}} = \dfrac{3y^2}{x^3}$

23. $\dfrac{4x^{-5}y^{-6}}{w^{-2}z^8} = \dfrac{4w^2}{x^5y^6z^8}$

24. $\dfrac{3^{-3}a^{-2}b^5}{c^{-3}d^{-4}} = \dfrac{b^5c^3d^4}{3^3a^2} = \dfrac{b^5c^3d^4}{27a^2}$

25. $156,340,200,000 = 1.563402\times10^{11}$

26. $179,632 = 1.79632\times10^5$

27. $0.00092 = 9.2\times10^{-4}$

28. $0.00000174 = 1.74\times10^{-6}$

29. $1.2\times10^5 = 120,000$

30. $6.034\times10^6 = 6,034,000$

31. $3\times10^6 = 3,000,000$

32. $2.5\times10^{-1} = 0.25$

33. $4.32\times10^{-5} = 0.0000432$

34. $6\times10^{-9} = 000000006\times10^{-9}$
$= 0.000000006$

35. $\dfrac{(28,0000,000)(5,000,000,000)}{7,000}$
$= \dfrac{(2.8\times10^7)(5\times10^9)}{7\times10^3}$
$= \dfrac{14\times10^{16}}{7\times10^3}$
$= 2\times10^{13}$

36. $(3.12\times10^5)(2.0\times10^6)(1.5\times10^8)$
$= 9.36\times10^{5+6+8}$
$= 9.36\times10^{19}$

37. $(1.6\times10^{-3})(3.0\times10^{-5})(2.0\times10^{-2})$
$= (1.6)(3.0)(2.0)\times10^{-3-5-2}$
$= 9.6\times10^{-10}$

38. $\dfrac{(0.00078)(0.000005)(0.00004)}{0.002}$
$= \dfrac{(7.8\times10^{-4})(5.0\times10^{-6})(4.0\times10^{-5})}{2.0\times10^{-3}}$
$= \dfrac{156\times10^{-15}}{2.0\times10^{-3}}$
$= 78\times10^{-12}$
$= 7.8\times10^{-11}$

39. $(3.5\times10^9)\times(0.20) = (3.5\times10^9)\times(2\times10^{-1})$
$= (3.5\times2)\times10^{9+(-1)}$
$= 7\times10^8$

The total cost is $\$7\times10^8$.

40. Seconds in one day $= 60\times60\times24$
$= 86,400$
$= 8.64\times10^4$

Cycles of radiation in one second
$= 9,192,631,770$
$= 9.19\times10^9$

Cycles of radiation in one day
$= (8.64\times10^4)(9.19\times10^9)$
$= 79.4016\times10^{13}$
$= 7.94\times10^{14}$

In one day, there are 7.94×10^{14} cycles.

41. $\dfrac{60}{1\times10^{-8}} = 60\times10^8 = 6\times10^9$

In one minute the computer can perform 6×10^9 operations.

42. $(2.8x^2 - 1.5x + 3.4) + (2.7x^2 + 0.5x - 5.7)$
$= (2.8 + 2.7)x^2 + (-1.5 + 0.5)x + (3.4 - 5.7)$
$= 5.5x^2 - x - 23$

43. $(4.1x^2 + 1.8x - 3.3) + (1.5x^2 - 2.9x - 0.7)$
$= (4.1 + 1.5)x^2 + (1.8 - 2.9)x + (-3.3 - 0.7)$
$= 5.6x^2 - 1.1x - 4$

44. $(x^3 + x^2 - 6x + 2) - (2x^3 - x^2 - 5x - 6)$
$= (x^3 + x^2 - 6x + 2) + (-2x^3 + x^2 + 5x + 6)$
$= (1-2)x^3 + (1+1)x^2 + (-6+5)x + (2+6)$
$= -x^3 + 2x^2 - x + 8$

45. $(4x^3 - x^2 - x + 3) - (-3x^3 + 2x^2 + 5x - 1)$
$= (4x^3 - x^2 - x + 3) + (3x^3 - 2x^2 - 5x + 1)$
$= (4+3)x^3 + (-1-2)x^2 + (-1-5)x + (3+1)$
$= 7x^3 - 3x^2 - 6x + 4$

46. $\left(\frac{3}{5}x^2y - \frac{1}{3}x + \frac{3}{4}\right) - \left(\frac{1}{2}x^2y + \frac{2}{7}x + \frac{1}{3}\right)$
$= \left(\frac{3}{5}x^2y - \frac{1}{3}x + \frac{3}{4}\right) + \left(-\frac{1}{2}x^2y - \frac{2}{7}x - \frac{1}{3}\right)$
$= \left(\frac{3}{5} - \frac{1}{2}\right)x^2y + \left(-\frac{1}{3} - \frac{2}{7}\right)x + \left(\frac{3}{4} - \frac{1}{3}\right)$
$= \frac{1}{10}x^2y - \frac{13}{21}x + \frac{5}{12}$

47. $\frac{1}{2}x^2 - \frac{3}{4}x + \frac{1}{5} - \left(\frac{1}{4}x^2 - \frac{1}{2}x + \frac{1}{10}\right)$
$= \left(\frac{1}{2}x^2 - \frac{3}{4}x + \frac{1}{5}\right) + \left(-\frac{1}{4}x^2 + \frac{1}{2}x - \frac{1}{10}\right)$
$= \left(\frac{1}{2} - \frac{1}{4}\right)x^2 + \left(-\frac{3}{4} + \frac{1}{2}\right)x + \left(\frac{1}{5} - \frac{1}{10}\right)$
$= \frac{1}{4}x^2 - \frac{1}{4}x + \frac{1}{10}$

48. $(3x^2 - 5x) + (8x - 3) - (x^2 + 2)$
$= (3x^2 - 5x) + (8x - 3) + (-x^2 - 2)$
$= (3-1)x^2 + (-5+8)x + (-3-2)$
$= 2x^2 + 3x - 5$

49. $(x^2 - 9) - (4x^2 + 5x) + (5x - 6)$
$= (x^2 - 9) + (-4x^2 - 5x) + (5x - 6)$
$= (1-4)x^2 + (-5+5)x + (-9-6)$
$= -3x^2 + 0x - 15$
$= -3x^2 - 15$

50. $(3x+1)(5x-1) = 15x^2 - 3x + 5x - 1$
$= 15x^2 + 2x - 1$

51. $(7x-2)(4x-3) = 28x^2 - 21x - 8x + 6$
$= 28x^2 - 29x + 16$

52. $(2x+3)(10x+9) = 20x^2 + 18x + 30x + 27$
$= 20x^2 + 48x + 27$

53. $5x(2x^2 - 6x + 3) = 10x^3 - 30x^2 + 15x$

54. $(3x^3 + 4x^2 - 5x + 8)(-5x)$
$= 3x^3(-5x) + 4x^2(-5x) - 5x(-5x) + 8(-5x)$
$= -15x^4 - 20x^3 + 25x^2 - 40x$

55. $(xy^2 + 5xy - 6)(-4xy^2)$
$= xy^2(-4xy^2) + 5xy(-4xy^2) - 6(-4xy^2)$
$= -4x^2y^4 - 20x^2y^3 + 24xy^2$

56. $(5a + 7b)(a - 3b) = 5a^2 - 15ab + 7ab - 21b^2$
$= 5a^2 - 8ab - 21b^2$

57. $(2x^2 - 3)(4x^2 - 5y) = 8x^4 - 10x^2y - 12x^2 + 15y$

58. $-3x^2y(5x^4y + 3x^2 - 2)$
$= -3x^2y(5x^4y) - 3x^2y(3x^2) - 3x^2y(-2)$
$= -15x^6y^2 - 9x^4y + 6x^2y$

59. $(4x+3)^2 = (4x)^2 + (2)(4x)(3) + (3)^2$
$= 16x^2 + 24x + 9$

60. $(a+5b)(a-5b) = (a)^2 - (5b)^2 = a^2 - 25b^2$

61. $(7x+6y)(7x-6y) = (7x)^2 - (6y)^2$
$= 49x^2 - 36y^2$

62. $(5a-2b)^2 = (5a)^2 - 2(5a)(2b) + (2b)^2$
$= 25a^2 - 20ab + 4b^2$

63. $(8x+9y)^2 = (8x)^2 + 2(8x)(9y) + (9y)^2$
$= 64x^2 + 144xy + 81y^2$

64. $(x^2 + 7x + 3)(4x - 1)$
$= 4x^3 + 28x^2 + 12x - x^2 - 7x - 3$
$= 4x^3 + 27x^2 + 5x - 3$

65. $(x-6)(2x-3)(x+4)$
$= (2x^2 - 15x + 18)(x+4)$
$= 2x^3 - 15x^2 + 18x + 8x^2 - 60x + 72$
$= 2x^3 - 7x^2 - 42x + 72$

66. $(12y^3 + 18x^2 + 24y) \div (6y) = \frac{12y^3}{6y} + \frac{18y^2}{6y} + \frac{24y}{6y}$
$= 2y^2 + 3y + 4$

67. $(30x^5+35x^4-90x^3)\div(5x^2)$
$$=\frac{30x^5}{5x^2}+\frac{35x^4}{5x^2}-\frac{90x^3}{5x^2}$$
$$=6x^3+7x^2-18x$$

68. $(16x^3-24x^2+32x)\div(4x)$
$$=\frac{16x^3}{4x}-\frac{24x^2}{4x}+\frac{32x}{4x}$$
$$=4x^2-6x+8$$

69. $(106x^6-24x^5+38x^4+26x^3)\div(2x^3)$
$$=\frac{106x^6}{2x^3}-\frac{24x^5}{2x^3}+\frac{38x^4}{2x^3}+\frac{26x^3}{2x^3}$$
$$=53x^3-12x^2+19x+13$$

70.
$$\begin{array}{r} 3x-2 \\ 5x+7\overline{\big)15x^2+11x-14} \\ \underline{15x^2+21x} \\ -10x-14 \\ \underline{-10x-14} \\ 0 \end{array}$$
$$\frac{15x^2+11x-14}{5x+7}=3x-2$$

71.
$$\begin{array}{r} 3x-7 \\ 4x+9\overline{\big)12x^2-x-62} \\ \underline{12x^2+27x} \\ -28x-63 \\ \underline{-28x-63} \\ 0 \end{array}$$
$$\frac{12x^2-x-63}{4x+9}=3x-7$$

72.
$$\begin{array}{r} 3x^2+2x+4 \\ 2x-1\overline{\big)6x^3+x^2+6x+5} \\ \underline{6x^3-3x^2} \\ 4x^2+6x \\ \underline{4x^2-2x} \\ 8x+5 \\ \underline{8x-4} \\ 9 \end{array}$$
$$(6x^3+x^2+6x+5)\div(2x-1)$$
$$=3x^2+2x+4+\frac{9}{2x-1}$$

73.
$$\begin{array}{r} 2x^2-5x+13 \\ x+2\overline{\big)2x^3-x^2+3x-1} \\ \underline{2x^3+4x^2} \\ -5x^2+3x \\ \underline{-5x^2-10x} \\ 13x-1 \\ \underline{13x+26} \\ -27 \end{array}$$
$$\frac{2x^3-x^2+3x-1}{x+2}=2x^2-5x+13-\frac{27}{x+2}$$

74.
$$\begin{array}{r} 5x+2 \\ 4x-1\overline{\big)20x^2+3x-2} \\ \underline{20x^2-5x} \\ 8x-2 \\ \underline{8x-2} \\ 0 \end{array}$$
$$(20x^2+3x-2)\div(4x-1)=5x+2$$

75.
$$\begin{array}{r} 3x-4 \\ 2x+3\overline{\big)6x^2+x-9} \\ \underline{6x^2+9x} \\ -8x-9 \\ \underline{-8x-12} \\ 3 \end{array}$$
$$(6x^2+x-9)\div(2x+3)=3x-4+\frac{3}{2x+3}$$

76.
$$\begin{array}{r} x^2+3x+8 \\ x-3\overline{\big)x^3+0x^2-x-24} \\ \underline{x^3-3x^2} \\ 3x^2-x \\ \underline{3x^2-9x} \\ 8x-24 \\ \underline{8x-24} \\ 0 \end{array}$$
$$(x^3-x-24)\div(x-3)=x^2+3x+8$$

77.
$$\begin{array}{r}
2x^2+4x+5 \\
x-2\overline{)2x^3+0x^2-3x+\ 1} \\
\underline{2x^3-4x^2} \\
4x^2-3x \\
\underline{4x^2-8x} \\
5x+\ 1 \\
\underline{5x-10} \\
11
\end{array}$$

$$(2x^3-3x+1)\div(x-2)=2x^2+4x+5+\frac{11}{x-2}$$

78. $(4^6)(4^{23})=4^{6+23}=4^{29}$

79. $\frac{6^{50}}{6^{35}}=6^{50-35}=6^{15}$

80.
$$\begin{aligned}(2x^3y^{-2})^{-2}&=2^{-2}(x^3)^{-2}(y^{-2})^{-2}\\&=2^{-2}x^{-6}y^4\\&=\frac{y^4}{2^2x^6}\\&=\frac{y^4}{4x^6}\end{aligned}$$

81. $\frac{-3ab^{-1}}{bc^{-2}}=\frac{-3ac^2}{b^{1+1}}=\frac{-3ac^2}{b^2}$

82.
$$\begin{aligned}\left(\frac{3x^0y^{-2}}{z^3}\right)^{-2}&=\left(\frac{3y^{-2}}{z^3}\right)^{-2}\\&=\frac{3^{-2}y^4}{z^{-6}}\\&=\frac{y^4z^6}{3^2}\\&=\frac{y^4z^6}{9}\end{aligned}$$

83. $0.000922=9.22\times10^{-4}$

84.
$$\begin{aligned}&(-6x^3-7x+8)-(-2x^3+5x-9)\\&=(-6x^3-7x+8)+(2x^3-5x+9)\\&=(-6+2)x^3+(-7-5)x+(8+9)\\&=-4x^3-12x+17\end{aligned}$$

85.
$$\begin{aligned}&-6x^2(2x^3+5x^2-2)\\&=-6x^2(2x^3)+(-6x^2)(5x^2)+(-6x^2)(-2)\\&=-12x^5-30x^4+12x^2\end{aligned}$$

86.
$$\begin{aligned}&(7x-3)(2x-1)(x+2)\\&=(14x^2-13x+3)(x+2)\\&=14x^3-13x^2+3x+28x^2-26x+6\\&=14x^3+15x^2-23x+6\end{aligned}$$

87.
$$\begin{array}{r}
2x^2+5x+3 \\
3x-2\overline{)6x^3+11x^2-\quad x-2} \\
\underline{6x^3-\ 4x^2} \\
15x^2-\quad x \\
\underline{15x^2-10x} \\
9x-2 \\
\underline{9x-6} \\
4
\end{array}$$

$$\frac{6x^3+11x^2-x-2}{3x-2}=2x^2+5x+3+\frac{4}{3x-2}$$

88.
$$\begin{aligned}\frac{2.8\times10^9}{3\times10^8}&=\frac{2.8}{3}\times10^{9-8}\\&\approx0.933\times10^1\\&=9.33\end{aligned}$$

The United States spent about \$9.33 per person.

89.
$$\begin{aligned}1.18\times10^9+1.6\times10^8&=1.18\times10^9+0.16\times10^9\\&=(1.18+0.16)\times10^9\\&=1.34\times10^9\end{aligned}$$

The total population is 1.34×10^9 people.

90.
$$\begin{aligned}(9.11\times10^{-28})(30,000)&=(9.11\times10^{-28})(3\times10^4)\\&=27.33\times10^{-24}\\&=2.733\times10^{-23}\end{aligned}$$

The mass is 2.733×10^{-23} gram.

91.
$$\begin{aligned}(3.4\times10^5)\times140&=(3.4\times10^5)\times(1.4\times10^2)\\&=(3.4\times1.4)\times10^{5+2}\\&=4.76\times10^7\end{aligned}$$

A gray whale will consume a total of 4.76×10^7 pounds of food.

92. $A=2x(2y+1)-xy=4xy+2x-xy=3xy+2x$

93. $A = 2x(x) - 4y(y) = 2x^2 - 4y^2$

How Am I Doing? Chapter 4 Test

1. $(3^{10})(3^{24}) = 3^{10+24} = 3^{34}$

2. $\dfrac{25^{18}}{25^{34}} = \dfrac{1}{25^{34-18}} = \dfrac{1}{25^{16}}$

3. $(8^4)^6 = 8^{4\cdot 6} = 8^{24}$

4. $$\begin{aligned}(-3xy^4)(-4x^3y^6) &= (-3)(-4)(x \cdot x^3)(y^4 \cdot y^6)\\ &= 12x^{1+3}y^{4+6}\\ &= 12x^4y^{10}\end{aligned}$$

5. $\dfrac{-35x^8y^{10}}{25x^5y^{10}} = -\dfrac{7x^{8-5}}{5y^{10-10}} = -\dfrac{7x^3}{5y^0} = -\dfrac{7x^3}{5}$

6. $(-5xy^6)^3 = (-5)^3x^3(y^6)^3 = -125x^3y^{18}$

7. $$\begin{aligned}\left(\frac{7a^7b^2}{3c^0}\right)^2 &= \left(\frac{7a^7b^2}{3}\right)^2\\ &= \frac{7^2(a^7)^2(b^2)^2}{3^2}\\ &= \frac{49a^{14}b^4}{9}\end{aligned}$$

8. $\dfrac{(3x^2)^3}{(6x)^2} = \dfrac{3^3(x^2)^3}{6^2x^2} = \dfrac{27x^6}{36x^2} = \dfrac{3x^{6-2}}{4} = \dfrac{3x^4}{4}$

9. $4^{-3} = \dfrac{1}{4^3} = \dfrac{1}{64}$

10. $6a^{-4}b^{-3}c^5 = \dfrac{6c^5}{a^4b^3}$

11. $\dfrac{3x^{-3}y^2}{x^{-4}y^{-5}} = \dfrac{3x^4y^2y^5}{x^3} = 3x^{4-3}y^{2+5} = 3xy^7$

12. $0.0005482 = 5.482 \times 10^{-4}$

13. $5.82 \times 10^8 = 582{,}000{,}000$

14. $$\begin{aligned}(4.0\times10^{-3})(3.0\times10^{-8})(2.0\times10^{4}) &= (4.0\times3.0\times2.0)\times10^{-3-8+4}\\ &= 24.0\times10^{-7}\\ &= 2.4\times10^{-6}\end{aligned}$$

15. $$\begin{aligned}(2x^2-3x-6)+(-4x^2+8x+6) &= (2-4)x^2+(-3+8)x+6-6\\ &= -2x^2+5x\end{aligned}$$

16. $$\begin{aligned}(3x^3-4x^2+3)-(14x^3-7x+11) &= (3x^3-4x^2+3)+(-14x^3+7x-11)\\ &= (3-14)x^3-4x^2+7x+(3-11)\\ &= -11x^3-4x^2+7x-8\end{aligned}$$

17. $$\begin{aligned}-7x^2(3x^3-4x^2+6x-2) &= -7x^2(3x^3)+(-7x^2)(-4x^2)+(-7x^2)(6x)+(-7x^2)(-2)\\ &= -21x^5+28x^4-42x^3+14x^2\end{aligned}$$

18. $$\begin{aligned}(5x^2y^2-6xy+2)(3x^2y) &= 5x^2y^2(3x^2y)-6xy(3x^2y)+2(3x^2y)\\ &= 15x^4y^3-18x^3y^2+6x^2y\end{aligned}$$

19. $$\begin{aligned}(5a-4b)(2a+3b) &= 10a^2+15ab-8ab-12b^2\\ &= 10a^2+7ab-12b^2\end{aligned}$$

20. $$\begin{aligned}(3x+2)(2x+1)(x-3) &= (6x^2+7x+2)(x-3)\\ &= 6x^3+7x^2+2x-18x^2-21x-6\\ &= 6x^3-11x^2-19x-6\end{aligned}$$

21. $$\begin{aligned}(7x^2+2y^2)^2 &= (7x^2)^2+2(7x^2)(2y^2)+(2y^2)^2\\ &= 49x^4+28x^2y^2+4y^4\end{aligned}$$

22. $$\begin{aligned}(5s-11t)(5s+11t) &= (5s)^2-(11t)^2\\ &= 25s^2-121t^2\end{aligned}$$

23. $$\begin{aligned}(3x-2)(4x^3-2x^2+7x-5) &= 3x(4x^3-2x^2+7x-5)-2(4x^3-2x^2+7x-5)\\ &= 12x^4-6x^3+21x^2-15x-8x^3+4x^2-14x+10\\ &= 12x^4-14x^3+25x^2-29x+10\end{aligned}$$

24. $$\begin{aligned}(3x^2-5xy)(x^2+3xy) &= 3x^4+9x^3y-5x^3y-15x^2y^2\\ &= 3x^4+4x^3y-15x^2y^2\end{aligned}$$

25. $$\begin{aligned}\frac{15x^6-5x^4+25x^3}{5x^3} &= \frac{15x^6}{5x^3}-\frac{5x^4}{5x^3}+\frac{25x^3}{5x^3}\\ &= 3x^3-x+5\end{aligned}$$

26.
$$\begin{array}{r} 2x^2-7x+4 \\ 4x+3\overline{)8x^3-22x^2-5x+12} \\ \underline{8x^3+6x^2} \\ -28x^2-5x \\ \underline{-28x^2-21x} \\ 16x+12 \\ \underline{16x+12} \\ 0 \end{array}$$

$$\frac{8x^3-22x^2-5x+12}{4x+3}=2x^2-7x+4$$

27.
$$\begin{array}{r} 2x^2+6x+12 \\ x-3\overline{)2x^3+0x^2-6x-36} \\ \underline{2x^3-6x^2} \\ 6x^2-6x \\ \underline{6x^2-18x} \\ 12x-36 \\ \underline{12x-36} \\ 0 \end{array}$$

$$\frac{2x^3-6x-36}{x-3}=2x^2+6x+12$$

28. $\frac{2.618\times10^{11}}{86}=0.03044\times10^{11}$
$=3.044\times10^9$

They would pump 3.04×10^9 barrels per year.

29. $d=rt$
$=2.49\times10^4(7)(24)$
$=418\times10^4$
$=4.18\times10^6$
In one week, the space probe would travel 4.18×10^6 miles.

Cumulative Test for Chapters 0–4

1. $\frac{11}{15}-\frac{7}{10}=\frac{22}{30}-\frac{21}{30}=\frac{1}{30}$

2. $(-3.7)\times(0.2)=-0.74$

3. $\left(-4\frac{1}{2}\right)\div\left(5\frac{1}{4}\right)=\left(-\frac{9}{2}\right)\div\left(\frac{21}{4}\right)$
$=-\frac{9}{2}\cdot\frac{4}{21}$
$=-\frac{\cancel{3}\cdot3\cdot\cancel{2}\cdot2}{\cancel{2}\cdot\cancel{3}\cdot7}$
$=-\frac{6}{7}$

4. $7x(3x-4)-5x(2x-3)-(3x)^2$
$=21x^2-28x-10x^2+15x-9x^2$
$=2x^2-13x$

5. 6% of 1842.5 = 0.06(1842.5) = 110.55

6. $3a^2+ab-4b^2=3(5)^2+(5)(-1)-4(-1)^2$
$=3(25)+(5)(-1)-4(1)$
$=75-5-4$
$=66$

7. $7x-3(4-2x)=14x-(3-x)$
$7x-12+6x=14x-3+x$
$13x-12=15x-3$
$-9=2x$
$-\frac{9}{2}=x$

8. $\frac{2}{3}x+6=4(x-11)$
$3\left(\frac{2}{3}x\right)+3(6)=3(4)(x-11)$
$2x+18=12x-132$
$18=10x-132$
$150=10x$
$15=x$

9. $2-5x>17$
$2-2-5x>17-2$
$-5x>15$
$\frac{-5x}{-5}<\frac{15}{-5}$
$x<-3$

10. $8(x+d)=3(y-d)$
$8x+8d=3y-3d$
$8x+11d=3y$
$11d=3y-8x$
$d=\frac{3y-8x}{11}$

11. $\frac{11,904}{x} = 0.96$
$11,904 = 0.96x$
$12,400 = x$
VBM has 12,400 employees.

12. $(0.06)(3320) = 199.20$
It would earn $199.20 in interest.

13. $(4x-5)(5x+1) = 20x^2 + 4x - 25x - 5$
$= 20x^2 - 21x - 5$

14. $(3x-5)^2 = (3x)^2 - 2(3x)(5) + 5^2$
$= 9x^2 - 30x + 25$

15. $(3x+2)(2x+1)(x-4)$
$= (3x+2)(2x^2 - 7x - 4)$
$= 6x^3 - 21x^2 - 12x + 4x^2 - 14x - 8$
$= 6x^3 - 17x^2 - 26x - 8$

16. $(-4x^2y^5)(5xy^3) = (-4)(5)x^{2+1}y^{5+3} = -20x^3y^8$

17. $\frac{14x^8y^3}{-21x^5y^{12}} = -\frac{2x^{8-5}}{3y^{12-3}} = -\frac{2x^3}{3y^9}$

18. $(-2x^3y^2z^0)^4 = (-2x^3y^2)^4$
$= (-2)^4(x^3)^4(y^2)^4$
$= 16x^{12}y^8$

19. $\frac{9x^{-3}y^{-4}}{w^2z^{-8}} = \frac{9z^8}{w^2x^3y^4}$

20. $42,500,000,000 = 4.25 \times 10^{10}$

21. $0.00056 = 5.6 \times 10^{-4}$

22. $\frac{(2.0\times10^{-12})(8.0\times10^{-20})}{4.0\times10^3} = \frac{16.0\times10^{-32}}{4.0\times10^3}$
$= 4.0\times10^{-35}$

23. $(x^3 - 3x^2 - 5x + 20) - (-4x^3 - 10x^2 + x - 30)$
$= (x^3 - 3x^2 - 5x + 20) + (4x^3 + 10x^2 - x + 30)$
$= (1+4)x^3 + (-3+10)x^2 + (-5-1)x + 20 + 30$
$= 5x^3 + 7x^2 - 6x + 50$

24. $-6xy^2(6x^2 - 3xy + 8y^2)$
$= -6xy^2(6x^2) + (-6xy^2)(-3xy) + (-6xy^2)(8y^2)$
$= -36x^3y^2 + 18x^2y^3 - 48xy^4$

25. $(2x^2 - 3x + 1)(3x - 5)$
$= (2x^2 - 3x + 1)(3x) + (2x^2 - 3x + 1)(-5)$
$= 6x^3 - 9x^2 + 3x - 10x^2 + 15x - 5$
$= 6x^3 - 19x^2 + 18x - 5$

26.
$$\begin{array}{r} x+5 \\ x-3\overline{)x^2+2x-12} \\ \underline{x^2-3x} \\ 5x-12 \\ \underline{5x-15} \\ 3 \end{array}$$

$\frac{x^2 + 2x - 12}{x-3} = x + 5 + \frac{3}{x-3}$

Chapter 5

5.1 Exercises

1. $3x^2$ and $5x^3$ are called <u>factors</u>.

3. The factoring is not complete because $6a^3 + 3a^2 - 9a$ contains a common factor of $3a$.

5. $3a^2 + 3a = 3a(a+1)$

Check: $3a(a+1) = 3a^2 + 3a$

7. $21ab - 14ab^2 = 7ab(3-2b)$

Check: $7ab(3-2b) = 21ab - 14ab^2$

9. $2\pi rh + 2\pi r^2 = 2\pi r(h+r)$

Check: $2\pi r(h+r) = 2\pi rh + 2\pi r^2$

11. $5x^3 + 25x^2 - 15x = 5x(x^2 + 5x - 3)$

Check: $5x(x^2 + 5x - 3) = 5x^3 + 25x^2 - 15x$

13. $12ab - 28bc + 20ac = 4(3ab - 7bc + 5ac)$

Check:

$4(3ab - 7bc + 5ac) = 12ab - 28bc + 20ac$

15. $16x^5 + 24x^3 - 32x^2 = 8x^2(2x^3 + 3x - 4)$

Check: $8x^2(2x^3 + 3x - 4) = 16x^5 + 24x^3 - 32x^2$

17. $14x^2y - 35xy - 63x = 7x(2xy - 5y - 9)$

Check: $7x(2xy - 5y - 9) = 14x^2y - 35xy - 63x$

19. $54x^2 - 45xy + 18x = 9x(6x - 5y + 2)$

Check: $9x(6x - 5y + 2) = 54x^2 - 45xy + 18x$

21. $3xy^2 - 2ay + 5xy - 2y = y(3xy - 2a + 5x - 2)$

Check:

$y(3xy - 2a + 5x - 2) = 3xy^2 - 2ay + 5xy - 2y$

23. $24x^2y - 40xy^2 = 8xy(3x - 5y)$

Check: $8xy(3x - 5y) = 24x^2y - 40xy^2$

25. $7x^3y^2 + 21x^2y^2 = 7x^2y^2(x+3)$

Check: $7x^2y^2(x+3) = 7x^3y^2 + 21x^2y^2$

27. $8x^4y + 24x^2y^2 - 32x^2y = 8x^2y(x^2 + 3y - 4)$

Check:

$8x^2y + (x^2 + 3y - 4) = 8x^4y + 24x^2y^2 - 32x^2y$

29. $7a(x + 2y) - b(x + 2y) = (x + 2y)(7a - b)$

31. $3x(x - 4) - 2(x - 4) = (x - 4)(3x - 2)$

33. $6b(2a - 3c) - 5d(2a - 3c) = (2a - 3c)(6b - 5d)$

35. $7c(b - a^2) - 5d(b - a^2) + 2f(b - a^2)$

$= (b - a^2)(7c - 5d + 2f)$

37. $2a(xy - 3) - 4(xy - 3) - z(xy - 3)$

$= (xy - 3)(2a - 4 - z)$

39. $4a^3(a - 3b) + (a - 3b) = (a - 3b)(4a^3 + 1)$

41. $(a + 2) - x(a + 2) = (a + 2)(1 - x)$

Cumulative Review

43. $I = rt,\ I_1 = I_2,\ r_1 = 6.72,\ t_1 = 2,\ t_2 = 6$

$6.72(2) = r_2(6)$

$r_2 = 2.24$

The rate at the slower speed is 2.24 feet per minute.

44. $16\% \text{ of } 5{,}350{,}000 = 0.16(5{,}350{,}000)$

$= 856{,}000$

856,000 metric tons were produced in Vietnam.

45. $49\% \text{ of } 5{,}350{,}000 = 0.49(5{,}350{,}000)$

$= 2{,}621{,}500$

2,621,500 metric tons were produced in Brazil.

46. $\left(\dfrac{856{,}000 \text{ metric tons}}{85{,}000{,}000 \text{ people}}\right) \cdot \left(\dfrac{2205 \text{ lb}}{1 \text{ metric ton}}\right)$

= about 22 lb per person

About 22 lb per person were produced in Vietnam.

47. $\left(\dfrac{2{,}621{,}500 \text{ metric tons}}{190{,}000{,}000}\right) \cdot \left(\dfrac{2205 \text{ lb}}{1 \text{ metric ton}}\right)$

= about 30 lb per person

About 30 lb per person were produced in Brazil.

Quick Quiz 5.1

1. $3x-4x^2+2xy = x(3-4x+2y)$

2. $20x^3-25x^2-5x = 5x(4x^2-5x-1)$

3. $8a(a+3b)-7b(a+3b) = (a+3b)(8a-7b)$

4. Answers may vary. Possible solution:
Determine that the largest integer that will divide into the coefficient of all terms is 36.
Determine that the variables common to all terms are a^2 and b^2.
Write the above common factors as the first part of the answer (the first factor).
Remove common factors, and what remains is the second part of the answer (the second factor).

5.2 Exercises

1. We must remove a common factor of 5 from the last two terms. This will give us:
$$3x^2-6xy+5x-10y = 3x(x-2y)+5(x-2y)$$
$$= (x-2y)(3x+5)$$

3. $ab-3a+4b-12 = a(b-3)+4(b-3)$
$= (b-3)(a+4)$
Check: $(b-3)(a+4) = ab-3a+4b-12$

5. $x^3-4x^2+3x-12 = x^2(x-4)+3(x-4)$
$= (x-4)(x^2+3)$
Check: $(x-4)(x^2+3) = x^3+3x-4x^2-12$
$= x^3-4x^2+3x-12$

7. $2ax+6bx-ay-3by = 2x(a+3b)-y(a+3b)$
$= (a+3b)(2x-y)$
Check: $(a+3b)(2x-y) = 2ax-ay+6bx-3by$
$= 2ax+6bx-ay-3by$

9. $3ax+bx-6a-2b = x(3a+b)-2(3a+b)$
$= (3a+b)(x-2)$
Check: $(3a+b)(x-2) = 3ax-6a+bx-2b$
$= 3ax+bx-6a-2b$

11. $5a+12bc+10b+6ac = 5a+10b+6ac+12bc$
$= 5(a+2b)+6c(a+2b)$
$= (a+2b)(5+6c)$
Check: $(a+2b)(5+6c) = 5a+6ac+10b+12bc$
$= 6a+12bc+10b+6ac$

13. $6c-12d+cx-2dx = 6(c-2d)+x(c-2d)$
$= (c-2d)(6+x)$
Check: $(c-2d)(6+x) = 6c+cx-12d-2dx$
$= 6c-12d+cx-2dx$

15. $y^2-2y-3y+6 = y(y-2)-3(y-2)$
$= (y-2)(y-3)$
Check: $(y-2)(y-3) = y^2-2y-3y+6$

17. $54-6y+9y-y^2 = 6(9-y)+y(9-y)$
$= (9-y)(6+y)$
Check: $(9-y)(6+y) = 54+9y-6y-y^2$
$= 54-6y+9y-y^2$

19. $6ax-y+2ay-3x = 6ax-3x+2ay-y$
$= 3x(2a-1)+y(2a-1)$
$= (2a-1)(3x+y)$
Check: $(2a-1)(3x+y) = 6ax-y+2ay-3x$

21. $2x^2+8x-3x-12 = 2x(x+4)-3(x+4)$
$= (x+4)(2x-3)$
Check: $(x+4)(2x-3) = 2x^2+8x-3x-12$

23. $t^3-t^2+t-1 = t^2(t-1)+(t-1) = (t-1)(t^2+1)$
Check: $(t-1)(t^2+1) = t^3-t^2+t-1$

25. $28x^2+8xy^2+21xw+6y^2w$
$= 4x(7x+2y^2)+3w(7x+2y^2)$
$= (7x+2y^2)(4x+3w)$
Check: $(7x+2y^2)(4x+3w)$
$= 28x^2+8xy^2+21xw+6y^2w$

27. Rearrange the terms so that factoring gives the same expression in parentheses.
$6a^2-12bd-8ad+9ab$
$= 6a^2-8ad+9ab-12bd$
$= 2a(3a-4d)+3b(3a-4d)$
$= (3a-4d)(2a+3b)$

Cumulative Review

29. x = % of total sales spent on travel.
$$\frac{x}{100\%} = \frac{119.1 \text{ billion}}{328.6 \text{ billion}}$$
$x = 0.362(100\%)$
$x = 36.2\%$
Travel sales account for 36.2% of the total.

30. Let x = average salary for an instructor.

$$\begin{aligned} x+(2x+11,200) &= 131,500 \\ x+2x+11,200 &= 131,500 \\ 3x &= 131,500-11,200 \\ 3x &= 120,300 \\ x &= 40,100 \end{aligned}$$

$2x + 11{,}200 = 2(40{,}100) + 11{,}200 = 91{,}400$
The average salaries are:
instructor: \$40,100 and professor: \$91,400

31. 2003–2004: $581-581(0.046) = 581-26.726 = 554.274$

2004–2005:
$$554.274+554.274(0.123) = 554.274+68.177 \approx 622$$
About 622 million metric tons or 622,000,000 metric tons were produced in 2005–2006.

Quick Quiz 5.2

1. $$\begin{aligned} 7ax+12a-14x-24 &= 7ax-14x+12a-24 \\ &= 7x(a-2)+12(a-2) \\ &= (7x+12)(a-2) \end{aligned}$$

2. $$\begin{aligned} 2xy^2-15+6x-5y^2 &= 2xy^2+6x-15-5y^2 \\ &= 2x(y^2+3)-5(3+y^2) \\ &= (y^2+3)(2x-5) \end{aligned}$$

3. $$\begin{aligned} 10xy-3x+40by-12b &= 10xy+40by-3x-12b \\ &= 10y(x+4b)-3(x+4b) \\ &= (10y-3)(x+4b) \end{aligned}$$

4. Answers may vary. Possible solution:
Start by grouping terms $10ax$ with $5ab$ and $2bx$ with b^2.
$(2x + b)$ can be factored out of both groups leaving the second factor to be $(5a + b)$.

5.3 Exercises

1. Find two numbers whose product is 6 and whose sum is 5.

3. x^2+2x+1; product: 1, sum: 2, + signs
$x^2+2x+1=(x+1)(x+1)$

5. $x^2+12x+35$; product: 35, sum: 12, + signs
$x^2+12x+35=(x+5)(x+7)$

7. x^2-4x+3; product: 3, sum: –4, – signs
$x^2-4x+3=(x-1)(x-3)$

9. $x^2-11x+28$; product: 28, sum: –11, – signs
$x^2-11x+28=(x-4)(x-7)$

11. $x^2+5x-24$; product: –24, sum 5
opposite signs with larger absolute value +
$x^2+5x-24=(x+8)(x-3)$

13. $x^2-13x-14$; product: –14, sum: –13,
opposite signs with larger absolute value –
$x^2-13x-14=(x-14)(x+1)$

15. $x^2+2x-35$; product: –35, sum: 2
opposite signs with larger absolute value +
$x^2+2x-35=(x+7)(x-5)$

17. $x^2-2x-24$; product: –24, sum: –2
opposite signs with larger absolute value –
$x^2-2x-24=(x-6)(x+4)$

19. $x^2+15x+36$; product: 36, sum: 15, + signs
$x^2+15x+36=(x+12)(x+3)$

21. $x^2-10x+24$; product: 24, sum: –10, – signs
$x^2-10x+24=(x-6)(x-4)$

23. $x^2+13x+30$; product: 30, sum: 13, + signs
$x^2+13x+30=(x+3)(x+10)$

25. x^2-6x+5; product: 5, sum: –6, – signs
$x^2-6x+5=(x-1)(x-5)$

27. $a^2+6a-16=(a+8)(a-2)$
Check:
$(a+8)(a-2)=a^2-2a+8a-16=a^2+6a-16$

29. $x^2-12x+32=(x-4)(x-8)$
Check:
$(x-4)(x-8)=x^2-8x-4x+32=x^2-12x+32$

31. $x^2+4x-21=(x+7)(x-3)$
Check:
$(x+7)(x-3)=x^2-3x+7x-21=x^2+4x-21$

33. $x^2+15x+56=(x+7)(x+8)$
Check:
$(x+7)(x+8)=x^2+8x+7x+56=x^2+15x+56$

35. $y^2+4y-45=(y+9)(y-5)$
Check:
$(y+9)(y-5)=y^2-5y+9y-45=y^2+4y-45$

37. $x^2+9x-36=(x+12)(x-3)$
Check: $(x+12)(x-3)=x^2-3x+12x-36$
$=x^2+9x-36$

39. $x^2-2xy-15y^2=(x-5y)(x+3y)$
Check: $(x-5y)(x+3y)=x^2+3xy-5xy-15y^2$
$=x^2-2xy-15y^2$

41. $x^2-16xy+63y^2=(x-9y)(x-7y)$
Check: $(x-9y)(x-7y)=x^2-7xy-9xy+63y^2$
$=x^2-16xy+63y^2$

43. $4x^2+24x+20=4(x^2+6x+5)=4(x+1)(x+5)$

45. $6x^2+18x+12=6(x^2+3x+2)=6(x+2)(x+1)$

47. $5x^2-30x+25=5(x^2-6x+5)=5(x-1)(x-5)$

49. $3x^2-6x-72=3(x^2-2x-24)=3(x-6)(x+4)$

51. $7x^2+21x-70=7(x^2+3x-10)$
$=7(x+5)(x-2)$

53. $3x^2-18x+15=3(x^2-6x+5)=3(x-5)(x-1)$

55. A = large rectangle area − small rectangle area
$A=10(12)-x(x+2)$
$=120-x^2-2x$
$=120-2x-x^2$
$=(10-x)(12+x)$

Cumulative Review

57.
$$A=P+Prt$$
$$A=P(1+rt)$$
$$\frac{A}{1+rt}=\frac{P(1+rt)}{1+rt}$$
$$\frac{A}{1+rt}=P$$

58. $2-3x\le 7$
$-3x\le 5$
$x\ge -\frac{5}{3}$

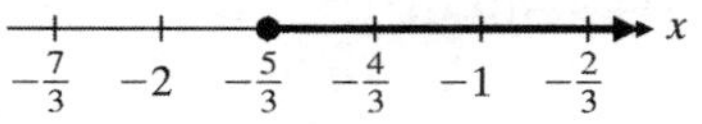

59. c = car's speed
$c=\frac{d}{t}=\frac{d}{2}$
c + 20 = train's speed
$\frac{d}{2}+20=\frac{d}{1.5}$
$1.5d+60=2d$
$60=0.5d$
$120=d$
The distance is 120 miles.

60. x = sales
$600+0.04(x)\ge 3800$
$0.04x\ge 3200$
$x\ge 80,000$
Carla must sell at least $80,000 of electronics.

61. $M=3$
$T=19+2M$
$T=19+2(3)$
$T=25$
The average temperature during April is 25°C.

62. $T=19+2M$, $T=29$
$29=19+2M$
$10=2M$
$5=M$
Five months after January will be June.

Quick Quiz 5.3

1. $x^2+17x+70$; product: 70, sum: 17, + signs
$x^2+17x+70=(x+7)(x+10)$

2. $x^2 - 14x + 48$; product: 48, sum: −14, − signs
$x^2 - 14x + 48 = (x-6)(x-8)$

3. $2x^2 - 4x - 96 = 2(x^2 - 2x - 48)$
product: −48, sum: −2
opposite signs with largest absolute value −
$2(x^2 - 2x - 48) = 2(x+6)(x-8)$

4. Answers may vary. Possible solution:
The first step is to factor out the greatest common factor of 4, leaving $4(x^2 - x - 30)$.
Next, write expression in factored form using variables n and m, $4(x + m)(x + n)$.
Next determine that the product of m and n is −30, and the sum is −1.
m and n may equal 5, −6.
Substitute values of m and n, then check.

5.4 Exercises

1. $4x^2 + 13x + 3$
Factorizations of 4: (2)(2) or (1)(4)
Factorization of 3: (1)(3)
Each factor will be positive. Find the factoring combination that yields a middle term of $13x$.
$4x^2 + 13x + 3 = (4x+1)(x+3)$

Check: $(4x+1)(x+3) = 4x^2 + 12x + x + 3$
$= 4x^2 + 13x + 3$

3. $5x^2 + 7x + 2$
Factorization of 5: (1)(5)
Factorization of 2: (1)(2)
Each factor will be positive. Find the factoring combination that yields a middle term of $7x$.
$5x^2 + 7x + 2 = (5x+2)(x+1)$
Check:
$(5x+2)(x+1) = 5x^2 + 5x + 2x + 2 = 5x^2 + 7x + 2$

5. $4x^2 + 5x - 6$
Factorizations of 4: (2)(2) or (1)(4)
Factorizations of 6: (1)(6) or (2)(3)
The constants in the factors will have opposite signs. Find the factoring combination that yields a middle term of $5x$.
$4x^2 + 5x - 6 = (4x-3)(x+2)$
Check:
$(4x-3)(x+2) = 4x^2 + 8x - 3x - 6 = 4x^2 + 5x - 6$

7. $2x^2 - 5x - 3$
Factorization of 2: (1)(2)
Factorization of 3: (1)(3)
The signs of the constants in the factors will have opposite signs. Find the factoring combination that yields a middle term of $-5x$.
$2x^2 - 5x - 3 = (x-3)(2x+1)$
Check:
$(x-3)(2x+1) = 2x^2 + x - 6x - 3 = 2x^2 - 5x - 3$

9. $9x^2 + 9x + 2$; grouping number: 18

$$\begin{aligned} 9x^2 + 9x + 2 &= 9x^2 + 3x + 6x + 2 \\ &= 3x(3x+1) + 2(3x+1) \\ &= (3x+1)(3x+2) \end{aligned}$$

Check: $(3x+1)(3x+2) = 9x^2 + 6x + 3x + 2$
$= 9x^2 + 9x + 2$

11. $15x^2 - 34x + 15$; grouping number: 225

$$\begin{aligned} 15x^2 - 34x + 15 &= 15x^2 - 25x - 9x + 15 \\ &= 5x(3x-5) - 3(3x-5) \\ &= (5x-3)(3x-5) \end{aligned}$$

Check: $(5x-3)(3x-5) = 15x^2 - 25x - 9x + 15$
$= 15x^2 - 34x + 15$

13. $2x^2 + 3x - 20$; grouping number: −40

$$\begin{aligned} 2x^2 + 3x - 20 &= 2x^2 + 8x - 5x - 20 \\ &= 2x(x+4) - 5(x+4) \\ &= (x+4)(2x-5) \end{aligned}$$

Check: $(x+4)(2x-5) = 2x^2 - 5x + 8x - 20$
$= 2x^2 + 3x - 20$

15. $8x^2 + 10x - 3$; grouping number: −24

$$\begin{aligned} 8x^2 + 10x - 3 &= 8x^2 + 12x - 2x - 3 \\ &= 4x(2x+3) - (2x+3) \\ &= (2x+3)(4x-1) \end{aligned}$$

Check: $(2x+3)(4x-1) = 8x^2 - 2x + 12x - 3$
$= 8x^2 + 10x - 3$

17. $6x^2 - 5x - 6$; grouping number: −36

$$\begin{aligned} 6x^2 - 5x - 6 &= 6x^2 - 9x + 4x - 6 \\ &= 3x(2x-3) + 2(2x-3) \\ &= (2x-3)(3x+2) \end{aligned}$$

19. $10x^2 + 3x - 1$; grouping number -10

$$\begin{aligned} 10x^2 + 3x - 1 &= 10x^2 + 5x - 2x - 1 \\ &= 5x(2x+1) - 1(2x+1) \\ &= (5x-1)(2x+1) \end{aligned}$$

21. $7x^2 - 5x - 18$; grouping number: -126

$$\begin{aligned} 7x^2 - 5x - 18 &= 7x^2 - 14x + 9x - 18 \\ &= 7x(x-2) + 9(x-2) \\ &= (7x+9)(x-2) \end{aligned}$$

23. $9y^2 - 13y + 4$; grouping number: 36

$$\begin{aligned} 9y^2 - 13y + 4 &= 9y^2 - 9y - 4y + 4 \\ &= 9y(y-1) - 4(y-1) \\ &= (y-1)(9y-4) \end{aligned}$$

25. $5a^2 - 13a - 6$; graphing number: -30

$$\begin{aligned} 5a^2 - 13a - 6 &= 5a^2 - 15a + 2a - 6 \\ &= 5a(a-3) + 2(a-3) \\ &= (a-3)(5a+2) \end{aligned}$$

27. $14x^2 + 17x - 6$; grouping number -84

$$\begin{aligned} 14x^2 + 17x - 6 &= 14x^2 + 21x - 4x - 6 \\ &= 7x(2x+3) - 2(2x+3) \\ &= (7x-2)(2x+3) \end{aligned}$$

29. $15x^2 + 4x - 4$; grouping number: -60

$$\begin{aligned} 15x^2 + 4x - 4 &= 15x^2 + 10x - 6x - 4 \\ &= 5x(3x+2) - 2(3x+2) \\ &= (3x+2)(5x-2) \end{aligned}$$

31. $12x^2 + 28x + 15$; grouping number: 180

$$\begin{aligned} 12x^2 + 28x + 15 &= 12x^2 + 18x + 10x + 15 \\ &= 6x(2x+3) + 5(2x+3) \\ &= (6x+5)(2x+3) \end{aligned}$$

33. $12x^2 - 16x - 3$; grouping number: -36

$$\begin{aligned} 12x^2 - 16x - 3 &= 12x^2 + 2x - 18x - 3 \\ &= 2x(6x+1) - 3(6x+1) \\ &= (6x+1)(2x-3) \end{aligned}$$

35. $3x^4 - 14x^2 - 5$; grouping number -15

$$\begin{aligned} 3x^4 - 14x^2 - 5 &= 3x^4 + x^2 - 15x^2 - 5 \\ &= x^2(3x^2+1) - 5(3x^2+1) \\ &= (x^2-5)(3x^2+1) \end{aligned}$$

37. $2x^2 + 11xy + 15y^2$; grouping number: 30

$$\begin{aligned} 2x^2 + 11xy + 15y^2 &= 2x^2 + 6xy + 5xy + 15y^2 \\ &= 2x(x+3y) + 5y(x+3y) \\ &= (x+3y)(2x+5y) \end{aligned}$$

39. $5x^2 + 16xy - 16y^2$; grouping number: -80

$$\begin{aligned} 5x^2 + 16xy - 16y^2 &= 5xy^2 + 20xy - 4xy - 16y^2 \\ &= 5x(x+4y) - 4y(x+4y) \\ &= (x+4y)(5x-4y) \end{aligned}$$

41.

$$\begin{aligned} 10x^2 - 25x - 15 &= 5(2x^2 - 5x - 3) \\ &= 5(2x^2 - 6x + x - 3) \\ &= 5[2x(x-3) + 1(x-3)] \\ &= 5(2x+1)(x-3) \end{aligned}$$

43.

$$\begin{aligned} 6x^3 + 9x^2 - 60x &= 3x(2x^2 + 3x - 20) \\ &= 3x(2x^2 + 8x - 5x - 20) \\ &= 3x[2x(x+4) - 5(x+4)] \\ &= 3x(2x-5)(x+4) \end{aligned}$$

45.

$$\begin{aligned} 5x^2 + 3x - 2 &= 5x^2 + 5x - 2x - 2 \\ &= 5x(x+1) - 2(x+1) \\ &= (x+1)(5x-2) \end{aligned}$$

47.

$$\begin{aligned} 12x^2 - 38x + 20 &= 2(6x^2 - 19x + 10) \\ &= 2(6x^2 - 15x - 4x + 10) \\ &= 2[3x(2x-5) - 2(2x-5)] \\ &= 2(2x-5)(3x-2) \end{aligned}$$

49.

$$\begin{aligned} 12x^3 - 20x^2 + 3x &= x(12x^2 - 20x + 3) \\ &= x(12x^2 - 18x - 2x + 3) \\ &= x[6x(2x-3) - (2x-3)] \\ &= x(2x-3)(6x-1) \end{aligned}$$

51.

$$\begin{aligned} 8x^2 + 16x - 10 &= 2(4x^2 + 8x - 5) \\ &= 2(4x^2 + 10x - 2x - 5) \\ &= 2[2x(2x+5) - (2x+5)] \\ &= 2(2x+5)(2x-1) \end{aligned}$$

Cumulative Review

53. $7x-3(4-2x)=2(x-3)-(5-x)$
$7x-12+6x=2x-6-5+x$
$13x-12=3x-11$
$13x-3x-12=3x-3x-11$
$10x-12=-11$
$10x-12+12=-11+12$
$10x=1$
$x=\frac{1}{10}$

54. $7x-3y=5$
$7x-3y+3y=5+3y$
$7x=5+3y$
$\frac{7x}{7}=\frac{5+3y}{7}$
$x=\frac{5+3y}{7}$

55. a. Add the heights of the 2005 bars.
6.1 + 4.8 + 4.4 + 2.3 + 1.8 = 19.4
19,400,000 travelers visited these states in 2005.

b. x = % of travelers to Hawaii
$x=\frac{4,400,000}{19,400,000}\times 100\%$
x = about 22.7%
About 22.7% visited Hawaii.

56. x = % of travelers to California in 2000
2000 total = 5.9 + 6.4 + 6.0 + 2.7 + 2.4 = 23.4
$x=\frac{6,400,000}{23,400,000}\times 100\%$
x = about 27.4%
About 27.4% visited California.

57. x = % of decrease in travelers to California
$x=\frac{6.4-4.8}{6.4}\times 100\%$
$x=25\%$
The percent decrease in the number of visitors to California from 2000 to 2005 was 25%.

58. x = % increase of visitors to New York from 2000 to 2005
$x=\frac{6.1-5.9}{5.9}\times 100\%$
x = about 3.4%
The percent decrease in the number of visitors to NY from 2000 to 2005 was about 3.4%.

Quick Quiz 5.4

1. $12x^2+16x-3$; grouping number −36
$12x^2+16x-3=12x^2+18x-2x-3$
$=6x(2x+3)-(2x+3)$
$=(6x-1)(2x+3)$

2. $10x^2-21x+9$; grouping number 90
$10x^2-21x+9=10x^2-15x-6x+9$
$=5x(2x-3)-3(2x-3)$
$=(5x-3)(2x-3)$

3. $6x^3-3x^2-30x=3x(2x^2-x-10)$
grouping number −20
$6x^3-3x^2-30x=3x[2x^2+4x-5x-10]$
$=3x[2x(x+2)-5(x+2)]$
$=3x(2x-5)(x+2)$

4. Answers may vary. Possible solution:
First step is to factor out common coefficients and variables.
$2x(5x^2+9xy-2y^2)$
Next step is to factor the inside expression using grouping.
Grouping number −10
$2x[5x^2+10xy-xy-2y^2]$
$=2x[5x(x+2y)-y(x+2y)]$
$=2x(5x-y)(x+2y)$
Lastly, check the solution by multiplying the factors.

How Am I Doing? Sections 5.1–5.4

1. $6xy-15z+21=3(2xy-5z+7)$

2. $30x^2-45xy-10x=5x(6x-9y-2)$

3. $7(4x-5)-b(4x-5)=(4x-5)(7-b)$

4. $2x(8y+3z)-5y(8y+3z)=(8y+3z)(2x-5y)$

5. $18+3x-6y-xy=3(6+x)-y(6+x)$
$=(6+x)(3-y)$

6. $15x-9xb+20w-12bw=3x(5-3b)+4w(5-3b)$
$=(5-3b)(3x+4w)$

7. $x^3-5x^2-3x+15=x^2(x-5)-3(x-5)$
$=(x-5)(x^2-3)$

8. $7a+21b+2ab+6b^2 = 7(a+3b)+2b(a+3b)$
$= (a+3b)(7+2b)$

9. $x^2-15x+56 = x^2-7x-8x+56$
$= x(x-7)-8(x-7)$
$= (x-7)(x-8)$

10. $x^2+12x-64 = x^2+16x-4x-64$
$= x(x+16)-4(x+16)$
$= (x+16)(x-4)$

11. $a^2+10ab+21b^2$; product: 21, sum: 10, + signs
$a^2+10ab+21b^2 = a^2+3ab+7ab+21b^2$
$= a(a+3b)+7b(a+3b)$
$= (a+7b)(a+3b)$

12. $7x^2-14x-245 = 7(x^2-2x-35)$
$= 7(x^2-7x+5x-35)$
$= 7[x(x-7)+5(x-7)]$
$= 7(x-7)(x+5)$

13. $12x^2+17x-5$; product: –60, sum: 17
$12x^2+17x-5 = 12x^2+20x-3x-5$
$= 4x(3x+5)-(3x+5)$
$= (4x-1)(3x+5)$

14. $3x^2-23x+14 = 3x^2-21x-2x+14$
$= 3x(x-7)-2(x-7)$
$= (x-7)(3x-2)$

15. $6x^2+17xy+12y^2 = 6x^2+9xy+8xy+12y^2$
$= 3x(2x+3y)+4y(2x+3y)$
$= (2x+3y)(3x+4y)$

16. $14x^3-20x^2-16x = 2x(7x^2-10x-8)$
$= 2x(7x^2-14x+4x-8)$
$= 2x[7x(x-2)+4(x-2)]$
$= 2x(x-2)(7x+4)$

5.5 Exercises

1. $9x^2-1 = (3x)^2-(1)^2 = (3x+1)(3x-1)$

3. $81x^2-16 = (9x)^2-(4)^2 = (9x+4)(9x-4)$

5. $x^2-49 = (x)^2-(7)^2 = (x+7)(x-7)$

7. $25x^2-81 = (5x)^2-(9)^2 = (5x+9)(5x-9)$

9. $x^2-25 = (x)^2-(5)^2 = (x+5)(x-5)$

11. $1-16x^2 = (1)^2-(4x)^2 = (1+4x)(1-4x)$

13. $16x^2-49y^2 = (4x)^2-(7y)^2$
$= (4x+7y)(4x-7y)$

15. $36x^2-169y^2 = (6x)^2-(13y)^2$
$= (6x+13y)(6x-13y)$

17. $100x^2-81 = (10x)^2-(9)^2 = (10x+9)(10x-9)$

19. $25a^2-81b^2 = (5a)^2-(9b)^2 = (5a+9b)(5a-9b)$

21. $9x^2+6x+1 = (3x)^2+2(3x)(1)+(1)^2 = (3x+1)^2$

23. $y^2-10y+25 = y^2-2(y)(5)+(5)^2 = (y-5)^2$

25. $36x^2-60x+25 = (6x)^2-2(6x)(5)+(5)^2$
$= (6x-5)^2$

27. $49x^2+28x+4 = (7x)^2+2(7x)(2)+2^2$
$= (7x+2)^2$

29. $x^2+14x+49 = (x)^2+2(x)(7)+(7)^2 = (x+7)^2$

31. $25x^2-40x+16 = (5x)^2-2(5x)(4)+4^2$
$= (5x-4)^2$

33. $81x^2+36xy+4y^2 = (9x)^2+2(9x)(2y)+(2y)^2$
$= (9x+2y)^2$

35. $9x^2-30xy+25y^2 = (3x)^2-2(3x)(5y)+(5y)^2$
$= (3x-5y)^2$

37. $16a^2+72ab+81b^2 = (4a)^2+2(4a)(9b)+(9b)^2$
$= (4a+9b)^2$

39. $49x^2-42xy+9y^2 = (7x)^2-2(7x)(3y)+(3y)^2$
$= (7x-3y)^2$

41. $9x^2+42x+49=(3x)^2+2(3x)(7)+7^2$
$=(3x+7)^2$

43. $144x^2-1=(12x)^2-(1)^2=(12x-1)(12x+1)$

45. $x^4-36=(x^2)^2-(6)^2=(x^2+6)(x^2-6)$

47. $4x^4-20x^2+25=(2x^2)^2-2(2x^2)(5)+(5)^2$
$=(2x^2-5)^2$

49. You cannot factor $9x^2+1$ because there are no combinations of the product of two binomials that give $9x^2+1$.

51. $16=4^2$
$56=(2)(4)(7)$
$c=7^2$
$c=49$
There is only one answer.

53. $16x^2-36=4(4x^2-9)=4(2x+3)(2x-3)$

55. $147x^2-3y^2=3(49x^2-y^2)$
$=3[(7x)^2-y^2]$
$=3(7x-y)(7x+y)$

57. $16x^2-16x+4=4(4x^2-4x+1)$
$=4[(2x)^2-2(2x)(1)+(1)^2]$
$=4(2x-1)^2$

59. $98x^2+84x+18=2(49x^2+42x+9)=2(7x+3)^2$

61. $x^2+16x+63=(x+9)(x+7)$

63. $2x^2+5x-3=(2x-1)(x+3)$

65. $16x^2-121=(4x)^2-(11)^2=(4x+11)(4x-11)$

67. $9x^2+42x+49=(3x+7)^2$

69. $36x^2-36x+9=9(4x^2-4x+1)$
$=9[(2x)^2-2(2x)(1)+(1)^2]$
$=9(2x+1)(2x-1)$

71. $2x^2-32x+126=2(x^2-16x+63)$
$=2(x-9)(x-7)$

Cumulative Review

73.
$$\begin{array}{r} x^2+3x+4 \\ x-2\overline{)\,x^3+x^2-2x-11} \\ \underline{x^3-2x^2} \\ 3x^2-2x \\ \underline{3x^2-6x} \\ 4x-11 \\ \underline{4x-8} \\ -3 \end{array}$$

$(x^3+x^2-2x-11)\div(x-2)=x^2+3x+4-\dfrac{3}{x-2}$

74.
$$\begin{array}{r} 2x^2+x-5 \\ 3x+4\overline{)\,6x^3+11x^2-11x-20} \\ \underline{6x^3+8x^2} \\ 3x^2-11x \\ \underline{3x^2+4x} \\ -15x-20 \\ \underline{-15x-20} \end{array}$$

$(6x^3+11x^2-11x-20)\div(3x+4)=2x^2+x-5$

75. Daily diet = 0.2(150) = 3 ounces
0.4(3) = 1.2 ounces of greens
0.35(3) = 1.05 ounces of bulk vegetables
0.25(3) = 0.75 ounce of fruit

76. Daily diet = 0.03(120) = 3.6 ounces
0.4(3.6) = 1.44 ounces of greens
0.35(3.6) = 1.26 ounces of bulk vegetables
0.25(3.6) = 0.9 ounce of fruit

Quick Quiz 5.5

1. $49x^2-81y^2=(7x)^2-(9y)^2$
$=(7x-9y)(7x+9y)$

2. $9x^2-48x+64=(3x)^2-2(3x)(8)+(8)^2$
$=(3x-8)^2$

3. $162x^2-200=2(81x^2-100)$
$=2[(9x)^2-(10)^2]$
$=2(9x-10)(9x+10)$

4. Answers may vary. Possible solution:
First factor out the common factor of 2.
$2(12x^2+60x+75)$
Next use grouping to factor inside expression.
Grouping number 900
$2(12x^2+30x+30x+75)$
$=2[6x(2x+5)+15(2x+5)]$
$=2(6x+15)(2x+5)$
Lastly, check by multiplying.

5.6 Exercises

1. $5a^2-3ab+8a=a(5a-3b+8)$
Check: $a(5a-3b+8)=5a^2-3ab+8a$

3. $16x^2-25y^2=(4x)^2-(5y)^2$
$=(4x+5y)(4x-5y)$
Check:
$(4x+5y)(4x-5y)=(4x)^2-(5y)^2=16x^2-25y^2$

5. $9x^2-12xy+4y^2=(3x-2y)^2$
Check: $(3x-2y)^2=9x^2+2(3x)(-2y)+(-2y)^2$
$=9x^2-12xy+4y^2$

7. $x^2+8x+15=(x+5)(x+3)$
Check:
$(x+5)(x+3)=x^2+3x+5x+15=x^2+8x+15$

9. $15x^2+7x-2=(5x-1)(3x+2)$
Check: $(5x-1)(3x+2)=15x^2+10x-3x-2$
$=15x^2+7x-2$

11. $ax-3cx+3ay-9cy=x(a-3c)+3y(a-3c)$
$=(a-3c)(x+3y)$
Check: $(x+3y)(a-3c)=ax-3cx+3ay-9cy$

13. $y^2+14y+49=(y)^2+2(y)(7)+(7)^2=(y+7)^2$

15. $4x^2-12x+9=(2x)^2-2(2x)(3)+3^2=(2x-3)^2$

17. $2x^2-11x+12=(2x-3)(x-4)$

19. $x^2-3xy-70y^2=(x-10y)(x+7y)$

21. $ax-5a+3x-15=a(x-5)+3(x-5)$
$=(a+3)(x-5)$

23. $16x-4x^3=4x(4-x^2)$
$=4x[(2)^2-(x)^2]$
$=4x(2-x)(2+x)$

25. $5x^3y^3-10x^2y^3+5xy^3=5xy^3(x^2-2x+1)$
$=5xy^3[x^2-2(x)(1)+1^2]$
$=5xy^3(x-1)^2$

27. $3xyz^2-6xyz-9xy=3xy(z^2-2z-3)$
$=3xy(z-3)(z+1)$

29. $3x^2+6x-105=3(x^2+2x-35)$
$=3(x+7)(x-5)$

31. $3x^3+3x^2-36x=3x(x^2+x-12)$
$=3x(x^2+4x-3x-12)$
$=3x[x(x+4)-3(x+4)]$
$=3x(x+4)(x-3)$

33. $7x^2-2x^4+4=-1(2x^4-7x^2-4)$
$=-1(2x^2+1)(x^2-4)$
$=-1(2x^2+1)(x+2)(x-2)$

35. $6x^2-3x+2$ is prime.

37. $5x^2+10xy-30y=5(x^2+2xy-6y)$

39. $30x^3+3x^2y-6xy^2=3x(10x^2+xy-2y^2)$
$=3x(2x+y)(5x-2)$

41. $24x^2-58x+30=2(12x^2-29x+15)$
$=2(12x^2-20x-9x+15)$
$=2[4x(3x-5)-3(3x-5)]$
$=2(3x-5)(4x-3)$

Cumulative Review

43. $x=\%$ increase in emissions 2000–2005
$$x=\frac{5322-2913}{2913}\times 100\%$$
$x=$ about 83%
The percent increase was about 83%.

44. x = % increase U.S. 2000–2005
y = % increase Europe 2000–2005

$$x = \frac{5957-5823}{5823}\times 100\% = 2.3\%$$

$$y = \frac{4675-4445}{4445}\times 100\% = 5.2\%$$

Europe has the larger, with about a 5% increase.

45. x = estimated emissions in 2010

$$83\% = \frac{x-5322}{5322}\times 100\%$$
$$0.83 = \frac{x-5322}{5322}$$
$$4401 = x-5322$$
$$9739.26 = x$$

China will emit about 9739 million metric tons.

46. x = estimated 2010 emissions in Europe

$$5.2\% = \frac{x-4675}{4675}\times 100\%$$
$$0.052 = \frac{x-4675}{4675}$$
$$241 = x-4675$$
$$4908.75 = x$$

Europe will emit about 4909 million metric tons.

Quick Quiz 5.6

1. $6x^2-17x+12$; product: 72, sum: -17, $-$ signs

$$\begin{aligned}6x^2-17x+12 &= 6x^2-8x-9x+12\\ &= 2x(3x-4)-3(3x-4)\\ &= (2x-3)(3x-4)\end{aligned}$$

2. $60x^2-9x-6 = 3(20x^2-3x-2)$
grouping number: -40

$$\begin{aligned}&= 3(20x^2+5x-8x-2)\\ &= 3[5x(4x+1)-2(4x+1)]\\ &= 3(5x-2)(4x+1)\end{aligned}$$

3. $25x^2+49$
prime

4. Answers may vary. Possible solution:
$2x^2+6xw-5x-15w$
The first step is to group terms and factor out common factors.
$2x(x+3w)-5(x+3w)$
Next, factor out $(x+3w)$.
$(2x-5)(x+3w)$
Finally, check by multiplying.

5.7 Exercises

1. $x^2-4x-21=0$
$(x+3)(x-7)=0$
$x+3=0$ $\quad$ $x-7=0$
$x=-3$ $\quad$ $x=7$
Check:
$$(-3)^2-4(-3)-21 \stackrel{?}{=} 0$$
$$9+12-21 \stackrel{?}{=} 0$$
$$0=0$$
$$(7)^2-4(7)-21 \stackrel{?}{=} 0$$
$$49-28-21 \stackrel{?}{=} 0$$
$$0=0$$

3. $x^2+14x+24=0$
$(x+12)(x+2)=0$
$x+12=0$ $\quad$ $x+2=0$
$x=-12$ $\quad$ $x=-2$
Check:
$$(-12)^2+14(-12)+24 \stackrel{?}{=} 0$$
$$144-168+24 \stackrel{?}{=} 0$$
$$0=0$$
$$(-2)^2+14(-2)+24 \stackrel{?}{=} 0$$
$$4-28+24 \stackrel{?}{=} 0$$
$$0=0$$

5. $2x^2-7x+6=0$
$2x^2-3x-4x+6=0$
$x(2x-3)-2(2x-3)=0$
$(2x-3)(x-2)=0$
$2x-3=0$ $\quad$ $x-2=0$
$x=\frac{3}{2}$ $\quad$ $x=2$
Check:
$$2\left(\frac{3}{2}\right)^2-7\left(\frac{3}{2}\right)+6 \stackrel{?}{=} 0$$
$$\frac{9}{2}-\frac{21}{2}+\frac{12}{2} \stackrel{?}{=} 0$$
$$0=0$$
$$2(2)^2-7(2)+6 \stackrel{?}{=} 0$$
$$8-14+6 \stackrel{?}{=} 0$$
$$0=0$$

7. $6x^2 - 13x = -6$

$6x^2 - 13x + 6 = 0$

$(3x-2)(2x-3) = 0$

$3x - 2 = 0$ $\quad$ $2x - 3 = 0$

$3x = 2$ $\quad$ $2x = 3$

$x = \frac{2}{3}$ $\quad$ $x = \frac{3}{2}$

Check:

$6\left(\frac{2}{3}\right)^2 - 13\left(\frac{2}{3}\right) \stackrel{?}{=} -6$ $\quad$ $6\left(\frac{3}{2}\right)^2 - 13\left(\frac{3}{2}\right) \stackrel{?}{=} -6$

$\frac{8}{3} - \frac{26}{3} \stackrel{?}{=} -6$ $\quad$ $\frac{27}{2} - \frac{39}{2} \stackrel{?}{=} -6$

$-6 = -6$ $\quad$ $-6 = -6$

9. $x^2 + 13x = 0$

$x(x+13) = 0$

$x = 0$ $\quad$ $x + 13 = 0$

$x = -13$

Check:

$(-13)^2 + (13)(-13) \stackrel{?}{=} 0$ $\quad$ $0^2 + 13(0) \stackrel{?}{=} 0$

$169 - 169 \stackrel{?}{=} 0$ $\quad$ $0 \stackrel{?}{=} 0$

$0 = 0$

11. $8x^2 = 72$

$8x^2 - 72 = 0$

$8(x^2 - 9) = 0$

$8(x-3)(x+3) = 0$

$x - 3 = 0$ $\quad$ $x + 3 = 0$

$x = 3$ $\quad$ $x = -3$

Check:

$8(-3)^2 \stackrel{?}{=} 72$ $\quad$ $8(3)^2 \stackrel{?}{=} 72$

$72 = 72$ $\quad$ $72 = 72$

13. $5x^2 + 3x = 8x$

$5x^2 - 5x = 0$

$5x(x-1) = 0$

$5x = 0$ $\quad$ $x - 1 = 0$

$x = 0$ $\quad$ $x = 1$

Check:

$5(0)^2 + 3(0) \stackrel{?}{=} 8(0)$ $\quad$ $5(1)^2 + 3(1) \stackrel{?}{=} 8(1)$

$0 + 0 \stackrel{?}{=} 0$ $\quad$ $5 + 3 \stackrel{?}{=} 8$

$0 = 0$ $\quad$ $8 = 8$

15. $6x^2 = 16x - 8$

$6x^2 - 16x + 8 = 0$

$2(3x^2 - 8x + 4) = 0$

$2(3x-2)(x-2) = 0$

$3x - 2 = 0$ $\quad$ $x - 2 = 0$

$3x = 2$ $\quad$ $x = 2$

$x = \frac{2}{3}$

Check:

$6\left(\frac{2}{3}\right)^2 \stackrel{?}{=} 16\left(\frac{2}{3}\right) - 8$ $\quad$ $6(2)^2 \stackrel{?}{=} 16(2) - 8$

$\frac{24}{9} \stackrel{?}{=} \frac{32}{3} - \frac{24}{3}$ $\quad$ $24 \stackrel{?}{=} 32 - 8$

$\frac{8}{3} = \frac{8}{3}$ $\quad$ $24 = 24$

17. $(x-5)(x+2) = -4(x+1)$

$x^2 + 2x - 5x - 10 = -4x - 4$

$x^2 + x - 6 = 0$

$(x+3)(x-2) = 0$

$x + 3 = 0$ $\quad$ $x - 2 = 0$

$x = -3$ $\quad$ $x = 2$

Check:

$(-3-5)(-3+2) \stackrel{?}{=} -4(-3+1)$

$(-8)(-1) \stackrel{?}{=} -4(-2)$

$8 = 8$

$(2-5)(2+2) \stackrel{?}{=} -4(2+1)$

$-3(4) \stackrel{?}{=} -4(3)$

$-12 = -12$

19. $4x^2 - 3x + 1 = -7x$

$4x^2 + 4x + 1 = 0$

$(2x+1)(2x+1) = 0$

$2x + 1 = 0$

$2x = -1$

$x = -\frac{1}{2}$

Check: $4\left(-\frac{1}{2}\right)^2 - 3\left(-\frac{1}{2}\right) + 1 \stackrel{?}{=} -7\left(-\frac{1}{2}\right)$

$1 + \frac{3}{2} + 1 \stackrel{?}{=} \frac{7}{2}$

$\frac{7}{2} = \frac{7}{2}$

21. $\frac{x^2}{3} - 5 + x = -5$

$x^2 - 15 + 3x = -15$

$x^2 + 3x = 0$

$x(x+3) = 0$

$x = 0 \qquad x + 3 = 0$

$x = -3$

Check:

$\frac{0^2}{3} - 5 + 0 \overset{?}{=} -5 \qquad \frac{(-3)^2}{3} - 5 - 3 \overset{?}{=} -5$

$-5 = -5 \qquad \frac{9}{3} - 8 \overset{?}{=} -5$

$-5 = -5$

23. $\frac{x^2 + 10x}{8} = -2$

$x^2 + 10x = -16$

$x^2 + 10x + 16 = 0$

$(x+8)(x+2) = 0$

$x + 8 = 0 \qquad x + 2 = 0$

$x = -8 \qquad x = -2$

Check:

$\frac{(-8)^2 + 10(-8)}{8} \overset{?}{=} -2 \qquad \frac{(-2)^2 + 10(-2)}{8} \overset{?}{=} -2$

$\frac{64 - 80}{8} \overset{?}{=} -2 \qquad \frac{4 - 20}{8} \overset{?}{=} -2$

$-2 = -2 \qquad -2 = -2$

25. $\frac{10x^2 - 25x}{12} = 5$

$10x^2 - 25x = 60$

$10x^2 - 25x - 60 = 0$

$5(2x^2 - 5x - 12) = 0$

$5(2x+3)(x-4) = 0$

$2x + 3 = 0 \qquad x - 4 = 0$

$x = -\frac{3}{2} \qquad x = 4$

Check:

$\frac{10\left(-\frac{3}{2}\right) - 25\left(-\frac{3}{2}\right)}{12} \overset{?}{=} 5 \qquad \frac{10(4)^2 - 25(4)}{12} \overset{?}{=} 5$

$\frac{\frac{90}{4} + \frac{150}{4}}{12} \overset{?}{=} 5 \qquad \frac{160 - 100}{12} \overset{?}{=} 5$

$\frac{60}{4} \overset{?}{=} 5 \qquad \frac{60}{4} \overset{?}{=} 5$

$5 = 5 \qquad 5 = 5$

27. An equation in the form $ax^2 + bx = 0$ can always be solved by factoring out x.

29. x = length

$\frac{1}{2}x + 3 = \text{width}$

$x\left(\frac{1}{2}x + 3\right) = 140$

$\frac{x^2}{2} + 3x = 140$

$x^2 + 6x = 280$

$x^2 + 6x - 280 = 0$

$(x+20)(x-14) = 0$

$x + 20 = 0 \qquad x - 14 = 0$

$x = -20 \qquad x = 14$

Width not negative $\qquad \frac{1}{2}x + 3 = 10$

The length is 14 m and the width is 10 m.

31. $G = \frac{x^2 - 3x + 2}{2}$, $x = 13$

$G = \frac{13^2 - 3(13) + 2}{2}$

$G = \frac{169 - 39 + 2}{2}$

$G = 66$

There are 66 possible groups.

33. $G = \frac{x^2 - 3x + 2}{2}$

$72 = x^2 - 3x + 2$

$0 = x^2 - 3x - 70$

$0 = (x-10)(x+7)$

$x - 10 = 0 \qquad x + 7 = 0$

$x = 10 \qquad x = -7$

Not possible

There are 10 students.

35. $S = -15t^2 + vt + h$

$v = 13 \qquad h = 6$

$S = -5t^2 + 13t + 6$

When the ball hits the ground $S = 0$.

$0 = -5t^2 + 13t + 6$

$0 = 5t^2 - 13t - 6$

$0 = (5t+2)(t-3)$

$5t+2=0$ $\quad t=-\frac{2}{5}$ $\qquad t-3=0$ $\quad t=3$

Time can't be negative. The ball will hit the ground at $t = 3$ sec.

$S=-5(2)^2+13(2)+6=-20+26+6=12$

After 2 seconds the ball is 12 m from the ground.

37. $x=70$

$$T=0.5[(70)^2-70] = 0.5(4900-70) = 2415 \text{ telephone calls}$$

These 70 people could make 2415 telephone calls between them.

39. H = number of handshakes

$H=0.5(17^2-17)$
$H=0.5(289-17)$
$H=136$

136 handshakes will take place.

Cumulative Review

41. $(2x^2y^3)(-5x^3y)=(2)(-5)x^{2+3}y^{3+1}=-10x^5y^4$

42. $(3a^4b^5)(4a^6b^8)=(3)(4)a^{4+6}b^{5+8}=12a^{10}b^{13}$

43. $\dfrac{21a^5b^{10}}{-14ab^{12}}=\dfrac{3a^{5-4}}{-2b^{12-10}}=-\dfrac{3a^4}{2b^2}$

44. $\dfrac{18x^3y^6}{54x^8y^{10}}=\dfrac{1}{3x^{8-3}y^{10-6}}=\dfrac{1}{3x^5y^4}$

Quick Quiz 5.7

1. $15x^2-8x+1=0$

grouping number 15

$$15x^2-8x+1=15x^2-5x-3x+1 = 5x(3x-1)-(3x-1) = (5x-1)(3x-1)$$

$5x-1=0$ $\quad 5x=1$ $\quad x=\frac{1}{5}$ $\qquad 3x-1=0$ $\quad 3x=1$ $\quad x=\frac{1}{3}$

2. $4+x(x-2)=7$

$4+x^2-2x=7$
$x^2-2x-3=0$
$(x-3)(x+1)=0$

$x-3=0$ $\quad x=3$ $\qquad x+1=0$ $\quad x=-1$

3. $4x^2=9x+9$

$4x^2-9x-9=0$

grouping number -36

$$4x^2-9x-9=4x^2-12x+3x-9 = 4x(x-3)+3(x-3) = (4x+3)(x-3)$$

$4x+3=0$ $\quad 4x=-3$ $\quad x=-\frac{3}{4}$ $\qquad x-3=0$ $\quad x=3$

4. Answers may vary. Possible solution:
Let x = width of rectangle, then length = $(2x + 3)$.

A = (width)(length)
$65=x(2x+3)$
$65=2x^2+3x$
$0=2x^2+3x-65$
$0=(2x+13)(x-5)$

$2x+13=0$ $\quad 2x=-13$ $\quad x=-\frac{13}{2}$ $\qquad x-5=0$ $\quad x=5$

x cannot be negative in this case, because it describes a length. So, $x = 5$.
length = 2(5) + 3 = 13 feet
width = 5 feet

Putting Your Skills to Work

1. $$PMT=\frac{PV}{\left[1-\frac{1}{(1+I)^N}\right]\Big/I} = \frac{25{,}000}{\left[1-\frac{1}{\left(1+\frac{0.09}{12}\right)^{60}}\right]\Big/\left(\frac{0.09}{12}\right)} = 518.96$$

Megan's monthly payment is \$520.

2. $5 \times 12 \times 520 = 31{,}200$
She will pay about \$31,200 over the next 5 years.

3. $31{,}200 - 25{,}000 = 6200$
She will pay about \$6200 in interest.

4. $PMT = \dfrac{23{,}000}{\left[1 - \dfrac{1}{\left(1+\frac{0.09}{12}\right)^{60}}\right] \Big/ \left(\frac{0.09}{12}\right)} = 477.44$

Megan's monthly payment would be \$480.

5. $480 \times 5 \times 12 = 28{,}800$
The total is about \$28,800.

6. Megan would pay \$25,000, since there is no rebate and no interest rate.

7. Option B would save Megan the most money, since \$25,000 < \$28,800.

8. Answers may vary.

Chapter 5 Review Problems

1. $12x^3 - 20x^2y = 4x^2(3x - 5y)$

2. $10x^3 - 35x^3y = 5x^3(2 - 7y)$

3. $24x^3y - 8x^2y^2 - 16x^3y^3 = 8x^2y(3x - y - 2xy^2)$

4. $45a^3b^3 + 30a^2b^3 - 15a^2b^4 = 15a^2b^3(3a + 2 - b)$

5. $3a^3 + 6a^2 - 9ab + 12a = 3a(a^2 + 2a - 3b + 4)$

6. $2x - 4y + 6z + 12 = 2(x - 2y + 3z + 6)$

7. $2a(a + 3b) - 5(a + 3b) = (a + 3b)(2a - 5)$

8. $15x^3y + 6xy^2 + 3xy = 3xy(5x^2 + 2y + 1)$

9. $2ax + 5a - 8x - 20 = 2ax - 8x + 5a - 20$
$= 2x(a - 4) + 5(a - 4)$
$= (2x + 5)(a - 4)$

10. $a^2 - 4ab + 7a - 28b = a(a - 4b) + 7(a - 4b)$
$= (a + 7)(a - 4b)$

11. $x^2y + 3y - 2x^2 - 6 = y(x^2 + 3) - 2(x^2 + 3)$
$= (x^2 + 3)(y - 2)$

12. $30ax - 15ay + 42x - 21y$
$= 3(10ax - 5ay + 14x - 7y)$
$= 3[5a(2x - y) + 7(2x - y)]$
$= 3(2x - y)(5a + 7)$

13. $15x^2 - 3x + 10x - 2 = 3x(5x - 1) + 2(5x - 1)$
$= (5x - 1)(3x + 2)$

14. $30w^2 - 18w + 5wz - 3z = 6w(5w - 3) + z(5w - 3)$
$= (5w - 3)(6w + z)$

15. $x^2 + 6x - 27 = (x + 9)(x - 3)$

16. $x^2 + 9x - 10 = (x + 10)(x - 1)$

17. $x^2 + 14x + 48 = (x + 6)(x + 8)$

18. $x^2 + 8xy + 15y^2 = (x + 3y)(x + 5y)$

19. $x^4 + 13x^2 + 42 = (x^2 + 6)(x^2 + 7)$

20. $x^4 - 2x^2 - 35 = (x^2 - 7)(x^2 + 5)$

21. $6x^2 + 30x + 36 = 6(x^2 + 5x + 6) = 6(x + 2)(x + 3)$

22. $3x^2 + 39x + 36 = 3(x^2 + 13x + 12)$
$= 3(x + 12)(x + 1)$

23. $2x^2 - 28x + 96 = 2(x^2 - 14x + 48)$
$= 2(x - 8)(x - 6)$

24. $4x^2 - 44x + 120 = 4(x^2 - 11x + 30)$
$= 4(x - 6)(x - 5)$

25. $4x^2 + 7x - 15 = (4x - 5)(x + 3)$

26. $12x^2 + 11x - 5 = 12x^2 - 4x + 15x - 5$
$= 4x(3x - 1) + 5(3x - 1)$
$= (3x - 1)(4x + 5)$

27. $15x^2 + 7x - 4 = (3x - 1)(5x + 4)$

28. $6x^2 - 13x + 6 = 6x^2 - 4x - 9x + 6$
$= 2x(3x-2) - 3(3x-2)$
$= (3x-2)(2x-3)$

29. $2x^2 - x - 3 = (2x-3)(x+1)$

30. $3x^2 + 2x - 8 = (3x-4)(x+2)$

31. $20x^2 + 48x - 5 = (10x-1)(2x+5)$

32. $20x^2 + 21x - 5 = 20x^2 - 4x + 25x - 5$
$= 4x(5x-1) + 5(5x-1)$
$= (5x-1)(4x+5)$

33. $6a^2 + 11a - 10 = (3a-2)(2a+5)$

34. $6a^2 - 19a + 10 = (3a-2)(2a-5)$

35. $6x^2 + 4x - 10 = 2(3x^2 + 2x - 5)$
$= 2(3x+5)(x-1)$

36. $6x^2 - 4x - 10 = 2(3x^2 - 2x - 5)$
$= 2(3x^2 - 5x + 3x - 5)$
$= 2[x(3x-5) + 1(3x-5)]$
$= 2(3x-5)(x+1)$

37. $4x^2 - 26x + 30 = 2(2x^2 - 13x + 15)$
$= 2(2x-3)(x-5)$

38. $4x^2 - 20x - 144 = 4(x^2 - 5x - 36)$
$= 4(x-9)(x+4)$

39. $16x^2 + 16x - 60 = 4(4x^2 + 4x - 15)$
$= 4(2x-3)(2x+5)$

40. $24x^2 + 6x - 3 = 3(8x^2 + 2x - 1)$
$= 3(4x-1)(2x+1)$

41. $12x^2 + xy - 6y^2 = (4x+3y)(3x-2y)$

42. $6x^2 + 5xy - 25y^2 = (3x-5y)(2x+5y)$

43. $49x^2 - y^2 = (7x)^2 - (y)^2 = (7x+y)(7x-y)$

44. $16x^2 - 36y^2 = 4(4x^2 - 9y^2)$
$= 4[(2x)^2 - (3y)^2]$
$= 4(2x-3y)(2x+3y)$

45. $36x^2 - 60x + 25 = (6x)^2 - 2(6x)(5) + (5)^2$
$= (6x-5)^2$

46. $16x^2 + 40x + 25 = (4x)^2 + 2(4x)(5) + 5^2$
$= (4x+5)^2$

47. $25x^2 - 36 = (5x)^2 - (6)^2 = (5x+6)(5x-6)$

48. $100x^2 - 9 = (10x)^2 - 3^2 = (10x-3)(10x+3)$

49. $y^2 - 36x^2 = y^2 - (6x)^2 = (y+6x)(y-6x)$

50. $9y^2 - 25x^2 = (3y)^2 - (5x)^2 = (3y+5x)(3y-5x)$

51. $36x^2 + 12x + 1 = (6x)^2 + 2(6x)(1) + 1^2 = (6x+1)^2$

52. $25x^2 - 20x + 4 = (5x)^2 - 2(5x)(2) + 2^2$
$= (5x-2)^2$

53. $16x^2 - 24xy + 9y^2 = (4x)^2 - 2(4x)(3y) + (3y)^2$
$= (4x-3y)^2$

54. $49x^2 - 28xy + 4y^2 = (7x)^2 - 2(7x)(2y) + (2y)^2$
$= (7x-2y)^2$

55. $2x^2 - 32 = 2(x^2 - 16)$
$= 2[(x)^2 - (4)^2]$
$= 2(x-4)(x+4)$

56. $3x^2 - 27 = 3(x^2 - 9)$
$= 3[(x)^2 - (3)^2]$
$= 3(x-3)(x+3)$

57. $28x^2 + 140x + 175 = 7(4x^2 + 20x + 25)$
$= 7[(2x)^2 + 2(2x)(5) + 5^2]$
$= 7(2x+5)^2$

58. $72x^2 - 192x + 128 = 8(9x^2 - 24x + 16)$
$= 8(3x-4)^2$

59. $4x^2 - 9y^2 = (2x)^2 - (3y)^2 = (2x+3y)(2x-3y)$

60. $x^2 + 6x + 9 = x^2 + 2(x)(3) + 3^2 = (x+3)^2$

61. $x^2 - 9x + 18 = (x-6)(x-3)$

62. $x^2 + 13x - 30 = (x+15)(x-2)$

63. $$\begin{aligned} 5x^2 - 13x - 6 &= 5x^2 - 15x + 2x - 6 \\ &= 5x(x-3) + 2(x-3) \\ &= (5x+2)(x-3) \end{aligned}$$

64. $$\begin{aligned} 9x^2 - 9x - 4 &= 9x^2 + 3x - 12x - 4 \\ &= 3x(3x+1) - 4(3x+1) \\ &= (3x-4)(3x+1) \end{aligned}$$

65. $24x - 60 = 12(2x - 5)$

66. $8x^2y^2 - 4xy = 4xy(2xy - 1)$

67. $$\begin{aligned} 50x^3y^2 + 30x^2y^2 - 10x^2y^2 &= 50x^3y^2 + 20x^2y^2 \\ &= 10x^2y^2(5x+2) \end{aligned}$$

68. $$\begin{aligned} &26a^3b - 13ab^3 + 52a^2b^4 \\ &= 13ab(2a^2 - b^2 + 4ab^3) \end{aligned}$$

69. $x^3 - 16x^2 + 64x = x(x^2 - 16x + 64) = x(x-8)^2$

70. $$\begin{aligned} 2x^2 + 40x + 200 &= 2(x^2 + 20x + 100) \\ &= 2[x^2 + 2(x)(10) + 10^2] \\ &= 2(x+10)^2 \end{aligned}$$

71. $3x^2 - 18x + 27 = 3(x^2 - 6x + 9) = 3(x-3)^2$

72. $$\begin{aligned} 25x^3 - 60x^2 + 36x &= x(25x^2 - 60x + 36) \\ &= x[(5x)^2 - 2(5x)(6) + 6^2] \\ &= x(5x-6)^2 \end{aligned}$$

73. $7x^2 - 9x - 10 = (7x+5)(x-2)$

74. $4x^2 - 13x - 12 = (4x+3)(x-4)$

75. $$\begin{aligned} 9x^3y - 4xy^3 &= xy(9x^2 - 4y^2) \\ &= xy(3x+2y)(3x-2y) \end{aligned}$$

76. $$\begin{aligned} &3x^3a^3 - 11x^4a^2 - 20x^5a \\ &= x^3a(3a^2 - 11xa - 20x^2) \\ &= x^3a(3a^2 + 4xa - 15xa - 20x^2) \\ &= x^3a[a(3a+4x) - 5x(3a+4x)] \\ &= x^3a(3a+4x)(a-5x) \end{aligned}$$

77. $$\begin{aligned} 12a^2 + 14ab - 10b^2 &= 2(6a^2 + 7ab - 5b^2) \\ &= 2(3a+5b)(2a-b) \end{aligned}$$

78. $$\begin{aligned} 121a^2 + 66ab + 9b^2 &= (11a)^2 + 2(11a)(3b) + (3b)^2 \\ &= (11a+3b)^2 \end{aligned}$$

79. $$\begin{aligned} 7a - 7 - ab + b &= 7(a-1) - b(a-1) \\ &= (a-1)(7-b) \end{aligned}$$

80. $$\begin{aligned} 18b - 42 + 3bc - 7c &= 18b + 3bc - 42 - 7c \\ &= 3b(6+c) - 7(6+c) \\ &= (3b-7)(6+c) \end{aligned}$$

81. $$\begin{aligned} 10b + 16 - 24x - 15bx &= 10b - 15bx + 16 - 24x \\ &= 5b(2-3x) + 8(2-3x) \\ &= (5b+8)(2-3x) \end{aligned}$$

82. $$\begin{aligned} 5xb - 35x + 4by - 28y &= 5x(b-7) + 4y(b-7) \\ &= (b-7)(5x+4y) \end{aligned}$$

83. $$\begin{aligned} 2a^2x - 15ax + 7x &= x(2a^2 - 15a + 7) \\ &= x(2a-1)(a-7) \end{aligned}$$

84. $$\begin{aligned} x^5 - 17x^3 + 16x &= x(x^4 - 17x^2 + 16) \\ &= x(x^2 - 16)(x^2 - 1) \\ &= x(x^2 - 4^2)(x^2 - 1^2) \\ &= x(x-4)(x+4)(x-1)(x+1) \end{aligned}$$

85. $$\begin{aligned} x^4 - 81y^{12} &= (x^2)^2 - (9y^6)^2 \\ &= (x^2 + 9y^6)(x^2 - 9y^6) \\ &= (x^2 + 9y^6)(x + 3y^3)(x - 3y^3) \end{aligned}$$

86. $$\begin{aligned} 6x^4 - x^2 - 15 &= 6x^4 - 10x^2 + 9x^2 - 15 \\ &= 2x^2(3x^2 - 5) + 3(3x^2 - 5) \\ &= (3x^2 - 5)(2x^2 + 3) \end{aligned}$$

87. $$\begin{aligned} 28yz - 16xyz + x^2yz &= yz(28 - 16x + x^2) \\ &= yz(14-x)(2-x) \end{aligned}$$

88. $12x^3+17x^2+6x = x(12x^2+17x+6)$
$= x(12x^2+8x+9x+6)$
$= x[4x(3x+2)+3(3x+2)]$
$= x(3x+2)(4x+3)$

89. $16w^2-2w-5 = (2w+1)(8w-5)$

90. $12w^2-12w+3 = 3(4w^2-4w+1)$
$= 3[(2w)^2-2(2w)(1)+1^2]$
$= 3(2w-1)^2$

91. $4y^3+10y^2-6y = 2y(2y^2+5y-3)$
$= 2y(2y-1)(y+3)$

92. $10y^2+33y-7 = 10y^2-2y+35y-7$
$= 2y(5y-1)+7(5y-1)$
$= (5y-1)(2y+7)$

93. $8y^{10}-16y^8 = 8y^8(y^2-2)$

94. $9x^4-144 = 9(x^4-16)$
$= 9[(x^2)^2-(4)^2]$
$= 9(x^2+4)(x^2-4)$
$= 9(x^2+4)(x+2)(x-2)$

95. $x^2-6x+12$ is prime.

96. $8x^2-19x-6$ is prime.

97. $8y^5+4y^3-60y = 4y(2y^4+y^2-15)$
$= 4y(2y^2-5)(y^2+3)$

98. $9xy^2+3xy-42x = 3x(3y^2+y-14)$
$= 3x(3y^2+7y-6y-14)$
$= 3x[y(3y+7)-2(3y+7)]$
$= 3x(3y+7)(y-2)$

99. $16x^4y^2-56x^2y+49$
$= (4x^2y)^2-2(4x^2y)(7)+(7)^2$
$= (4x^2y-7)^2$

100. $128x^3y-2xy = 2xy(64x^2-1)$
$= 2xy[(8x)^2-1^2]$
$= 2xy(8x-1)(8x+1)$

101. $2ax+5a-10b-4bx = 2ax+5a-4bx-10b$
$= a(2x+5)-2b(2x+5)$
$= (2x+5)(a-2b)$

102. $2x^3-9+x^2-18x = 2x^3+x^2-18x-9$
$= x^2(2x+1)-9(2x+1)$
$= (2x+1)(x^2-9)$
$= (2x+1)(x^2-3^2)$
$= (2x+1)(x-3)(x+3)$

103. $x^2+x-20=0$
$(x+5)(x-4)=0$
$x+5=0$ $\quad$ $x-4=0$
$x=-5$ $\quad$ $x=4$

104. $2x^2+11x-6=0$
$2x^2+12x-x-6=0$
$2x(x+6)-1(x+6)=0$
$(2x-1)(x+6)=0$
$2x-1=0$ $\quad$ $x+6=0$
$2x=1$ $\quad$ $x=-6$
$x=\frac{1}{2}$

105. $7x^2=15x+x^2$
$7x^2-x^2-15x=0$
$6x^2-15x=0$
$3x(2x-5)=0$
$3x=0$ $\quad$ $2x-5=0$
$x=0$ $\quad$ $2x=5$
$x=\frac{5}{2}$

106. $5x^2-x=4x^2+12$
$x^2-x=12$
$x^2-x-12=0$
$(x-4)(x+3)=0$
$x-4=0$ $\quad$ $x+3=0$
$x=4$ $\quad$ $x=-3$

107. $2x^2+9x-5=0$
$(2x-1)(x+5)=0$
$2x-1=0$ $\quad$ $x+5=0$
$x=\frac{1}{2}$ $\quad$ $x=-5$

108. $x^2+11x+24=0$
$(x+8)(x+3)=0$
$x+8=0 \qquad x+3=0$
$x=-8 \qquad x=-3$

109. $x^2+14x+45=0$
$(x+9)(x+5)=0$
$x+9=0 \qquad x+5=0$
$x=-9 \qquad x=-5$

110. $5x^2=7x+6$
$5x^2-7x-6=0$
$(5x+3)(x-2)=0$
$5x+3=0 \qquad x-2=0$
$5x=-3 \qquad x=2$
$x=-\frac{3}{5}$

111. $3x^2+6x=2x^2-9$
$x^2+6x+9=0$
$(x+3)^2=0$
$x+3=0$
$x=-3$

112. $4x^2+9x-9=0$
$(4x-3)(x+3)=0$
$4x-3=0 \qquad x+3=0$
$4x=3 \qquad x=-3$
$x=\frac{3}{4}$

113. $5x^2-11x+2=0$
$(5x-1)(x-2)=0$
$5x-1=0 \qquad x-2=0$
$x=\frac{1}{5} \qquad x=2$

114. b = base of triangle
a = altitude = b + 5
area $=\frac{1}{2}ba$
$25=\frac{1}{2}b(b+5)$
$25=\frac{1}{2}(b^2+5b)$
$50=b^2+5b$
$0=b^2+5b-50$
$0=(b+10)(b-5)$
$b+10=0 \qquad b-5=0$
$b=-10$ not possible $\qquad b=5$
The base = 5 inches and the altitude = 5 + 5 = 10 inches.

115. w = width of rectangle
l = length = $(2w-4)$
Area $=wl=w(2w-4)$
$20=w(2w-4)$
$30=2w^2-4w$
$0=2w^2-4w-30$
$0=2(w^2-2w-15)$
$0=2(w-5)(w+3)$
$w-5=0 \qquad w+3=0$
$w=5$ feet $\qquad w=-3$ not possible
width = 5
length = 2(5) − 4 = 6
The width is 5 feet and the length is 6 feet.

116. $h=-16t^2+80t+96$
$0=-16t^2+80t+96$
$0=-16(t^2-5t-6)$
$0=-16(t-6)(t+1)$
$t-6=0 \qquad t+1=0$
$t=6 \qquad t=-1$
Since the time must be positive, t = 6 seconds.

117. $480=-5x^2+100x$
$5x^2-100x+480=0$
$5(x^2-20x+96)=0$
$5(x-12)(x-8)=0$
$x-12=0$ or $x-8=0$
$x=12 \qquad x=8$
The current is 12 amperes or 8 amperes.

How Am I Doing? Chapter 5 Test

1. $x^2+12x-28=(x+14)(x-2)$

2. $16x^2-81=(4x)^2-9^2=(4x+9)(4x-9)$

3. $10x^2+27x+5=10x^2+2x+25x+5$
$=2x(5x+1)+5(5x+1)$
$=(5x+1)(2x+5)$

4. $9a^2-30a+25=(3a)^2-2(3a)(5)+(5)^2$
$=(3a-5)^2$

5. $7x - 9x^2 + 14xy = x(7 - 9x + 14y)$

6. $10xy + 15by - 8x - 12b = 5y(2x + 3b) - 4(2x + 3b)$
$= (2x + 3b)(5y - 4)$

7. $6x^3 - 20x^2 + 16x = 2x(3x^2 - 10x + 8)$
$= 2x(3x^2 - 6x - 4x + 8)$
$= 2x[3x(x - 2) - 4(x - 2)]$
$= 2x(x - 2)(3x - 4)$

8. $5a^2c - 11ac + 2c = c(5a^2 - 11a + 2)$
$= c(5a - 1)(a - 2)$

9. $81x^2 - 100 = (9x)^2 - (10)^2 = (9x + 10)(9x - 10)$

10. $9x^2 - 15x + 4 = (3x - 1)(3x - 4)$

11. $20x^2 - 45 = 5(4x^2 - 9)$
$= 5[(2x)^2 - 3^2]$
$= 5(2x + 3)(2x - 3)$

12. $36x^2 + 1 = (6x)^2 + (1)^2$
It is prime.

13. $3x^3 + 11x^2 + 10x = x(3x^2 + 11x + 10)$
$= x(3x + 5)(x + 2)$

14. $60xy^2 - 20x^2y - 45y^3$
$= -5y(-12xy + 4x^2 + 9y^2)$
$= -5y(4x^2 - 12xy + 9y^2)$
$= -5y[(2x)^2 - 2(2x)(3y) + (3y)^2]$
$= -5y(2x - 3y)^2$

15. $81x^2 - 1 = (9x)^2 - (1)^2 = (9x + 1)(9x - 1)$

16. $81y^4 - 1 = (9y^2 + 1)(9y^2 - 1)$
$= (9y^2 + 1)(3y + 1)(3y - 1)$

17. $2ax + 6a - 5x - 15 = 2a(x + 3) - 5(x + 3)$
$= (x + 3)(2a - 5)$

18. $aw^2 - 8b + 2bw^2 - 4a = aw^2 - 4a + 2bw^2 - 8b$
$= a(w^2 - 4) + 2b(w^2 - 4)$
$= (w^2 - 4)(a + 2b)$
$= (w - 2)(w + 2)(a + 2b)$

19. $3x^2 - 3x - 90 = 3(x^2 - x - 30) = 3(x - 6)(x + 5)$

20. $2x^3 - x^2 - 15x = x(2x^2 - x - 15)$
$= x(2x^2 - 6x + 5x - 15)$
$= x[2x(x - 3) + 5(x - 3)]$
$= x(x - 3)(2x + 5)$

21. $x^2 + 14x + 45 = 0$
$(x + 9)(x + 5) = 0$
$x + 9 = 0$ $x + 5 = 0$
$x = -9$ $x = -5$

22. $14 + 3x(x + 2) = -7x$
$14 + 3x^2 + 6x = -7x$
$3x^2 + 13x + 14 = 0$
$(3x + 7)(x + 2) = 0$
$3x + 7 = 0$ $x + 2 = 0$
$3x = -7$ $x = -2$
$x = -\frac{7}{3}$

23. $2x^2 + x - 10 = 0$
$(2x + 5)(x - 2) = 0$
$2x + 5 = 0$ $x - 2 = 0$
$x = -\frac{5}{2}$ $x = 2$

24. $x^2 - 3x - 28 = 0$
$(x - 7)(x + 4) = 0$
$x - 7 = 0$ $x + 4 = 0$
$x = 7$ $x = -4$

25. x = width
$2x - 1$ = length
$x(2x - 1) = 91$
$2x^2 - x = 91$
$2x^2 - x - 91 = 0$
$(2x + 13)(x - 7) = 0$
$2x + 13 = 0$ $x - 7 = 0$
$2x = -13$ $x = 7$
$x = -\frac{13}{2}$
Since the width cannot be negative, $x = 7$ and $2x - 1 = 2(7) - 1 = 13$.
The width = 7 miles, and the length = 13 miles.

Cumulative Test Chapters 0–5

1. $x = \frac{60}{300} = 0.20$
60 is 20% of 300.

2. $0.12(4.5) = x$
$0.54 = x$
0.54 is 12% of 4.5.

3. $4.3 - 2(3.4 - 7.2) - 6.1 = 4.3 - 2(-3.8) - 6.1$
$= 4.3 + 7.6 - 6.1$
$= 5.8$

4. $(-2x^3y^4)(-4xy^6) = (-2)(-4)x^{3+1}y^{4+6} = 8x^4y^{10}$

5. $(-3)^4 = (-3)(-3)(-3)(-3) = 81$

6. $(9x - 4)(3x + 2) = 27x^2 + 18x - 12x - 8$
$= 27x^2 + 6x - 8$

7. $(2x^2 - 6x + 1)(x - 3)$
$= 2x^3 - 6x^2 + x - 6x^2 + 18x - 3$
$= 2x^3 - 12x^2 + 19x - 3$

8. $5x - 6 \le 7x + 3$
$5x - 7x \le 3 + 6$
$-2x \le 9$
$x \ge -\frac{9}{2}$

9. $3x - (7 - 5x) = 3(4x - 5)$
$3x - 7 + 5x = 12x - 15$
$-7 + 8x = 12x - 15$
$8 = 4x$
$2 = x$

10. $\frac{1}{2}x - 3 = \frac{1}{4}(3x + 3)$
$4\left(\frac{1}{2}x - 3\right) = 4\left[\frac{1}{4}(3x + 3)\right]$
$2x - 12 = 3x + 3$
$-12 = x + 3$
$-15 = x$

11. $3(q + 1) = 8q - p$
$3q + 3 = 8q - p$
$p + 3 = 5q$
$\frac{p + 3}{5} = q$

12. $6x^2 - 5x + 1 = 6x^2 - 3x - 2x + 1$
$= 3x(2x - 1) - 1(2x - 1)$
$= (2x - 1)(3x - 1)$

13. $6x^2 + 5x - 4 = (3x + 4)(2x - 1)$

14. $9x^2 + 3x - 2 = 9x^2 + 6x - 3x - 2$
$= 3x(3x + 2) - 1(3x + 2)$
$= (3x + 2)(3x - 1)$

15. $121x^2 - 64y^2 = (11x)^2 - (8y)^2$
$= (11x + 8y)(11x - 8y)$

16. $4x + 120 - 80x^2 = -80x^2 + 4x + 120$
$= -4(20x^2 - x - 30)$
$= -4(20x^2 + 24x - 25x - 30)$
$= -4[4x(5x + 6) - 5(5x + 6)]$
$= -4(5x + 6)(4x - 5)$

17. $x^2 + 5x + 9$ is prime.

18. $16x^3 + 40x^2 + 25x = x(16x^2 + 40x + 25)$
$= x[(4x)^2 + 2(4x)(5) + 5^2]$
$= x(4x + 5)^2$

19. $16x^4 - b^4 = (4x^2)^2 - (b^2)^2$
$= (4x^2 - b^2)(4x^2 + b^2)$
$= [(2x)^2 - (b)^2](4x^2 + b^2)$
$= (2x - b)(2x + b)(4x^2 + b^2)$

20. $2ax - 4bx + 3a - 6b = 2x(a - 2b) + 3(a - 2b)$
$= (a - 2b)(2x + 3)$

21. $15x^2 + 6x - 9 = 3(5x^2 + 2x - 3)$
$= 3(5x - 3)(x + 1)$

22. $x^2 + 5x - 36 = 0$
$(x - 4)(x + 9) = 0$
$x - 4 = 0$ $\quad$ $x + 9 = 0$
$x = 4$ $\quad$ $x = -9$

23. $3x^2 - 11x + 10 = 0$
$(3x - 5)(x - 2) = 0$
$3x - 5 = 0$ $\quad$ $x - 2 = 0$
$x = \frac{5}{3}$ $\quad$ $x = 2$

24. x = length
$x - 5$ = width
$A = LW$
$300 = x(x-5)$
$300 = x^2 - 5x$
$0 = x^2 - 5x - 300$
$0 = (x-20)(x+15)$
$x - 20 = 0$ $\quad$ $x + 15 = 0$
$x = 20$ $\quad$ $x = -15$
Not possible
$x - 5 = 15$
The length is 20 feet and the width is 15 feet.

Chapter 6

6.1 Exercises

1. $\frac{3a-9b}{a-3b}=\frac{3(a-3b)}{a-3b}=3$

3. $\frac{6x+18}{x^2+3x}=\frac{6(x+3)}{x(x+3)}=\frac{6}{x}$

5. $\frac{9x^2+6x+1}{1-9x^2}=\frac{(3x+1)(3x+1)}{(1+3x)(1-3x)}=\frac{3x+1}{1-3x}$

7. $\frac{3a^2b(a-2b)}{6ab^2}=\frac{a(a-2b)}{2b}$

9. $\frac{x^2+x-2}{x^2-x}=\frac{(x+2)(x-1)}{x(x-1)}=\frac{x+2}{x}$

11. $\frac{x^2-3x-10}{3x^2+5x-2}=\frac{(x-5)(x+2)}{(3x-1)(x+2)}=\frac{x-5}{3x-1}$

13. $\frac{x^2+4x-21}{x^3-49x}=\frac{(x+7)(x-3)}{x(x^2-49)}$
$=\frac{(x+7)(x-3)}{x(x+7)(x-7)}$
$=\frac{x-3}{x(x-7)}$

15. $\frac{3x^2-11x-4}{x^2+x-20}=\frac{(3x+1)(x-4)}{(x+5)(x-4)}=\frac{3x+1}{x+5}$

17. $\frac{3x^2-8x+5}{4x^2-5x+1}=\frac{(3x-5)(x-1)}{(4x-1)(x-1)}=\frac{3x-5}{4x-1}$

19. $\frac{5x^2-27x+10}{5x^2+3x-2}=\frac{(5x-2)(x-5)}{(5x-2)(x+1)}=\frac{x-5}{x+1}$

21. $\frac{12-3x}{5x^2-20x}=\frac{-3(-4+x)}{5x(x-4)}=\frac{-3}{5x}=-\frac{3}{5x}$

23. $\frac{2x^2-7x-15}{25-x^2}=\frac{(2x+3)(x-5)}{(5-x)(5+x)}$
$=\frac{-2x-3}{x+5}$
$=\frac{-2x-3}{5+x}$

25. $\frac{(3x+4)^2}{9x^2+9x-4}=\frac{(3x+4)(3x+4)}{(3x+4)(3x-1)}=\frac{3x+4}{3x-1}$

27. $\frac{2x^2+9x-18}{30-x-x^2}=\frac{2x^2+9x-18}{-(x^2+x-30)}$
$=\frac{(2x-3)(x+6)}{-(x+6)(x-5)}$
$=\frac{2x-3}{-x+5}$
$=\frac{2x-3}{5-x}$

29. $\frac{a^2+2ab-3b^2}{2a^2+5ab-3b^2}=\frac{(a+3b)(a-b)}{(2a-b)(a+3b)}=\frac{a-b}{2a-b}$

Cumulative Review

31. $(3x-7)^2=(3x)^2-2(3x)(7)+(7)^2$
$=9x^2-42x+49$

32. $(7x+6y)(7x-6y)=(7x)^2-(6y)^2$
$=49x^2-36y^2$

33. $(2x+3)(x-4)(x-2)$
$=(2x+3)(x^2-6x+8)$
$=2x^3-12x^2+16x+3x^2-18x+24$
$=2x^3-9x^2-2x+24$

34. $(x^2-3x-5)(2x^2+x+3)$
$=2x^4+x^3+3x^2-6x^3-3x^2-9x-10x^2$
$-5x-15$
$=2x^4-5x^3-10x^2-14x-15$

35. $\frac{4\frac{7}{8}}{3} = \frac{\frac{39}{8}}{\frac{3}{1}} = \frac{39}{8} \div \frac{3}{1} = \frac{39}{8} \cdot \frac{1}{3} = \frac{13}{8} = 1\frac{5}{8}$

Each lot will be $1\frac{5}{8}$ acres.

36. $17.5 \text{ lb} \cdot \frac{22 \text{ min}}{\text{lb}} = 385$

The cooking time is 385 minutes, or 6 hours, 25 minutes.

37. Let x = number of Spanish speakers in Spain
$x = 22{,}500{,}000 + 17{,}000{,}000$
$= 39{,}500{,}000$
39,500,000 people speak Spanish in Spain.

38. Let x = number of Spanish speakers in Mexico
$x = (3)22{,}500{,}000 + 12{,}500{,}000$
$= 80{,}000{,}000$
80,000,000 people speak Spanish in Mexico.

Quick Quiz 6.1

1. $$\frac{x^3+3x^2}{x^3-2x^2-15x} = \frac{x^2(x+3)}{x(x^2-2x-15)} = \frac{x^2(x+3)}{x(x+3)(x-5)} = \frac{x}{x-5}$$

2. $$\frac{6-2ab}{ab^2-3b} = \frac{2(3-ab)}{b(ab-3)} = \frac{2(3-ab)}{-b(3-ab)} = -\frac{2}{b}$$

3. $$\frac{8x^2+6x-5}{16x^2+40x+25} = \frac{8x^2-4x+10x-5}{16x^2+20x+20x+25} = \frac{4x(2x-1)+5(2x-1)}{4x(4x+5)+5(4x+5)} = \frac{(4x+5)(2x-1)}{(4x+5)(4x+5)} = \frac{2x-1}{4x+5}$$

4. Answers may vary. Possible solution: Completely factoring both numerator and denominator is the only way to see what factors are shared, and may consequently be eliminated. In this case, it can be seen that $(x - y)$ is a common factor.

6.2 Exercises

1. Before multiplying rational expressions, we should always first try to factor the numerator and denominator completely and divide out any common factors.

3. $$\frac{3x+6}{x-3} \cdot \frac{x^2+2x-15}{x^2+4x+4} = \frac{3(x+2)}{x-3} \cdot \frac{(x+5)(x-3)}{(x+2)(x+2)} = \frac{3(x+5)}{x+2}$$

5. $$\frac{24x^3}{4x^2-36} \cdot \frac{2x^2+6x}{16x^2} = \frac{3(8x^3)}{4(x^2-9)} \cdot \frac{2x(x+3)}{2 \cdot 8x^2} = \frac{3 \cdot 8x^3}{4(x+3)(x-3)} \cdot \frac{2x(x+3)}{2 \cdot 8x^2} = \frac{3x^2}{4(x-3)}$$

7. $$\frac{x^2+3x-10}{x^2+x-20} \cdot \frac{x^2-3x-4}{x^2+4x+3} = \frac{(x-2)(x+5)}{(x-4)(x+5)} \cdot \frac{(x-4)(x+1)}{(x+3)(x+1)} = \frac{x-2}{x+3}$$

9. $$\frac{x+6}{x-8} \div \frac{x+5}{x^2-6x-16} = \frac{x+6}{x-8} \cdot \frac{x^2-6x-16}{x+5} = \frac{x+6}{x-8} \cdot \frac{(x-8)(x+2)}{x+5} = \frac{(x+6)(x+2)}{x+5}$$

11. $$(5x+4) \div \frac{25x^2-16}{5x^2+11x-12} = \frac{5x+4}{1} \cdot \frac{5x^2+11x-12}{25x^2-16} = \frac{5x+4}{1} \cdot \frac{(x+3)(5x-4)}{(5x+4)(5x-4)} = x+3$$

13. $\frac{3x^2+12xy+12y^2}{x^2+4xy+3y^2} \div \frac{4x+8y}{x+y}$
$= \frac{3(x+2y)(x+2y)}{(x+3y)(x+y)} \div \frac{4(x+2y)}{x+y}$
$= \frac{3(x+2y)(x+2y)}{(x+3y)(x+y)} \cdot \frac{x+y}{4(x+2y)}$
$= \frac{3(x+2y)}{4(x+3y)}$

15. $\frac{(x+5)^2}{3x^2-7x+2} \cdot \frac{x^2-4x+4}{x+5}$
$= \frac{(x+5)(x+5)}{(3x-1)(x-2)} \cdot \frac{(x-2)(x-2)}{x+5}$
$= \frac{(x+5)(x-2)}{3x-1}$

17. $\frac{x^2+x-30}{10-2x} \div \frac{x^2+4x-12}{5x+15}$
$= \frac{x^2+x-30}{10-2x} \cdot \frac{5x+15}{x^2+4x-12}$
$= \frac{(x+6)(x-5)}{-2(x-5)} \cdot \frac{5(x+3)}{(x+6)(x-2)}$
$= -\frac{5(x+3)}{2(x-2)}$

19. $\frac{y^2+4y-12}{y^2+2y-24} \cdot \frac{y^2-16}{y^2+2y-8}$
$= \frac{(y+6)(y-2)}{(y+6)(y-4)} \cdot \frac{(y+4)(y-4)}{(y+4)(y-2)}$
$= 1$

21. $\frac{x^2+7x+12}{2x^2+9x+4} \div \frac{x^2+6x+9}{2x^2-x-1}$
$= \frac{x^2+7x+12}{2x^2+9x+4} \cdot \frac{2x^2-x-1}{x^2+6x+9}$
$= \frac{(x+4)(x+3)}{(2x+1)(x+4)} \cdot \frac{(2x+1)(x-1)}{(x+3)(x+3)}$
$= \frac{x-1}{x+3}$

Cumulative Review

23. $6x^2+3x-18 = 5x-2+6x^2$
$3x-18 = 5x-2$
$-18 = 2x-2$
$-16 = 2x$
$-8 = x$

24. $(7x^2-x-1)(x-3)$
$= (7x^2-x-1)(x)+(7x^2-x-1)(-3)$
$= 7x^3-x^2-x-21x^2+3x+3$
$= 7x^3-22x^2+2x+3$

25. Area of road = 8981(90)
Area of sidewalk = 8981(10.5)
Difference in area = 8981(90) − 8981(10.5)
= 8981(79.5)
Difference in cost = $8991(79.5)x = \$713,989.5x$

26. Harold: length = x, width = x
George: length = $x + 3$, width = $x - 2$
$A = LW$
$36 = (x+3)(x-2)$
$36 = x^2+x-6$
$0 = x^2+x-42$
$0 = (x+7)(x-6)$
$x+7=0$ $\quad$ $x-6=0$
$x=-7$ $\quad$ $x=6$
Not possible
The dimensions of their gardens are as follows:
Harold: 6 ft by 6 ft
George: 9 ft by 4 ft

Quick Quiz 6.2

1. $\frac{2x-10}{x-4} \cdot \frac{x^2+5x+4}{x^2-4x-5} = \frac{2(x-5)}{x-4} \cdot \frac{(x+4)(x+1)}{(x-5)(x+1)}$
$= \frac{2(x+4)}{x-4}$

2. $\frac{3x^2-13x-10}{3x^2+2x} \cdot \frac{x^2-25x}{x^2-25}$
$= \frac{(3x+2)(x-5)}{x(3x+2)} \cdot \frac{x(x-25)}{(x-5)(x+5)}$
$= \frac{x-25}{x+5}$

3. $\frac{2x^2-18}{3x^2+3x} \div \frac{x^2+6x+9}{x^2+4x+3}$
$= \frac{2x^2-18}{3x^2+3x} \cdot \frac{x^2+4x+3}{x^2+6x+9}$
$= \frac{2(x-3)(x+3)}{3x(x+1)} \cdot \frac{(x+3)(x+1)}{(x+3)(x+3)}$
$= \frac{2(x-3)}{3x}$

4. Answers may vary. Possible solution: The first step is to change the operation from division to multiplication, by changing the operator and inverting the second fraction. Secondly, all terms must be factored completely. Next, common factors in the numerators and denominators may be canceled. Lastly the multiplication operation is performed.

6.3 Exercises

1. The LCD would be a product that contains each factor. However, any repeated factor in any one denominator must be repeated the greatest number of times it occurs in any one denominator. So the LCD would be $(x+5)(x+3)^2$.

3. $\frac{x}{x+5}+\frac{2x+1}{5+x}=\frac{x+2x+1}{x+5}=\frac{3x+1}{x+5}$

5. $\frac{3x}{x+3}-\frac{x+5}{x+3}=\frac{3x-(x+5)}{x+3}$
$=\frac{3x-x-5}{x+3}$
$=\frac{2x-5}{x+3}$

7. $\frac{8x+3}{5x+7}-\frac{6x+10}{5x+7}=\frac{8x+3-(6x+10)}{5x+7}$
$=\frac{8x+3-6x-10}{5x+7}$
$=\frac{2x-7}{5x+7}$

9. $3a^2b^3,\ ab^2$
$\text{LCD}=3a^2b^3$

11. $18x^2y^5=2\cdot 3^2x^2y^5$
$30x^3y^3=2\cdot 3\cdot 5x^3y^3$
$\text{LCD}=2\cdot 3^2\cdot 5x^3y^5=90x^3y^5$

13. $2x-6=2(x-3)$
$9x-27=9(x-3)$
$\text{LCD}=2\cdot 9(x-3)=18(x-3)$

15. $x+3=x+3$
$x^2-9=(x+3)(x-3)$
$\text{LCD}=(x+3)(x-3)=x^2-9$

17. $3x^2+14x-5=(3x-1)(x+5)$
$9x^2-6x+1=(3x-1)(3x-1)$
$\text{LCD}=(x+5)(3x-1)^2$

19. $\frac{7}{ab}+\frac{3}{b}=\frac{7}{ab}+\frac{3}{b}\cdot\frac{a}{a}=\frac{7}{ab}+\frac{3a}{ab}=\frac{7+3a}{ab}$

21. $\frac{3}{x+7}+\frac{8}{x^2-49}=\frac{3}{x+7}\cdot\frac{x-7}{x-7}+\frac{8}{(x+7)(x-7)}$
$=\frac{3x-21+8}{(x+7)(x-7)}$
$=\frac{3x-13}{(x+7)(x-7)}$

23. $\frac{4y}{y+1}+\frac{y}{y-1}=\frac{4y}{y+1}\cdot\frac{y-1}{y-1}+\frac{y}{y-1}\cdot\frac{y+1}{y+1}$
$=\frac{4y^2-4y}{(y+1)(y-1)}+\frac{y^2+y}{(y+1)(y-1)}$
$=\frac{4y^2-4y+y^2+y}{(y+1)(y-1)}$
$=\frac{5y^2-3y}{(y+1)(y-1)}$

25. $\frac{6}{5a}+\frac{5}{3a+2}=\frac{6}{5a}\cdot\frac{3a+2}{3a+2}+\frac{5}{3a+2}\cdot\frac{5a}{5a}$
$=\frac{6(3a+2)+5\cdot 5a}{5a(3a+2)}$
$=\frac{18a+12+25a}{5a(3a+2)}$
$=\frac{43a+12}{5a(3a+2)}$

27. $\frac{2}{3xy}+\frac{1}{6yz}=\frac{2}{3xy}\cdot\frac{2z}{2z}+\frac{1}{6yz}\cdot\frac{x}{x}$
$=\frac{4z}{6xyz}+\frac{x}{6xyz}$
$=\frac{4z+x}{6xyz}$

29. $\frac{5x+6}{x-3}-\frac{x-2}{2x-6}=\frac{5x+6}{x-3}\cdot\frac{2}{2}-\frac{x-2}{2(x-3)}$
$=\frac{2(5x+6)-(x-2)}{2(x-3)}$
$=\frac{10x+12-x+2}{2(x-3)}$
$=\frac{9x+14}{2(x-3)}$

31. $\frac{3x}{x^2-25}-\frac{2}{x+5}=\frac{3x}{(x+5)(x-5)}-\frac{2}{x+5}\cdot\frac{x-5}{x-5}$
$=\frac{3x-2(x-5)}{(x+5)(x-5)}$
$=\frac{3x-2x+10}{(x+5)(x-5)}$
$=\frac{x+10}{(x+5)(x-5)}$

33. $\frac{a+3b}{2}-\frac{a-b}{5}=\frac{a+3b}{2}\cdot\frac{5}{5}-\frac{a-b}{5}\cdot\frac{2}{2}$
$=\frac{5(a+3b)-2(a-b)}{2\cdot 5}$
$=\frac{5a+15b-2a+2b}{10}$
$=\frac{3a+17b}{10}$

35. $\frac{8}{2x-3}-\frac{6}{x+2}=\frac{8}{2x-3}\cdot\frac{x+2}{x+2}-\frac{6}{x+2}\cdot\frac{2x-3}{2x-3}$
$=\frac{8x+16}{(2x-3)(x+2)}-\frac{12x-18}{(2x-3)(x+2)}$
$=\frac{8x+16-(12x-18)}{(2x-3)(x+2)}$
$=\frac{8x+16-12x+18}{(2x-3)(x+2)}$
$=\frac{-4x+34}{(2x-3)(x+2)}$

37. $\frac{x}{x^2+2x-3}-\frac{x}{x^2-5x+4}$
$=\frac{x}{(x+3)(x-1)}-\frac{x}{(x-4)(x-1)}$
$=\frac{x}{(x+3)(x-1)}\cdot\frac{x-4}{x-4}-\frac{x}{(x-4)(x-1)}\cdot\frac{x+3}{x+3}$
$=\frac{x^2-4x}{(x+3)(x-1)(x-4)}-\frac{x^2+3x}{(x+3)(x-1)(x-4)}$
$=\frac{x^2-4x-(x^2+3x)}{(x+3)(x-1)(x-4)}$
$=\frac{x^2-4x-x^2-3x}{(x+3)(x-1)(x-4)}$
$=\frac{-7x}{(x+3)(x-1)(x-4)}$

39. $\frac{3}{x^2+9x+20}+\frac{1}{x^2+10x+24}$
$=\frac{3}{(x+4)(x+5)}\cdot\frac{x+6}{x+6}+\frac{1}{(x+4)(x+6)}\cdot\frac{x+5}{x+5}$
$=\frac{3x+18+x+5}{(x+4)(x+5)(x+6)}$
$=\frac{4x+23}{(x+4)(x+5)(x+6)}$

41. $\frac{3x-8}{x^2-5x+6}+\frac{x+2}{x^2-6x+8}$
$=\frac{3x-8}{(x-2)(x-3)}+\frac{x+2}{(x-4)(x-2)}$
$=\frac{3x-8}{(x-2)(x-3)}\cdot\frac{x-4}{x-4}+\frac{x+2}{(x-4)(x-2)}\cdot\frac{x-3}{x-3}$
$=\frac{3x^2-20x+32}{(x-2)(x-3)(x-4)}+\frac{x^2-x-6}{(x-2)(x-3)(x-4)}$
$=\frac{3x^2-20x+32+x^2-x-6}{(x-2)(x-3)(x-4)}$
$=\frac{4x^2-21x+26}{(x-2)(x-3)(x-4)}$
$=\frac{(x-2)(4x-13)}{(x-2)(x-3)(x-4)}$
$=\frac{4x-13}{(x-3)(x-4)}$

43. $\frac{6x}{y-2x}-\frac{5x}{2x-y}=\frac{6x}{-(2x-y)}-\frac{5x}{2x-y}$
$=\frac{-6x-5x}{2x-y}$
$=\frac{-11x}{2x-y}$
$=\frac{11x}{y-2x}$

45. $\frac{3y}{8y^2+2y-1}-\frac{5y}{2y^2-9y-5}$
$=\frac{3y}{(4y-1)(2y+1)}-\frac{5y}{(2y+1)(y-5)}$
$=\frac{3y}{(4y-1)(2y+1)}\cdot\frac{y-5}{y-5}-\frac{5y}{(2y+1)(y-5)}\cdot\frac{4y-1}{4y-1}$
$=\frac{3y^2-15y}{(4y-1)(2y+1)(y-5)}-\frac{20y^2-5y}{(4y-1)(2y+1)(y-5)}$
$=\frac{3y^2-15y-(20y^2-5y)}{(4y-1)(2y+1)(y-5)}$
$=\frac{3y^2-15y-20y^2+5y}{(4y-1)(2y+1)(y-5)}$
$=\frac{-17y^2-10y}{(4y-1)(2y+1)(y-5)}$

47. $\frac{4y}{y^2+4y+3}+\frac{2}{y+1}$
$=\frac{4y}{(y+3)(y+1)}+\frac{2}{y+1}$
$=\frac{4y}{(y+3)(y+1)}+\frac{2}{y+1}\cdot\frac{y+3}{y+3}$
$=\frac{4y}{(y+3)(y+1)}+\frac{2y+6}{(y+3)(y+1)}$
$=\frac{4y+2y+6}{(y+3)(y+1)}$
$=\frac{6y+6}{(y+3)(y+1)}$
$=\frac{6(y+1)}{(y+3)(y+1)}$
$=\frac{6}{y+3}$

Cumulative Review

49. $\frac{1}{3}(x-2)+\frac{1}{2}(x+3)=\frac{1}{4}(3x+1)$
$12\left[\frac{1}{3}(x-2)+\frac{1}{2}(x+3)\right]=12\left(\frac{1}{4}\right)(3x+1)$
$4(x-2)+6(x+3)=3(3x+1)$
$4x-8+6x+18=9x+3$
$10x+10=9x+3$
$x+10=3$
$x=-7$

50. $5ax=2(ay-3bc)$
$5ax=2ay-6bc$
$5ax+6bc=2ay$
$\frac{5ax+6bc}{2a}=\frac{2ay}{2a}$
$\frac{5ax+6bc}{2a}=y$

51. $x-\frac{1}{5}x>\frac{1}{2}+\frac{1}{10}x$
$10\left(x-\frac{1}{5}x\right)>10\left(\frac{1}{2}+\frac{1}{10}x\right)$
$10x-2x>5+x$
$8x>5+x$
$7x>5$
$x>\frac{5}{7}$

52. $(3x^3y^4)^4=3^4x^{3\cdot4}y^{4\cdot4}=81x^{12}y^{16}$

53. Let x = number of days.
$1.50(2x)>50$
$3x>50$
$x>\frac{50}{3}$
$x>16\frac{2}{3}$
It is cheaper if you use the subway at least 17 days per month.

54. 5.5% − 0.03% = 5.47%
0.0547(5,200,000) = 284,440
284,440 more people spoke Swedish than Sámi.

Quick Quiz 6.3

1. $\frac{3}{x^2-2x-8}+\frac{2}{x-4}$
$=\frac{3}{(x-4)(x+2)}+\frac{2}{x-4}\cdot\frac{x+2}{x+2}$
$=\frac{3+2x+4}{(x-4)(x+2)}$
$=\frac{7+2x}{(x-4)(x+2)}$

2. $\frac{2x+y}{xy}-\frac{b-y}{by}=\frac{2x+y}{xy}\cdot\frac{b}{b}-\frac{b-y}{by}\cdot\frac{x}{x}$
$=\frac{2bx+by}{bxy}-\frac{bx-xy}{bxy}$
$=\frac{2bx+by-bx+xy}{bxy}$
$=\frac{bx+by+xy}{bxy}$

3. $\frac{2}{x^2-9}+\frac{3}{x^2+7x+12}$
$=\frac{2}{(x-3)(x+3)}\cdot\frac{x+4}{x+4}+\frac{3}{(x+4)(x+3)}\cdot\frac{x-3}{x-3}$
$=\frac{2x+8+3x-9}{(x+4)(x+3)(x-3)}$
$=\frac{5x-1}{(x+4)(x+3)(x-3)}$

4. Answers may vary. Possible solution: First factor each denominator completely. The LCD will be the product containing each different factor. If a factor occurs more than once in any one denominator, the LCD will contain that factor repeated the greatest number of times that it occurs in any one denominator.

How Am I Doing? Sections 6.1–6.3

1. $\frac{8x-48}{x^2-6x}=\frac{8(x-6)}{x(x-6)}=\frac{8}{x}$

2. $\frac{2x^2-7x-15}{x^2-12x+35}=\frac{(2x+3)(x-5)}{(x-5)(x-7)}=\frac{2x+3}{x-7}$

3. $\frac{y^2+2y+1}{2x^2-2x^2y^2}=\frac{(y+1)(y+1)}{-2x^2(y^2-1)}$
$=\frac{(y+1)(y+1)}{-2x^2(y-1)(y+1)}$
$=\frac{y+1}{2x^2(1-y)}$

4. $\frac{5x^2-23x+12}{5x^2+7x-6}=\frac{(5x-3)(x-4)}{(5x-3)(x+2)}=\frac{x-4}{x+2}$

5. $\frac{12a^2}{2x+10}\cdot\frac{8x+40}{16a^3}=\frac{12a^2}{2(x+5)}\cdot\frac{8(x+5)}{16a^3}$
$=\frac{6a^2}{x+5}\cdot\frac{x+5}{2a^3}$
$=\frac{6a^2}{2a^3}$
$=\frac{3}{a}$

6. $\frac{x-5}{x^2+5x-14}\cdot\frac{x^2+12x+35}{15-3x}$
$=\frac{x-5}{(x+7)(x-2)}\cdot\frac{(x+7)(x+5)}{3(5-x)}$
$=-\frac{x+5}{3(x-2)}$

7. $\frac{x^2-9}{2x+6}\div\frac{2x^2-5x-3}{4x^2-1}$
$=\frac{x^2-9}{2x+6}\cdot\frac{4x^2-1}{2x^2-5x-3}$
$=\frac{(x+3)(x-3)}{2(x+3)}\cdot\frac{(2x+1)(2x-1)}{(2x+1)(x-3)}$
$=\frac{2x-1}{2}$

8. $\frac{3a^2+7a+2}{4a^2+11a+6}\div\frac{6a^2-13a-5}{16a^2-9}$
$=\frac{3a^2+7a+2}{4a^2+11a+6}\cdot\frac{16a^2-9}{6a^2-13a-5}$
$=\frac{(3a+1)(a+2)}{(4a+3)(a+2)}\cdot\frac{(4a+3)(4a-3)}{(2a-5)(3a+1)}$
$=\frac{4a-3}{2a-5}$

9. $$\frac{x-3y}{xy}-\frac{4a-y}{ay}=\frac{x-3y}{xy}\cdot\frac{a}{a}-\frac{4a-y}{ay}\cdot\frac{x}{x}$$
$$=\frac{a(x-3y)-x(4a-y)}{axy}$$
$$=\frac{ax-3ay-4ax+xy}{axy}$$
$$=\frac{xy-3ax-3ay}{axy}$$

10. $$\frac{7}{2x-4}+\frac{-14}{x^2-4}$$
$$=\frac{7}{2(x-2)}\cdot\frac{x+2}{x+2}+\frac{-14}{(x+2)(x-2)}\cdot\frac{2}{2}$$
$$=\frac{7(x+2)-14\cdot 2}{2(x+2)(x-2)}$$
$$=\frac{7x+14-28}{2(x+2)(x-2)}$$
$$=\frac{7x-14}{2(x+2)(x-2)}$$
$$=\frac{7(x-2)}{2(x+2)(x-2)}$$
$$=\frac{7}{2(x+2)}$$

11. $$\frac{2x}{x^2+10x+21}+\frac{x-3}{x+7}$$
$$=\frac{2x}{(x+3)(x+7)}+\frac{x-3}{x+7}\cdot\frac{x+3}{x+3}$$
$$=\frac{2x+(x-3)(x+3)}{(x+3)(x+7)}$$
$$=\frac{2x+x^2-9}{(x+3)(x+7)}$$
$$=\frac{x^2+2x-9}{(x+3)(x+7)}$$

12. $$\frac{4}{x^2-2x-3}-\frac{5x}{x^2+5x+4}$$
$$=\frac{4}{(x-3)(x+1)}\cdot\frac{x+4}{x+4}-\frac{5x}{(x+4)(x+1)}\cdot\frac{x-3}{x-3}$$
$$=\frac{4x+16-5x^2+15x}{(x-3)(x+1)(x+4)}$$
$$=\frac{-5x^2+19x+16}{(x-3)(x+1)(x+4)}$$

6.4 Exercises

1. $$\frac{\frac{3}{x}}{\frac{2}{x^2}+\frac{5}{x}}=\frac{\frac{3}{x}}{\frac{2}{x^2}+\frac{5}{x}}\cdot\frac{x^2}{x^2}=\frac{\frac{3}{x}\cdot x^2}{\frac{2}{x^2}\cdot x^2+\frac{5}{x}\cdot x^2}=\frac{3x}{2+5x}$$

3. $$\frac{\frac{3}{a}+\frac{3}{b}}{\frac{3}{ab}}=\frac{\frac{3}{a}+\frac{3}{b}}{\frac{3}{ab}}\cdot\frac{ab}{ab}$$
$$=\frac{\frac{3}{a}\cdot ab+\frac{3}{b}\cdot ab}{\frac{3}{ab}\cdot ab}$$
$$=\frac{3b+3a}{3}$$
$$=\frac{3(b+a)}{3}$$
$$=b+a$$

5. $$\frac{\frac{x}{6}-\frac{1}{3}}{\frac{2}{3x}+\frac{5}{6}}=\frac{\frac{x}{6}(6x)-\frac{1}{3}(6x)}{\frac{2}{3x}(6x)+\frac{5}{6}(6x)}=\frac{x^2-2x}{4+5x}$$

7. $$\frac{\frac{7}{5x}-\frac{1}{x}}{\frac{3}{5}+\frac{2}{x}}=\frac{\frac{7}{5x}(5x)-\frac{1}{x}(5x)}{\frac{3}{5}(5x)+\frac{2}{x}(5x)}=\frac{7-5}{3x+10}=\frac{2}{3x+10}$$

9. $$\frac{\frac{5}{x}+\frac{3}{y}}{3x+5y}=\frac{\frac{5}{x}+\frac{3}{y}}{3x+5y}\cdot\frac{xy}{xy}$$
$$=\frac{\frac{5}{x}\cdot xy+\frac{3}{y}\cdot xy}{(3x+5y)xy}$$
$$=\frac{5y+3x}{(3x+5y)xy}$$
$$=\frac{1}{xy}$$

11. $$\frac{4-\frac{1}{x^2}}{2+\frac{1}{x}}=\frac{4-\frac{1}{x^2}}{2+\frac{1}{x}}\cdot\frac{x^2}{x^2}$$
$$=\frac{4\cdot x^2-\frac{1}{x^2}\cdot x^2}{2x^2+\frac{1}{x}\cdot x^2}$$
$$=\frac{4x^2-1}{2x^2+x}$$
$$=\frac{(2x+1)(2x-1)}{x(2x+1)}$$
$$=\frac{2x-1}{x}$$

13. $\dfrac{\frac{2}{x+6}}{\frac{2}{x-6}-\frac{2}{x^2-36}} = \dfrac{\frac{2}{x+6}}{\frac{2}{x-6}-\frac{2}{(x-6)(x+6)}} \cdot \dfrac{(x-6)(x+6)}{(x-6)(x+6)}$

$= \dfrac{\frac{2(x-6)(x+6)}{(x+6)}}{\frac{2(x-6)(x+6)}{(x-6)} - \frac{2(x-6)(x+6)}{(x-6)(x+6)}}$

$= \dfrac{2(x-6)}{2(x+6)-2}$

$= \dfrac{2(x-6)}{2[(x+6)-1]}$

$= \dfrac{x-6}{x+6-1}$

$= \dfrac{x-6}{x+5}$

15. $\dfrac{a+\frac{3}{a}}{\frac{a^2+2}{3a}} = \dfrac{a+\frac{3}{a}}{\frac{a^2+2}{3a}} \cdot \dfrac{3a}{3a} = \dfrac{a(3a)+\left(\frac{3}{a}\right)(3a)}{\left(\frac{a^2+2}{3a}\right)(3a)} = \dfrac{3a^2+9}{a^2+2}$

17. $\dfrac{\frac{3}{x-3}}{\frac{1}{x^2-9}+\frac{2}{x+3}}$

$= \dfrac{\frac{3}{x-3}}{\frac{1}{x^2-9}+\frac{2}{x+3}} \cdot \dfrac{(x+3)(x-3)}{(x+3)(x-3)}$

$= \dfrac{\left(\frac{3}{x-3}\right)\frac{(x+3)(x-3)}{1}}{\frac{1}{(x+3)(x-3)}\frac{(x+3)(x-3)}{1} + \left(\frac{2}{x+3}\right)\frac{(x+3)(x-3)}{1}}$

$= \dfrac{3(x+3)}{1+2(x-3)}$

$= \dfrac{3(x+3)}{1+2x-6}$

$= \dfrac{3x+9}{2x-5}$

19. $\dfrac{\frac{2}{y-1}+2}{\frac{2}{y+1}-2} = \dfrac{\frac{2}{y-1}+2}{\frac{2}{y+1}-2} \cdot \dfrac{(y-1)(y+1)}{(y-1)(y+1)}$

$= \dfrac{\frac{2(y-1)(y+1)}{y-1} + 2(y-1)(y+1)}{\frac{2(y-1)(y+1)}{(y+1)} - 2(y-1)(y+1)}$

$= \dfrac{2(y+1)+2(y^2-1)}{2(y-1)-2(y^2-1)}$

$= \dfrac{2y+2+2y^2-2}{2y-2-2y^2+2}$

$= \dfrac{2y(y+1)}{2y(-y+1)}$

$= \dfrac{y+1}{-y+1}$

21. No expression in any denominator can be allowed to be zero, since division by zero is undefined. So −3, 5, and 0 are not allowable replacements for the variable *x*.

Cumulative Review

23.
$$P = 2(l+w)$$
$$P = 2l+2w$$
$$P-2l = 2w$$
$$\frac{P-2l}{2} = w$$

24.
$$7+x < 11+5x$$
$$7 < 11+4x$$
$$-4 < 4x$$
$$\frac{-4}{4} < \frac{4x}{4}$$
$$-1 < x \text{ or } x > -1$$

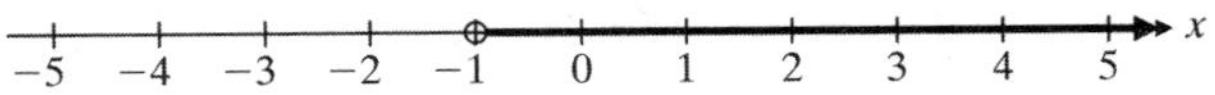

25.
$$2x-9 = \frac{x}{2}$$
$$2(2x-9) = 2\left(\frac{x}{2}\right)$$
$$4x-18 = x$$
$$3x = 18$$
$$x = 6$$

The number is 6.

26. Increase = 5.5 − 3.5 = 2.0

$\frac{2.0}{3.5} \approx 0.57 = 57\%$ increase

This is an increase of approximately 57%.

Quick Quiz 6.4

1. $\frac{\frac{a}{4b}-\frac{1}{3}}{\frac{5}{4b}-\frac{4}{a}} = \frac{\frac{a}{4b}-\frac{1}{3}}{\frac{5}{4b}-\frac{4}{a}} \cdot \frac{12ab}{12ab}$

$= \frac{\frac{a}{4b}\cdot 12ab - \frac{1}{3}\cdot 12ab}{\frac{5}{46}\cdot 12ab - \frac{4}{a}\cdot 12ab}$

$= \frac{3a^2 - 4ab}{15a - 48b}$

$= \frac{a(3a-4b)}{3(5a-16b)}$

2. $\frac{a+b}{\frac{1}{a}+\frac{1}{b}} = \frac{a \cdot ab + b \cdot ab}{\frac{1}{a}\cdot ab + \frac{1}{b}\cdot ab}$

$= \frac{a^2b + ab^2}{b+a}$

$= \frac{ab(a+b)}{b+a}$

$= ab$

3. $\frac{\frac{10}{x^2-25}}{\frac{3}{x+5}+\frac{2}{x-5}} = \frac{\frac{10}{(x+5)(x-5)}}{\frac{3}{x+5}+\frac{2}{x-5}} \cdot \frac{(x+5)(x-5)}{(x+5)(x-5)}$

$= \frac{\frac{10(x+5)(x-5)}{(x+5)(x-5)}}{\frac{3(x+5)(x-5)}{x+5} + \frac{2(x+5)(x-5)}{x-5}}$

$= \frac{10}{3(x-5)+2(x+5)}$

$= \frac{10}{3x-15+2x+10}$

$= \frac{10}{5x-5}$

$= \frac{10}{5(x-1)}$

$= \frac{2}{x-1}$

4. Answers may vary. Possible solution:
The first step is to find the LCD.

$x - 3 = \quad (x-3)$

$2x - 6 = 2 \cdot (x-3)$

$x + 5 = \quad (x+5)$

$\text{LCD} = 2 \quad (x-3)(x+5)$

Next, multiply the numerator and denominator of the complex fraction by the LCD. Lastly, cancel common factors to eliminate the denominators of each individual fraction.

6.5 Exercises

1. $\frac{7}{x}+\frac{3}{4}=\frac{-2}{x}$

$4x\left(\frac{7}{x}\right)+4x\left(\frac{3}{4}\right)=4x\left(-\frac{2}{x}\right)$

$28+3x=-8$

$3x=-36$

$x=-12$

Check: $\frac{7}{-12}+\frac{3}{4} \stackrel{?}{=} \frac{-2}{-12}$

$\frac{7}{-12}+\frac{9}{12} \stackrel{?}{=} \frac{1}{6}$

$\frac{2}{12} \stackrel{?}{=} \frac{1}{6}$

$\frac{1}{6}=\frac{1}{6}$

3. $\frac{-1}{4x}+\frac{3}{2}=\frac{5}{x}$

$4x\left(-\frac{1}{4x}\right)+4x\left(\frac{3}{2}\right)=4x\left(\frac{5}{x}\right)$

$-1+6x=20$

$6x=21$

$x=\frac{21}{6}$

$x=\frac{7}{2}$ or $3\frac{1}{2}$ or 3.5

Check: $-\frac{1}{4\left(\frac{7}{2}\right)}+\frac{3}{2} \stackrel{?}{=} \frac{5}{\frac{7}{2}}$

$-\frac{1}{14}+\frac{3}{2} \stackrel{?}{=} \frac{10}{7}$

$-\frac{1}{14}+\frac{21}{14} \stackrel{?}{=} \frac{10}{7}$

$\frac{20}{14} \stackrel{?}{=} \frac{10}{7}$

$\frac{10}{7}=\frac{10}{7}$

5.
$$\frac{5x+3}{3x}=\frac{7}{3}-\frac{9}{x}$$
$$3x\left(\frac{5x+3}{3x}\right)=3x\left(\frac{7}{3}\right)-3x\left(\frac{9}{x}\right)$$
$$5x+3=7x-27$$
$$3=2x-27$$
$$30=2x$$
$$15=x$$
Check: $\frac{5(15)+3}{3(15)}\stackrel{?}{=}\frac{7}{3}-\frac{9}{15}$
$$\frac{78}{45}\stackrel{?}{=}\frac{35}{15}-\frac{9}{15}$$
$$\frac{26}{15}=\frac{26}{15}$$

7.
$$\frac{x+5}{3x}=\frac{1}{2}$$
$$6x\cdot\frac{x+5}{3x}=6x\cdot\frac{1}{2}$$
$$2(x+5)=3x$$
$$2x+10=3x$$
$$-x=-10$$
$$x=10$$
Check: $\frac{10+5}{3(10)}\stackrel{?}{=}\frac{1}{2}$
$$\frac{15}{30}\stackrel{?}{=}\frac{1}{2}$$
$$\frac{1}{2}=\frac{1}{2}$$

9.
$$\frac{6}{3x-5}=\frac{3}{2x}$$
$$2x(3x-5)\left(\frac{6}{3x-5}\right)=2x(3x-5)\left(\frac{3}{2x}\right)$$
$$12x=9x-15$$
$$3x=-15$$
$$x=-5$$
Check: $\frac{6}{3(-5)-5}\stackrel{?}{=}\frac{3}{2(-5)}$
$$-\frac{6}{20}\stackrel{?}{=}-\frac{3}{10}$$
$$-\frac{3}{10}=-\frac{3}{10}$$

11.
$$\frac{2}{2x+5}=\frac{4}{x-4}$$
$$(2x+5)(x-4)\left(\frac{2}{2x+5}\right)=(2x+5)(x-4)\left(\frac{4}{x-4}\right)$$
$$2(x-4)=4(2x+5)$$
$$2x-8=8x+20$$
$$-8=6x+20$$
$$-28=6x$$
$$-\frac{28}{6}=x$$
$$-\frac{14}{3}=x \text{ or } x=-4\frac{2}{3}$$
Check: $\frac{2}{2\left(-\frac{14}{3}\right)+5}\stackrel{?}{=}\frac{4}{-\frac{14}{3}-4}$
$$\frac{2}{-\frac{28}{3}+\frac{15}{3}}\stackrel{?}{=}\frac{4}{-\frac{14}{3}-\frac{12}{3}}$$
$$\frac{2}{-\frac{13}{3}}\stackrel{?}{=}\frac{4}{-\frac{26}{3}}$$
$$-\frac{6}{13}=-\frac{6}{13}$$

13.
$$\frac{2}{x}+\frac{x}{x+1}=1$$
$$x(x+1)\left(\frac{2}{x}\right)+x(x+1)\left(\frac{x}{x+1}\right)=x(x+1)(1)$$
$$(x+1)(2)+x^2=x^2+x$$
$$2x+2+x^2=x^2+x$$
$$2x+2=x$$
$$2=-x$$
$$-2=x$$
Check: $\frac{2}{-2}+\frac{-2}{-2+1}\stackrel{?}{=}1$
$$-1+2\stackrel{?}{=}1$$
$$1=1$$

15.
$$\frac{85-4x}{x}=7-\frac{3}{x}$$
$$x\left(\frac{85-4x}{x}\right)=7x-x\left(\frac{3}{x}\right)$$
$$85-4x=7x-3$$
$$85=11x-3$$
$$88=11x$$
$$8=x$$

Check: $\frac{85-4(8)}{8} \stackrel{?}{=} 7-\frac{3}{8}$

$\frac{53}{8} \stackrel{?}{=} \frac{56}{8}-\frac{3}{8}$

$\frac{53}{8} = \frac{53}{8}$

17.

$$\begin{aligned} \frac{1}{x+4}-2 &= \frac{3x-2}{x+4} \\ (x+4)\left(\frac{1}{x+4}\right)-(x+4)\cdot 2 &= (x+4)\cdot\frac{3x-2}{x+4} \\ 1-(2x+8) &= 3x-2 \\ 1-2x-8 &= 3x-2 \\ -5x &= 5 \\ x &= -1 \end{aligned}$$

Check: $\frac{1}{-1+4}-2 \stackrel{?}{=} \frac{3(-1)-2}{-1+4}$

$\frac{1}{3}-\frac{6}{3} \stackrel{?}{=} -\frac{5}{3}$

$-\frac{5}{3} = -\frac{5}{3}$

19.

$$\begin{aligned} \frac{2}{x-6}-5 &= \frac{2(x-5)}{x-6} \\ (x-6)\left(\frac{2}{x-6}\right)-5(x-6) &= (x-6)\left[\frac{2(x-5)}{x-6}\right] \\ 2-5x+30 &= 2x-10 \\ -5x+32 &= 2x-10 \\ 42 &= 7x \\ 6 &= x \end{aligned}$$

Check: $\frac{2}{6-6}-5 \stackrel{?}{=} \frac{2(6-5)}{6-6}$

$\frac{2}{0}-5 \stackrel{?}{=} \frac{2}{0}$

$x = 6$ makes the denominators zero so it is an extraneous solution. There is no solution.

21.

$$\begin{aligned} \frac{2}{x+1}-\frac{1}{x-1} &= \frac{2x}{x^2-1} \\ \frac{2}{x+1}-\frac{1}{x-1} &= \frac{2x}{(x+1)(x-1)} \\ (x+1)(x-1)\left(\frac{2}{x+1}\right)-(x+1)(x-1)\left(\frac{1}{x-1}\right) &= (x+1)(x-1)\left(\frac{2x}{(x+1)(x-1)}\right) \\ (x-1)(2)-(x+1) &= 2x \\ 2x-2-x-1 &= 2x \\ x-3 &= 2x \\ -3 &= x \end{aligned}$$

Check: $\dfrac{2}{-3+1}-\dfrac{1}{-3-1} \stackrel{?}{=} \dfrac{2(-3)}{(-3)^2-1}$

$$-1+\frac{1}{4} \stackrel{?}{=} -\frac{6}{8}$$

$$-\frac{3}{4} = -\frac{3}{4}$$

23.

$$\frac{y+1}{y^2+2y-3} = \frac{1}{y+3}-\frac{1}{y-1}$$

$$\frac{y+1}{(y+3)(y-1)} = \frac{1}{y+3}-\frac{1}{y-1}$$

$$(y+3)(y-1)\left(\frac{y+1}{(y+3)(y-1)}\right) = (y+3)(y-1)\left(\frac{1}{y+3}\right)-(y+3)(y-1)\left(\frac{1}{y-1}\right)$$

$$y+1 = y-1-y-3$$

$$y+1 = -4$$

$$y = -5$$

Check: $\dfrac{-5+1}{(-5)^2+2(-5)-3} \stackrel{?}{=} \dfrac{1}{-5+3}-\dfrac{1}{-5-1}$

$$\frac{-4}{12} \stackrel{?}{=} -\frac{1}{2}+\frac{1}{6}$$

$$-\frac{4}{12} = -\frac{4}{12}$$

25.

$$\frac{2x}{x+4}-\frac{8}{x-4} = \frac{2x^2+32}{x^2-16}$$

$$\frac{2x}{x+4}-\frac{8}{x-4} = \frac{2x^2+32}{(x+4)(x-4)}$$

$$(x+4)(x-4)\left(\frac{2x}{x+4}\right)-(x+4)(x-4)\left(\frac{8}{x-4}\right) = (x+4)(x-4)\left(\frac{2x^2+32}{(x+4)(x-4)}\right)$$

$$(x-4)(2x)-(x+4)(8) = 2x^2+32$$

$$2x^2-8x-8x-32 = 2x^2+32$$

$$2x^2-16x-32 = 2x^2+32$$

$$-16x-32 = 32$$

$$-16x = 64$$

$$x = -4$$

Since $x = -4$ makes the first denominator 0, $\dfrac{2x}{x+4}$, $x = -4$ is extraneous and there is no solution.

27.

$$\frac{4}{x^2-1}+\frac{7}{x+1}=\frac{5}{x-1}$$
$$\frac{4}{(x+1)(x-1)}+\frac{7}{x+1}=\frac{5}{x-1}$$
$$(x+1)(x-1)\frac{4}{(x+1)(x-1)}+(x+1)(x-1)\frac{7}{x+1}=(x+1)(x-1)\frac{5}{x-1}$$
$$4+(x-1)(7)=(x+1)(5)$$
$$4+7x-7=5x+5$$
$$2x-3=5$$
$$2x=8$$
$$x=4$$

Check: $\frac{4}{(4)^2-1}+\frac{7}{4+1}\stackrel{?}{=}\frac{5}{4-1}$

$$\frac{4}{15}+\frac{7}{5}\stackrel{?}{=}\frac{5}{3}$$
$$\frac{25}{15}\stackrel{?}{=}\frac{5}{3}$$
$$\frac{5}{3}=\frac{5}{3}$$

29.

$$\frac{x+11}{x^2-5x+4}+\frac{3}{x-1}=\frac{5}{x-4}$$
$$\frac{x+11}{(x-4)(x-1)}+\frac{3}{x-1}=\frac{5}{x-4}$$
$$(x-4)(x-1)\left(\frac{x+11}{(x-4)(x-1)}\right)+(x-4)(x-1)\left(\frac{3}{x-1}\right)=(x-4)(x-1)\left(\frac{5}{x-4}\right)$$
$$x+11+(x-4)(3)=(x-1)(5)$$
$$x+11+3x-12=5x-5$$
$$4x-1=5x-5$$
$$-1=x-5$$
$$4=x$$

Since $x = 4$ makes the denominator zero, for example $\frac{5}{x-4}$, $x = 4$ is extraneous and there is no solution.

31. $x - 2 \neq 0$, so $x \neq 2$
$x - 4 \neq 0$, so $x \neq 4$
$x^2-6x+8\neq 0$
$(x-2)(x-4)\neq 0$
$x-2\neq 0$ $\quad$ $x-4\neq 0$
$x\neq 2$ $\quad$ $x\neq 4$
Extraneous solutions $x = 2$, $x = 4$

Cumulative Review

33. $8x^2-2x-1=(4x+1)(2x-1)$

34. $5(x-2)=8-(3+x)$
$5x-10=8-3-x$
$6x=15$
$x=\frac{5}{2}$ or $2\frac{1}{2}$ or 2.5

35. Let w = width.
length = $2w - 8$
Perimeter = 2(length) + 2(width)
$44=2(2w-8)+2w$
$44=4w-16+2w$
$44=6w-16$
$-6w=-60$
$w=10$
$2w - 8 = 2(10) - 8 = 20 - 8 = 12$
The dimensions are as follows:
width = 10 in.
length = 12 in.

36. Multiples of \$8.23: $\frac{115,000}{1000}=115$
$115(8.23) = 946.45$
The monthly payment will be \$946.45.

Quick Quiz 6.5

1. $\frac{3}{4x}-\frac{5}{6x}=2-\frac{1}{2x}$
$12x\cdot\frac{3}{4x}-12x\cdot\frac{5}{6x}=12x\cdot 2-12x\cdot\frac{1}{2x}$
$9-10=24x-6$
$6-1=24x$
$x=\frac{5}{24}$

2. $\frac{x}{x-1}-\frac{2}{x}=\frac{1}{x-1}$
$x(x-1)\cdot\frac{x}{x-1}-x(x-1)\cdot\frac{2}{x}=x(x-1)\cdot\frac{1}{x-1}$
$x^2-2(x-1)=x$
$x^2-2x+2=x$
$x^2-3x+2=0$
$(x-2)(x-1)=0$

$x \neq 1$ because it makes the denominator 0, for example $\frac{x}{x-1}$. $x = 2$ is the solution.

3. $$\frac{6}{x^2-2x-8}+\frac{5}{x+2}=\frac{1}{x-4}$$
$$(x-4)(x+2)\cdot\frac{6}{(x-4)(x+2)}+(x-4)(x+2)\cdot\frac{5}{x+2}=(x-4)(x+2)\cdot\frac{1}{x-4}$$
$$6+5(x-4)=x+2$$
$$6+5x-20=x+2$$
$$5x-14=x+2$$
$$4x=16$$
$$x=4$$

$x \neq 4$ because it makes the denominator 0, $\frac{1}{x-4}$, so there is no solution..

4. Answers may vary. Possible solution:
To find the LCD first factor each denominator, then multiply one instance of each factor.
$$\begin{aligned} x^2-9 &= \quad (x-3)(x+3) \\ 3x-9 &= 3\ (x-3) \\ 2x+6 &= 2 \qquad\quad (x+3) \\ 2x^2-18 &= 2\quad (x-3)(x+3) \\ \text{LCD} &= 2\cdot 3\cdot (x-3)(x+3) \end{aligned}$$

6.6 Exercises

1. $$\frac{4}{9}=\frac{8}{x}$$
$$4\cdot x=9\cdot 8$$
$$4x=72$$
$$x=18$$

3. $$\frac{x}{17}=\frac{12}{5}$$
$$x\cdot 5=17\cdot 12$$
$$5x=204$$
$$x=\frac{204}{5} \text{ or } 40\frac{4}{5} \text{ or } 40.8$$

5. $$\frac{9.1}{8.4}=\frac{x}{6}$$
$$9.1\cdot 6=8.4\cdot x$$
$$54.6=8.4x$$
$$x=6.5$$

7. $$\frac{7}{x}=\frac{40}{130}$$
$$x\cdot 40=7\cdot 130$$
$$40x=910$$
$$x=\frac{910}{40}=22.75=22\frac{3}{4}$$

9. d = distance apart

$\frac{d}{5.5}=\frac{15}{\frac{3}{4}}$

$\frac{3}{4}d=82.5$

$d=110$

The two cities are 110 miles apart.

11. a. $500(1.3)=650$

Robyn received 650 New Zealand dollars.

b. $500(1.3-1.15)=75$

She would have received 75 New Zealand dollars less.

13. x = speed limit in miles per hour

$\frac{x}{90}=\frac{62}{100}$

$100x=5580$

$x=55.8\approx 56$

The limit is 56 miles per hour.

15. x = miles from base of mountain

$\frac{x}{\frac{3}{4}}=\frac{136}{3\frac{1}{2}}$

$3\frac{1}{2}x=\frac{3}{4}(136)$

$\frac{7}{2}x=102$

$x=29.1$

He is 29 miles away.

17. $x=20,\ y=29,\ m=13$

$\frac{n}{m}=\frac{y}{x}$

$\frac{n}{13}=\frac{29}{20}$

$20n=377$

$n=\frac{377}{20}=18\frac{17}{20}$ inches

19. $\frac{y}{n}=\frac{x}{m}$

$\frac{y}{40}=\frac{175}{35}$

$35y=7000$

$y=200$ meters

21. $\frac{a}{g}=\frac{d}{k}$

$\frac{5}{7}=\frac{8}{k}$

$5k=56$

$k=\frac{56}{5}$

$k=11\frac{1}{5}$ ft

23. $\frac{k}{d}=\frac{h}{b}$

$\frac{k}{32}=\frac{24}{20}$

$20k=32\cdot 24$

$k=\frac{768}{20}=38\frac{2}{5}$ m

25. x = length

$\frac{30}{x}=\frac{5}{8}$

$5x=8(30)$

$x=\frac{240}{5}$

$x=48$ inches

27. x = flower height

$\frac{x}{13}=\frac{5}{3}$

$3x=65$

$x=\frac{65}{3}\approx 22$ inches

Total height = 13 + 22 = 35 inches

29. x = amount of acceleration in 11 seconds

$\frac{x}{11}=\frac{3}{2}$

$2x=33$

$x=16.5$

$45+16.5=61.5$

He will be traveling at 61.5 miles per hour.

31.

	D	R	$T=\frac{D}{R}$
Helicopter	1050	s	$\frac{1050}{s}$
Airline	1250	$s+40$	$\frac{1250}{s+40}$

$$\frac{1250}{s+40}=\frac{1050}{s}$$
$$1050s = 1050s + 42{,}000$$
$$200s = 42{,}000$$
$$s = 210$$
$s + 40 = 250$
The speeds are as follows:
commuter airline, 250 kilometers/hr;
helicopter, 210 kilometers/hr

33. a. $\frac{\$0.79}{7.5 \text{ oz}} = \$0.11/\text{oz}$
It costs \$0.11/oz.

b. $\frac{\$1.49}{16 \text{ oz}} = \$0.09/\text{oz}$
It costs \$0.09/oz.

c. x = price of 40 oz bucket
$$\frac{x}{40}=\frac{1.49}{16}$$
$$16x = 59.6$$
$$x = 3.73$$
The price would be \$3.73.

35. x = time to rake together
$$\frac{1}{6}+\frac{1}{8}=\frac{1}{x}$$
$$24x\left(\frac{1}{6}\right)+24x\left(\frac{1}{8}\right)=24x\left(\frac{1}{x}\right)$$
$$4x+3x=24$$
$$7x=24$$
$$x=\frac{24}{7}=3\frac{3}{7}$$
It will take $3\frac{3}{7}$ hours or 3 hours 26 minutes.

Cumulative Review

37. $0.000892465 = 8.92465\times 10^{-4}$

38. $5.82\times 10^{8} = 582{,}000{,}000$

39. $\frac{x^{-3}y^{-2}}{z^4w^{-8}}=\frac{w^8}{x^3y^2z^4}$

40. $\left(\frac{2}{3}\right)^{-3}=\left(\frac{3}{2}\right)^{3}=\frac{3^3}{2^3}=\frac{27}{8}$ or $3\frac{3}{8}$

Quick Quiz 6.6

1. $\frac{16.5}{2.1}=\frac{x}{7}$
$$2.1\cdot x = 16.5\cdot 7$$
$$2.1x = 115.5$$
$$x = 55$$

2. x = expected on-time departures
$$\frac{164}{205}=\frac{x}{215}$$
$$x=\frac{164}{205}(215)$$
$$x=172$$
There are 172 expected on-time departures.

3. x = height of tree
$$\frac{x}{34}=\frac{6}{8.4}$$
$$x=\frac{6}{8.5}(34)$$
$$x=24$$
The tree is 24 ft tall.

4. Answers may vary. Possible solution:
One of the fractions needs to be inverted in order for the equation to be an accurate statement.

Putting Your Skills to Work

1. 138 + 142 + 157 + 158 + 160 + 165 + 170 + 183 + 192 + 203 + 210 + 214 = 2092
She accumulated \$2092 in finance charges.

2. $\frac{\$2092}{12}\approx \$174.33$
Her average finance charge was \$174.33.

3. Answers may vary.

4. Answers may vary.

Chapter 6 Review Problems

1. $\frac{bx}{bx-by}=\frac{bx}{b(x-y)}=\frac{x}{x-y}$

2. $\frac{4x-4y}{5y-5x}=\frac{4(x-y)}{-5(x-y)}=-\frac{4}{5}$

3. $\frac{2x^2+5x-3}{2x^2-9x+4}=\frac{(2x-1)(x+3)}{(2x-1)(x-4)}=\frac{x+3}{x-4}$

4. $\dfrac{3x^2+7x+2}{3x^2+13x+4}=\dfrac{(3x+1)(x+2)}{(3x+1)(x+4)}=\dfrac{x+2}{x+4}$

5. $\dfrac{x^2-9}{x^2-10x+21}=\dfrac{(x+3)(x-3)}{(x-7)(x-3)}=\dfrac{x+3}{x-7}$

6. $\dfrac{2x^2+18x+40}{3x+15}=\dfrac{2(x+5)(x+4)}{3(x+5)}=\dfrac{2(x+4)}{3}$

7. $\dfrac{x^3-4x^2}{x^3-x^2-12x}=\dfrac{x^2(x-4)}{x(x-4)(x+3)}=\dfrac{x}{x+3}$

8. $\dfrac{2x^2+7x-15}{25-x^2}=\dfrac{2x^2+10x-3x-15}{25-x^2}$
$=\dfrac{2x(x+5)-3(x+5)}{(5-x)(5+x)}$
$=\dfrac{(2x-3)(x+5)}{(5-x)(5+x)}$
$=\dfrac{2x-3}{5-x}$

9. $\dfrac{2x^2-2xy-24y^2}{2x^2+5xy-3y^2}=\dfrac{2(x^2-xy-12y^2)}{2x^2+5xy-3y^2}$
$=\dfrac{2(x-4y)(x+3y)}{(2x-y)(x+3y)}$
$=\dfrac{2(x-4y)}{2x-y}$

10. $\dfrac{4-y^2}{3y^2+5y-2}=\dfrac{(2+y)(2-y)}{(3y-1)(y+2)}=\dfrac{2-y}{3y-1}$

11. $\dfrac{5x^3-10x^2}{25x^4+5x^3-30x^2}=\dfrac{5x^2(x-2)}{5x^2(5x+6)(x-1)}$
$=\dfrac{x-2}{(5x+6)(x-1)}$

12. $\dfrac{16x^2-4y^2}{4x-2y}=\dfrac{4(2x+y)(2x-y)}{2(2x-y)}$
$=2(2x+y)$
$=4x+2y$

13. $\dfrac{2x^2+6x}{3x^2-27}\cdot\dfrac{x^2+3x-18}{4x^2-4x}$
$=\dfrac{2x(x+3)}{3(x+3)(x-3)}\cdot\dfrac{(x+6)(x-3)}{4x(x-1)}$
$=\dfrac{x+6}{6(x-1)}$

14. $\dfrac{y^2+8y+16}{5y^2+20y}\div\dfrac{y^2+7y+12}{2y^2+5y-3}$
$=\dfrac{y^2+8y+16}{5y^2+20y}\cdot\dfrac{2y^2+5y-3}{y^2+7y+12}$
$=\dfrac{(y+4)(y+4)}{5y(y+4)}\cdot\dfrac{(2y-1)(y+3)}{(y+4)(y+3)}$
$=\dfrac{2y-1}{5y}$

15. $\dfrac{2y^2+3y-2}{2y^2+y-1}\div\dfrac{2y^2+y-1}{2y^2-3y-2}$
$=\dfrac{2y^2+3y-2}{2y^2+y-1}\cdot\dfrac{2y^2-3y-2}{2y^2+y-1}$
$=\dfrac{(2y-1)(y+2)}{(2y-1)(y+1)}\cdot\dfrac{(2y+1)(y-2)}{(2y-1)(y+1)}$
$=\dfrac{(y+2)(2y+1)(y-2)}{(y+1)^2(2y-1)}$

16. $\dfrac{6y^2+13y-5}{9y^2+3y}\div\dfrac{4y^2+20y+25}{12y^2}$
$=\dfrac{6y^2+13y-5}{9y^2+3y}\cdot\dfrac{12y^2}{4y^2+20y+25}$
$=\dfrac{(3y-1)(2y+5)}{3y(3y+1)}\cdot\dfrac{12y^2}{(2y+5)(2y+5)}$
$=\dfrac{4y(3y-1)}{(3y+1)(2y+5)}$

17. $\dfrac{3xy^2+12y^2}{2x^2-11x+5}\div\dfrac{2xy+8y}{8x^2+2x-3}$
$=\dfrac{3xy^2+12y^2}{2x^2-11x+5}\cdot\dfrac{8x^2+2x-3}{2xy+8y}$
$=\dfrac{3y^2(x+4)}{(2x-1)(x-5)}\cdot\dfrac{(2x-1)(4x+3)}{2y(x+4)}$
$=\dfrac{3y(4x+3)}{2(x-5)}$

18. $\frac{11}{x-2}\cdot\frac{2x^2-8}{44}=\frac{11}{x-2}\cdot\frac{2(x+2)(x-2)}{44}=\frac{x+2}{2}$

19. $\frac{x^2-5xy-24y^2}{2x^2-2xy-24y^2}\cdot\frac{4x^2+4xy-24y^2}{x^2-10xy+16y^2}$
$=\frac{(x-8y)(x+3y)}{2(x-4y)(x+3y)}\cdot\frac{4(x+3y)(x-2y)}{(x-8y)(x-2y)}$
$=\frac{2(x+3y)}{x-4y}$

20. $\frac{2x^2+10x+2}{8x-8}\cdot\frac{3x-3}{4x^2+20x+4}$
$=\frac{2(x^2+5x+1)}{8(x-1)}\cdot\frac{3(x-1)}{4(x^2+5x+1)}$
$=\frac{3}{16}$

21. $\frac{6}{y+2}+\frac{2}{3y}=\frac{6}{y+2}\cdot\frac{3y}{3y}+\frac{2}{3y}\cdot\frac{y+2}{y+2}$
$=\frac{18y+2(y+2)}{3y(y+2)}$
$=\frac{18y+2y+4}{3y(y+2)}$
$=\frac{20y+4}{3y(y+2)}$
$=\frac{4(5y+1)}{3y(y+2)}$

22. $3+\frac{2}{x+1}+\frac{1}{x}=\frac{3}{1}\cdot\frac{x(x+1)}{x(x+1)}+\frac{2}{x+1}\cdot\frac{x}{x}+\frac{1}{x}\cdot\frac{x+1}{x+1}$
$=\frac{3x(x+1)+2x+(x+1)}{x(x+1)}$
$=\frac{3x^2+3x+2x+x+1}{x(x+1)}$
$=\frac{3x^2+6x+1}{x(x+1)}$

23. $\frac{7}{x+2}+\frac{3}{x-4}=\frac{7}{x+2}\cdot\frac{x-4}{x-4}+\frac{3}{x-4}\cdot\frac{x+2}{x+2}$
$=\frac{7x-28}{(x+2)(x-4)}+\frac{3x+6}{(x+2)(x-4)}$
$=\frac{7x-28+3x+6}{(x+2)(x-4)}$
$=\frac{10x-22}{(x+2)(x-4)}$

24. $\frac{2}{x^2-9}+\frac{x}{x+3}=\frac{2}{(x+3)(x-3)}+\frac{x}{x+3}\cdot\frac{x-3}{x-3}$
$=\frac{2+x^2-3x}{(x+3)(x-3)}$
$=\frac{(x-2)(x-1)}{(x+3)(x-3)}$

25. $\frac{x}{y}+\frac{3}{2y}+\frac{1}{y+2}$
$=\frac{x}{y}\cdot\frac{2(y+2)}{2(y+2)}+\frac{3}{2y}\cdot\frac{y+2}{y+2}+\frac{1}{y+2}\cdot\frac{2y}{2y}$
$=\frac{2x(y+2)+3y+6+2y}{2y(y+2)}$
$=\frac{2xy+4x+5y+6}{2y(y+2)}$

26. $\frac{4}{a}+\frac{2}{b}+\frac{3}{a+b}$
$=\frac{4}{a}\cdot\frac{b(a+b)}{b(a+b)}+\frac{2}{b}\cdot\frac{a(a+b)}{a(a+b)}+\frac{3}{a+b}\cdot\frac{ab}{ab}$
$=\frac{4ab+4b^2}{ab(a+b)}+\frac{2a^2+2ab}{ab(a+b)}+\frac{3ab}{ab(a+b)}$
$=\frac{4ab+4b^2+2a^2+2ab+3ab}{ab(a+b)}$
$=\frac{2a^2+9ab+4b^2}{ab(a+b)}$
$=\frac{(2a+b)(a+4b)}{ab(a+b)}$

27. $\frac{3x+1}{3x}-\frac{1}{x}=\frac{3x+1}{3x}-\frac{1}{x}\cdot\frac{3}{3}$
$=\frac{3x+1-3}{3x}$
$=\frac{3x-2}{3x}$

28. $\frac{x+4}{x+2}-\frac{1}{2x}=\frac{x+4}{x+2}\cdot\frac{2x}{2x}-\frac{1}{2x}\cdot\frac{x+2}{x+2}$
$=\frac{2x^2+8x}{2x(x+2)}-\frac{x+2}{2x(x+2)}$
$=\frac{2x^2+8x-(x+2)}{2x(x+2)}$
$=\frac{2x^2+8x-x-2}{2x(x+2)}$
$=\frac{2x^2+7x-2}{2x(x+2)}$

29. $\frac{27}{x^2-81}+\frac{3}{2(x+9)}$
$=\frac{27}{(x+9)(x-9)}\cdot\frac{2}{2}+\frac{3}{2(x+9)}\cdot\frac{x-9}{x-9}$
$=\frac{54+3x-27}{2(x+9)(x-9)}$
$=\frac{3x+27}{2(x+9)(x-9)}$
$=\frac{3(x+9)}{2(x+9)(x-9)}$
$=\frac{3}{2(x-9)}$

30. $\frac{1}{x^2+7x+10}-\frac{x}{x+5}$
$=\frac{1}{(x+2)(x+5)}-\frac{x}{x+5}\cdot\frac{x+2}{x+2}$
$=\frac{1-x^2-2x}{(x+2)(x+5)}$

31. $\frac{\frac{4}{3y}-\frac{2}{y}}{\frac{1}{2y}+\frac{1}{y}}=\frac{\frac{4}{3y}-\frac{2}{y}}{\frac{1}{2y}+\frac{1}{y}}\cdot\frac{6y}{6y}$
$=\frac{\frac{4}{3y}\cdot 6y-\frac{2}{y}\cdot 6y}{\frac{1}{2y}\cdot 6y+\frac{1}{y}\cdot 6y}$
$=\frac{8-12}{3+6}$
$=\frac{-4}{9}$
$=-\frac{4}{9}$

32. $\frac{\frac{5}{x}+\frac{1}{2x}}{\frac{x}{4}+x}=\frac{\frac{5}{x}+\frac{1}{2x}}{\frac{x}{4}+\frac{x}{1}}\cdot\frac{4x}{4x}$
$=\frac{\frac{5}{x}\cdot 4x+\frac{1}{2x}\cdot 4x}{\frac{x}{4}\cdot 4x+\frac{x}{1}\cdot 4x}$
$=\frac{20+2}{x^2+4x^2}$
$=\frac{22}{5x^2}$

33. $\frac{w-\frac{4}{w}}{1+\frac{2}{w}}=\frac{w-\frac{4}{w}}{1+\frac{2}{w}}\cdot\frac{w}{w}$
$=\frac{w^2-4}{w+2}$
$=\frac{(w+2)(w-2)}{w+2}$
$=w-2$

34. $\frac{1-\frac{w}{w-1}}{1+\frac{w}{1-w}}=\frac{1-\frac{w}{w-1}}{1+\frac{w}{-(w-1)}}=\frac{1-\frac{w}{w-1}}{1-\frac{w}{w-1}}=1$

35. $\frac{1+\frac{1}{y^2-1}}{\frac{1}{y+1}-\frac{1}{y-1}}=\frac{1+\frac{1}{(y+1)(y-1)}}{\frac{1}{y+1}-\frac{1}{y-1}}\cdot\frac{(y+1)(y-1)}{(y+1)(y-1)}$
$=\frac{(y+1)(y-1)+1}{y-1-(y+1)}$
$=\frac{y^2-1+1}{y-1-y-1}$
$=\frac{y^2}{-2}$
$=-\frac{y^2}{2}$

36. $\frac{\frac{1}{y}+\frac{1}{x+y}}{1+\frac{2}{x+y}}=\frac{\frac{1}{y}+\frac{1}{x+y}}{1+\frac{2}{x+y}}\cdot\frac{y(x+y)}{y(x+y)}$
$=\frac{x+y+y}{y(x+y)+2y}$
$=\frac{x+2y}{xy+y^2+2y}$
$=\frac{x+2y}{y(x+y+2)}$

37. $\frac{\frac{1}{a+b}-\frac{1}{a}}{b}=\frac{\frac{a-(a+b)}{a(a+b)}}{b}$
$=\frac{a-a-b}{a(a+b)}\cdot\frac{1}{b}$
$=\frac{-b}{a(a+b)}\cdot\frac{1}{b}$
$=-\frac{1}{a(a+b)}$

38. $$\frac{\frac{2}{a+b}-\frac{3}{b}}{\frac{1}{a+b}}=\frac{\frac{2}{a+b}-\frac{3}{b}}{\frac{1}{a+b}}\cdot\frac{b(a+b)}{b(a+b)}$$
$$=\frac{2b-3(a+b)}{b}$$
$$=\frac{2b-3a-3b}{b}$$
$$=\frac{-3a-b}{b}\text{ or }-\frac{3a+b}{b}$$

39. $$\frac{x+5y}{x-6y}\div\left(\frac{1}{5y}-\frac{1}{x+5y}\right)$$
$$=\frac{x+5y}{x-6y}\div\left(\frac{1}{5y}\cdot\frac{x+5y}{x+5y}-\frac{1}{x+5y}\cdot\frac{5y}{5y}\right)$$
$$=\frac{x+5y}{x-6y}\div\left[\frac{x+5y}{5y(x+5y)}-\frac{5y}{5y(x+5y)}\right]$$
$$=\frac{x+5y}{x-6y}\div\frac{x+5y-5y}{5y(x+5y)}$$
$$=\frac{x+5y}{x-6y}\div\frac{x}{5y(x+5y)}$$
$$=\frac{x+5y}{x-6y}\cdot\frac{5y(x+5y)}{x}$$
$$=\frac{5y(x+5y)^2}{x(x-6y)}$$

40. $$\left(\frac{1}{x+2y}-\frac{1}{x-y}\right)\div\frac{2x-4y}{x^2-3xy+2y^2}$$
$$=\left(\frac{1}{x+2y}-\frac{1}{x-y}\right)\cdot\frac{x^2-3xy+2y^2}{2x-4y}$$
$$=\frac{x-y-(x+2y)}{(x+2y)(x-y)}\cdot\frac{(x-y)(x-2y)}{2(x-2y)}$$
$$=\frac{(x-y-x-2y)(x-y)(x-2y)}{(x+2y)(x-y)(2)(x-2y)}$$
$$=\frac{(-3y)(x-y)(x-2y)}{2(x+2y)(x-y)(x-2y)}$$
$$=\frac{-3y}{2(x+2y)}$$

41. $$\frac{8a-1}{6a+8}=\frac{3}{4}$$
$$\frac{8a-1}{2(3a+4)}=\frac{3}{4}$$
$$4(3a+4)\left[\frac{8a-1}{2(3a+4)}\right]=4(3a+4)\left(\frac{3}{4}\right)$$
$$2(8a-1)=3(3a+4)$$
$$16a-2=9a+12$$
$$7a-2=12$$
$$7a=14$$
$$a=2$$

42. $$\frac{8}{a-3}=\frac{12}{a+3}$$
$$(a-3)(a+3)\frac{8}{a-3}=(a-3)(a+3)\frac{12}{a+3}$$
$$8(a+3)=12(a-3)$$
$$8a+24=12a-36$$
$$60=4a$$
$$15=a$$

43. $$\frac{2x-1}{x}-\frac{1}{2}=-2$$
$$2x\left(\frac{2x-1}{x}\right)-2x\left(\frac{1}{2}\right)=-2(2x)$$
$$4x-2-x=-4x$$
$$7x=2$$
$$x=\frac{2}{7}$$

44. $$\frac{5-x}{x}-\frac{7}{x}=-\frac{3}{4}$$
$$4x\left(\frac{5-x}{x}\right)-4x\left(\frac{7}{x}\right)=4x\left(-\frac{3}{4}\right)$$
$$4(5-x)-28=-3x$$
$$20-4x-28=-3x$$
$$-4x-8=-3x$$
$$-8=x$$

45. $$\frac{5}{2}-\frac{2y+7}{y+6}=3$$
$$2(y+6)\left(\frac{5}{2}\right)-2(y+6)\left(\frac{2y+7}{y+6}\right)=2(y+6)(3)$$
$$5y+30-4y-14=6y+36$$
$$y+16=6y+36$$
$$-20=5y$$
$$-4=y$$

46.
$$\frac{5}{4}-\frac{1}{2x}=\frac{1}{x}+2$$
$$4x\left(\frac{5}{4}\right)-4x\left(\frac{1}{2x}\right)=4x\left(\frac{1}{x}\right)+4x(2)$$
$$5x-2=4+8x$$
$$-2=4+3x$$
$$-6=3x$$
$$-2=x$$

47.
$$\frac{7}{8x}-\frac{3}{4}=\frac{1}{4x}+\frac{1}{2}$$
$$8x\left(\frac{7}{8x}\right)-8x\left(\frac{3}{4}\right)=8x\left(\frac{1}{4x}\right)+8x\left(\frac{1}{2}\right)$$
$$7-6x=2+4x$$
$$7=2+10x$$
$$5=10x$$
$$\frac{1}{2}=x$$

48.
$$\frac{1}{3x}+2=\frac{5}{6x}-\frac{1}{2}$$
$$6x\left(\frac{1}{3x}\right)+6x(2)=6x\left(\frac{5}{6x}\right)-6x\left(\frac{1}{2}\right)$$
$$2+12x=5-3x$$
$$15x=3$$
$$x=\frac{1}{5}$$

49.
$$\frac{3}{y-3}=\frac{3}{2}+\frac{y}{y-3}$$
$$2(y-3)\left(\frac{3}{y-3}\right)=2(y-3)\left(\frac{3}{2}\right)+2(y-3)\left(\frac{y}{y-3}\right)$$
$$6=3(y-3)+2y$$
$$6=3y-9+2y$$
$$6=5y-9$$
$$15=5y$$
$$3=y$$

Since $y = 3$ causes a denominator in the original equation to equal 0, there is no solution.

50.
$$\frac{3x}{x^2-4}-\frac{2}{x+2}=-\frac{4}{x-2}$$
$$(x+2)(x-2)\cdot\frac{3x}{(x+2)(x-2)}-(x+2)(x-2)\cdot\frac{2}{x+2}=(x+2)(x-2)\cdot\frac{-4}{x-2}$$
$$3x-2(x-2)=-4(x+2)$$
$$3x-2x+4=-4x-8$$
$$5x=-12$$
$$x=-\frac{12}{5}\text{ or }-2\frac{2}{5}\text{ or }-2.4$$

51.
$$\frac{9}{2}-\frac{7y-4}{y+2}=-\frac{1}{4}$$
$$4(y+2)\left(\frac{9}{2}\right)-4(y+2)\left(\frac{7y-4}{y+2}\right)=4(y+2)\left(-\frac{1}{4}\right)$$
$$18(y+2)-4(7y-4)=-(y+2)$$
$$18y+36-28y+16=-y-2$$
$$-10y+52=-y-2$$
$$52=9y-2$$
$$54=9y$$
$$6=y$$

52.
$$\frac{3y-1}{3y}-\frac{6}{5y}=\frac{1}{y}-\frac{4}{15}$$
$$15y\left(\frac{3y-1}{3y}\right)-15y\left(\frac{6}{5y}\right)=15y\left(\frac{1}{y}\right)-15y\left(\frac{4}{15}\right)$$
$$5(3y-1)-3(6)=15-4y$$
$$15y-5-18=15-4y$$
$$19y=38$$
$$y=2$$

53.
$$\frac{y+18}{y^2-16}=\frac{y}{y+4}-\frac{y}{y-4}$$
$$\frac{y+18}{(y+4)(y-4)}=\frac{y}{y+4}-\frac{y}{y-4}$$
$$(y+4)(y-4)\cdot\frac{y+18}{(y+4)(y-4)}=(y+4)(y-4)\cdot\frac{y}{y+4}-(y+4)(y-4)\cdot\frac{y}{y-4}$$
$$y+18=(y-4)(y)-(y+4)(y)$$
$$y+18=y^2-4y-y^2-4y$$
$$y+18=-8y$$
$$18=-9y$$
$$-2=y$$

54.
$$\frac{4}{x^2-1}=\frac{2}{x-1}+\frac{2}{x+1}$$
$$(x+1)(x-1)\cdot\frac{4}{(x+1)(x-1)}=(x+1)(x-1)\cdot\frac{2}{x-1}+(x+1)(x-1)\cdot\frac{2}{x+1}$$
$$4=(x+1)(2)+(x-1)(2)$$
$$4=2x+2+2x-2$$
$$4=4x$$
$$1=x$$

$x = 1$ is not an extraneous solution. There is no solution.

55.
$$\frac{3y+1}{y^2-y}-\frac{3}{y-1}=\frac{4}{y}$$
$$y(y-1)\cdot\frac{3y+1}{y(y-1)}-y(y-1)\cdot\frac{3}{y-1}=y(y-1)\cdot\frac{4}{y}$$
$$\frac{3y+1}{y(y-1)}-\frac{3}{y-1}\cdot\frac{y}{y}=\frac{4}{y}\cdot\frac{y-1}{y-1}$$
$$3y+1-3y=4(y-1)$$
$$3y+1-3y=4y-4$$
$$-4y=-5$$
$$y=\frac{5}{4}\text{ or }1\frac{1}{4}\text{ or }1.25$$

56.
$$\frac{3}{y-2}+\frac{4}{3y+2}=\frac{1}{2-y}$$
$$(y-2)(3y+2)\cdot\frac{3}{y-2}+(y-2)(3y+2)\cdot\frac{4}{3y+2}=(y-2)(3y+2)\cdot\frac{1}{2-y}$$
$$9y+6+4y-8=-3y-2$$
$$10y=0$$
$$y=0$$

57.
$$\frac{x}{4}=\frac{7}{10}$$
$$x\cdot 10=4\cdot 7$$
$$10x=28$$
$$x=\frac{28}{10}$$
$$x=\frac{14}{5}=2\frac{4}{5}=2.8$$

58.
$$\frac{8}{5}=\frac{2}{x}$$
$$8\cdot x=5\cdot 2$$
$$8x=10$$
$$x=\frac{5}{4}=1.25=1.3$$

59.
$$\frac{33}{10}=\frac{x}{8}$$
$$33\cdot 8=10\cdot x$$
$$264=10x$$
$$26.4=x$$

60.
$$\frac{16}{x}=\frac{24}{9}$$
$$16\cdot 9=x\cdot 24$$
$$144=24x$$
$$6=x$$

61. $\frac{13.5}{0.6}=\frac{360}{x}$
$13.5x=0.6(360)$
$13.5x=216$
$x=16$

62. $\frac{2\frac{1}{2}}{3\frac{1}{4}}=\frac{7}{x}$
$\frac{\frac{5}{2}}{\frac{13}{4}}=\frac{7}{x}$
$\frac{5}{2}x=\frac{91}{4}$
$x=\frac{91}{10}$ or $9\frac{1}{10}$ or 9.1

63. x = gallons to cover 400 square feet
$\frac{x}{400}=\frac{5}{240}$
$240x=2000$
$x=\frac{2000}{240}=8\frac{1}{3}$
8.3 gallons of paint are needed.

64. $\frac{3}{100}=\frac{5}{x}$
$3x=5(100)$
$3x=500$
$\frac{3x}{3}=\frac{500}{3}$
$x=\frac{500}{3}=166\frac{2}{3}$
She can make 167 cookies.

65. x = gallons of gas for trip
$\frac{10}{360}=\frac{x}{1400}$
$x=\frac{10}{360}(1400)$
$x=38.88$
Leila will use 38.9 gallons.

66. x = distance from Houston to Dallas
$\frac{4}{640}=\frac{1.5}{x}$
$x=\frac{640}{4}(1.5)$
$x=240$
Distance from Houston to Dallas is 240 miles.

67.

	D	R	$T=\frac{D}{R}$
Train	180	$s+20$	$\frac{180}{s+20}$
Car	120	s	$\frac{120}{s}$

$\frac{180}{s+20}=\frac{120}{s}$
$180s=120s+2400$
$60s=2400$
$s=40$
$s+20=60$
Car's speed is 40 mph. Train's speed is 60 mph.

68. x = time for both people to prepare baked goods when working together
$\frac{1}{3}+\frac{1}{5}=\frac{1}{x}$
$15x\cdot\frac{1}{3}+15x\cdot\frac{1}{5}=15x\cdot\frac{1}{x}$
$5x+3x=15$
$8x=15$
$x=\frac{15}{8}$
$x=1\frac{7}{8}$

Working together it will take them $1\frac{7}{8}$ hours or 1 hour, 53 minutes.

69. $\frac{5.75}{3}=\frac{x}{95}$
$(5.75)(95)=3x$
$546.25=3x$
$\frac{546.25}{3}=\frac{3x}{3}$
$182.1=x$
The peak of the canyon is 182 feet tall.

70. x = height of building
$\frac{x}{450}=\frac{8}{3}$
$3x=3600$
$x=1200$
The office building is 1200 feet tall.

71. x = time in hours for both people to wash windows when working together

$$\frac{1}{4}+\frac{1}{6}=\frac{1}{x}$$
$$12x\cdot\frac{1}{4}+12x\cdot\frac{1}{6}=12x\cdot\frac{1}{x}$$
$$3x+2x=12$$
$$5x=12$$
$$x=2\frac{2}{5}$$

Working together it will take them $2\frac{2}{5}$ hours or 2 hours, 24 minutes.

72. x = time to plow together

$$\frac{1}{20}+\frac{1}{30}=\frac{1}{x}$$
$$60x\left(\frac{1}{20}\right)+60x\left(\frac{1}{30}\right)=60x\left(\frac{1}{x}\right)$$
$$3x+2x=60$$
$$5x=60$$
$$x=12$$

Working together it will take them 12 hours.

73. $$\frac{a^2+2a-8}{6a^2-3a^3}=\frac{(a+4)(a-2)}{-3a^2(a-2)}=-\frac{a+4}{3a^2}$$

74. $$\frac{4a^3+20a^2}{2a^2+13a+15}=\frac{4a^2(a+5)}{(2a+3)(a+5)}=\frac{4a^2}{2a+3}$$

75. $$\frac{x^2-y^2}{x^2+4xy+3y^2}\cdot\frac{x^2+xy-6y^2}{x^2+xy-2y^2}$$
$$=\frac{(x+y)(x-y)}{(x+y)(x+3y)}\cdot\frac{(x+3y)(x-2y)}{(x+2y)(x-y)}$$
$$=\frac{x-2y}{x+2y}$$

76. $$\frac{x^2-6xy-16y^2}{x^2+4xy-21y^2}\cdot\frac{x^2-8xy+15y^2}{x^2-3xy-10y^2}$$
$$=\frac{(x-8y)(x+2y)}{(x+7y)(x-3y)}\cdot\frac{(x-3y)(x-5y)}{(x-5y)(x+2y)}$$
$$=\frac{x-8y}{x+7y}$$

77. $$\frac{x}{x+3}+\frac{9x+18}{x^2+3x}=\frac{x}{x+3}\cdot\frac{x}{x}+\frac{9x+18}{x(x+3)}$$
$$=\frac{x^2+9x+18}{x(x+3)}$$
$$=\frac{(x+6)(x+3)}{x(x+3)}$$
$$=\frac{x+6}{x}$$

78. $$\frac{x-30}{x^2-5x}+\frac{x}{x-5}=\frac{x-30}{x(x-5)}+\frac{x}{x-5}\cdot\frac{x}{x}$$
$$=\frac{x-30+x^2}{x(x-5)}$$
$$=\frac{x^2+x-30}{x(x-5)}$$
$$=\frac{(x+6)(x-5)}{x(x-5)}$$
$$=\frac{x+6}{x}$$

79. $$\frac{a+b}{ax+ay}-\frac{a+b}{bx+by}=\frac{a+b}{a(x+y)}\cdot\frac{b}{b}-\frac{a+b}{b(x+y)}\cdot\frac{a}{a}$$
$$=\frac{b(a+b)-a(a+b)}{ab(x+y)}$$
$$=\frac{ab+b^2-a^2-ab}{ab(x+y)}$$
$$=\frac{b^2-a^2}{ab(x+y)}$$

80. $$\frac{\frac{5}{3x}+\frac{2}{9x}}{\frac{3}{x}+\frac{8}{3x}}=\frac{\frac{5}{3x}+\frac{2}{9x}}{\frac{3}{x}+\frac{8}{3x}}\cdot\frac{9x}{9x}$$
$$=\frac{\frac{5(9x)}{3x}+\frac{2(9x)}{9x}}{\frac{3(9x)}{x}+\frac{8(9x)}{3x}}$$
$$=\frac{15+2}{27+24}$$
$$=\frac{17}{51}$$
$$=\frac{1}{3}$$

81. $\dfrac{\frac{4}{5y}-\frac{8}{y}}{y+\frac{y}{5}}=\dfrac{\frac{4}{5y}-\frac{8}{y}}{y+\frac{y}{5}}\cdot\dfrac{5y}{5y}$

$=\dfrac{\frac{4(5y)}{5y}-\frac{8(5y)}{y}}{y(5y)+\frac{y(5y)}{5}}$

$=\dfrac{4-40}{5y^2+y^2}$

$=\dfrac{-36}{6y^2}$

$=-\dfrac{6}{y^2}$

82. $\dfrac{x-3y}{x+2y}\div\left(\dfrac{2}{y}-\dfrac{12}{x+3y}\right)$

$=\dfrac{\frac{x-3y}{x+2y}}{\frac{2}{y}-\frac{12}{x+3y}}$

$=\dfrac{\frac{x-3y}{x+2y}}{\frac{2}{y}-\frac{12}{x+3y}}\cdot\dfrac{y(x+2y)(x+3y)}{y(x+2y)(x+3y)}$

$=\dfrac{y(x-3y)(x+3y)}{2(x+2y)(x+3y)-12y(x+2y)}$

$=\dfrac{y(x-3y)(x+3y)}{2(x+2y)[x+3y-6y]}$

$=\dfrac{y(x-3y)(x+3y)}{2(x+2y)(x-3y)}$

$=\dfrac{y(x+3y)}{2(x+2y)}$

83. $\dfrac{7}{x+2}=\dfrac{4}{x-4}$

$7(x-4)=4(x+2)$

$7x-28=4x+8$

$3x=36$

$x=12$

84. $\dfrac{2x-1}{3x-8}=\dfrac{5}{8}$

$8(2x-1)=5(3x-8)$

$16x-8=15x-40$

$x=-32$

85. $2+\dfrac{4}{b-1}=\dfrac{4}{b^2-b}$

$2b(b-1)+b(b-1)\left(\dfrac{4}{b-1}\right)=b(b-1)\left(\dfrac{4}{b(b-1)}\right)$

$2b^2-2b+4b=4$

$2b^2+2b=4$

$2b^2+2b-4=0$

$2(b^2+b-2)=0$

$2(b+2)(b-1)=0$

$b+2=0$　　　$b-1=0$

$b=-2$　　　$b=1$

Check: $2+\dfrac{4}{1-1}\stackrel{?}{=}\dfrac{4}{1^2-1}$

$2+\dfrac{4}{0}\stackrel{?}{=}\dfrac{4}{0}$

$b = 1$ does not check.

Solution: $b = -2$

86. x = weight of 40 m of wire

$\dfrac{x}{40}=\dfrac{8}{30}$

$30x=8(40)$

$30x=320$

$x=10.7$

It weighs 10.7 kg.

87. x = distance driven in 7 hours

$\dfrac{x}{7}=\dfrac{270}{4.5}$

$4.5x=7(270)$

$4.5x=1890$

$x=420$

She drove 420 miles.

How Am I Doing? Chapter 6 Test

1. $\dfrac{2ac+2ad}{3a^2c+3a^2d}=\dfrac{2a(c+d)}{3a^2(c+d)}=\dfrac{2}{3a}$

2. $\dfrac{8x^2-2x^2y^2}{y^2+4y+4}=\dfrac{2x^2(2-y)(2+y)}{(y+2)^2}=\dfrac{2x^2(2-y)}{y+2}$

3. $\dfrac{x^2+2x}{2x-1}\cdot\dfrac{10x^2-5x}{12x^3+24x^2}=\dfrac{x(x+2)}{2x-1}\cdot\dfrac{5x(2x-1)}{12x^2(x+2)}$

$=\dfrac{5}{12}$

4. $\frac{x+2y}{12y^2}\cdot\frac{4y}{x^2+xy-2y^2}=\frac{x+2y}{12y^2}\cdot\frac{4y}{(x+2y)(x-y)}$
$=\frac{1}{3y(x-y)}$

5. $\frac{2a^2-3a-2}{a^2+5a+6}\div\frac{a^2-5a+6}{a^2-9}$
$=\frac{2a^2-3a-2}{a^2+5a+6}\cdot\frac{a^2-9}{a^2-5a+6}$
$=\frac{(2a+1)(a-2)}{(a+2)(a+3)}\cdot\frac{(a+3)(a-3)}{(a-2)(a-3)}$
$=\frac{2a+1}{a+2}$

6. $\frac{1}{a^2-a-2}+\frac{3}{a-2}=\frac{1}{(a-2)(a+1)}+\frac{3}{a-2}\cdot\frac{a+1}{a+1}$
$=\frac{1+3(a+1)}{(a-2)(a+1)}$
$=\frac{1+3a+3}{(a-2)(a+1)}$
$=\frac{3a+4}{(a-2)(a+1)}$

7. $\frac{x-y}{xy}-\frac{a-y}{ay}=\frac{x-y}{xy}\cdot\frac{a}{a}-\frac{a-y}{ay}\cdot\frac{x}{x}$
$=\frac{ax-ay-ax+xy}{axy}$
$=\frac{xy-ay}{axy}$
$=\frac{y(x-a)}{axy}$
$=\frac{x-a}{ax}$

8. $\frac{3x}{x^2-3x-18}-\frac{x-4}{x-6}$
$=\frac{3x}{(x-6)(x+3)}-\frac{x-4}{x-6}$
$=\frac{3x}{(x-6)(x+3)}-\frac{x-4}{x-6}\cdot\frac{x+3}{x+3}$
$=\frac{3x}{(x-6)(x+3)}-\frac{x^2-x-12}{(x-6)(x+3)}$
$=\frac{3x-(x^2-x-12)}{(x-6)(x+3)}$
$=\frac{3x-x^2+x+12}{(x-6)(x+3)}$
$=\frac{-x^2+4x+12}{(x-6)(x+3)}$
$=\frac{-(x-6)(x+2)}{(x-6)(x+3)}$
$=-\frac{x+2}{x+3}$

9. $\frac{\frac{x}{3y}-\frac{1}{2}}{\frac{4}{3y}-\frac{2}{x}}=\frac{\frac{x}{3y}-\frac{1}{2}}{\frac{4}{3y}-\frac{2}{x}}\cdot\frac{6xy}{6xy}$
$=\frac{\frac{x}{3y}\cdot 6xy-\frac{1}{2}\cdot 6xy}{\frac{4}{3y}\cdot 6xy-\frac{2}{x}\cdot 6xy}$
$=\frac{2x^2-3xy}{8x-12y}$
$=\frac{x(2x-3y)}{4(2x-3y)}$
$=\frac{x}{4}$

10. $\frac{\frac{6}{b}-4}{\frac{5}{bx}-\frac{10}{3x}}=\frac{\frac{6}{b}-4}{\frac{5}{bx}-\frac{10}{3x}}\cdot\frac{3bx}{3bx}$
$=\frac{\frac{6(3bx)}{b}-4(3bx)}{\frac{5(3bx)}{bx}-\frac{10(3bx)}{3x}}$
$=\frac{18x-12bx}{15-10b}$
$=\frac{6x(3-2b)}{5(3-2b)}$
$=\frac{6x}{5}$

11. $\frac{2x^2+3xy-9y^2}{4x^2+13xy+3y^2}=\frac{(2x-3y)(x+3y)}{(4x+y)(x+3y)}=\frac{2x-3y}{4x+y}$

12. $\frac{1}{x+4}-\frac{2}{x^2+6x+8}$

$=\frac{1}{x+4}-\frac{2}{(x+4)(x+2)}$

$=\frac{1}{x+4}\cdot\frac{x+2}{x+2}-\frac{2}{(x+4)(x+2)}$

$=\frac{x+2}{(x+4)(x+2)}-\frac{2}{(x+4)(x+2)}$

$=\frac{x+2-2}{(x+4)(x+2)}$

$=\frac{x}{(x+4)(x+2)}$

13.

$$\frac{4}{3x}-\frac{5}{2x}=5-\frac{1}{6x}$$

$$6x\left(\frac{4}{3x}\right)-6x\left(\frac{5}{2x}\right)=6x(5)-6x\left(\frac{1}{6x}\right)$$

$$8-15=30x-1$$

$$-7=30x-1$$

$$-6=30x$$

$$\frac{-6}{30}=x$$

$$-\frac{1}{5}=x$$

Check: $\frac{4}{3\left(-\frac{1}{5}\right)}-\frac{5}{2\left(-\frac{1}{5}\right)}\stackrel{?}{=}5-\frac{1}{6\left(-\frac{1}{5}\right)}$

$$-\frac{20}{3}+\frac{25}{2}\stackrel{?}{=}5+\frac{5}{6}$$

$$-\frac{40}{6}+\frac{75}{6}\stackrel{?}{=}\frac{30}{6}+\frac{5}{6}$$

$$\frac{35}{6}=\frac{35}{6}$$

14.

$$\frac{x-3}{x-2}=\frac{2x^2-15}{x^2+x-6}-\frac{x+1}{x+3}$$

$$\frac{x-3}{x-2}=\frac{2x^2-15}{(x+3)(x-2)}-\frac{x+1}{x+3}$$

$$(x-2)(x+3)\left(\frac{x-3}{x-2}\right)=(x-2)(x+3)\left[\frac{2x^2-15}{(x+3)(x-2)}\right]-(x-2)(x+3)\left(\frac{x+1}{x+3}\right)$$

$$(x+3)(x-3)=2x^2-15-(x-2)(x+1)$$

$$x^2-9=2x^2-15-(x^2-x-2)$$

$$x^2-9=2x^2-15-x^2+x+2$$

$$x^2-9=x^2+x-13$$

$$-9=x-13$$

$$4=x$$

Check: $\frac{4-3}{4-2} \stackrel{?}{=} \frac{2(4)^2-15}{4^2+4-6} - \frac{4+1}{4+3}$

$\frac{1}{2} \stackrel{?}{=} \frac{17}{14} - \frac{5}{7}$

$\frac{7}{14} \stackrel{?}{=} \frac{17}{14} - \frac{10}{14}$

$\frac{7}{14} = \frac{7}{14}$

15. $3 - \frac{7}{x+3} = \frac{x-4}{x+3}$

$(x+3)(3) - (x+3)\left(\frac{7}{x+3}\right) = (x+3)\left(\frac{x-4}{x+3}\right)$

$3x + 9 - 7 = x - 4$

$2x = -6$

$x = -3$

$x = -3$ is extraneous. There is no solution.

16. $\frac{3}{3x-5} = \frac{7}{5x+4}$

$3(5x+4) = 7(3x-5)$

$15x + 12 = 21x - 35$

$12 = 6x - 35$

$47 = 6x$

$\frac{47}{6} = x$

Check: $\frac{3}{3\left(\frac{47}{6}\right)-5} \stackrel{?}{=} \frac{7}{5\left(\frac{47}{6}\right)+4}$

$\frac{3}{\frac{111}{6}} \stackrel{?}{=} \frac{7}{\frac{259}{6}}$

$\frac{18}{111} \stackrel{?}{=} \frac{42}{259}$

$\frac{6}{37} = \frac{6}{37}$

17. $\frac{9}{x} = \frac{13}{5}$

$45 = 13x$

$\frac{45}{13} = x$

18. $\frac{9.3}{2.5} = \frac{x}{10}$

$\frac{10(9.3)}{2.5} = x$

$37.2 = x$

19. x = on-time flights

$\frac{x}{200} = \frac{113}{150}$

$150x = 113(200)$

$150x = 22{,}600$

$x = 151$

151 flights can be expected to be on time.

20. x = cost of wood for 92 days

$\frac{x}{92} = \frac{100}{25}$

$25x = 9200$

$x = 368$

It will cost \$368.

21. $\frac{\text{height}}{\text{shadow}}$: $\frac{6}{7} = \frac{87}{x}$

$6x = 609$

$x = 101.5$

The rope bridge should be 102 feet.

Cumulative Test for Chapters 0–6

1. 0.25% of $2.57 = (0.0025)(2.57)$
$= 0.006$ centimeters

2. x = % of agricultural exports to Canada

$x = \frac{10{,}570{,}000{,}000}{62{,}958{,}000{,}000} \cdot 100\%$

$x = 16.8\%$

3. Difference = 7% − 6.5% = 0.5%
Savings = 0.005(22,500) = \$112.50

4. $5(x-3) - 2(4-2x) = 7(x-1) - (x-2)$

$5x - 15 - 8 + 4x = 7x - 7 - x + 2$

$9x - 23 = 6x - 5$

$3x - 23 = -5$

$3x = 18$

$x = 6$

5. $A = \pi r^2 h$

$\frac{A}{\pi r^2} = \frac{\pi r^2 h}{\pi r^2}$

$\frac{A}{\pi r^2} = h$

6. $4(2-x)<3$
$$8-4x<3$$
$$-4x<-5$$
$$\frac{-4x}{-4}>\frac{-5}{-4}$$
$$x>\frac{5}{4}$$

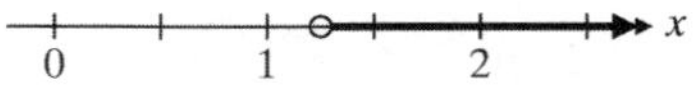

7.
$$\frac{3}{5}x-2\le\frac{1}{2}x+\frac{3}{5}$$
$$10\left(\frac{3}{5}x\right)-10(2)\le 10\left(\frac{1}{2}x\right)+10\left(\frac{3}{5}\right)$$
$$6x-20\le 5x+6$$
$$x-20\le 6$$
$$x\le 26$$

8.
$$3ax+3bx-2ay-2by=3x(a+b)-2y(a+b)$$
$$=(a+b)(3x-2y)$$

9.
$$8a^3-38a^2b-10ab^2=2a(4a^2-19ab-5b^2)$$
$$=2a(4a+b)(a-5b)$$

10.
$$\frac{-16a^3b^{-1}}{4a^{-2}b^5}=\frac{-4a^{3+2}}{b^{5+1}}=\frac{-4a^5}{b^6}$$

11.
$$\frac{4x^2-25}{2x^2+9x-35}=\frac{(2x+5)(2x-5)}{(2x-5)(x+7)}=\frac{2x+5}{x+7}$$

12.
$$\frac{x^2-4}{x^2-25}\cdot\frac{3x^2-14x-5}{3x^2+6x}$$
$$=\frac{(x+2)(x-2)}{(x+5)(x-5)}\cdot\frac{(3x+1)(x-5)}{3x(x+2)}$$
$$=\frac{(x-2)(3x+1)}{3x(x+5)}$$

13.
$$\frac{2x^2-9x+9}{8x-12}\div\frac{x^2-3x}{2x}$$
$$=\frac{2x^2-9x+9}{8x-12}\cdot\frac{2x}{x^2-3x}$$
$$=\frac{(2x-3)(x-3)}{4(2x-3)}\cdot\frac{2x}{x(x-3)}$$
$$=\frac{1}{2}$$

14.
$$\frac{5}{2x+4}+\frac{3}{x-3}=\frac{5}{2(x+2)}\cdot\frac{x-3}{x-3}+\frac{3}{x-3}\cdot\frac{2(x+2)}{2(x+2)}$$
$$=\frac{5x-15+6x+12}{2(x+2)(x-3)}$$
$$=\frac{11x-3}{2(x+2)(x-3)}$$

15.
$$\frac{8}{c^2-4}-\frac{2}{c^2-5c+6}$$
$$=\frac{8}{(c+2)(c-2)}\cdot\frac{c-3}{c-3}-\frac{2}{(c-2)(c-3)}\cdot\frac{c+2}{c+2}$$
$$=\frac{8(c-3)}{(c+2)(c-2)(c-3)}-\frac{2(c+2)}{(c-2)(c-3)(c+2)}$$
$$=\frac{8c-24-(2c+4)}{(c+2)(c-2)(c-3)}$$
$$=\frac{6c-28}{(c+2)(c-2)(c-3)}$$

16.
$$\frac{3x-2}{3x+2}=2$$
$$(3x+2)\left(\frac{3x-2}{3x+2}\right)=(3x+2)(2)$$
$$3x-2=6x+4$$
$$-6=3x$$
$$-2=x$$

17.
$$\frac{x-3}{x}=\frac{x+2}{x+3}$$
$$x(x+2)=(x-3)(x+3)$$
$$x^2+2x=x^2-9$$
$$2x=-9$$
$$x=-\frac{9}{2}\text{ or }-4\frac{1}{2}\text{ or }-4.5$$

18.
$$\frac{\frac{1}{x-3}+\frac{5}{x^2-9}}{\frac{6x}{x+3}}=\frac{\frac{1}{x-3}+\frac{5}{(x+3)(x-3)}}{\frac{6x}{x+3}}\cdot\frac{(x+3)(x-3)}{(x+3)(x-3)}$$
$$=\frac{x+3+5}{6x(x-3)}$$
$$=\frac{x+8}{6x(x-3)}$$

19. $\dfrac{\frac{3}{a}+\frac{2}{b}}{\frac{5}{a^2}-\frac{2}{b^2}} = \dfrac{\frac{3}{a}+\frac{2}{b}}{\frac{5}{a^2}-\frac{2}{b^2}} \cdot \dfrac{a^2b^2}{a^2b^2}$

$= \dfrac{\frac{3}{a}\cdot a^2b^2 + \frac{2}{b}\cdot a^2b^2}{\frac{5}{a^2}\cdot a^2b^2 - \frac{2}{b^2}\cdot a^2b^2}$

$= \dfrac{3ab^2 + 2a^2b}{5b^2 - 2a^2}$

20. $\dfrac{2x+1}{3} = \dfrac{4}{5}$

$5(2x+1) = 3(4)$

$10x + 5 = 12$

$10x = 7$

$x = \dfrac{7}{10}$ or 0.7

21. $\dfrac{\text{mi}}{\text{in.}}$: $\dfrac{130}{2\frac{1}{2}} = \dfrac{x}{4}$

$\left(2\dfrac{1}{2}\right)x = 130.4$

$\dfrac{5}{2}x = 520$

$x = 208$

She has 208 miles to drive.

22. x = phone calls to make 110 sales

$\dfrac{x}{110} = \dfrac{22}{5}$

$5x = 2420$

$x = 484$

He must make 484 phone calls.

Chapter 7

7.1 Exercises

1. The x-coordinate of the origin is 0.

3. (5, 1) is an ordered pair because the order is important. The graphs of (5, 1) and (1, 5) are different.

5.

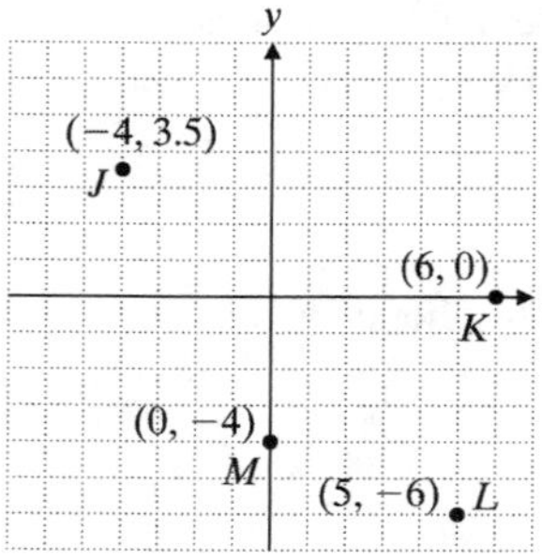

7. R: (−3, −5)

S: $\left(-4\frac{1}{2}, 0\right)$

X: (3, −5)

Y: $\left(2\frac{1}{2}, 6\right)$

9. (−4, −1), (−3, −2), (−2, −3), (−1, −5), (0, −3), (2, −1)

11. $y = 4x + 7$

a. (0,), $y = 4(0) + 7 = 0 + 7 = 7$
(0, 7)

b. (2,), $y = 4(2) + 7 = 8 + 7 = 15$
(2, 15)

13. $y = -6x + 5$

a. (−1,), $y = -6(-1) + 5 = 6 + 5 = 11$
(−1, 11)

b. (3,), $y = -6(3) + 5 = -18 + 5 = -13$
(3, −13)

15. $3x - 4y = 11$

a. (−3,):

$$\begin{aligned} 3(-3) - 4y &= 11 \\ -9 - 4y &= 11 \\ -4y &= 20 \\ y &= -5 \end{aligned}$$

(−3, −5)

b. (5,):

$$\begin{aligned} 3x - 4(1) &= 11 \\ 3x - 4 &= 11 \\ 3x &= 15 \\ x &= 5 \end{aligned}$$

(5, 1)

17. $3x + 2y = -6$

a. (−2,):

$$\begin{aligned} 3(-2) + 2y &= -6 \\ -6 + 2y &= -6 \\ 2y &= 0 \\ y &= 0 \end{aligned}$$

(−2, 0)

b. (, 3)

$$\begin{aligned} 3x + 2(3) &= -6 \\ 3x + 6 &= -6 \\ 3x &= -12 \\ x &= -4 \end{aligned}$$

(−4, 3)

19. $x + \frac{1}{3}y = 5$

a. (, 18), $x = 5 - \frac{1}{3}(18) = 5 - 6 = -1$

(−1, 18)

b. $\left(\frac{8}{3}, \ \right)$, $y = 3\left(5 - \frac{8}{3}\right) = 3\left(\frac{7}{3}\right) = 7$

$\left(\frac{8}{3}, 7\right)$

21. Locate the grid for Lynbrook, NY: B5

23. Locate the grid for Athol, Mass: E1

25. Look for the grid for Hartford, CT: D3

27. a. (0, 882), (1, 803), (2, 746), (3, 767), (4, 705), (5, 615)

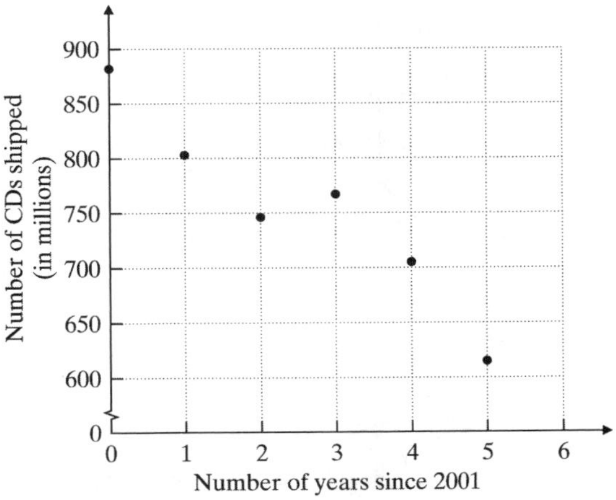

b. The number of CDs shipped decreased overall between 2001 and 2006 with a slight increase from 2003 to 2004.

29. a. (2003, 3.2), (2004, 3.6), (2005, 3.9), (2006, 4.2)

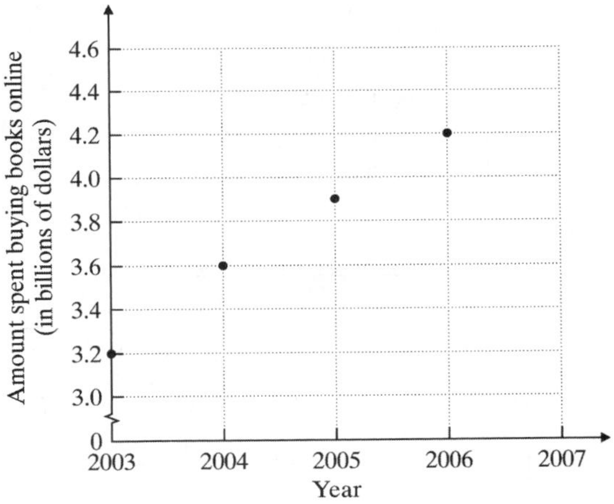

b. An estimated $4.5 billion was spent buying books on line in 2007.

Cumulative Review

31. $\frac{8x}{x+y}+\frac{8y}{x+y}=\frac{8x+8y}{x+y}=\frac{8(x+y)}{x+y}=8$

32. $\frac{x^2-y^2}{x^2+4xy+3y^2}\cdot\frac{x^2-9y^2}{x^2-6xy+9y^2}$

$=\frac{(x+y)(x-y)}{(x+y)(x+3y)}\cdot\frac{(x+3y)(x-3y)}{(x-3y)(x-3y)}$

$=\frac{x-y}{x-3y}$

Quick Quiz 7.1

1.

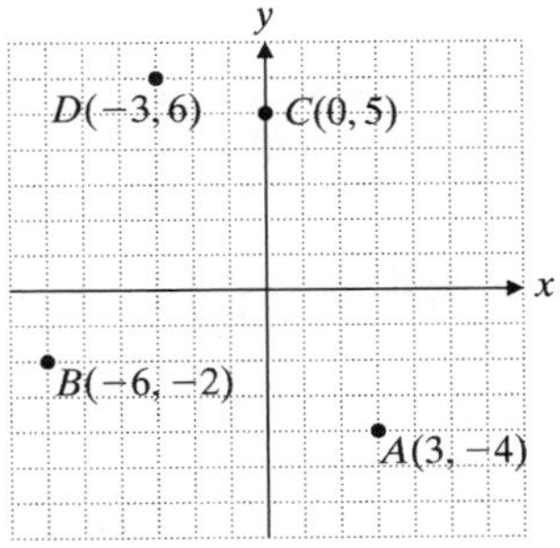

2. $y=-5x-7$

a. $(-2,\ \ )$, $y=-5(-2)-7=10-7=3$
$(-2, 3)$

b. $(3,\ \ )$, $y=-5(3)-7=-15-7=-22$
$(3, -22)$

c. $(0,\ \ )$, $y=-5(0)-7=0-7=-7$
$(0, -7)$

3. $4x-3y=-12$

$-3y=-12-4x$

$y=\frac{-12-4x}{-3}$

or

$4x=-12+3y$

$x=\frac{-12+3y}{4}$

a. $(3,\ \ )$, $y=\frac{-12-4(3)}{-3}=\frac{-24}{-3}=8$

$(3, 8)$

b. $(\ \ ,-8)$, $x=\frac{-12+3(-8)}{4}=\frac{-36}{4}=-9$

$(-9, -8)$

c. $(\ \ , 10),\ x = \dfrac{-12+3(10)}{4} = \dfrac{18}{4} = \dfrac{9}{2}$

$\left(\dfrac{9}{2}, 10\right)$

4. Answers may vary. Possible solution: Isolate x on the left side of the equation. Substitute the given value for y and solve for x.

7.2 Exercises

1. No, replacing x by −2 and y by 5 in $2x + 5y = 0$ does not result in a true statement so (−2, 5) is not a solution.

3. The x-intercept of a line is the point where the line crosses the <u>x-axis</u>.

5. $y = -2x + 1$

x	$y = -2x + 1$	y
0	$-2(0) + 1 = 0 + 1$	1
−2	$-2(-2) + 1 = -4 + 1$	5
1	$-2(1) + 1 = -2 + 1$	−1

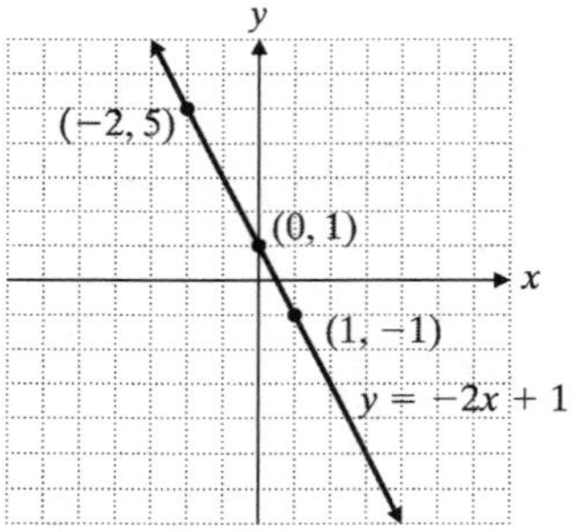

7. $y = x - 4$

x	$y = x - 4$	y
0	$y = 0 - 4$	−4
2	$y = 2 - 4$	−2
4	$y = 4 - 4$	0

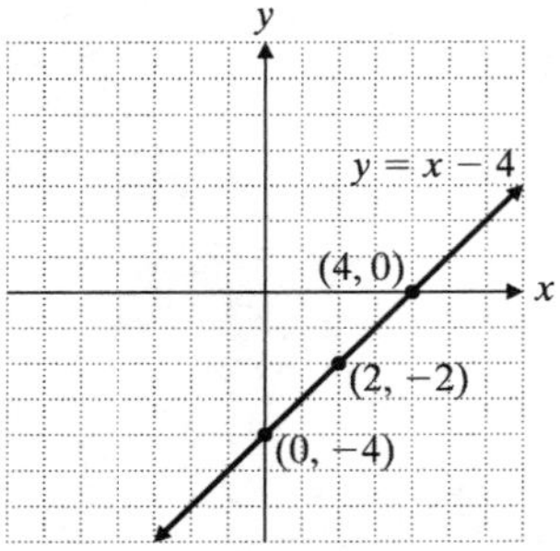

9. $y = -2x + 3$

x	$y = -2x + 3$	y
2	$y = -2(2) + 3 = -4 + 3$	−1
4	$y = -2(4) + 3 = -8 + 3$	−5
0	$y = -2(0) + 3 = 0 + 3$	3

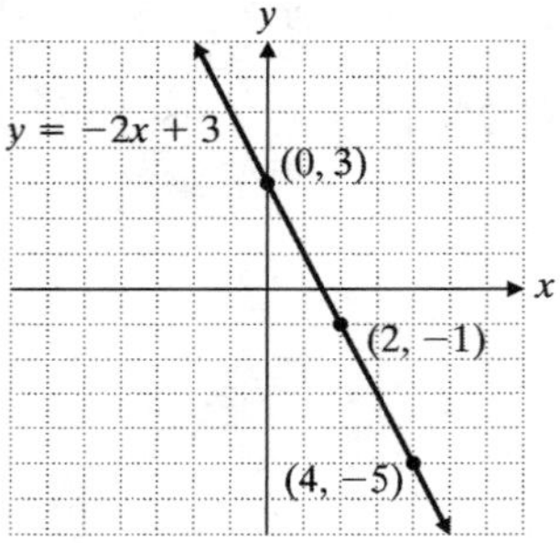

11. $y = 3x + 2$

x	$y = 3x + 2$	y
−1	$y = 3(-1) + 2 = -3 + 2$	−1
0	$y = 3(0) + 2 = 0 + 2$	2
1	$y = 3(1) + 2 = 3 + 2$	5

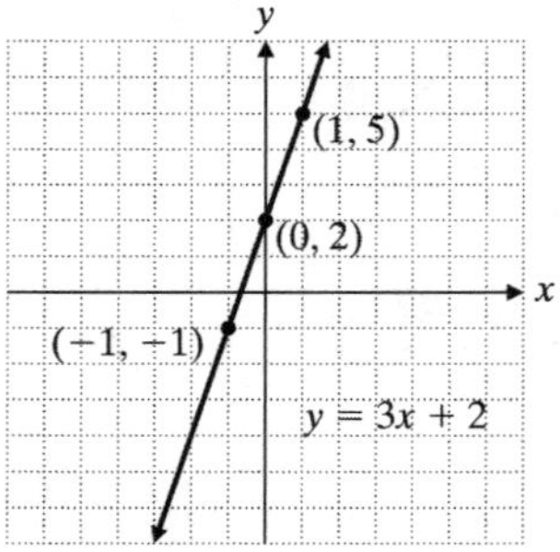

13. $3x - 2y = 0$

$2y = 3x$

$y = \frac{3x}{2}$

x	$y = \frac{3x}{2}$	y
−2	$y = \frac{3(-2)}{2} = \frac{-6}{2}$	−3
0	$y = \frac{3(0)}{2} = \frac{0}{2}$	0
2	$y = \frac{3(2)}{2} = \frac{6}{2}$	3

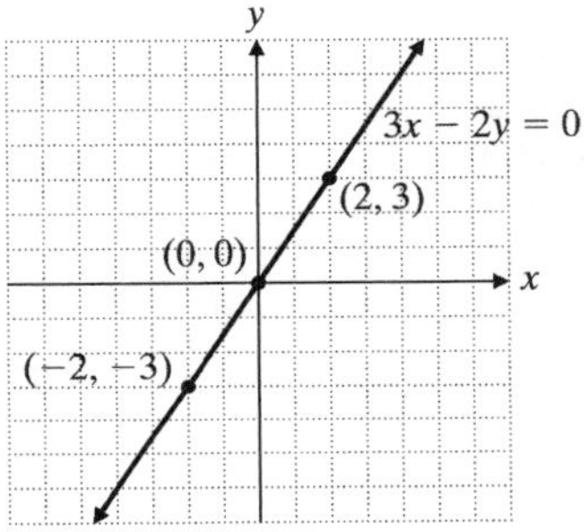

15. $y = -\frac{3}{4}x + 3$

x	$y = -\frac{3}{4}x + 3$	y
−4	$y = -\frac{3}{4}(-4) + 3 = 3 + 3$	6
0	$y = -\frac{3}{4}(0) + 3 = 0 + 3$	3
4	$y = -\frac{3}{4}(4) + 3 = -3 + 3$	0

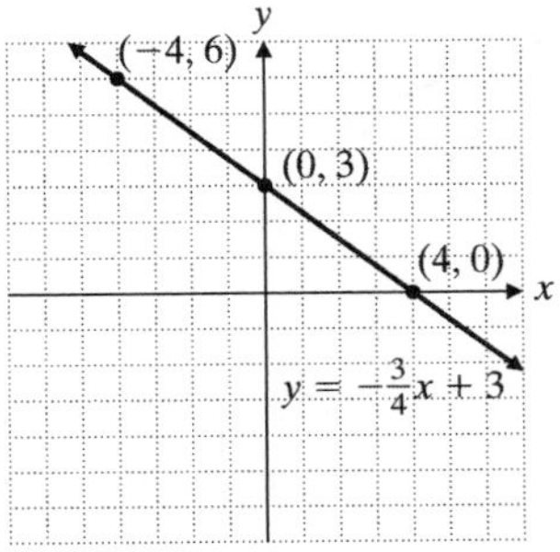

17. $4x + 3y = 12$

$3y = -4x + 12$

$y = -\frac{4}{3}x + 4$

x	$y = -\frac{4}{3}x + 4$	y
0	$y = -\frac{4}{3}(0) + 4 = 0 + 4$	4
3	$y = -\frac{4}{3}(3) + 4 = -4 + 4$	0
6	$y = -\frac{4}{3}(6) + 4 = -8 + 4$	−4

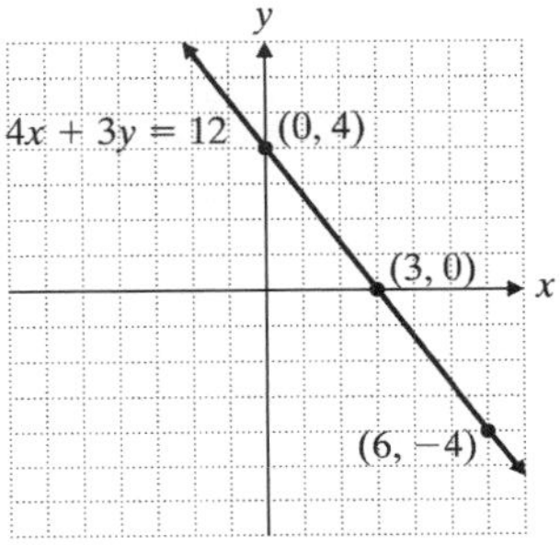

19. $y = 6 - 2x$

Let $x = 0$.

$y = 6 - 2(0) = 6 - 0 = 6$

Let $y = 0$.

$0 = 6 - 2x$

$2x = 6$

$x = 3$

Let $x = 1$.

$y = 6 - 2(1) = 6 - 2 = 4$

x	y
0	6
1	4
3	0

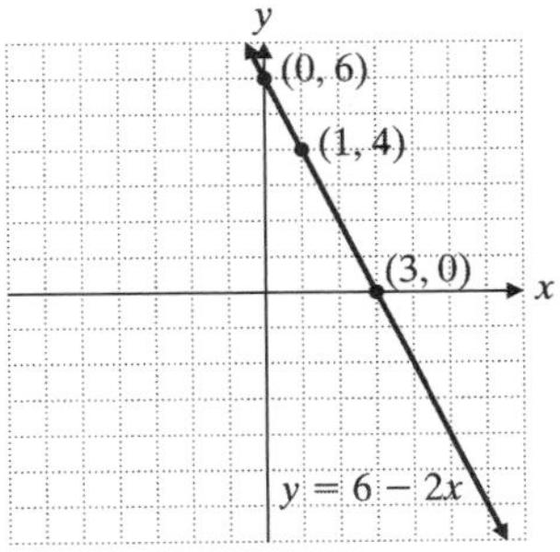

21. $x + 3 = 6y$
Let $x = 0$.
$0+3=6y$
$\frac{1}{2}=y$
Let $y = 0$.
$x+3=6(0)$
$x=-3$
Let $x = 3$.
$3+3=6y$
$1=y$

x	y
−3	0
0	$\frac{1}{2}$
3	1

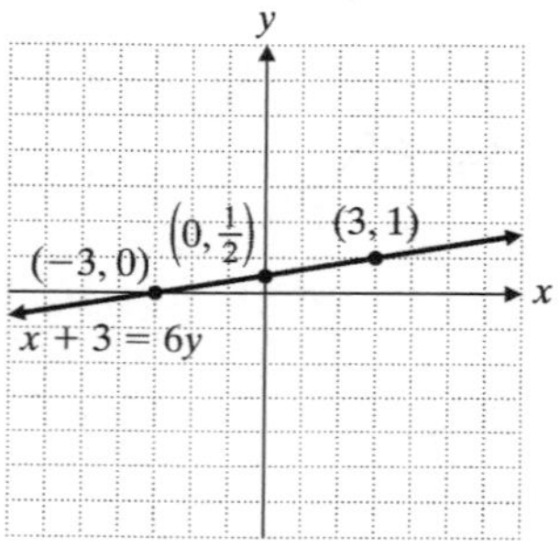

23. $y-2=3y$
$-2=2y$
$-1=y$
This is a horizontal line 1 unit below the axis.

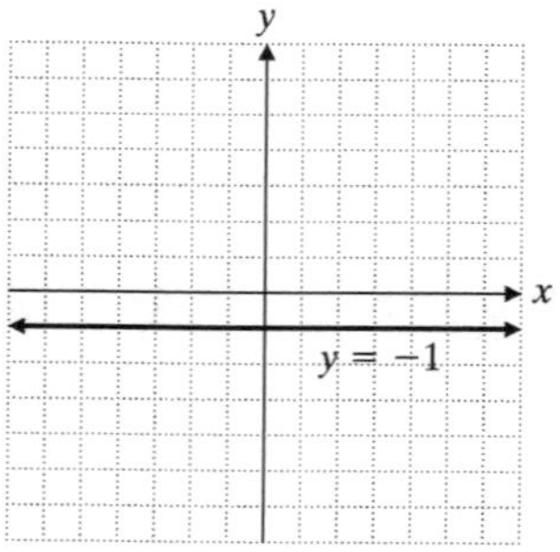

25. $2x+9=5x$
$-3x+9=0$
$-3x=-9$
$x=3$
This is a vertical line 3 units to the right of the axis.

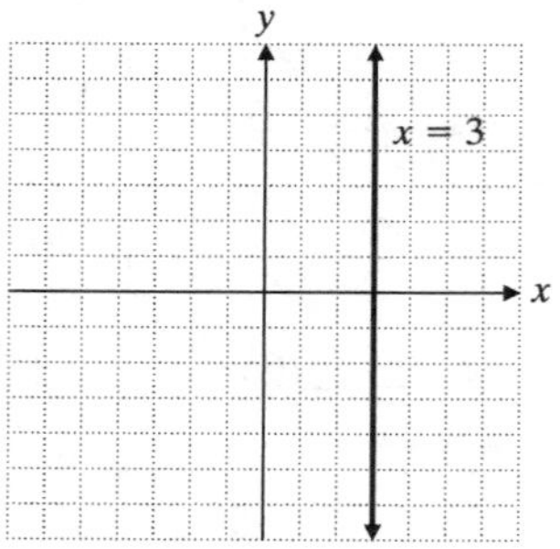

27. $2x+5y-2=-12$
$2x+5y=-10$
$5y=-2x-10$
$y=-\frac{2}{5}x-2$

x	$y=-\frac{2}{5}x-2$	y
−5	$y=-\frac{2}{5}(-5)-2=2-2$	0
0	$y=-\frac{2}{5}(0)-2=0-2$	−2
5	$y=-\frac{2}{5}(5)-2=-2-2$	−4

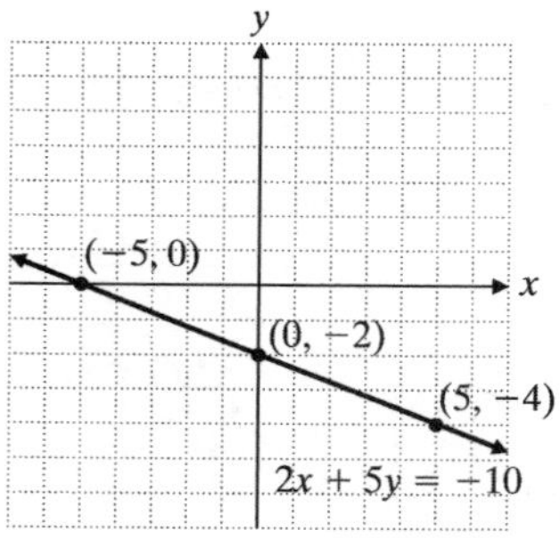

29. $C = 8m$

m	0	15	30	45	60	75
C	0	120	240	360	480	600

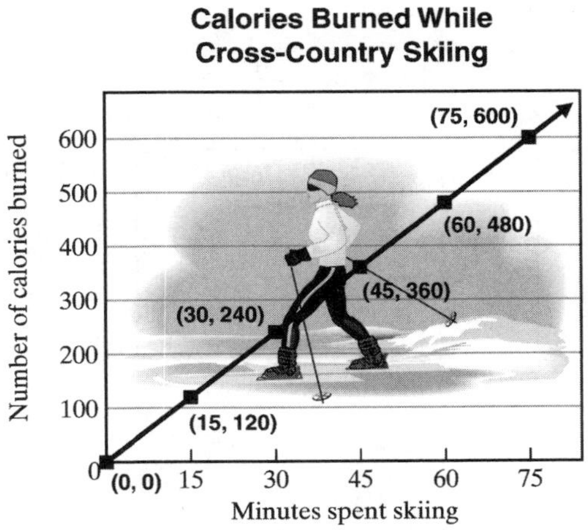

31. $S = 11t + 395$

t	$S = 11t + 395$	S
0	$S = 11(0) + 395 = 0 + 395$	395
4	$S = 11(4) + 395 = 44 + 395$	439
8	$S = 11(8) + 395 = 88 + 395$	483
16	$S = 11(16) + 395 = 176 + 395$	571

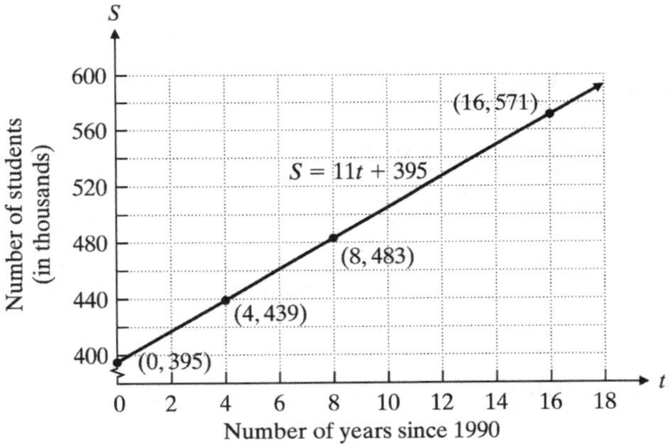

Cumulative Review

33.

$$\frac{x}{x+4}+\frac{x-1}{x-4}=2$$

$$(x-4)(x+4)\cdot\frac{x}{x+4}+(x-4)(x+4)\left(\frac{x-4}{x-4}\right)=(x-4)(x+4)(2)$$

$$(x-4)(x)+(x+4)(x-1)=2(x+4)(x-4)$$

$$x^2-4x+x^2+3x-4=2x^2-32$$

$$-x=-28$$

$$x=28$$

34. $4-3x \le 18$
$-3x \le 14$
$x \ge -\frac{14}{3}$

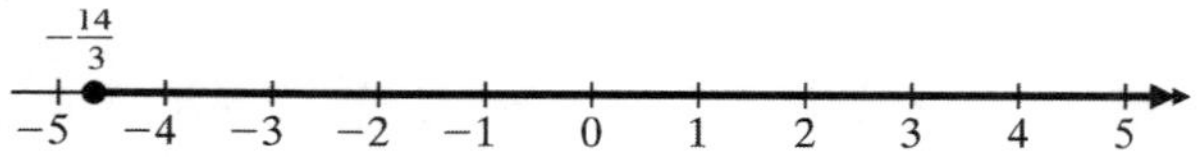

Quick Quiz 7.2

1. $y = -2x + 4$

x	$y = -2x + 4$	y
0	$y = -2(0) + 4 = 0 + 4$	4
2	$y = -2(2) + 4 = -4 + 4$	0
3	$y = -2(3) + 4 = -6 + 4$	−2

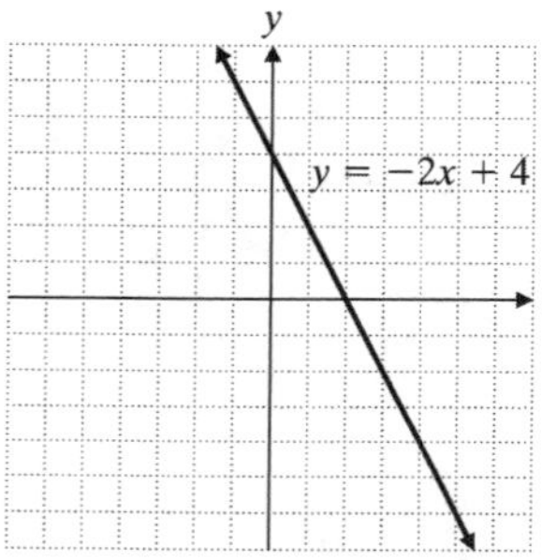

2. $y = \frac{2}{5}x - 3$

x	$y = \frac{2}{5}x - 3$	y
−5	$y = \frac{2}{5}(-5) - 3 = -2 - 3$	−5
0	$y = \frac{2}{5}(0) - 3 = 0 - 3$	−3
5	$y = \frac{2}{5}(5) - 3 = 2 - 3$	−1

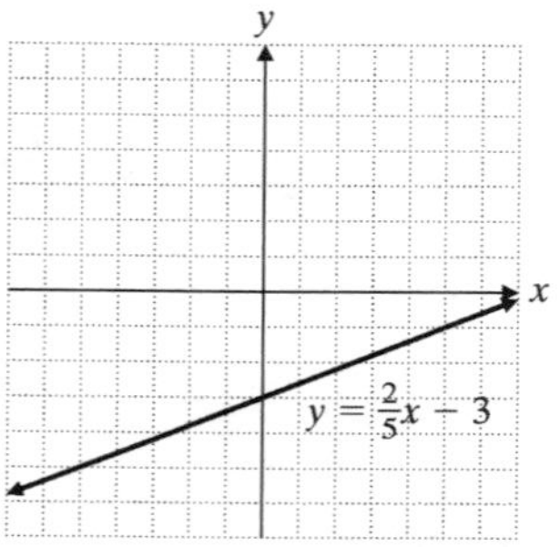

3. $4y + 1 = x + 9$
$x = 4y - 8$ or $y = \frac{1}{4}x + 2$

x	$x = 4y - 8$	y
−4	$x = 4(1) - 8 = 4 - 8$	1
0	$x = 4(2) - 8 = 8 - 8$	2
4	$x = 4(3) - 8 = 12 - 8$	3

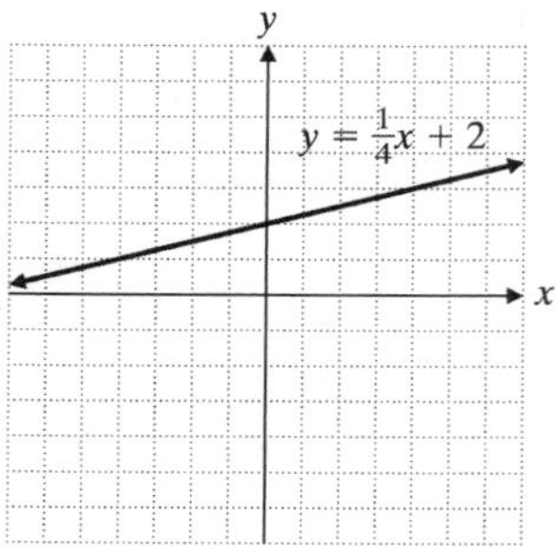

4. Answers may vary. Possible solution: The most important ordered pair gives the y-intercept. With the y-intercept and the slope (covered in the next section), the line may be graphed.

7.3 Exercises

Use $m = \frac{y_2 - y_1}{x_2 - x_1}$ in Exercises 1–13.

1. You cannot find the slope of the line passing through (5, −12) and (5, −6) because division by zero is impossible so the slope is undefined.

3. (4, 1) and (6, 7)
$m = \frac{7-1}{6-4} = \frac{6}{2} = 3$

5. (5, 10) and (6, 5)
$m = \frac{5-10}{6-5} = \frac{-5}{1} = -5$

7. (−2, 1) and (3, 4)
$m = \frac{4-1}{3-(-2)} = \frac{3}{5}$

9. (−6, −5) and (2, −7)
$m = \frac{-7-(-5)}{2-(-6)} = \frac{-2}{8} = -\frac{1}{4}$

11. $(-3, 0)$ and $(0, -4)$

$$m = \frac{-4-0}{0-(-3)} = \frac{-4}{3} = -\frac{4}{3}$$

13. $\left(\frac{3}{4}, -4\right)$ and $(2, -8)$

$$m = \frac{-8-(-4)}{2-\frac{3}{4}} = \frac{-4}{\frac{5}{4}} = -\frac{16}{5}$$

Use $y = mx + b$ in Exercises 15–33.

15. $y = 8x + 9$; $m = 8$, $b = 9$, y-intercept $(0, 9)$

17. $3x + y - 4 = 0$, $y = -3x - 4$, $m = -3$, $b = 4$, y-intercept $(0, 4)$

19. $y = -\frac{8}{7}x + \frac{3}{4}$; $m = -\frac{8}{7}$, $b = \frac{3}{4}$, y-intercept $\left(0, \frac{3}{4}\right)$

21. $y = -6x$; $m = -6$, $b = 0$, y-intercept $(0, 0)$

23. $6x + y = \frac{4}{5}$

$y = -6x + \frac{4}{5}$; $m = -6$, $b = \frac{4}{5}$,

y-intercept $\left(0, \frac{4}{5}\right)$

25. $3x + 4y - 6 = 0$

$$4y = -3x + 6$$
$$y = -\frac{3}{4}x + \frac{3}{2}$$

$m = -\frac{3}{4}$

$b = \frac{3}{2}$, y-intercept $\left(0, \frac{3}{2}\right)$

27. $7x - 3y = 4$

$y = \frac{7}{3}x - \frac{4}{3}$; $m = \frac{7}{3}$, $b = -\frac{4}{3}$,

y-intercept $\left(0, -\frac{4}{3}\right)$

29. $m = \frac{3}{5}$, y-intercept $(0, 3)$, $b = 3$

$y = \frac{3}{5}x + 3$

31. $m = 4$, y-intercept $(0, -5)$, $b = -5$
$y = 4x - 5$

33. $m = -\frac{5}{4}$, y-intercept $\left(0, -\frac{3}{4}\right)$, $b = -\frac{3}{4}$

$y = -\frac{5}{4}x - \frac{3}{4}$

35. $m = \frac{3}{4}$, $b = -4$

$y = \frac{3}{4}x - 4$

Graph $(0, -4)$. Find another point using the slope. Begin at $(0, -4)$. Go up 3 units and right 4 units: $(4, -1)$.

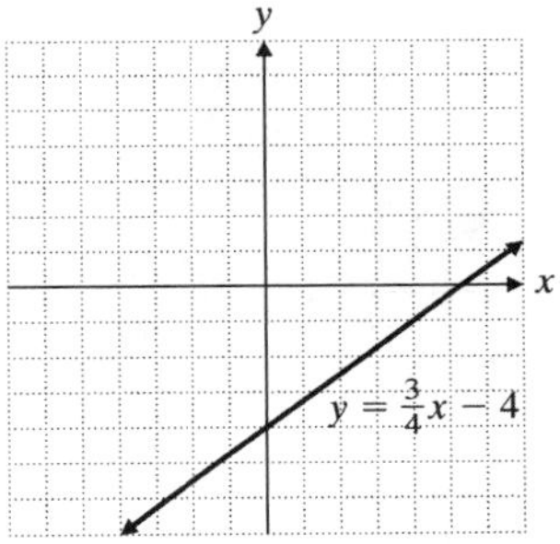

37. $m = -\frac{5}{3}$, $b = 2$

$y = -\frac{5}{3}x + 2$

Graph $(0, 2)$. Find another point using the slope. Begin at $(0, 2)$. Go down 5 units and right 3 units: $(3, -3)$

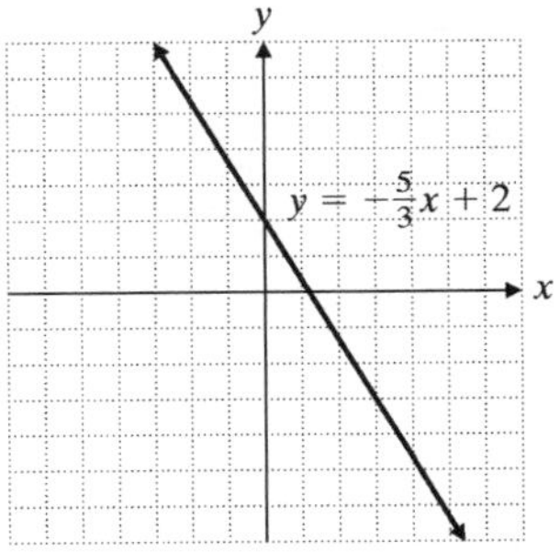

39. $y = \frac{2}{3}x + 2$

$m = \frac{2}{3}$, $b = 2$, y-intercept $(0, 2)$

Find another point using the slope: (3, 4).

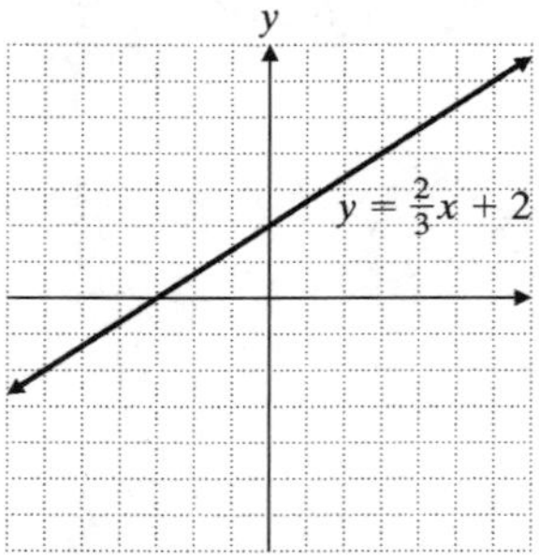

41. $y + 2x = 3$

$y = -2x + 3$

$m = -2$ or $\frac{-2}{1}$, $b = 3$, y-intercept $(0, 3)$

Find another point using the slope: (1, 1)

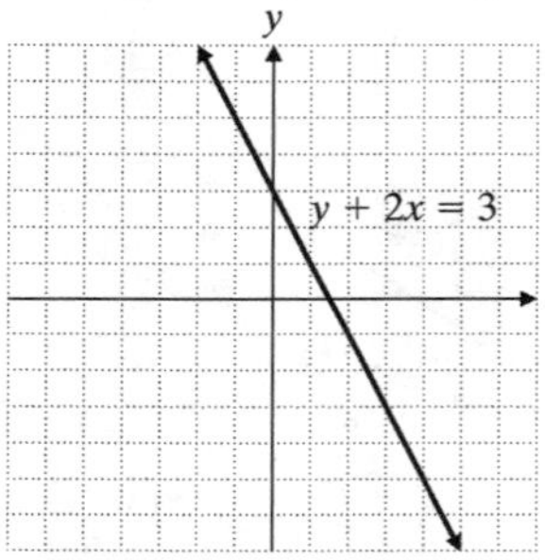

43. $y = 2x$

$m = 2$, $b = 0$, y-intercept $(0, 0)$

Find another point using the slope: (1, 2)

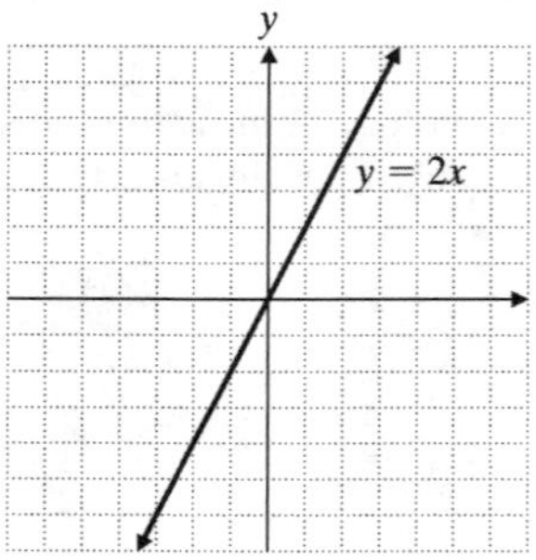

45. $m_1 = \frac{5}{6}$

a. $m_2 = m_1 = \frac{5}{6}$

b. $m_2 = -\frac{1}{m_1} = -\frac{6}{5}$

47. $m_1 = 6$

a. $m_2 = m_1 = 6$

b. $m_2 = -\frac{1}{m_1} = -\frac{1}{6}$

49. $y = \frac{2}{3}x + 6$, $m_1 = \frac{2}{3}$

a. $m_2 = m_1 = \frac{2}{3}$

b. $m_2 = -\frac{1}{m_1} = -\frac{3}{2}$

51. Yes; after graphing the points, one can see they possibly all lie on the same line. Notice the slope between each set of points is equal.

$m_1 = \frac{6-(-4)}{18-3} = \frac{10}{15} = \frac{2}{3}$, $m_2 = \frac{0-6}{9-18} = \frac{-6}{-9} = \frac{2}{3}$,

$m_3 = \frac{0-(-4)}{9-3} = \frac{4}{6} = \frac{2}{3}$

From the graph, $b = -6$.

$y = \frac{2}{3}x - 6$ or $2x - 3y = 18$

53. $y = 5(7x + 125)$

a. $y = 5 \cdot 7x + 5 \cdot 125$

$y = 35x + 625$

b. $m = 35$, $b = 625$, y-intercept $(0, 625)$

c. The slope is the amount of increase in the number of home health aides in the U.S. per year.

Cumulative Review

55.

$$3x + 8 > 2x + 12$$
$$3x - 2x + 8 > 2x - 2x + 12$$
$$x + 8 > 12$$
$$x > 4$$

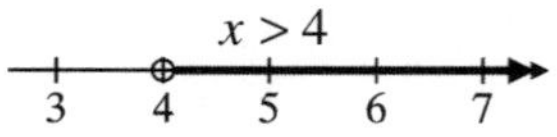

56. $\frac{1}{4}x+3>\frac{2}{3}x+2$

$$12\left(\frac{1}{4}x+3\right)>12\left(\frac{2}{3}x+2\right)$$
$$3x+12\cdot 3>4\cdot 2x+12\cdot 2$$
$$3x+36>8x+24$$
$$-5x>-12$$
$$x<\frac{12}{5}$$

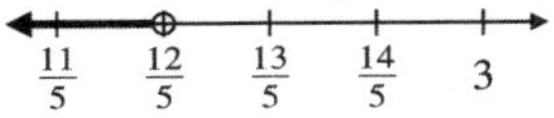

57. $\frac{1}{2}(x+2)\le\frac{1}{3}x+5$

$$6\left[\frac{1}{2}(x+2)\right]\le 6\left(\frac{1}{3}x+5\right)$$
$$3(x+2)\le 2x+30$$
$$3x+6\le 2x+30$$
$$x\le 24$$

58. $7x-2(x+3)\le 4(x-7)$

$$7x-2x-6\le 4x-28$$
$$5x-6\le 4x-28$$
$$x-6\le -28$$
$$x\le -22$$

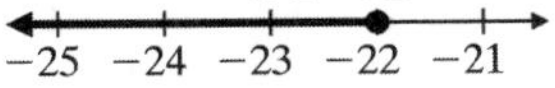

Quick Quiz 7.3

1. $(-2, 5), (-6, 3)$

$$m=\frac{3-5}{-6-(-2)}=\frac{-2}{-4}=\frac{1}{2}$$

2. $6x-2y-3=0$

$$2y=6x-3$$
$$y=3x-\frac{3}{2}$$

$m=3,\ b=-\frac{3}{2}$, y-intercept $\left(0,-\frac{3}{2}\right)$

3. $m=-\frac{5}{7}$

$$b=-5$$
$$y=mx+b$$
$$y=-\frac{5}{7}x-5$$

4. Answers may vary. Possible solution: First identify that the slope of the line of the equation given is $\frac{7}{11}$. The slope of a line perpendicular to the line of the equation given will be the negative reciprocal of the slope of line of the equation given. This is written as follows: $m_1m_2=-1$ or $m_1=-\frac{1}{m_2}$

How Am I Doing? Sections 7.1–7.3

1.

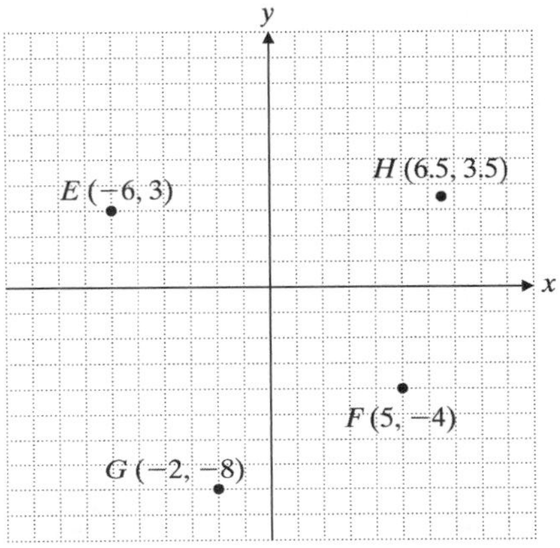

2. $A(3, 9)$
$B(7, -7)$
$C(-7, -8)$
$D(-4, 3)$

3. $y=-7x+3$

$y=-7(4)+3=-25,\ (4, -25)$
$y=-7(0)+3=3,\ (0, 3)$
$y=-7(-2)+3=17,\ (-2, 17)$

4. $6x-5y=-30$

$$5y=6x+30$$
$$y=\frac{6}{5}x+6$$

x	$y=\frac{6}{5}x+6$	y
0	$y=\frac{6}{5}(0)+6=0+6$	6
−5	$y=\frac{6}{5}(-5)+6=-6+6$	0
−10	$y=\frac{6}{5}(-10)+6=-12+6$	−6

5. $y = -4x - 3$

x	$y = -4x - 3$	y
-2	$y = -4(-2) - 3 = 8 - 3$	5
-1	$y = -4(-1) - 3 = 4 - 3$	1
0	$y = -4(0) - 3 = 0 - 3$	-3

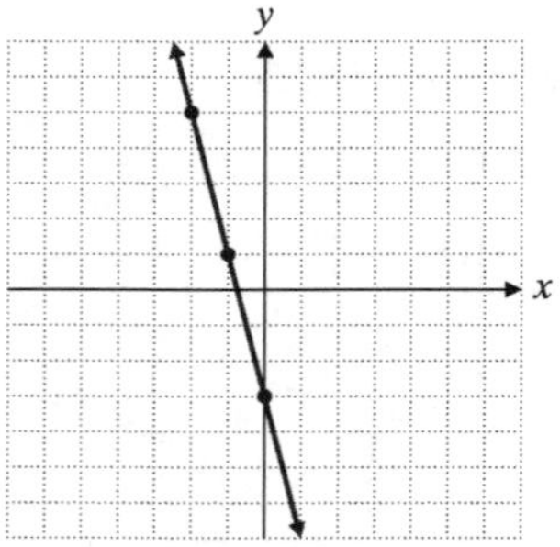

6. $y = \frac{3}{4}x - 1$

x	$y = \frac{3}{4}x - 1$	y
-4	$y = \frac{3}{4}(-4) - 1 = -3 - 1$	-4
0	$y = \frac{3}{4}(0) - 1 = 0 - 1$	-1
4	$y = \frac{3}{4}(4) - 1 = 3 - 1$	2

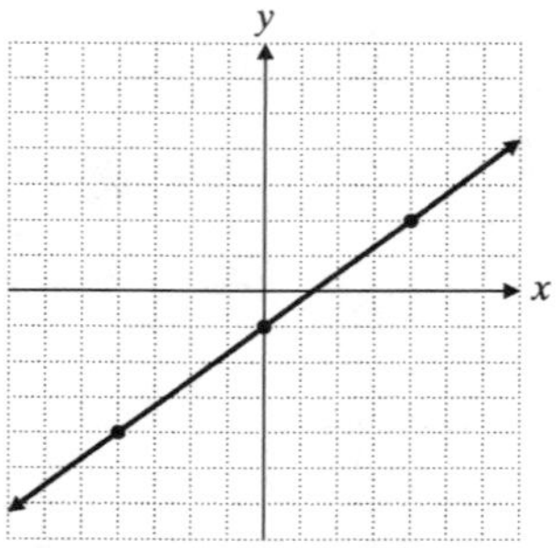

7. $3x - 5y = 0$

$$5y = 3x$$

$$y = \frac{3}{5}x$$

x	$y = \frac{3}{5}x$	y
-5	$y = \frac{3}{5}(-5)$	-3
0	$y = \frac{3}{5}(0)$	0
5	$y = \frac{3}{5}(5)$	3

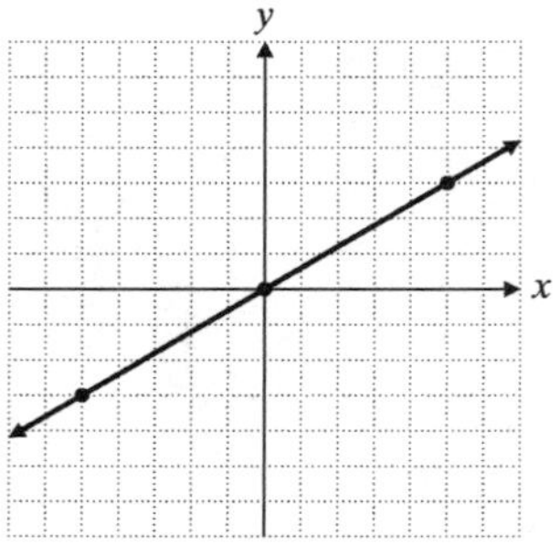

8. $-5x + 2 = y + 3$

$$-5x - 1 = y$$

x	$y = -5x - 1$	y
-1	$y = -5(-1) - 1 = 5 - 1$	4
0	$y = -5(0) - 1 = 0 - 1$	-1
1	$y = -5(1) - 1 = -5 - 1$	-6

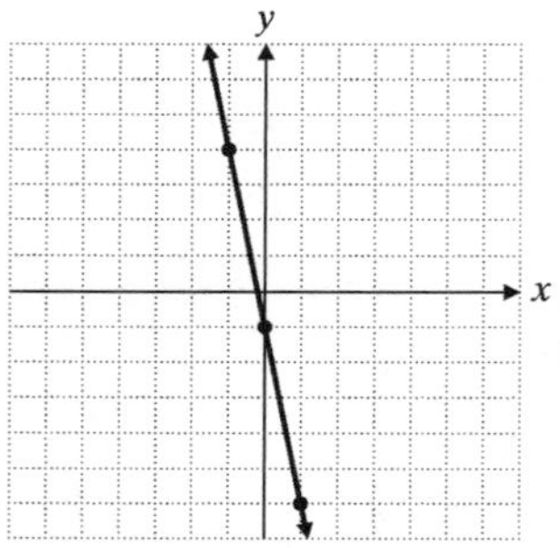

9. $(-2, 5), (0, 1)$

$$m = \frac{1 - 5}{0 - (-2)} = \frac{-4}{2} = -2$$

10. $-12x+6y+8=0$
$$6y=12x-8$$
$$y=\frac{12x-8}{6}$$
$$y=2x-\frac{4}{3}$$
$m=2$
$b=-\frac{4}{3}$, y-intercept $\left(0, -\frac{4}{3}\right)$

11. $m_1=\frac{3}{4}$
$m_2=-\frac{1}{m_1}=-\frac{4}{3}$

12. $3y-5x+10=0$
$$3y=5x-10$$
$$y=\frac{5x-10}{3}$$
$$y=\frac{5}{3}x-\frac{10}{3}$$
$m_{||}=\frac{5}{3}$

7.4 Exercises

Use $m=\frac{y_2-y_1}{x_2-x_1}$ and $y=mx+b$ as needed in Exercises 1–37.

1. $m=4$, $(-3, 0)$
$y=mx+b$
$0=4(-3)+b \Rightarrow b=12$
$y=4x+12$

3. $m=-2$, $(3, 5)$
$y=mx+b$
$5=(-2)3+b$
$b=11$
$y=-2x+11$

5. $m=-3$, $\left(\frac{1}{2}, 2\right)$
$y=mx+b$
$2=-3\left(\frac{1}{2}\right)+b \Rightarrow b=\frac{7}{2}$
$y=-3x+\frac{7}{2}$

7. $m=\frac{1}{4}$, $(4, 5)$
$y=mx+b$
$5=\frac{1}{4}(4)+b$
$b=4$
$y=\frac{1}{4}x+4$

9. $(3, -12)$ and $(-4, 2)$
$m=\frac{2-(-12)}{-4-3}=\frac{14}{-7}=-2$
$y=mx+b$
$2=-2(-4)+b \Rightarrow b=-6$
$y=-2x-6$

11. $(2, -6)$ and $(-1, 6)$
$m=\frac{6-(-6)}{-1-2}=\frac{12}{-3}=-4$
$y=mx+b$
$6=-4(-1)+b \Rightarrow b=2$
$y=-4x+2$

13. $(3, 5)$ and $(-1, -15)$
$m=\frac{-15-5}{-1-3}=\frac{-20}{-4}=5$
$y=mx+b$
$5=5(3)+b \Rightarrow b=-10$
$y=5x-10$

15. $\left(1, \frac{5}{6}\right)$ and $\left(3, \frac{3}{2}\right)$
$m=\frac{\frac{3}{2}-\frac{5}{6}}{3-1}=\frac{\frac{4}{6}}{2}=\frac{1}{3}$
$y=mx+b$
$\frac{5}{6}=\frac{1}{3}(1)+b \Rightarrow b=\frac{1}{2}$
$y=\frac{1}{3}x+\frac{1}{2}$

17. $m=-3$, $(-1, 3)$
$y=mx+b$
$3=-3(-1)+b$
$b=0$
$y=-3x$

19. (2, −3) and (−1, 6)

$m = \frac{6-(-3)}{-1-2} = \frac{9}{-3} = -3$

$y = mx + b$

$3 = -3(2) + b \Rightarrow b = 3$

$y = -3x + 3$

21. $b = 1,\ m = -\frac{2}{3}$

$y = -\frac{2}{3}x + 1$

23. $b = -4,\ m = \frac{2}{3}$

$y = \frac{2}{3}x - 4$

25. $b = 0,\ m = -\frac{2}{3}$

$y = -\frac{2}{3}x$

27. $b = -2$, $m = 0$

$y = -2$

29. $m = 0$, (7, −2)

The line is horizontal so $y = -2$.

31. (4, −6) perpendicular to x-axis so m is undefined.

$x = 4$ for all values of y.

$x = 4$

33. (0, 5) parallel to $y = \frac{1}{3}x + 4$.

$5 = \frac{1}{3}(0) + b$

$b = 5$

$y = \frac{1}{3}x + 5$

35. Perpendicular to $y = 2x - 9 \Rightarrow m_1 = 2$

$m_2 = -\frac{1}{m_1} = -\frac{1}{2}$

$y = mx + b$, (2, 3)

$3 = -\frac{1}{2}(2) + b \Rightarrow b = 4$

$y = -\frac{1}{2}x + 4$

37. (0, 227) and (10, 251)

$m = \frac{251-227}{10-0} = \frac{24}{10} = 2.4,\ b = 227$

$y = mx + b$

$y = 2.4x + 227$

Cumulative Review

39.

$$\frac{3}{t} - \frac{2}{t-1} = \frac{4}{t}$$

$$t(t-1)\left(\frac{3}{t}\right) - t(t-1)\left(\frac{2}{t-1}\right) = t(t-1)\left(\frac{4}{t}\right)$$

$$3(t-1) - 2t = 4(t-1)$$

$$3t - 3 - 2t = 4t - 4$$

$$t - 3 = 4t - 4$$

$$1 = 3t$$

$$\frac{1}{3} = t$$

40.

$$\frac{4x^2-25}{4x^2-20x+25} \cdot \frac{x^2-9}{2x^2+x-15}$$

$$= \frac{(2x+5)(2x-5)}{(2x-5)(2x-5)} \cdot \frac{(x+3)(x-3)}{(2x-5)(x+3)}$$

$$= \frac{(2x+5)(x-3)}{(2x-5)^2}$$

41. 2nd week: Price = 80 − 0.15(80) = \$68

3rd week: Price = 68 − 0.1(68) = \$61.20

42. x = minutes beyond 200

$0.21x + 50 = 68.90$

$0.21x = 18.90$

$x = 90$

Total time = 200 + 90 = 290 minutes

Quick Quiz 7.4

1. (3, −5), $m = \frac{2}{3}$

$y = mx + b$

$-5 = \frac{2}{3}(3) + b$

$-7 = b$

$y = \frac{2}{3}x - 7$

2. (–2, 7), (–4, –5)

$m = \dfrac{-5-7}{-4-(-2)} = \dfrac{-12}{-2} = 6$

$m = 6$

$y = mx + b$

$-5 = 6(-4) + b$

$19 = b$

$y = 6x + 19$

3. (4, 5), (4, –2)

$m = \dfrac{-2-5}{4-4} = \dfrac{-7}{0} =$ undefined

This is a vertical line $x = 4$.

4. Answers may vary. Possible solution: Zero slope indicates a horizontal line. Since the line is horizontal, and it passes through (–2, –3) the equation must be $y = -3$.

7.5 Exercises

1. No. All points in one region will be solutions to the inequality, while points in the other region will not be solutions. Thus testing any point will give the same result, as long as the test point is not on the boundary line.

3. $y < 2x - 4$

Graph $y = 2x - 4$ with a dashed line.

Test point: (2, –2)

$-2 < 2(2) - 4$

$-2 < 0$ True

Shade the region containing (2, –2).

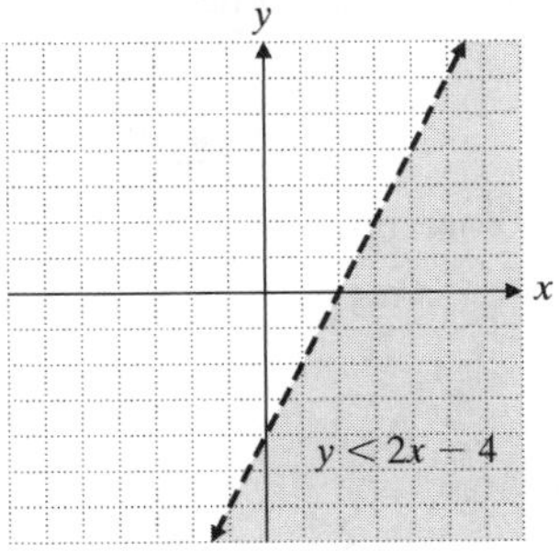

5. $2x - 3y < 6$

Graph $2x - 3y = 6$ using a dashed line.

Test point: (0, 0)

$2(0) - 3(0) < 6$

$0 < 6$ True

Shade the region containing (0, 0).

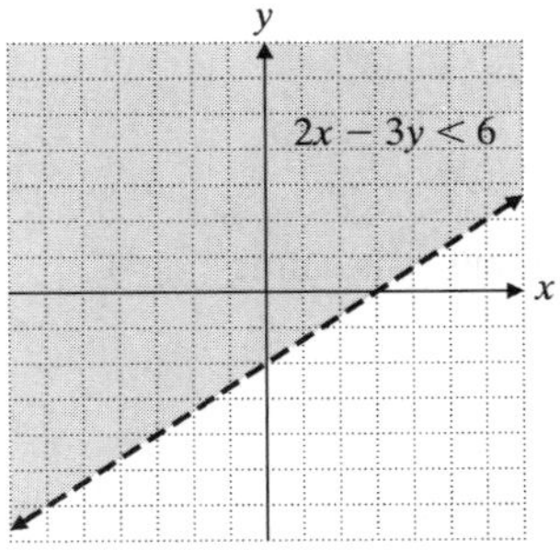

7. $2x - y \geq 3$

Graph $2x - y = 3$ with a solid line.

Test point: (0, 0)

$2(0) - (0) \geq 3$

$0 \geq 3$ False

Shade the region not containing (0, 0).

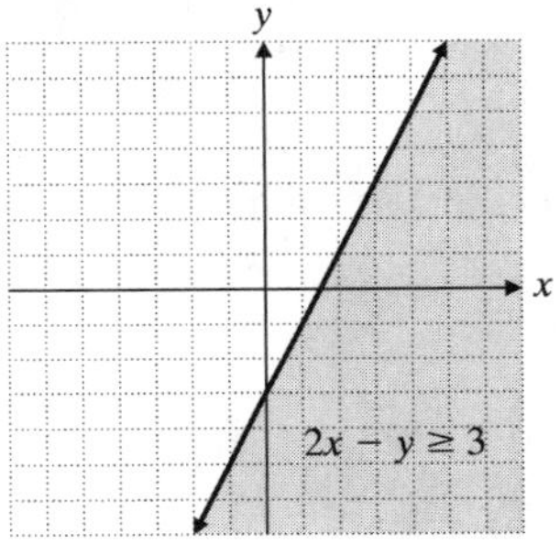

9. $y \geq 4x$

Graph $y = 4x$ with a solid line.

Test point: (1, 0)

$0 \geq 4(1)$

$0 \geq 4$ False

Shade the region not containing (1, 0).

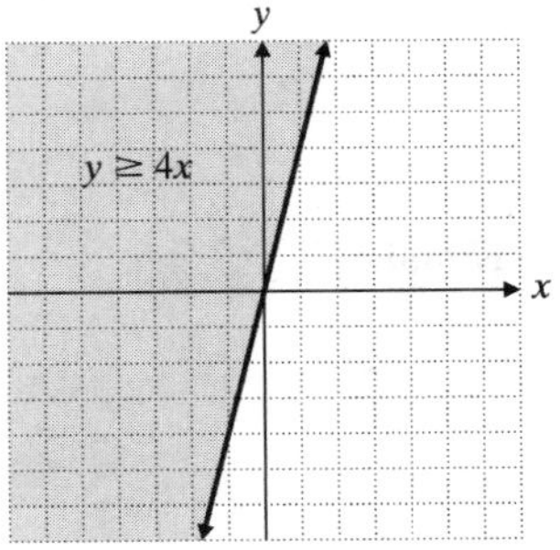

11. $y < -\dfrac{1}{2}x$

Graph $y = -\dfrac{1}{2}x$ with a dashed line.

Test point: (1, 0)

$0 < -\frac{1}{2}(1)$

$0 < -\frac{1}{2}$ False

Shade the region not containing (1, 0).

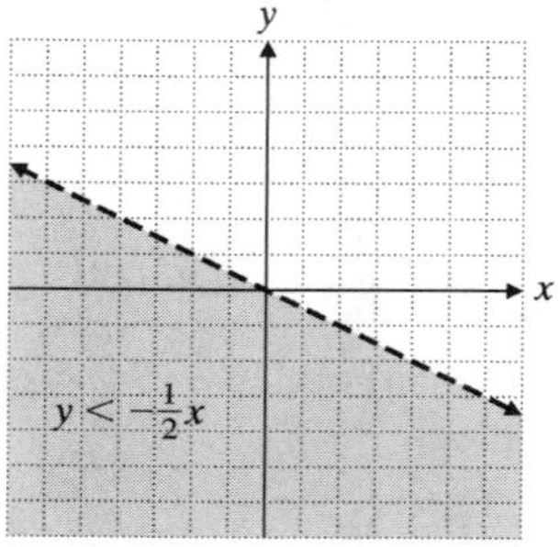

13. $x \geq 2$

Graph $x = 2$ with a solid line.

Test point: (0, 0)

$0 \geq 2$ False

Shade the region not containing (0, 0).

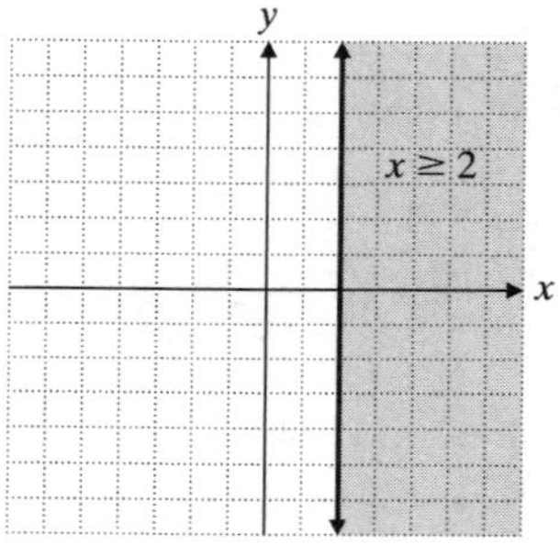

15. $2x - 3y + 6 \geq 0$

Graph $2x - 3y + 6 = 0$ or $2x - 3y = -6$ with a solid line.

Test point: (0, 0)

$2(0) - 3(0) + 6 \geq 0$

$6 \geq 0$ True

Shade the region containing (0, 0).

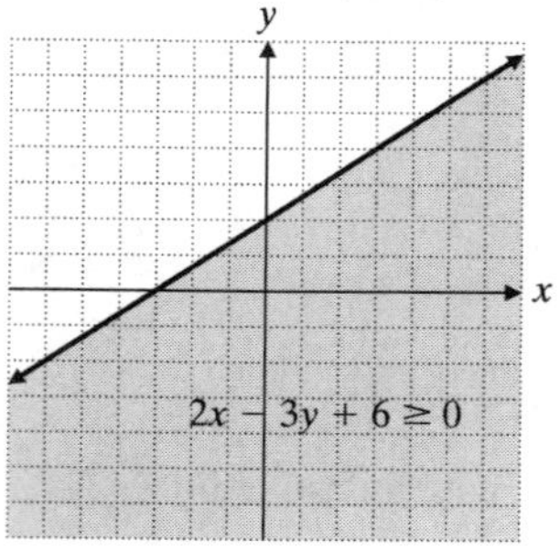

17. $2x > -3y$

Graph $2x = -3y$ with a dashed line.

Test point: (1, 1)

$2(1) > -3(1)$

$2 > -3$ True

Shade the region containing (1, 1).

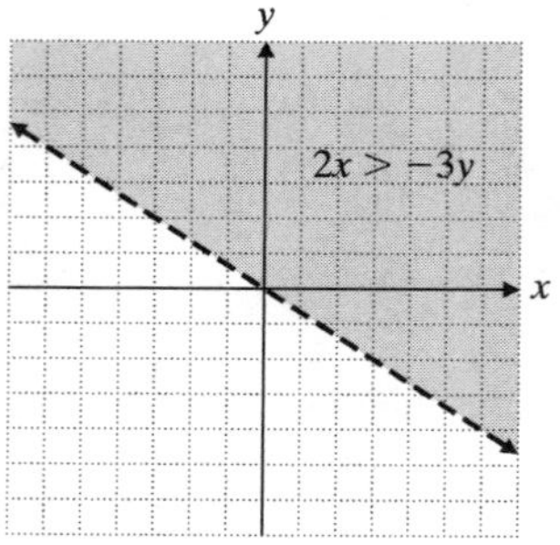

19. $2x > 3 - y$

Graph $2x = 3 - y$ or $y = -2x + 3$ with a dashed line.

Test point: (0, 0)

$2(0) > 3 - (0)$

$0 > 3$ False

Shade the region not containing (0, 0).

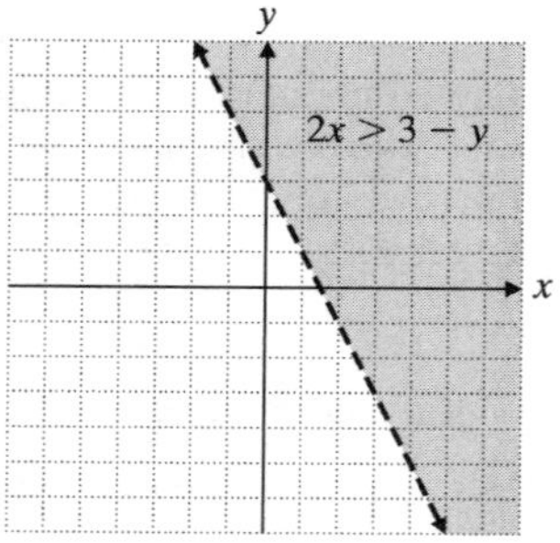

21. $x > -2y$

Graph $x = -2y$ with a dashed line.

Test point: (1, 0)

$(1) > -2(0)$

$1 > 0$ True

Shade the region containing (1, 0).

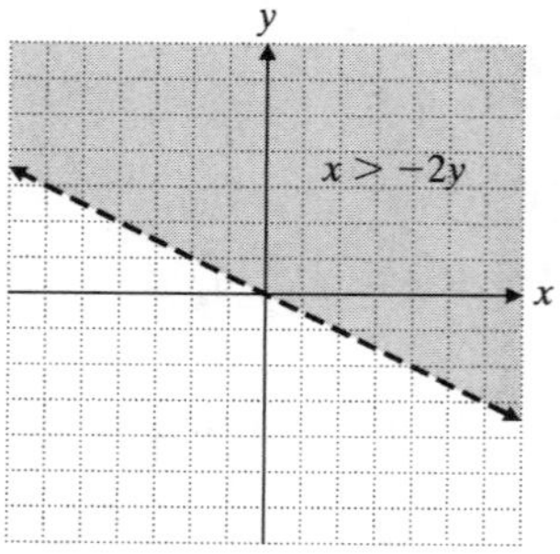

Cumulative Review

23.
$$3x^2+2x-8=0$$
$$3x^2+6x-4x-8=0$$
$$3x(x+2)-4(x+2)=0$$
$$(3x-4)(x+2)=0$$
$$3x-4=0 \qquad x+2=0$$
$$x=\frac{4}{3} \qquad x=-2$$

24. (5, −2) and (10, 0)
$$m=\frac{0-(-2)}{10-5}=\frac{2}{5}$$
$$y=mx+b$$
$$0=\frac{2}{5}(10)+b$$
$$b=-4$$
$$y=\frac{2}{5}x-4$$

Quick Quiz 7.5

1. Use a dashed line. If the inequality has a > or a < symbol, the points on the line itself are not included. This is indicated by a dashed line.

2. $3y \le -7x$
   Graph $3y=-7x$ with a solid line.
   Test point: (1, 1)
   $3(1) \le -7(1)$
   $3 \le -7$ False
   Shade the region not containing (1, 1).

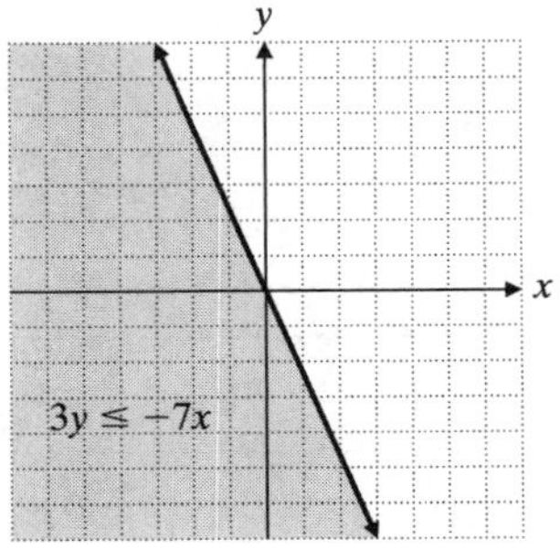

3. $-5x+2y>-3$
   Graph $-5x+2y=-3$ with a dashed line.
   Test point: (0, 0)
   $-5(0)+2(0)>-3$
   $0>-3$ True
   Shade the region containing (0, 0).

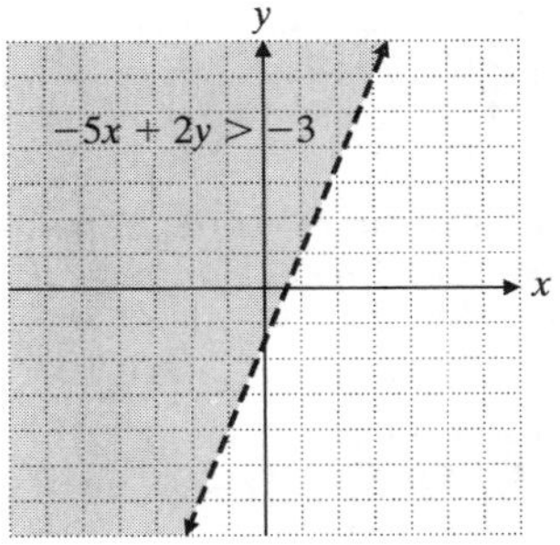

4. Answers may vary. Possible solution:
   The inequality is first graphed without shading. The test point coordinates are then substituted into the inequality. If the result of the substitution results in a true statement, the area where the test point lies is shaded. If false, the opposite area is shaded.

7.6 Exercises

1. You can describe a function using a table of values, an algebraic equation, or a graph.

3. The domain of a function is the set of possible values of the independent variable.

5. If a vertical line can intersect the graph more than once, the relation is not a function. If no such line exists, then the relation is a function.

7. $\left\{\left(\frac{2}{5}, 4\right), \left(3, \frac{2}{5}\right), \left(-3, \frac{2}{5}\right), \left(\frac{2}{5}, -1\right)\right\}$

   a. Domain $=\left\{-3, \frac{2}{5}, 3\right\}$

   Range $=\left\{-1, \frac{2}{5}, 4\right\}$

   b. Not a function because two different ordered pairs have the same first coordinate.

9. {(6, 2.5), (3, 1.5), (0, 0.5)}

   a. The domain consists of all the first coordinates and the range consists of all the second coordinates.
   Domain = {0, 3, 6}
   Range = {0.5, 1.5, 2.5}

   b. Function, because no two ordered pairs have the same first coordinate.

11. a. {(12, 1), (14, 3), (1, 12), (9, 12)}
Domain = {1, 9, 12, 14}
Range = {1, 3, 12}

b. Function, because no two ordered pairs have the same first coordinate.

13. a. {(3, 75), (5, 95), (3, 85), (7, 100)}
Domain = {3, 5, 7}
Range = {75, 85, 95, 100}

b. Not a function because two different ordered pairs have the same first coordinate.

15. $y = x^2 + 3$

x	$y = x^2 + 3$	y
−2	$y = (-2)^2 + 3 = 4 + 3$	7
−1	$y = (-1)^2 + 3 = 1 + 3$	4
0	$y = 0^2 + 3 = 0 + 3$	3
1	$y = 1^2 + 3 = 1 + 3$	4
2	$y = 2^2 + 3 = 4 + 3$	7

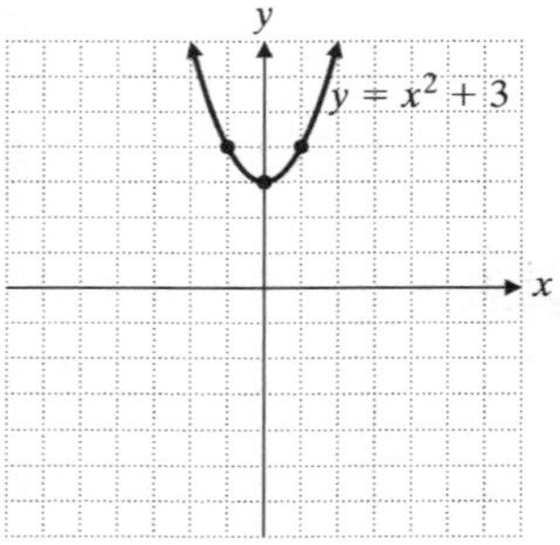

17. $y = 2x^2$

x	$y = 2x^2$	y
−2	$y = 2(-2)^2 = 2(4)$	8
−1	$y = 2(-1)^2 = 2(1)$	2
0	$y = 2(0)^2 = 2(0)$	0
1	$y = 2(1)^2 = 2(1)$	2
2	$y = 2(2)^2 = 2(4)$	8

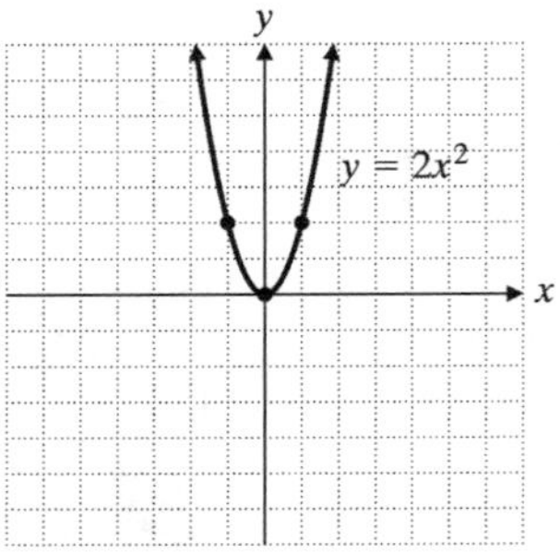

19. $x = -2y^2$

y	$x = -2y^2$	x
−2	$x = -2(-2)^2 = -2(4)$	−8
−1	$x = -2(-1)^2 = -2(1)$	−2
0	$x = -2(0)^2 = -2(0)$	0
1	$x = -2(1)^2 = -2(1)$	−2
2	$x = -2(2)^2 = -2(4)$	−8

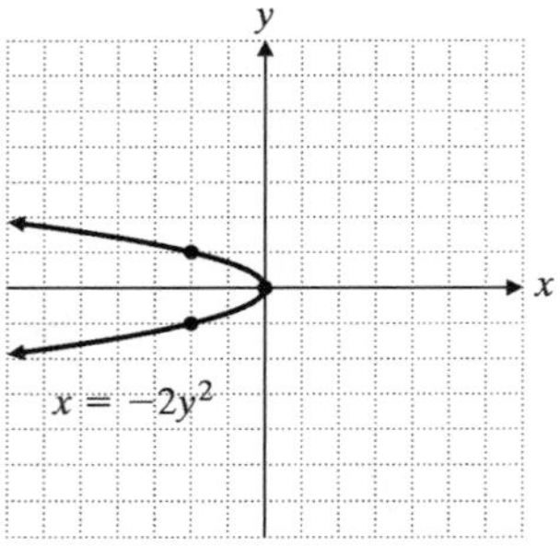

21. $x = y^2 - 4$

y	$x = y^2 - 4$	x
−2	$x = (-2)^2 - 4 = 4 - 4$	0
−1	$x = (-1)^2 - 4 = 1 - 4$	−3
0	$x = (0)^2 - 4 = 0 - 4$	−4
1	$x = 1^2 - 4 = 1 - 4$	−3
2	$x = 2^2 - 4 = 4 - 4$	0

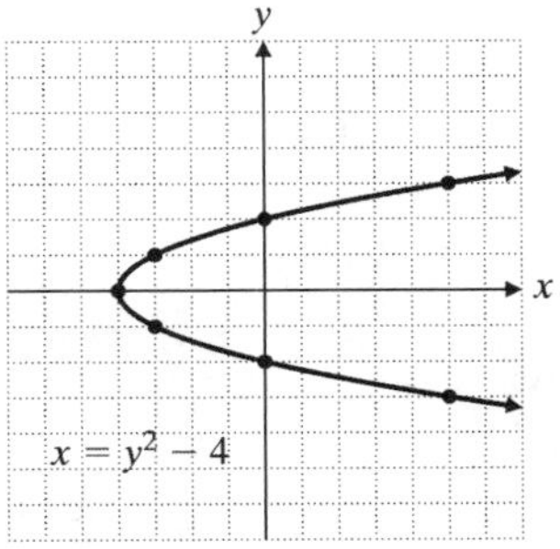

23. $y = \frac{2}{x}$

x	$y = \frac{2}{x}$	y
−4	$y = \frac{2}{-4}$	$-\frac{1}{2}$
−2	$y = \frac{2}{-2}$	−1
−1	$y = \frac{2}{-1}$	−2
1	$y = \frac{2}{1}$	2
2	$y = \frac{2}{1}$	1
4	$y = \frac{2}{4}$	$\frac{1}{2}$

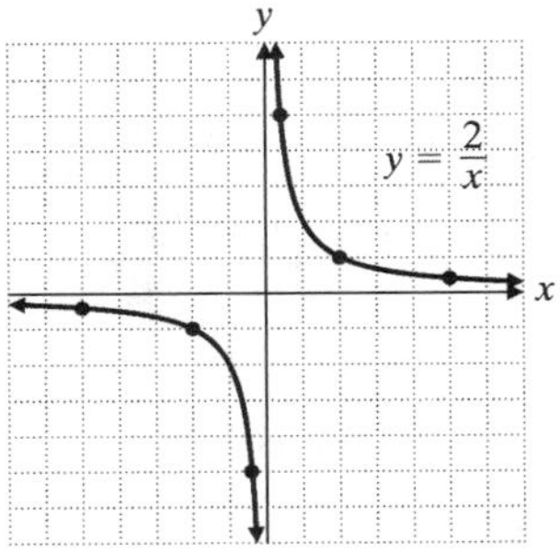

25. $y = \frac{4}{x^2}$

x	$y = \frac{4}{x^2}$	y
−4	$y = \frac{4}{(-4)^2} = \frac{4}{16}$	$\frac{1}{4}$
−2	$y = \frac{4}{(-2)^2} = \frac{4}{4}$	1
−1	$y = \frac{4}{(-1)^2} = \frac{4}{1}$	4
1	$y = \frac{4}{1^2} = \frac{4}{1}$	4
2	$y = \frac{4}{2^2} = \frac{4}{4}$	1
4	$y = \frac{4}{4^2} = \frac{4}{16}$	$\frac{1}{4}$

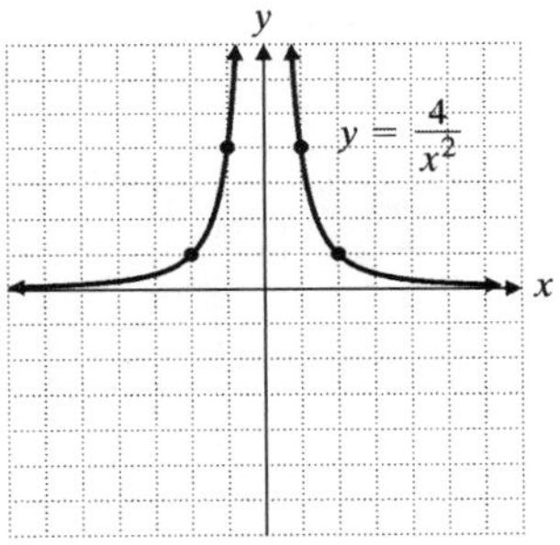

27. $x = (y+1)^2$

y	$x = (y+1)^2$	x
−3	$x = (-3+1)^2 = (-2)^2$	4
−2	$x = (-2+1)^2 = (-1)^2$	1
−1	$x = (-1+1)^2 = 0^2$	0
0	$x = (0+1)^2 = 1^2$	1
1	$x = (1+1)^2 = 2^2$	4

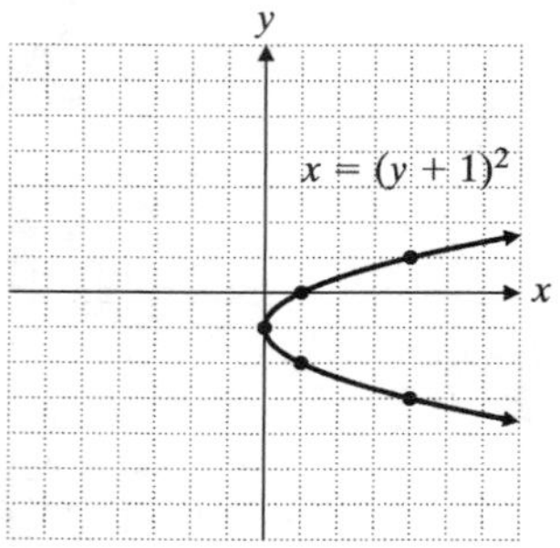

29. $y = \dfrac{4}{x-2}$

x	$y = \frac{4}{x-2}$	y
0	$y = \frac{4}{0-2} = \frac{4}{-2}$	−2
1	$y = \frac{4}{1-2} = \frac{4}{-1}$	−4
$\frac{3}{2}$	$y = \frac{4}{\frac{3}{2}-2} = \frac{4}{-\frac{1}{2}}$	−8
$\frac{5}{2}$	$y = \frac{4}{\frac{5}{2}-2} = \frac{4}{\frac{1}{2}}$	8
3	$y = \frac{4}{3-2} = \frac{4}{1}$	4
4	$y = \frac{4}{4-2} = \frac{4}{2}$	2

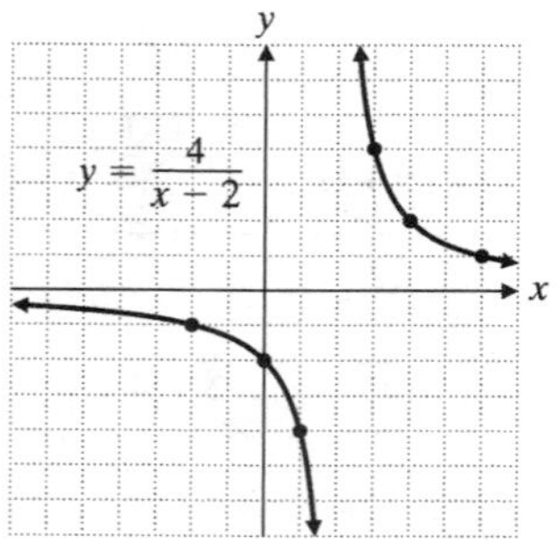

31. Function, passes the vertical line test.

33. Not a function, fails the vertical line test.

35. Function, passes the vertical line test.

37. Not a function, fails the vertical line test.

39. $f(x) = 2 - 3x$

a. $f(-8) = 2 - 3(-8) = 2 + 24 = 26$

b. $f(0) = 2 - 3(0) = 2 - 0 = 2$

c. $f(2) = 2 - 3(2) = 2 - 6 = -4$

41. $f(x) = 2x^2 - x + 3$

a. $f(0) = 2(0)^2 - 0 + 3 = 0 - 0 + 3 = 3$

b. $f(-3) = 2(-3)^2 - (-3) + 3 = 18 + 3 + 3 = 24$

c. $f(2) = 2(2)^2 - (2) + 3 = 8 - 2 + 3 = 9$

43. $f(x) = 0.02x^2 + 0.08x + 31.6$

$f(0) = 0.02(0)^2 + 0.08(0) + 31.6 = 31.6$

$f(4) = 0.02(4)^2 + 0.08(4) + 31.6 = 32.24$

$f(10) = 0.02(10)^2 + 0.08(10) + 31.6 = 34.4$

The curve slopes more steeply for larger values of x. The growth rate is increasing as x gets larger.

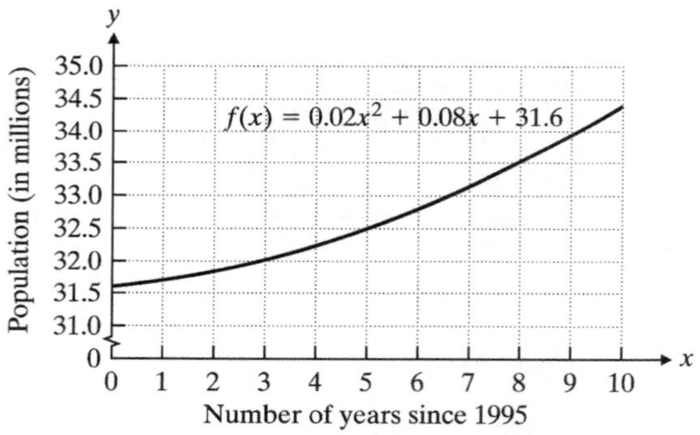

Cumulative Review

45. $(2x^2 - 2x + 1) + (-6x^2 - x + 7)$

$= 2x^2 - 6x^2 - 2x - x + 1 + 7$

$= -4x^2 - 3x + 8$

46. $(-3x^2 + 4x + 1) - (x^2 + 2x - 5)$

$= -3x^2 - x^2 + 4x - 2x + 1 + 5$

$= -4x^2 + 2x + 6$

47. $(4x^2 - 2x + 3)(x + 2)$

$= (4x^2 - 2x + 3)(x) + (4x^2 - 2x + 3)(2)$

$= 4x^3 - 2x^2 + 3x + 8x^2 - 4x + 6$

$= 4x^3 + 6x^2 - x + 6$

48. $$\begin{array}{r} 2x+4 \\ 3x-1\overline{)6x^2+10x-5} \\ \underline{6x^2-2x} \\ 12x-5 \\ \underline{12x-4} \\ -1 \end{array}$$

$$(6x^2+10x-5)\div(3x-1)=2x+4+\frac{-1}{3x-1}$$

Quick Quiz 7.6

1. No; two different ordered pairs have the same first coordinate.

2. $f(x)=3x^2-4x+2$

   a. $f(-3)=3(-3)^2-4(-3)+2$
   $=27+12+2$
   $=41$

   b. $f(4)=3(4)^2-4(4)+2=48-16+2=34$

3. $g(x)=\frac{7}{x-3}$

   a. $g(2)=\frac{7}{2-3}=\frac{7}{-1}=-7$

   b. $g(-5)=\frac{7}{-5-3}=\frac{7}{-8}=-\frac{7}{8}$

4. Answers may vary. Possible solution: Because duplicate elements are recorded only once in the domain and in the range.

Putting Your Skills to Work

1. $FV=PMT\times\frac{(1+I)^N-1}{I}$
   $=175\times\frac{(1+0.0067)^{480}-1}{0.0067}$
   $\approx 620{,}000$

   Louis's retirement account will be about $620,000.

2. $175\times 480=84{,}000$
   He will contribute $84,000.

3. $620{,}000-84{,}000=536{,}000$
   The account earned about $536,000 in interest.

4. Answers may vary.

5. Answers may vary.

6. Answers may vary.

Chapter 7 Review Problems

1. A: (2, −3), B: (−1, 0), C: (3, 2), D: (−3, −3)

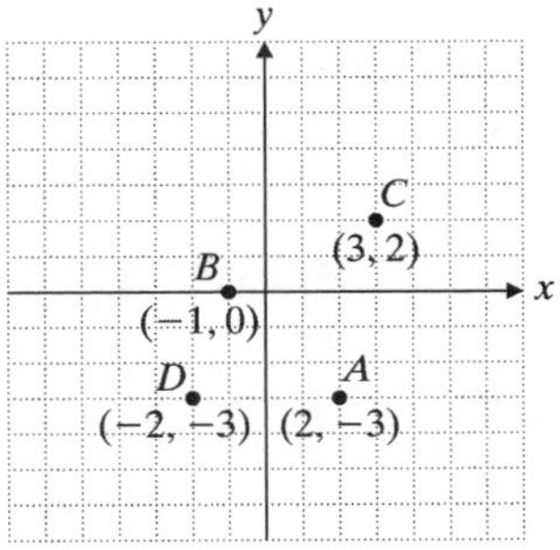

2. E: (4, 4), F: (0, 3), G: (1, −4), H: (−4, −1)

3. $y=7-3x$

   a. (0,): $y=7-3(0)$
   $y=7-0$
   $y=7$: (0, 7)

   b. (, 10): $10=7-3x$
   $3x=-3$
   $x=-1$: (−1, 10)

4. $2x+5y=12$

   a. (1,): $2(1)+5y=12$
   $5y=10$
   $y=2$: (1, 2)

   b. (, 4): $2x+5(4)=12$
   $2x=-8$
   $x=-4$: (−4, 4)

5. $x=6$

   a. (, −1): $x=6$: (6, −1)

   b. (, 3): $x=6$: (6, 3)

6. $3y = 2x + 6$

x	$3y = 2x + 6$	y
-3	$3y = 2(-3)+6$ $3y = 0$	0
0	$3y = 2(0)+6$ $3y = 6$	2
3	$3y = 2(3)+6$ $3y = 12$	4

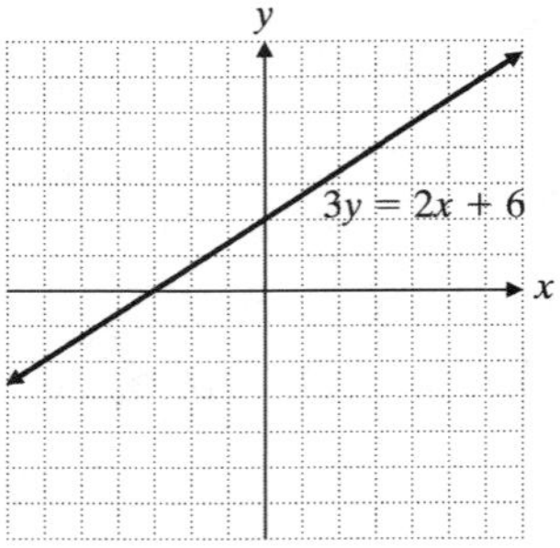

7. $5y + x = -15$
$x = -5y - 15$

y	$x = -5y - 15$	x
-2	$x = -5(-2)-15$ $x = 10-15$	-5
-3	$x = -5(-3)-15$ $x = 15-15$	0
-4	$x = -5(-4)-15$ $x = 20-15$	5

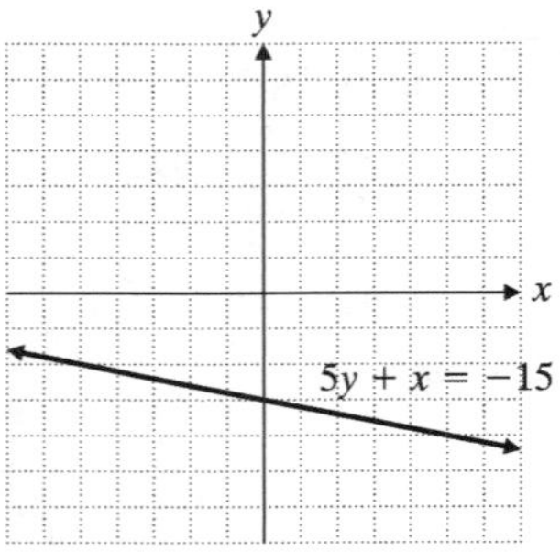

8. $2y + 4x = -8 + 2y$
$4x = -8$
$x = -2$

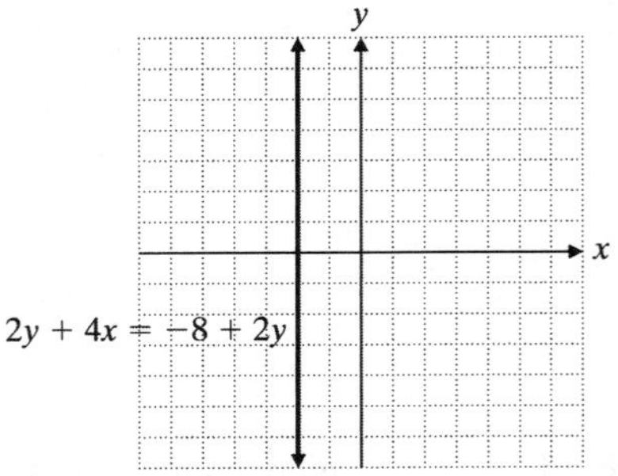

9. $(5, -3)$ and $\left(2, -\frac{1}{2}\right)$

$$m = \frac{y_2 - y_1}{x_2 - x_1} = \frac{-\frac{1}{2}-(-3)}{2-5} = \frac{\frac{5}{2}}{-3} = -\frac{5}{6}$$

10. $9x - 11y + 15 = 0$

$$9x + 15 = 11y$$
$$\frac{9}{11}x + \frac{15}{11} = y$$
$$y = \frac{9}{11}x + \frac{15}{11}$$

$m = \frac{9}{11}$, $b = \frac{15}{11}$, y-intercept $\left(0, \frac{15}{11}\right)$

11. $m = -\frac{1}{2}$, $b = 3$

$y = mx + b$

$y = -\frac{1}{2}x + 3$

12. $y = \frac{3}{5}x - 2$, $m_1 = \frac{3}{5}$

$$m_2 = -\frac{1}{m_1} = -\frac{5}{3}$$

13. $y = -\frac{1}{2}x + 3$

$m = -\frac{1}{2}$, $b = 3$, y-intercept $(0, 3)$

Find another point using the slope (2, 2).

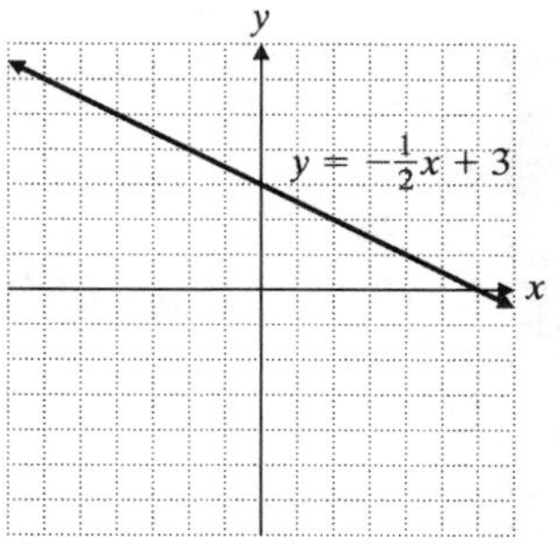

14. $2x - 3y = -12$

$-3y = -2x - 12$

$y = \frac{2}{3}x + 4$

$m = \frac{2}{3}$, $b = 4$, y-intercept (0, 4)

Find another point using the slope (3, 6).

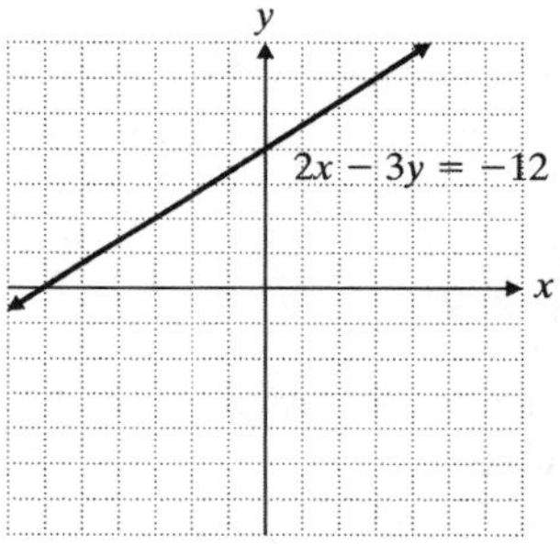

15. $6x - 2y = 6 + 6x$

$-2y = 6$

$y = -3$

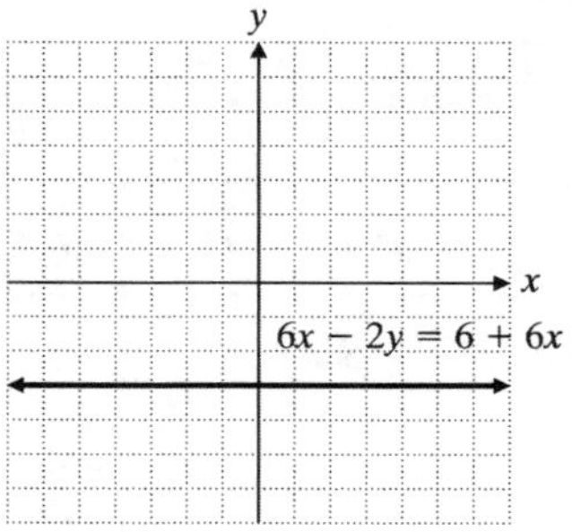

16. $m = 1$, (2, 5)

$y = mx + b$

$5 = 2 + b$

$3 = b$

$y = x + 3$

17. (3, –4), $m = -6$

$y = mx + b$

$-4 = -6(3) + b$

$-4 = -18 + b$

$b = 14$

$y = -6x + 14$

18. $m = -\frac{1}{3}$, (–1, 4)

$y = mx + b$

$4 = -\frac{1}{3}(-1) + b$

$4 = \frac{1}{3} + b$

$\frac{11}{3} = b$

$y = -\frac{1}{3}x + \frac{11}{3}$

19. (3, 7) and (–6, 7)

$m = \frac{y_2 - y_1}{x_2 - x_1} = \frac{7-7}{-6-3} = 0$

Horizontal line with $y = 7$

$y = 7$

20. $y = -\frac{2}{3}x + 4$, $m_1 = -\frac{2}{3}$

$m_2 = m_1 = -\frac{2}{3}$

21. $-2x + 6y = 3$

$6y = 2x + 3$

$y = \frac{2x+3}{6}$

$y = \frac{1}{3}x + \frac{1}{2}$

$m_1 = \frac{1}{3}$

$m_2 = -\frac{1}{m_1} = -3$

22. (0, –3) and (3, –1)

$m = \frac{-3-(-1)}{0-3} = \frac{2}{3}$, $b = -3$

$y = \frac{2}{3}x - 3$

23. (0, 1) and (1, −2)

$m = \dfrac{-2-1}{1-0} = -3,\ b = 1$

$y = -3x + 1$

24. $x = 5$

25. $y < \dfrac{1}{3}x + 2$

Graph $y = \dfrac{1}{3}x + 2$ with a dashed line.

Test point: (0, 0)

$0 < \dfrac{1}{3}(0) + 2$

$0 < 2$ True

Shade the region containing (0, 0).

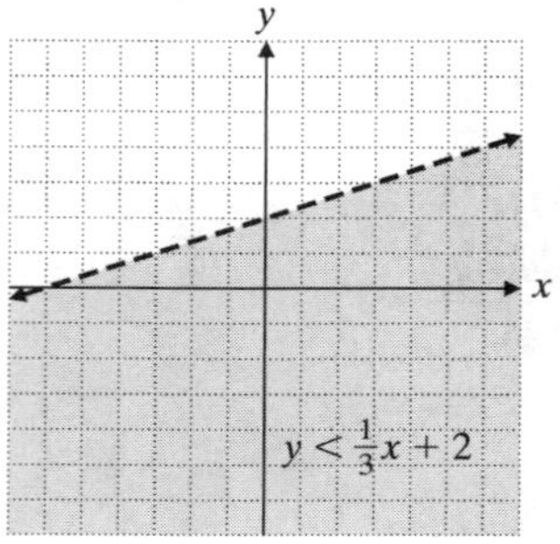

26. $3y + 2x \geq 12$

Graph $3y + 2x = 12$ with a solid line.

Test point: (0, 0)

$3(0) + 2(0) \geq 12$

$0 \geq 12$ False

Shade the region not containing (0, 0).

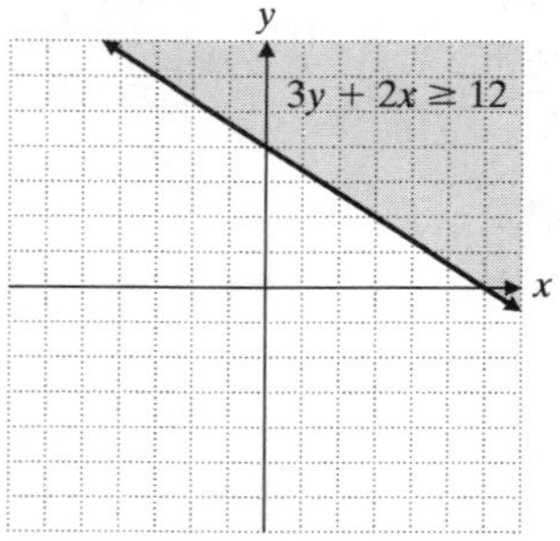

27. $x \leq 2$

Graph $x = 2$ with a solid line.

Test point: (0, 0)

$0 \leq 2$ True

Shade the region containing (0, 0).

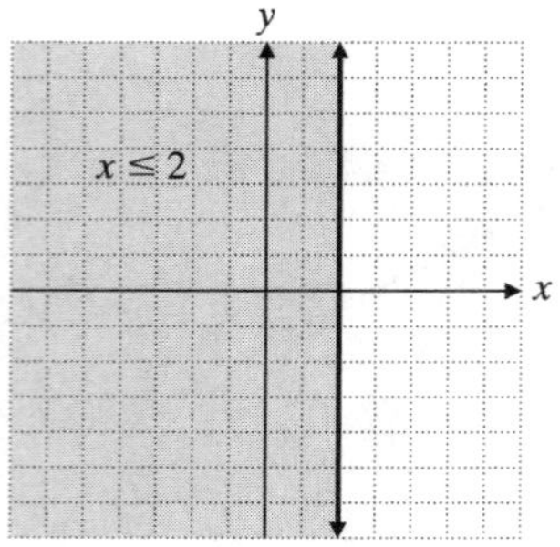

28. The domain consists of all the first coordinates and the range consists of all the second coordinates.

{(5, −6), (−6, 5), (−5, 5), (−6, −6)}

Domain: {−6, −5, 5}

Range: {−6, 5}

Not a function, because two different ordered pairs have the same first coordinate.

29. The domain consists of all the first coordinates and the range consists of all the second coordinates.

{(2, −3), (5, −3), (6, 4), (−2, 4)}

Domain: {−2, 2, 5, 6}

Range: {−3, 4}

Function, because no two ordered pairs have the same first coordinate.

30. Function, passes the vertical line test.

31. Not a function, fails the vertical line test.

32. Function, passes the vertical line test.

33. $y = x^2 - 5$

x	$y = x^2 - 5$	y
−2	$y = (-2)^2 - 5 = 4 - 5$	−1
−1	$y = (-1)^2 - 5 = 1 - 5$	−4
0	$y = 0^2 - 5 = 0 - 5$	−5
1	$y = 1^2 - 5 = 1 - 5$	−4
2	$y = 2^2 - 5 = 4 - 5$	−1

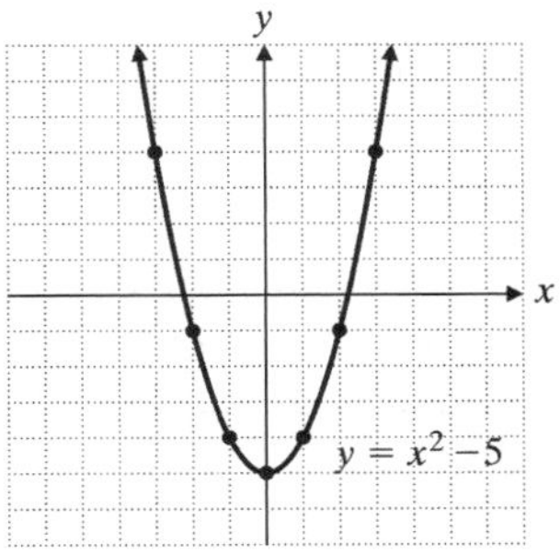

34. $x = y^2 + 3$

y	$x = y^2 + 3$	x
−2	$x = (-2)^2 + 3 = 4 + 3$	7
−1	$x = (-1)^2 + 3 = 1 + 3$	4
0	$x = 0^2 + 3 = 0 + 3$	3
1	$x = 1^2 + 3 = 1 + 3$	4
2	$x = 2^2 + 3 = 4 + 3$	7

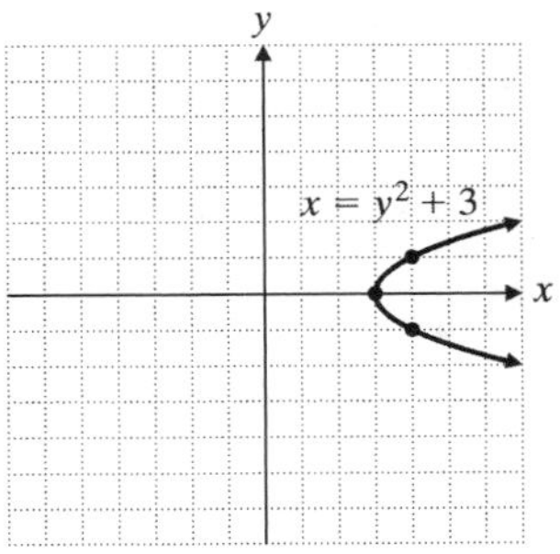

35. $y = (x - 3)^2$

x	$y = (x-3)^2$	y
1	$y = (1-3)^2 = (-2)^2$	4
2	$y = (2-3)^2 = (-1)^2$	1
3	$y = (3-3)^2 = 0^2$	0
4	$y = (4-3)^2 = 1^2$	1
5	$y = (5-3)^2 = 2^2$	4

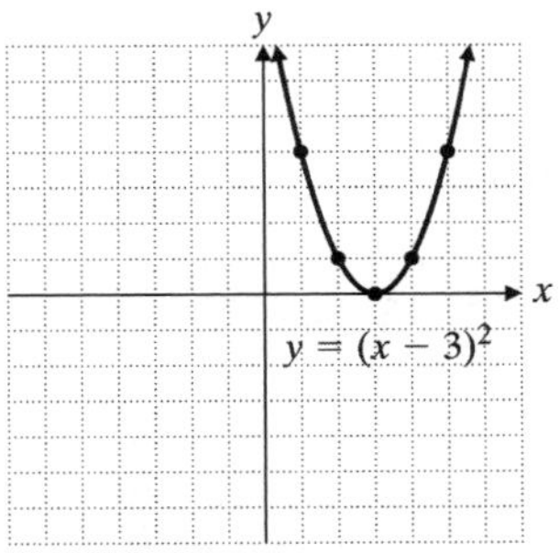

36. $f(x) = 7 - 6x$

a. $f(0) = 7 - 6(0) = 7 - 0 = 7$

b. $f(-4) = 7 - 6(-4) = 7 + 24 = 31$

37. $g(x) = -2x^2 + 3x + 4$

a. $g(-1) = -2(-1)^2 + 3(-1) + 4 = -2 - 3 + 4 = -1$

b. $g(3) = -2(3)^2 + 3(3) + 4 = -18 + 9 + 4 = -5$

38. $f(x) = 8x - 3$

a. $f(-1) = 8(-1) - 3 = -8 - 3 = -11$

b. $f\left(\frac{1}{2}\right) = 8\left(\frac{1}{2}\right) - 3 = 4 - 3 = 1$

39. $f(x) = \frac{2}{x+4}$

a. $f(-2) = \frac{2}{-2+4} = \frac{2}{2} = 1$

b. $f(6) = \frac{2}{6+4} = \frac{2}{10} = \frac{1}{5}$

40. $f(x) = x^2 - 2x + \frac{3}{x}$

a. $f(-1) = (-1)^2 - 2(-1) + \frac{3}{(-1)} = 1 + 2 - 3 = 0$

b. $f(3) = (3)^2 - 2(3) + \frac{3}{3} = 9 - 6 + 1 = 4$

Use $y = 30 + 0.09x$ in Exercises 41–46.

41. $y = 30 + 0.09(2000) = 210$
Their monthly bill would be \$210.

42. $y = 30 + 0.09(1600) = 174$
Their monthly bill would be \$174.

43. $y = 0.09x + 30$
$y = mx + b$, $b = 30$, y-intercept (0, 30)
It tells us that if no electricity is used, the cost is a minimum of \$30.

44. $y = 0.09x + 30$
$y = mx + b$, $m = 0.09$
It tells us that the bill increases \$0.09 for each kilowatt-hour of use.

45. $147 = 0.09x + 30$
$117 = 0.09x$
$x = 1300$
They used 1300 kilowatt-hours.

46. $246 = 0.09x + 30$
$216 = 0.09x$
$x = 2400$
They used 2400 kilowatt-hours.

47. $y = 1.15x + 65.6$
$y = 1.15(10) + 65.6 = 77.1$
In 2000, 77.1 million barrels were consumed.

48. $y = 1.15x + 65.6$
$y = 1.15(25) + 65.6 = 94.35$
In 2015, 94.35 million barrels were consumed.

49. The slope is 1.15. The slope tells us the rate of change per year. In this case it tells us that for each additional year, the world uses 1.15 million more barrels of petroleum.

50. The y-intercept is (0, 65.6). This tells us that when $x = 0$, which is the year 1990, a total of 65.6 million barrels of petroleum were consumed worldwide.

51. $96.65 = 1.15x + 65.6$
$27 = x$
27 years after 1990 = 2017

52. $100.1 = 1.15x + 65.6$
$30 = x$
30 years after 1990 = 2020

53. $5x + 3y = -15$
$3y = -5x - 15$
$y = -\frac{5}{3} - 5$

x	$y = -\frac{5}{3}x - 5$	y
0	$y = -\frac{5}{3}(0) - 5$	-5
-3	$y = -\frac{5}{3}(-3) - 5$	0

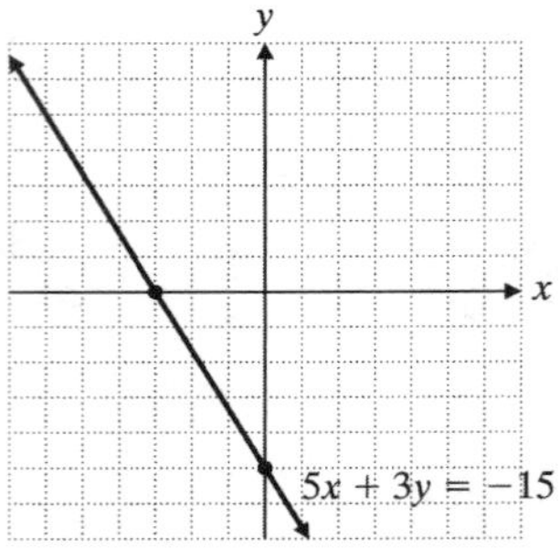

54. $y = \frac{3}{4}x - 3$

x	$y = \frac{3}{4}x - 3$	y
0	$y = \frac{3}{4}(0) - 3$	-3
4	$y = \frac{3}{4}(4) - 3$	0

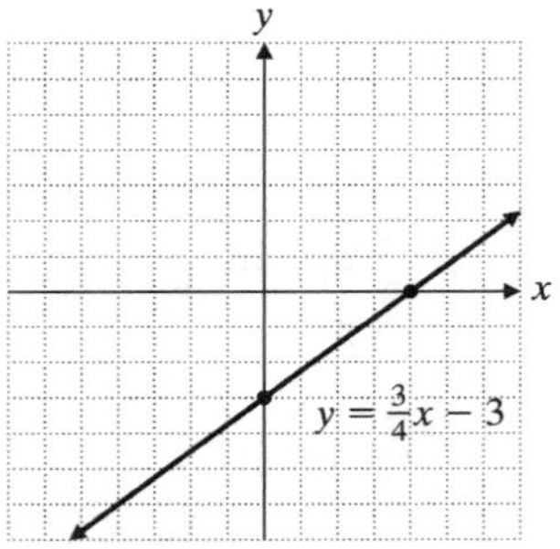

55. $(2, -7)$ and $(-3, -5)$
$y = \frac{-5-(-7)}{-3-2} = \frac{2}{-5} = -\frac{2}{5}$

56. $7x+6y-10=0$
$$6y-10=-7x$$
$$6y=-7x+10$$
$$y=-\frac{7}{6}x+\frac{5}{3}$$
$m=-\frac{7}{6}$, $b=\frac{5}{3}$, y-intercept $\left(0, \frac{5}{3}\right)$

57. $m=\frac{2}{3}$, $(3, -5)$
$$y=mx+b$$
$$-5=\frac{2}{3}(3)+b \Rightarrow b=-7$$
$$y=\frac{2}{3}x-7$$

58. $y < -2x + 1$
Graph $y = -2x + 1$ with a dashed line.
Test point: (0, 0)
$0 < -2(0)+1$
$0 < 1$ True
Shade the region containing (0, 0).

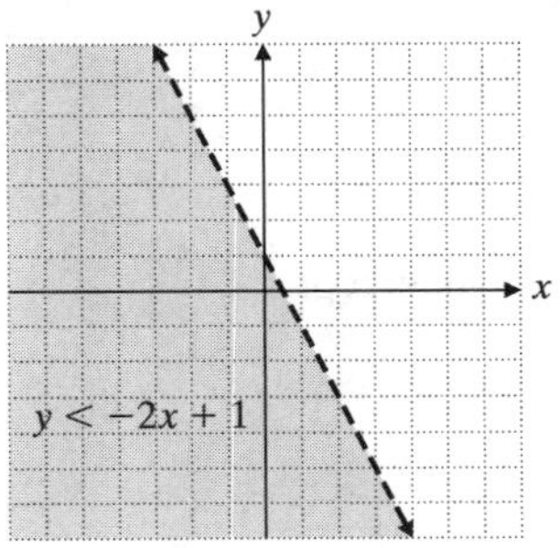

59. $x = 1994 - 1994 = 0$
$y = -269(0) + 17{,}020$
$y = 17{,}020{,}000$
17,020,000 people in 1994

$x = 2000 - 1994 = 6$
$y = -269(6) + 17{,}020$
$y = 15{,}406{,}000$
15,406,000 people in 2000

$x = 2008 - 1994 = 14$
$y = -269(14) + 17{,}020$
$y = 13{,}254{,}000$
13,254,000 people in 2008

60. Plot the points and connect (0, 17,020), (6, 15,406), (14, 13,254).

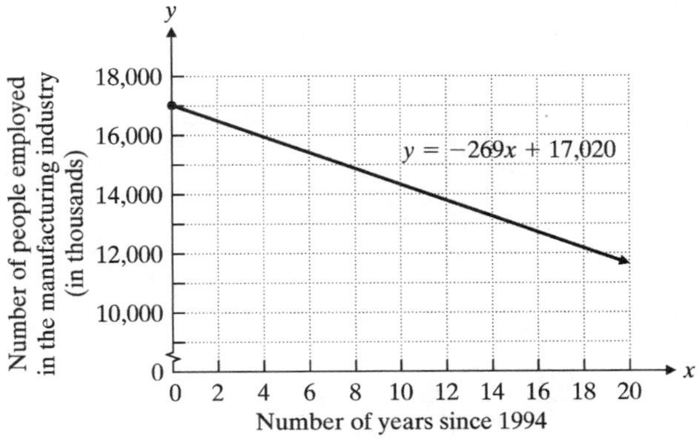

61. The slope is −269. The slope tells us that the number of people employed in manufacturing decreases each year by 269 thousand. In other words, employment in manufacturing goes down 269,000 people each year.

62. The y-intercept is (0, 17,020). This tells us that in the year of 1994, the number of manufacturing jobs was 17,020 thousand which is 17,020,000.

63. $y = -269x + 17{,}020$
$$11{,}640 = -269x + 17{,}020$$
$$20 = x$$
20 years after 1994 is 2014.

64. $y = -269x + 17{,}020$
$$10{,}295 = -269x + 17{,}020$$
$$25 = x$$
25 years after 1994 is 2019.

How Am I Doing? Chapter 7 Test

1. B: (6, 1), C: (−4, −3), D: (−3, 0), E: (5, −2)

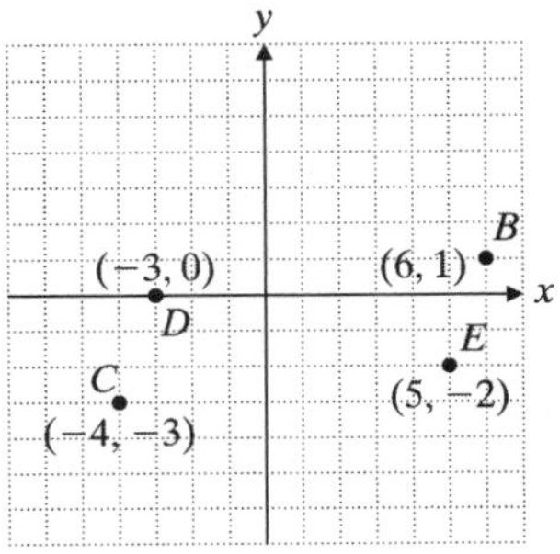

2. $6x - 3 = 5x - 2y$

$x + 2y = 3$

Find the x- and y-intercepts.

Let $x = 0$.

$0 + 2y = 3$

$y = \frac{3}{2}$

$\left(0, \frac{3}{2}\right)$

Let $y = 0$.

$x + 2(0) = 3$

$x = 3$

$(3, 0)$

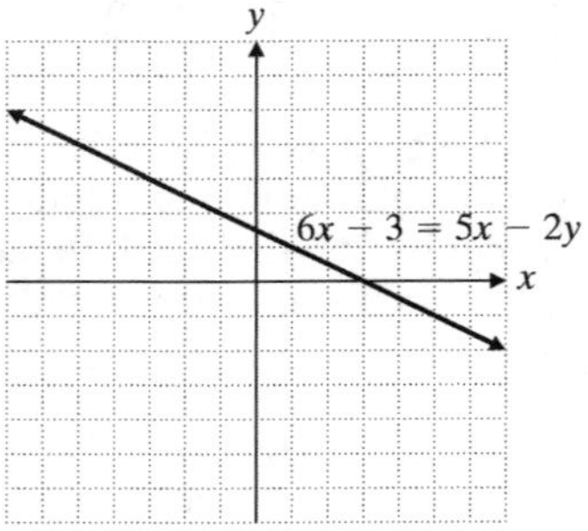

3. $8x + 2y = -4$

$2y = -8x - 4$

$y = -4x - 2$

x	$y = -4x - 2$	y
0	$y = 4(0) - 2$	−2
$-\frac{1}{2}$	$y = -4\left(-\frac{1}{2}\right) - 2$	0
−1	$y = -4(-1) - 2$	2

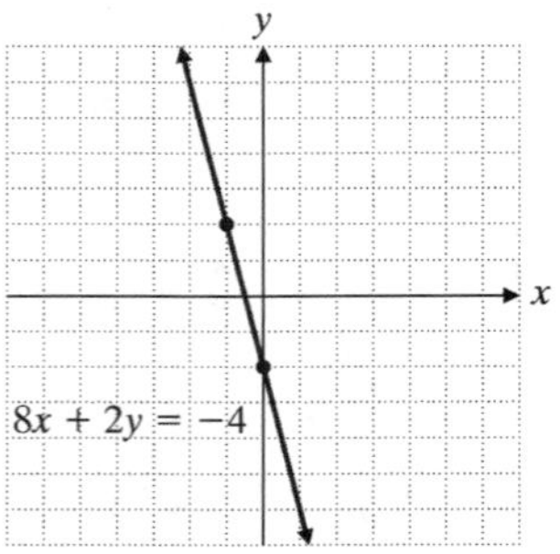

4. $y = \frac{2}{3}x - 4$

x	$y = \frac{2}{3}x - 4$	y
−3	$y = \frac{2}{3}(-3) - 4$	−6
0	$y = \frac{2}{3}(0) - 4$	−4
6	$y = \frac{2}{3}(6) - 4$	0

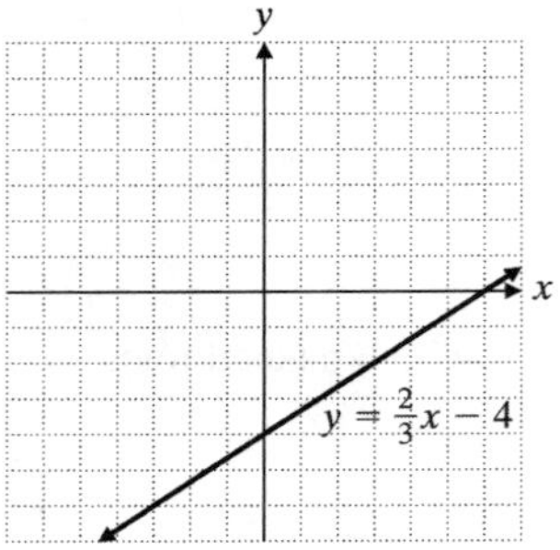

5. $3x + 2y - 5 = 0$

$2y = -3x + 5$

$y = -\frac{3}{2}x + \frac{5}{2}$

$m = -\frac{3}{2},\ b = \frac{5}{2},$ y-intercept $\left(0, \frac{5}{2}\right)$

6. $(8, 6)$ and $(-3, -5)$

$m = \frac{y_2 - y_1}{x_2 - x_1} = \frac{-5-6}{-3-8} = \frac{-11}{-11} = 1$

7. $(4, -2),\ m = \frac{1}{2}$

$y = mx + b$

$-2 = \frac{1}{2}(4) + b \Rightarrow b = -4$

$y = \frac{1}{2}x - 4$ or $x - 2y = 8$

8. $(-3, 11)$ and $(-6, 11)$

$m = \frac{11-11}{-6-(-3)} = 0$

9. (5, −4) and (−3, 8)

$$m=\frac{y_2-y_1}{x_2-x_1}=\frac{8-(-4)}{-3-5}=\frac{12}{-8}=-\frac{3}{2}$$
$$y=mx+b$$
$$-4=\left(-\frac{3}{2}\right)(5)+b$$
$$-8=-15+2b$$
$$\frac{7}{2}=b$$
$$y=-\frac{3}{2}x+\frac{7}{2}$$

10. (2, 7) and (2, −2)

$$m=\frac{y_2-y_1}{x_2-x_1}=\frac{-2-7}{2-2}=\frac{-9}{0}\text{ undefined}$$

Vertical line: $x = 2$

11. $4y \le 3x$

Graph $4y = 3x$ with a solid line.

Test point: (1, 1)

$4(1) \le 3(1)$

$4 \le 3$ False

Shade the region not containing (1, 1).

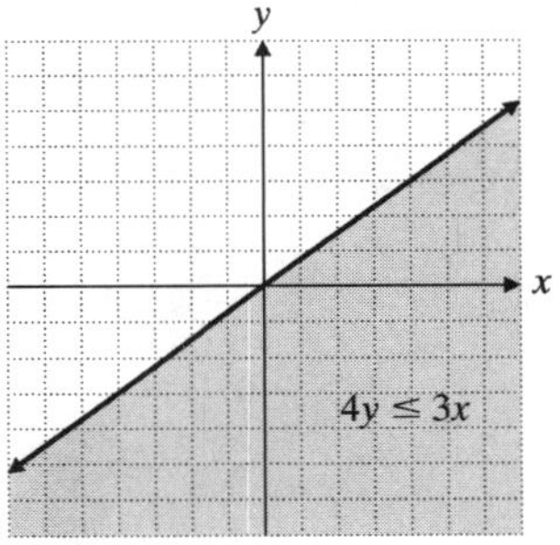

12. $-3x - 2y > 10$

Graph $-3x - 2y = 10$ with a dashed line.

Test point: (0, 0)

$-3(0) - 2(0) > 10$

$0 > 10$ False

Shade the region not containing (0, 0).

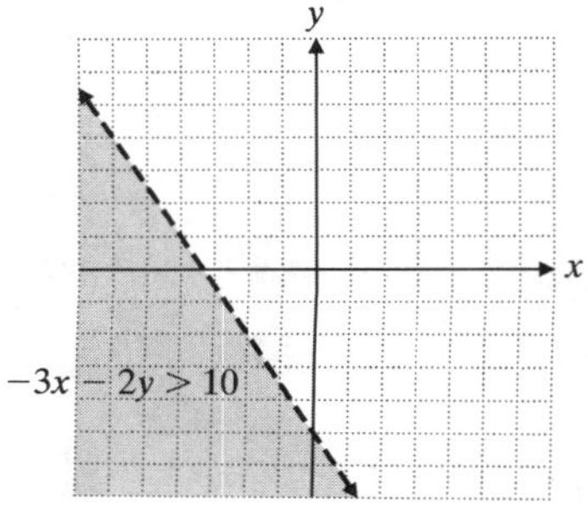

13. {(2, −8), (3, −7), (2, 5)}

No; two different ordered pairs have the same first coordinate.

14. Function, passes the vertical line test.

15. $y = 2x^2 - 3$

x	$y=2x^2-3$	y
−2	$y=2(-2)^2-3$	5
−1	$y=2(-1)^2-3$	−12
0	$y=2(0)^2-3$	−3
1	$y=2(1)^2-3$	−1
2	$y=2(2)^2-3$	5

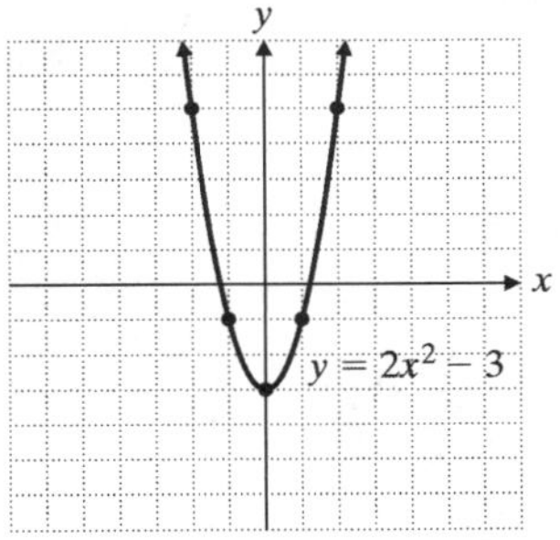

16. $f(x) = -x^2 - 2x - 3$

a. $f(0) = -(0)^2 - 2(0) - 3 = 0 - 0 - 3 = -3$

b. $f(-2) = -(-2)^2 - 2(-2) - 3 = -4 + 4 - 3 = -3$

17. $g(x)=\dfrac{3}{x-4}$

a. $g(3)=\dfrac{3}{3-4}=\dfrac{3}{-1}=-3$

b. $g(-11)=\dfrac{3}{-11-4}=\dfrac{3}{-15}=-\dfrac{1}{5}$

Cumulative Test for Chapters 0–7

1. $(-3x^3y^4z)^3 = (-3)^3x^{3\cdot 3}y^{4\cdot 3}z^3 = -27$

2. $\dfrac{-16x^{-3}y^4}{24x^5y^{-6}} = \dfrac{-2y^{4+6}}{3x^{5+3}} = \dfrac{-2y^{10}}{3x^8}$

3. $$\begin{aligned} 2-3(4-x) &= x-(3-x) \\ 2-12+3x &= x-3+x \\ -10+3x &= 2x-3 \\ -10+x &= -3 \\ x &= 7 \end{aligned}$$

4. $$\begin{aligned} \frac{600}{R} &= \frac{600}{R+50} + \frac{500(R+20)}{R^2+50R} \\ 5(R+50)\left(\frac{600}{R}\right) &= R(R+50)\left[\frac{600}{R+50} + \frac{500(R+20)}{R^2+50R}\right] \\ 600(R+50) &= 600R + 500(R+20) \\ 600R + 30,000 &= 600R + 500R + 10,000 \\ 20,000 &= 500R \\ 40 &= R \end{aligned}$$

5. $$\begin{aligned} x &= 8-2^4 \div 8 - 10 \\ x &= 8-16 \div 8 - 10 \\ x &= 8-2-10 \\ x &= -4 \end{aligned}$$

6. $$\begin{aligned} A &= lw + lh \\ A - lh &= lw \\ \frac{A-lh}{l} &= w \text{ or } w = \frac{A}{l} - h \end{aligned}$$

7. $$\begin{aligned} 50a^2 - 98b^2 &= 2(25a^2 - 49b^2) \\ &= 2[(5a)^2 - (7b)^2] \\ &= 2(5a+7b)(5a-7b) \end{aligned}$$

8. $$\begin{aligned} 9x^2 - 9x - 4 &= 9x^2 + 3x - 12x - 4 \\ &= 3x(3x+1) - 4(3x+1) \\ &= (3x-4)(3x+1) \end{aligned}$$

9. $$\begin{aligned} \frac{x-\frac{1}{x}}{\frac{1}{2}+\frac{1}{2x}} &= \frac{x-\frac{1}{x}}{\frac{1}{2}+\frac{1}{2x}} \cdot \frac{2x}{2x} \\ &= \frac{x \cdot 2x - \frac{1}{x} \cdot 2x}{\frac{1}{2} \cdot 2x + \frac{1}{2x} \cdot 2x} \\ &= \frac{2x^2-2}{x+1} \\ &= \frac{2(x+1)(x-1)}{x+1} \\ &= 2(x-1) \end{aligned}$$

10. (5, 3), (2, −3)

$$m = \frac{y_2 - y_1}{x_2 - x_1} = \frac{-3-3}{2-5} = 2$$

$$y = mx + b$$
$$-3 = 2(2) + b$$
$$-7 = b$$
$$y = 2x - 7$$

11. (7, −4) and slope is undefined.
$x = 7$

12. (3, −5), $m = -\frac{2}{3}$

$$y = mx + b$$
$$-5 = \left(-\frac{2}{3}\right)(3) + b$$
$$-5 = -2 + b$$
$$-3 = b$$
$$y = -\frac{2}{3}x - 3$$

13. (−8, −3) and (11, −3)

$$m = \frac{y_2 - y_1}{x_2 - x_1} = \frac{-3-(-3)}{11-(-8)} = \frac{0}{19} = 0$$

14. $4x - 3y = 6$

$$-3y = 6 - 4x$$
$$y = \frac{6-4x}{-3}$$
$$y = \frac{4}{3}x - 2$$
$$m = \frac{4}{3}$$

15. $y = \frac{2}{3}x - 4$

x	$y = \frac{2}{3}x - 4$	y
−3	$y = \frac{2}{3}(-3) - 4$	−6
0	$y = \frac{2}{3}(0) - 4$	−4
6	$y = \frac{2}{3}(6) - 4$	0

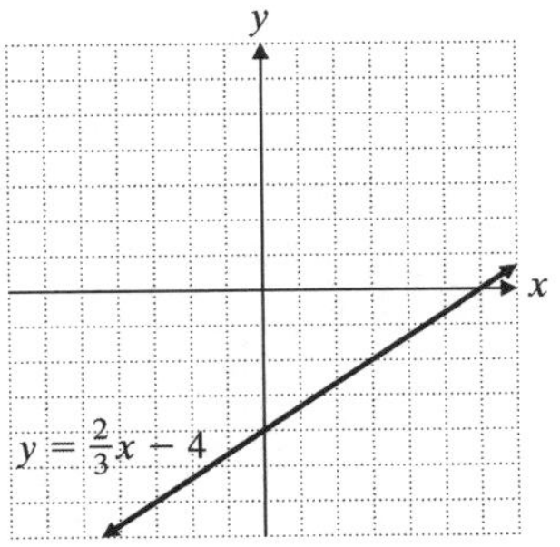

16. $3x + 8 = 5x$

$$8 = 2x$$
$$4 = x$$

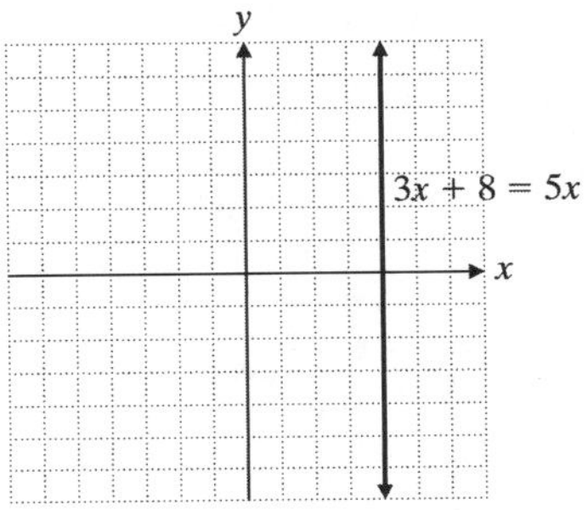

17. $2x + 5y \le -10$
Graph $2x + 5y = -10$ with a solid line.
Test point: (0, 0)

$$2(0) + 5(0) \le -10$$
$$0 \le -10 \quad \text{False}$$

Shade the region not containing (0, 0).

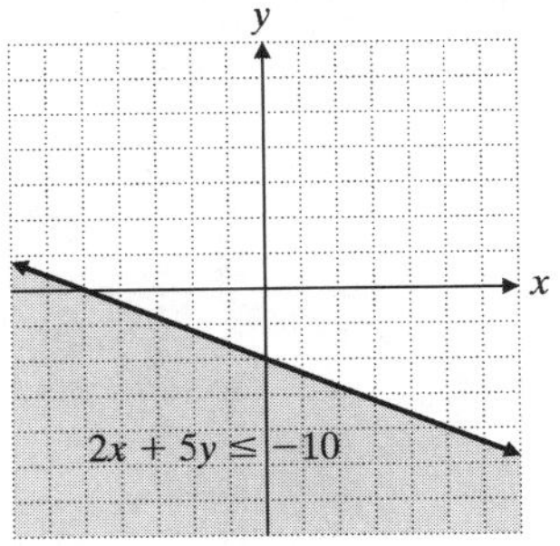

18. $x = y^2 + 3$

y	$x = y^2 + 3$	x
−2	$x = (-2)^2 + 3$	7
−1	$x = (-1)^2 + 3$	4
0	$x = (0)^2 + 3$	3
1	$x = 1^2 + 3$	4
2	$x = 2^2 + 3$	7

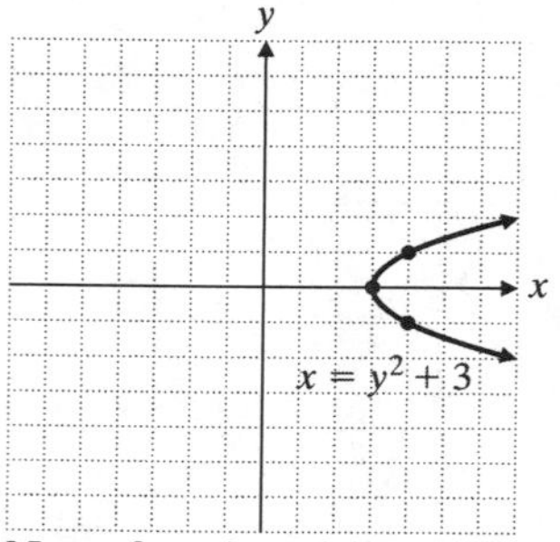

Not a function, fails the vertical line test.

19. $\{(3, 10), (10, -3), (0, -3), (5, 10)\}$
This is a function because no two ordered pairs have the same first coordinate.

20. $f(x) = 3x^2 - 2x + \frac{4}{x}$

a. $f(-1) = 3(-1)^2 - 2(-1) + \frac{4}{-1} = 3 + 2 - 4 = 1$

b. $f(2) = 3(2)^2 - 2(2) + \frac{4}{2} = 12 - 4 + 2 = 10$

Chapter 8

8.1 Exercises

1. The lines $y = 2x - 5$ and $y = 2x + 6$ are parallel with y-intercepts at -5 and 6 respectively. There is no solution. The system is inconsistent.

3. If two lines have different slopes you can conclude they <u>intersect</u> and the system has <u>one</u> solution.

5. If two lines have different slopes but the same y-intercept, the lines intersect at the y-intercept which is the solution to the system.

7. $x - y = 3$
   $x + y = 5$
   Graph both equations on the same coordinate system.
   The lines intersect at (4, 1).

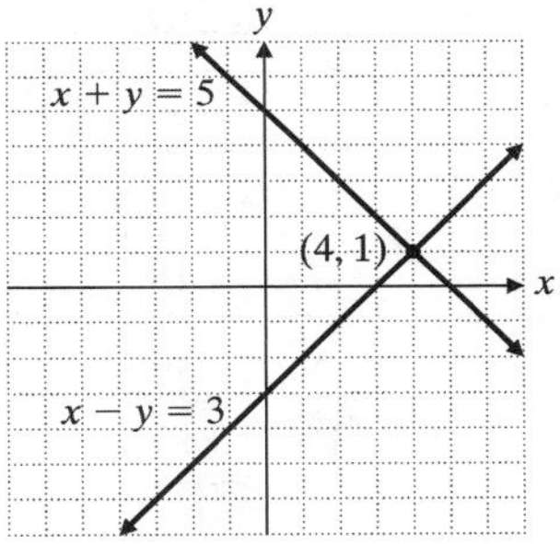

9. $y = -3x$
   $y = 2x + 5$
   Graph both equations on the same coordinate system.
   The lines intersect at (−1, 3).

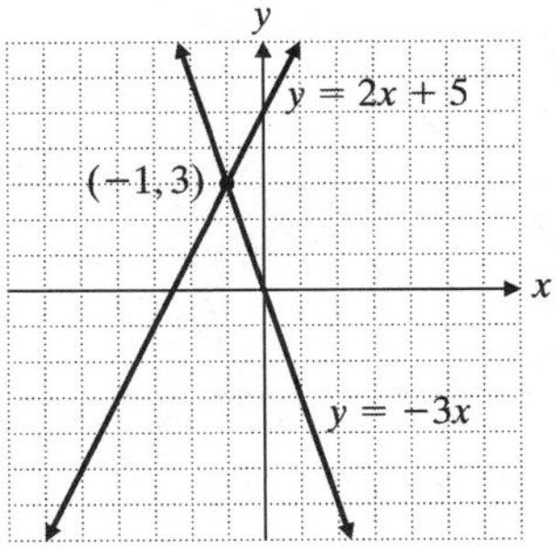

11. $4x + y = 5$
    $3x - 2y = 12$
    Graph both equations on the same coordinate system.
    The lines intersect at (2, −3).

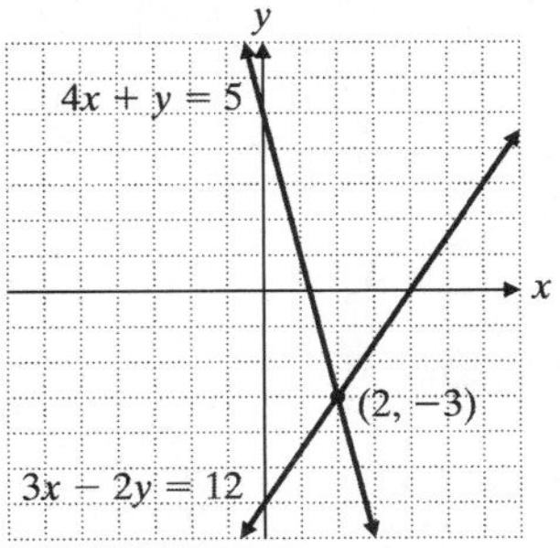

13. $-2x + y - 3 = 0$
    $4x + y + 3 = 0$
    Graph both equations on the same coordinate system.
    The lines intersect at (−1, 1).

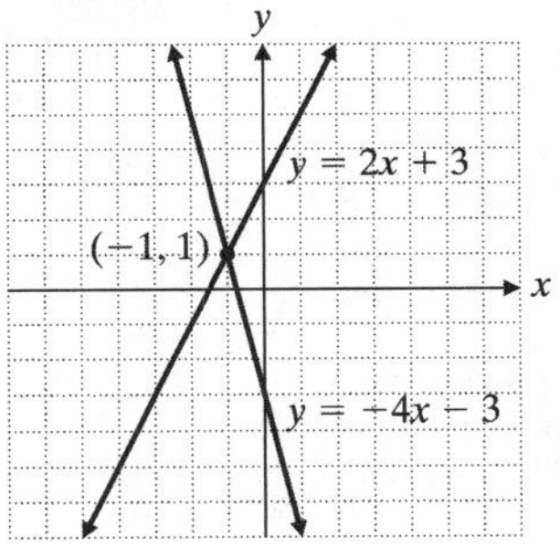

15. $3x - 2y = -18$
    $2x + 3y = 14$
    Graph both equations on the same coordinate system.
    The lines intersect at (−2, 6).

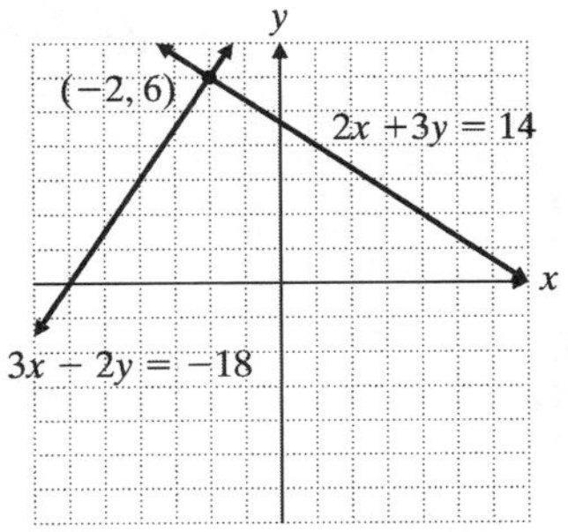

17. $y = \frac{3}{4}x + 7$

$y = -\frac{1}{2}x + 2$

Graph both equations on the same coordinate system.

The lines intersect at (–4, 4).

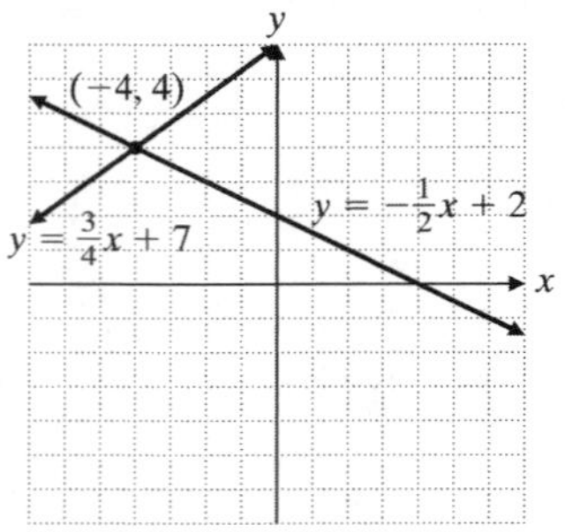

19. $3x - 2y = -4$

$-9x + 6y = -9$

Graph both equations on the same coordinate system. Notice that the lines are parallel.

No solution; inconsistent system

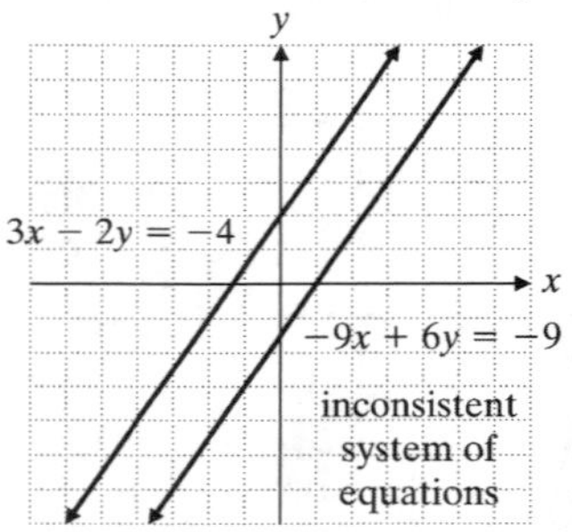

21. $y - 2x - 6 = 0$

$\frac{1}{2}y - 3 = x$

Graph both equations on the same coordinate system. Notice that the equations represent the same line.

Infinite number of solutions, dependent equations

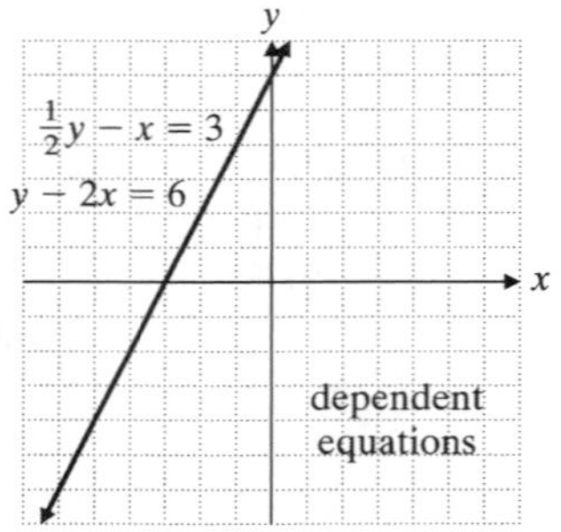

23. $y = \frac{1}{2}x - 2$

$y = \frac{2}{3}x - 1$

Graph both equations on the same coordinate system.

The lines intersect at (–6, –5).

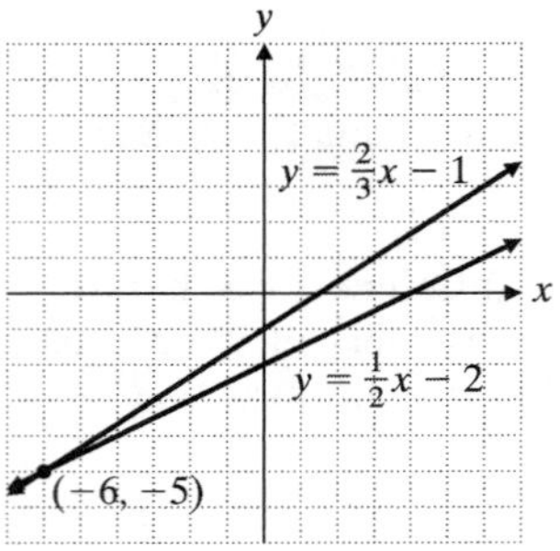

25. a. Newtown: $y = 7x$
Tryus: $y = 3x + 36$

b. $y = 7x$

x	y
1	7
4	28
8	56
12	84

$y = 3x + 36$

x	y
1	39
4	48
8	60
12	72

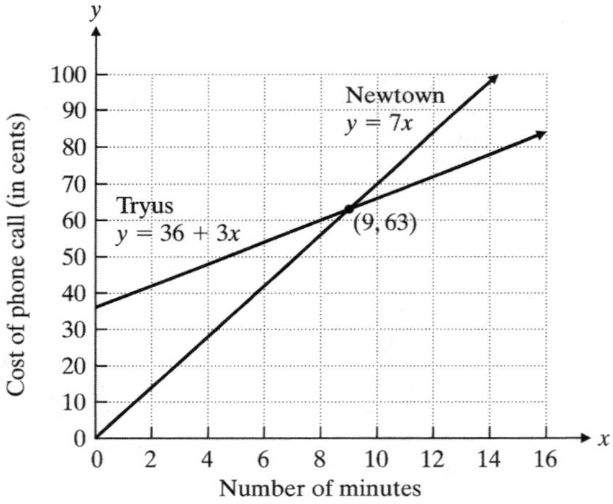

c. Since the intersection is at (9, 63), the cost is the same at 9 minutes.

d. Notice that the graph for Tyrus Telephone Company is below that of Newtown for 10 minutes or more. Therefore, choose Tyrus to spend less.

27. $88x + 57y = 683.10$
$95x - 48y = 7460.64$
(47.52, −61.38)

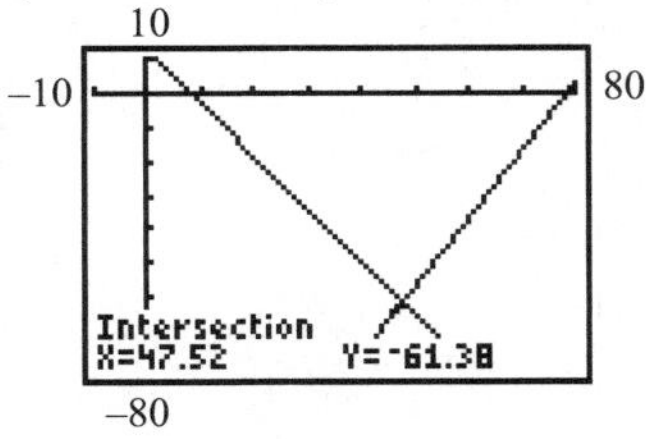

Cumulative Review

29. x = number of people in 2002
$\frac{8}{3}x = 400{,}000$
$x = 150{,}000$
150,000 people visited Machu Picchu in 2002.

30. $\frac{2000}{2500} \cdot 100\% = 80\%$
80% of the daily limit actually visit Machu Picchu daily.

31. 2500(365) = 912,500 maximum people
2000(365) = 730,000 average people
The maximum is 912,500 people and the average is 730,000 people.

32. x = number of people
$125{,}000 \le 120x$
$x \ge 1042$ people
$\frac{1042}{2500} \cdot 100\% = 41.7\%$
They need to transport at least 1042 people which is about 41.7%.

Quick Quiz 8.1

1. $x - 3y = 6$
$4x + 3y = 9$
Graph both equations on the same coordinate system.
The lines intersect at (3, −1).

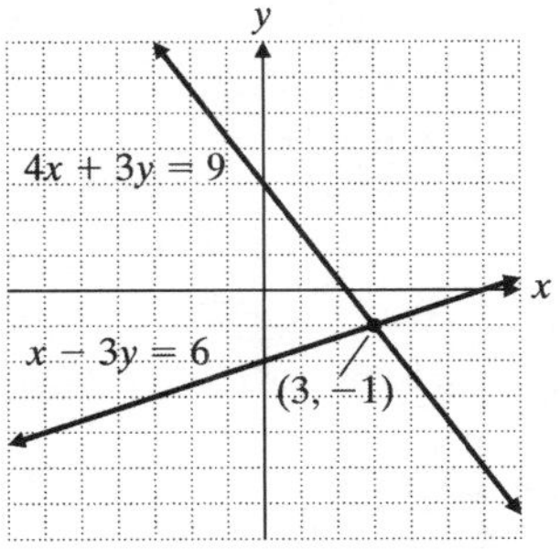

2. $y = \frac{3}{5}x + 2$
$m = \frac{3}{5}$, y-intercept (0, 2)
$5y - 3x = 35$
$5y = 3x + 35$
$y = \frac{3}{5}x + 7$
$m = 7$, y-intercept (0, 7)
The slope of each line is $\frac{3}{5}$. The y-intercept of the first line is (0, 2). The y-intercept of second line is (0, 7). Since the lines have the same slopes but different y-intercepts, they are parallel lines. They do not intersect. Therefore there is no solution. It is an inconsistent system of equations.

3. The two equations when graphed yield the same line. The equations are dependent. There is an infinite number of solutions.

4. Answers may vary. Possible solution:
The two equations represent lines that are parallel lines. There is therefore no solution.

8.2 Exercises

1. $4x + 3y = 9$ (1)
$x = 3y + 6$ (2)
Substitute $3y + 6$ for x in (1).
$4(3y + 6) + 3y = 9$
$12y + 24 + 3y = 9$
$15y + 24 = 9$
$15y = -15$
$y = -1$
Substitute −1 for y in (2).
$x = 3(-1) + 6 = 3$
(3, −1)

Check:
$4(3)+3(-1) \stackrel{?}{=} 9$
$9=9$
$3 \stackrel{?}{=} 3(-1)+6$
$3=3$

3. $2x+y=4$ (1)
$2x-y=0$ (2)
Solve (1) for y.
$y=-2x+4$ (3)
Substitute $-2x+4$ for y in (2).
$2x-(-2x+4)=0$
$2x+2x-4=0$
$4x-4=0$
$4x=4$
$x=1$
Substitute 1 for x in (3).
$y=-2(1)+4=2$
(1, 2)
Check:
$2(1)-2 \stackrel{?}{=} 0$
$0=0$
$2(1)+2 \stackrel{?}{=} 4$
$4=4$

5. $5x+2y=5$ (1)
$3x+y=4$ (2)
Solve (2) for y.
$y=-3x+4$ (3)
Substitute $-3x+4$ for y in (1).
$5x+2(-3x+4)=5$
$5x-6x+8=5$
$-x=-3$
$x=3$
Substitute 3 for x in (3).
$y=-3(3)+4=-5$
(3, −5)
Check:
$5(3)+2(-5) \stackrel{?}{=} 5$
$5=5$
$3(3)+(-5) \stackrel{?}{=} 4$
$4=4$

7. $4x-3y=-9$ (1)
$x-y=-3$ (2)
Solve (2) for x.
$x=y-3$ (3)
Substitute $y-3$ for x in (1).
$4(y-3)-3y=-9$
$4y-12-3y=-9$
$y=3$
Substitute 3 for y in (3).
$x=3-3=0$
(0, 3)
Check:
$4(0)-3(3) \stackrel{?}{=} -9$
$-9=-9$
$0-3 \stackrel{?}{=} -3$
$-3=-3$

9. $2a-3b=0$ (1)
$3a+b=22$ (2)
Solve (2) for b.
$b=22-3a$ (3)
Substitute $32-3a$ for b in (1).
$2a-3(22-3a)=0$
$2a-66+9a=0$
$a=6$
Substitute 6 for a in (3).
$b=22-3(6)$
$b=4$
(6, 4)
Check:
$2(6)-3(4) \stackrel{?}{=} 0$
$0=0$
$3(6)+4 \stackrel{?}{=} 22$
$22=22$

11. $3x-y=3$ (1)
$x+3y=11$ (2)
Solve (2) for x.
$x=-3y+11$ (3)
Substitute $-3y+11$ for x in (1).
$3(-3y+11)-y=3$
$-9y+33-y=3$
$-10y+33=3$
$-10y=-30$
$y=3$
Substitute 3 for y in (3).
$x=-3(3)+11$
$x=2$
(2, 3)
Check:
$3(2)-3 \stackrel{?}{=} 6$
$3=3$
$2+3(3) \stackrel{?}{=} 11$
$11=11$

13. $p+2q-4=0$ (1)
$7p-q-3=0$ (2)
Solve (1) for p.
$p=4-2q$ (3)
Substitute $4-2q$ for p in (2).

$$7(4-2q)-q-3=0$$
$$28-14q-q-3=0$$
$$-15q=-25$$
$$q=\frac{5}{3}$$

Substitute $\frac{5}{3}$ for q in (3).

$$p=4-2\left(\frac{5}{3}\right)=\frac{12}{3}-\frac{10}{3}=\frac{2}{3}$$

$\left(\frac{2}{3},\frac{5}{3}\right)$

Check:

$$\frac{2}{3}+2\left(\frac{5}{3}\right)-4 \stackrel{?}{=} 0$$
$$\frac{2}{3}+\frac{10}{3}-\frac{12}{3} \stackrel{?}{=} 0$$
$$0=0$$
$$7\left(\frac{2}{3}\right)-\frac{5}{3}-3 \stackrel{?}{=} 0$$
$$\frac{14}{3}-\frac{5}{3}-\frac{9}{3} \stackrel{?}{=} 0$$
$$0=0$$

15. $3x-y-9=0$ (1)
$8x+5y-1=0$ (2)
Solve (1) for y.
$3x-9=y$ (3)
Substitute $3x-9$ for y in (2).

$$8x+5(3x-9)-1=0$$
$$8x+15x-45-1=0$$
$$23x-46=0$$
$$x=2$$

Substitute 2 for x in (3).
$y=3(2)-9=-3$
$(2, -3)$
Check:

$$8(2)+5(-3)-1 \stackrel{?}{=} 0$$
$$16-15-1 \stackrel{?}{=} 0$$
$$0=0$$
$$3(2)-(-3)-9 \stackrel{?}{=} 0$$
$$6+3-9 \stackrel{?}{=} 0$$
$$0=0$$

17. $\frac{5}{3}x+\frac{1}{3}y=-3$ (1)
$-2x+3y=24$ (2)
Multiply (1) by 3.
$5x+y=-9$ (3)
Solve (3) for y.
$y=-5x-9$ (4)
Substitute $-5x-9$ for y in (2).

$$-2x+3(-5x-9)=24$$
$$-2x-15x-27=24$$
$$-17x=51$$
$$x=-3$$

Substitute -3 for x in (4).
$y=-5(-3)-9=6$
$(-3, 6)$

19. $4x+5y=2$ (1)
$\frac{1}{5}x+y=\frac{-7}{5}$ (2)
Multiply (2) by 5.
$x+5y=-7$
Solve for x.
$x=-5y-7$ (3)
Substitute $-5y-7$ for x in (1).

$$4(-5y-7)+5y=2$$
$$-20y-28+5y=2$$
$$-15y=30$$
$$y=-2$$

Substitute -2 for y in (3).
$x=-5(-2)-7=3$
$(3, -2)$

21. $\frac{4}{7}x-\frac{2}{7}y=2$
Multiply by 7.
$4x+2y=14$ (1)
$3x+y=13$ (2)
Solve (2) for y.
$y=-3x+13$ (3)
Substitute $-3x+13$ for y in (1).

$$4x+2(-3x+13)=14$$
$$4x-6x+26=14$$
$$-2x+26=14$$
$$-2x=-12$$
$$x=6$$

Substitute 6 for x in (3).
$y=-3(6)+13$
$y=-5$
$(6, -5)$

23. $3x+y=-1$ (1)
$-2x-y=-1$ (2)
Solve (2) for y.
$-y=-1+2x$
$y=1-2x$ (3)
Substitute $1-2x$ for y in (1).

$3x+(1-2x)=-1$
$3x+1-2x=-1$
$x=-2$
Substitute −2 for x in (3).
$y = 1 - 2(-2)$
$y = 5$
$(-2, 5)$

25. $3x - 4y = 36$ (1)
$y+3=3(x+1)$
$y+3=3x+3$
$y=3x$ (2)
Substitute $3x$ for y in (1).
$3x-4(3x)=36$
$3x-12x=36$
$-9x=36$
$x=-4$
Substitute −4 for x in (2).
$y = 3(-4) = -12$
$(-4, -12)$

27. $2x=4(2y+2)$
$3(x-3y)+2=17$

$2x=8y+8$
$3x-9y+2=17$

$x=4y+4$ (1)
$3x-9y=15$ (2)
Substitute $4y + 4$ for x in (2).
$3(4y+4)-9y=15$
$12y+12-9y=15$
$3y=3$
$y=1$
Substitute 1 for y in (1).
$x = 4(1) + 4 = 8$
$(8, 1)$

29. You would need 3 (7) equations to solve for 3 (7) unknowns. Substitution can reduce the system to one equation with one unknown. Then substituting back successively gives another unknown, thus requiring one equation for each unknown.

31. The solution to a system must satisfy <u>both</u> equations.

33. If the graphs are parallel, the system has no solutions because there are no points in common.

35. a. LEP: $y = 90x + 50$
AP: $y = 80x + 100$

b. $90x+50=80x+100$
$10x=50$
$x=5$
$y = 90(5) + 50 = 500$
$(5, 500)$
It would take 5 times and cost $500.

c. $90x+50-(80x+100)=80$
$10x-50=80$
$10x=130$
$x=13$
They expect plowing 13 times and he should hire Adirondack Plowing.

Cumulative Review

36. $7x+11y=19$
$11y=-7x+19$
$y=-\frac{7}{11}x+\frac{19}{11}$
$m=-\frac{7}{11}$, $b=\frac{19}{11}$, y-intercept $\left(0, \frac{19}{11}\right)$

37. $6x + 3y \geq 9$
Graph $6x + 3y = 9$ with a solid line.
Test point: (0, 0)
$6(0)+3(0)\geq 9$
$0\geq 9$ False
Shade the region not containing (0, 0).

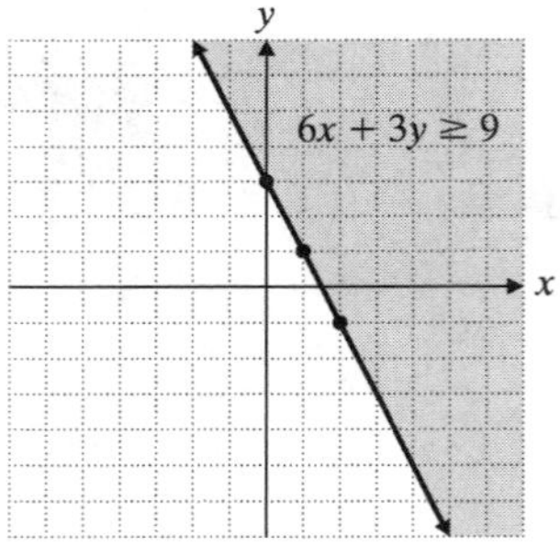

38. x = cost of gasoline per day
$x=\frac{60\text{ miles}}{\text{day}}\cdot\frac{\text{gallon}}{24\text{ miles}}\cdot\frac{4.20\text{ dollars}}{\text{gallon}}$
$x=\frac{10.50\text{ dollars}}{\text{day}}$
It costs Darrin $10.50 per day.

39. $\frac{64.82-23.33}{64.82}\cdot 100\% = 64\%$

x = amount spent by 18–24 year olds on Halloween
$x = (0.10 + 1)64.82$
$x = \$71.30$
About 64% was spent on decorations and candy. The average 18-24-year-old spent about \$71.30.

Quick Quiz 8.2

1. $10x+3y=8$ (1)
$2x+y=2$ (2)
Solve (2) for y.
$y = 2 - 2x$ (3)
Substitute $2 - 2x$ for y in (1).
$10x+3(2-2x)=8$
$10x+6-6x=8$
$4x=2$
$x=\frac{1}{2}$
Substitute $\frac{1}{2}$ for x in (3).
$y=2-2\left(\frac{1}{2}\right)$
$y=1$
$\left(\frac{1}{2}, 1\right)$
Check:
$10\left(\frac{1}{2}\right)+3(1) \stackrel{?}{=} 8$
$8=8$
$2\left(\frac{1}{2}\right)+1 \stackrel{?}{=} 2$
$2=2$

2. $-3x-y=-9$ (1)
$-x+2y=-10$ (2)
Solve (2) for x.
$-x=-10-2y$
$x=10+2y$ (3)
Substitute $10 + 2y$ for x in (1).
$-3(10+2y)-y=-9$
$-30-6y-y=-9$
$-7y=21$
$y=-3$
Substitute -3 for y in (3).
$x = 10 + 2(-3)$
$x = 4$
$(4, -3)$
Check:
$-3(4)-(-3) \stackrel{?}{=} -9$
$-9=-9$
$-4+2(-3) \stackrel{?}{=} -10$
$-10=-10$

3. $-14x-4y+20=0$ (1)
$5x-y+5=0$ (2)
Solve (2) for y.
$-y=-5x-5$
$y=5x+5$ (3)
Substitute $5x + 5$ for y in (1).
$-14x-4(5x+5)+20=0$
$-14x-20x-20+20=0$
$-34x=0$
$x=0$
Substitute 0 for x in (3).
$y = 5(0) + 5$
$y = 5$
$(0, 5)$
Check:
$-14(0)-4(5)+20 \stackrel{?}{=} 0$
$0=0$
$5(0)-5+5 \stackrel{?}{=} 0$
$0=0$

4. Answers may vary. Possible solution:
Label the equations (1) and (2). Solve (1) for y and label this equation (3). Substitute value of y in (3) into (2) and solve (2) for x. Substitute the found value of x back into (3) and solve for y. Check that the values of x and y satisfy both (1) and (2).

8.3 Exercises

1. $3x+2y=5$ (1)
$5x-y=3$ (2)
Multiply (2) by 2 and then add the equations to eliminate y.

3. $4x-3y=10$ (1)
$5x+4y=0$ (2)
Multiply (1) by 4 and (2) by 3 and then add the equations to eliminate y.

5. $-x+y=-3$ (1)
$\underline{-2x-y=6}$ (2)
$-3x=3$
$x=-1$
Substitute -1 for x in (1).

$-(-1)+y=-3$
$y=-4$
$(-1, -4)$
Check:
$-2(-1)-(-4) \stackrel{?}{=} 6$
$6=6$
$-(-1)+(-4) \stackrel{?}{=} -3$
$-3=-3$

7. $2x+3y=1$ (1)
$x-2y=4$ (2)
Multiply (2) by −2.
$2x+3y=1$
$\underline{-2x+4y=-8}$
$7y=-7$
$y=-1$
Substitute −1 for y in (2).
$x-2(-1)=4$
$x=2$
$(2, -1)$
Check:
$2(2)+3(-1) \stackrel{?}{=} 1$
$1=1$
$2-2(-1) \stackrel{?}{=} 4$
$4=4$

9. $6x-y=-7$ (1)
$6x+2y=5$ (2)
Multiply (2) by −1 to eliminate x.
$6x-y=-7$
$\underline{-6x-2y=-5}$
$-3y=-12$
$y=4$
Substitute 4 for y in (1).
$6x-4=-7$
$x=\frac{-7+4}{6}$
$x=-\frac{1}{2}$
$\left(-\frac{1}{2}, 4\right)$
Check:
$6\left(-\frac{1}{2}\right)-4 \stackrel{?}{=} -7$
$-7=-7$
$6\left(-\frac{1}{2}\right)+2(4) \stackrel{?}{=} 5$
$5=5$

11. $5x-15y=9$ (1)
$-x+10y=1$ (2)
Multiply (2) by 5.
$5x-15y=9$
$\underline{-5x+50y=5}$
$35y=14$
$y=\frac{2}{5}$
Substitute $\frac{2}{5}$ for y in (2).
$-x+10\left(\frac{2}{5}\right)=1$
$-x=-3$
$x=3$
$\left(3, \frac{2}{5}\right)$
Check:
$5(3)-15\left(\frac{2}{5}\right) \stackrel{?}{=} 9$
$9=9$
$-3+10\left(\frac{2}{5}\right) \stackrel{?}{=} 1$
$1=1$

13. $2x+5y=2$ (1)
$3x+y=3$ (2)
Multiply (2) by −5.
$2x+5y=2$
$\underline{-15x-5y=-15}$
$-13x \quad =-13$
$x=1$
Substitute 1 for x in (2).
$3(1)+y=3$
$y=0$
$(1, 0)$
Check:
$2(1)+5(0) \stackrel{?}{=} 2$
$2=2$
$3(1)+0 \stackrel{?}{=} 3$
$3=3$

15. $8x+6y=-2$ (1)
$10x-9y=-8$ (2)
Multiply (1) by 3 and (2) by 2.
$24x+18y=-6$
$\underline{20x-18y=-16}$
$44x \quad =-22$
$x=-\frac{1}{2}$

Substitute $-\frac{1}{2}$ for x in (1).

$$8\left(-\frac{1}{2}\right)+6y=-2$$
$$6y=2$$
$$y=\frac{1}{3}$$

$\left(-\frac{1}{2}, \frac{1}{3}\right)$

Check:

$$10\left(-\frac{1}{2}\right)-9\left(\frac{1}{3}\right) \stackrel{?}{=} -8$$
$$-8=-8$$
$$8\left(-\frac{1}{2}\right)+6\left(\frac{1}{3}\right) \stackrel{?}{=} -2$$
$$-2=-2$$

17. $2x+3y=-8$ (1)
$5x+4y=-34$ (2)
Multiply (1) by −5 and (2) by 2.

$$\begin{aligned} -10x-15y &= 40 \\ 10x+8y &= -68 \\ \hline -7y &= -28 \\ y &= 4 \end{aligned}$$

Substitute 4 for y in (1).

$$2x+3(4)=-8$$
$$2x=-20$$
$$x=-10$$

(−10, 4)

Check:

$$5(-10)+4(4) \stackrel{?}{=} -34$$
$$-34=-34$$
$$2(-10)+3(4) \stackrel{?}{=} -8$$
$$-8=-8$$

19. $4x+3y=1$ (1)
$2x-5y=-19$ (2)
Multiply (2) by −2 to eliminate x.

$$\begin{aligned} 4x+3y &= 1 \\ -4x+10y &= 38 \\ \hline 13y &= 39 \\ y &= 3 \end{aligned}$$

Substitute 3 for y in (1).

$$4x+3(3)=1$$
$$4x=1-9$$
$$x=-2$$

(−2, 3)

Check:

$$4(-2)+3(3) \stackrel{?}{=} 1$$
$$1=1$$
$$2(-2)-5(3) \stackrel{?}{=} -19$$
$$-19=-19$$

21. $12x-6y=-2$ (1)
$-9x-7y=-10$ (2)
Multiply (1) by 3 and (2) by 4.

$$\begin{aligned} 36x-18y &= -6 \\ -36x-28y &= -40 \\ \hline -46y &= -46 \\ y &= 1 \end{aligned}$$

Substitute 1 for y in (1).

$$12x-6(1)=-2$$
$$12x=4$$
$$x=\frac{1}{3}$$

$\left(\frac{1}{3}, 1\right)$

Check:

$$-9\left(\frac{1}{3}\right)-7(1) \stackrel{?}{=} -10$$
$$-10=-10$$
$$12\left(\frac{1}{3}\right)-6(1) \stackrel{?}{=} -2$$
$$-2=-2$$

23. $\frac{1}{4}x-\frac{3}{4}y=2$ (1)
$2x+3y=1$ (2)
Multiply equation (1) by 4.

25. $\frac{1}{2}x+\frac{2}{3}y=\frac{1}{3}$ (1)
$\frac{3}{4}x-\frac{4}{5}y=2$ (2)
Multiply equation (1) by 6 and equation (2) by 20.

27. $x+\frac{5}{4}y=\frac{9}{4}$ (1)
$\frac{2}{5}x-y=\frac{3}{5}$ (2)
Multiply (1) by 4 and (2) by 5.

$$\begin{aligned} 4x+5y &= 9 \\ 2x-5y &= 3 \\ \hline 6x \quad &= 12 \\ x &= 2 \end{aligned}$$

Substitute 2 for x in (1).

$2+\frac{5}{4}y=\frac{9}{4}$
$\frac{5}{4}y=\frac{1}{4}$
$y=\frac{1}{5}$
$\left(2, \frac{1}{5}\right)$
Check:
$\frac{2}{5}(2)-\frac{1}{5} \stackrel{?}{=} \frac{3}{5}$
$\frac{3}{5}=\frac{3}{5}$
$2+\frac{5}{4}\left(\frac{1}{5}\right) \stackrel{?}{=} \frac{9}{4}$
$\frac{9}{4}=\frac{9}{4}$

29. $\frac{x}{6}+\frac{y}{2}=-\frac{1}{2}$ (1)
$x-9y=21$ (2)
Multiply (1) by −6.
$-x-3y=3$
$\underline{x-9y=21}$
$-12y=24$
$y=-2$
Substitute −2 for y in (2).
$x-9(-2)=21$
$x=3$
(3, −2)
Check:
$\frac{3}{6}+\frac{-2}{2} \stackrel{?}{=} -\frac{1}{2}$
$\frac{1}{2}-\frac{2}{2} \stackrel{?}{=} -\frac{1}{2}$
$-\frac{1}{2}=-\frac{1}{2}$
$3-9(-2) \stackrel{?}{=} 21$
$21=21$

31. $\frac{1}{3}x+10y=-2$ (1)
$-3x+10y=-12$ (2)
Multiply (1) by 9 to eliminate x.
$3x+90y=-18$
$\underline{-3x+10y=-12}$
$100y=-30$
$y=-\frac{3}{10}$
Substitute $-\frac{3}{10}$ for y in (1).
$\frac{1}{3}x+10\left(-\frac{3}{10}\right)=-2$
$x=3$
$\left(3, -\frac{3}{10}\right)$
Check:
$\frac{1}{3}(3)+10\left(-\frac{3}{10}\right) \stackrel{?}{=} -2$
$-2=-2$
$-3(3)+10\left(-\frac{3}{10}\right) \stackrel{?}{=} -12$
$-12=-12$

33. $\frac{2}{3}x+\frac{3}{5}y=-\frac{1}{5}$ (1)
$\frac{1}{4}x+\frac{1}{3}y=\frac{1}{4}$ (2)
Multiply (1) by 15 and (2) by 12.
$10x+9y=-3$ (3)
$\underline{3x+4y=3}$ (4)
Multiply (3) by 3 and (4) by −10.
$30x+27y=-9$
$\underline{-30x-40y=-30}$
$-13y=-39$
$y=3$
Substitute 3 for y in (2).
$\frac{1}{4}x+\frac{1}{3}(3)=\frac{1}{4}$
$\frac{1}{4}x+1=\frac{1}{4}$
$x+4=1$
$x=-3$
(−3, 3)
Check:
$\frac{2}{3}(-3)+\frac{3}{5}(3) \stackrel{?}{=} -\frac{1}{5}$
$-2+\frac{9}{5} \stackrel{?}{=} -\frac{1}{5}$
$-\frac{1}{5}=-\frac{1}{5}$
$\frac{1}{4}(-3)+\frac{1}{3}(3) \stackrel{?}{=} \frac{1}{4}$
$-\frac{3}{4}+1 \stackrel{?}{=} \frac{1}{4}$
$\frac{1}{4}=\frac{1}{4}$

35. $0.5x - 0.3y = 0.1$ (1)
$5x + 3y = 6$ (2)
Multiply equation (1) by 10.
$5x - 3y = 1$
$5x + 3y = 6$

37. $4x + 0.5y = 9$ (1)
$0.2x - 0.05y = 1$ (2)
Multiply equation (1) by 10 and equation (2) by 100.
$40x + 5y = 90$
$20x - 5y = 100$

39. $0.2x + 0.3y = 0.4$ (1)
$0.5x + 0.4y = 0.3$ (2)
Multiply (1) by 10 and (2) by 10.
$2x + 3y = 4$ (3)
$5x + 4y = 3$ (4)
Multiply (3) by 5 and (4) by −2.
$$\begin{array}{r} 10x + 15y = 20 \\ \underline{-10x - 8y = -6} \\ 7y = 14 \\ y = 2 \end{array}$$
Substitute 2 for y in (3).
$2x + 3(2) = 4$
$2x = -2$
$x = -1$
(−1, 2)

41. $0.04x - 0.03y = 0.05$ (1)
$0.05x + 0.08y = -0.76$ (2)
Multiply (1) and (2) by 100.
$4x - 3y = 5$ (3)
$5x + 8y = -76$ (4)
Multiply (3) by 5 and (4) by −4.
$$\begin{array}{r} 20x - 15y = 25 \\ \underline{-20x - 32y = 304} \\ -47y = 329 \\ y = -7 \end{array}$$
Substitute −7 for y in (3).
$4x - 3(-7) = 5$
$4x = -16$
$x = -4$
(−4, −7)

43. $-0.6x - 0.08y = -4$ (1)
$3x + 2y = 4$ (2)
Multiply (1) by 100 and (2) by 20.
$$\begin{array}{r} -60x - 8y = -400 \\ \underline{60x + 40y = 80} \\ 32y = -320 \\ y = -10 \end{array}$$
Substitute −10 for y in (2).
$3x + 2(-10) = 4$
$3x = 24$
$x = 8$
(8, −10)

45. $3x - y = -8$ (1)
$-9x + 2y = 18$ (2)
Multiply (1) by 3 to eliminate x.
$$\begin{array}{r} 9x - 3y = -24 \\ \underline{-9x + 2y = 18} \\ -y = -6 \\ y = 6 \end{array}$$
Substitute 6 for y in (1).
$3x - 6 = -8$
$x = -\frac{2}{3}$
$\left(-\frac{2}{3}, 6\right)$

47. $\frac{1}{3}x - \frac{1}{4}y = 0$ (1)
$\frac{1}{6}x - \frac{1}{2}y = -6$ (2)
Multiply (1) by 12 and (2) by −6.
$$\begin{array}{r} 4x - 3y = 0 \\ \underline{-x + 3y = 36} \\ 3x \quad = 36 \\ x = 12 \end{array}$$
Substitute 12 for x in (1).
$\frac{1}{3}(12) - \frac{1}{4}y = 0$
$-\frac{1}{4}y = -4$
$y = 16$
(12, 16)

49. $0.05x - 0.04y = -0.08$ (1)
$0.2x + 0.3y = 0.6$ (2)
Multiply (1) by 100 and (2) by 10.
$5x - 4y = -8$ (3)
$2x + 3y = 6$ (4)
Multiply (3) by 2 and (4) by −5.

$$\begin{aligned} 10x - 8y &= -16 \\ -10x - 15y &= -30 \\ \hline -23y &= -46 \\ y &= 2 \end{aligned}$$

Substitute 2 for y in (3).

$$\begin{aligned} 5x - 4(2) &= -8 \\ 5x &= 0 \\ x &= 0 \end{aligned}$$

(0, 2)

51. $3(3x+2) = y + 5x \quad (1)$

$4 = -2(x - y) \quad (2)$

Expand both equations.

$$\begin{aligned} 9x + 6 &= y + 5x \\ 4 &= -2x + 2y \end{aligned}$$

Rearrange.

$$\begin{aligned} 4x - y &= -6 \quad (1) \\ 2x - 2y &= -4 \quad (2) \end{aligned}$$

Multiply (2) by −1 to eliminate x.

$$\begin{aligned} 4x - y &= -6 \\ -4x + 4y &= 8 \\ \hline 3y &= 2 \\ y &= \frac{2}{3} \end{aligned}$$

Substitute $\frac{2}{3}$ for y in (1).

$$\begin{aligned} 4x - \frac{2}{3} &= -6 \\ x &= -\frac{4}{3} \end{aligned}$$

$\left(-\frac{4}{3}, \frac{2}{3}\right)$

Check: $4\left(-\frac{4}{3}\right) - \frac{2}{3} \stackrel{?}{=} -6$ $\quad -6 = -6$

$2\left(-\frac{4}{3}\right) - 2\left(\frac{2}{3}\right) \stackrel{?}{=} -4$ $\quad -4 = -4$

Cumulative Review

52. $d = rt \Rightarrow t = \frac{d}{r}$

Bobby's time for 100 laps < Carlos' time for 96 laps

$$\begin{aligned} \frac{100 \times 4.75}{r} &< \frac{96 \times 4.75}{95} \\ \frac{475}{r} &< 4.8 \\ 475 &< 4.8r \\ 98.96 &< r \end{aligned}$$

Bobby will have to average more than 99 mph.

53. x = the number of future games to be won

$$45 + x = 0.5(45 + 55 + 62)$$
$$45 + x = 81$$
$$x = 36$$

They need to win 36 of the next 62 games.

54. $64x^2 - 4 = 4(16x^2 - 1)$
$$= 4[(4x)^2 - 1^2]$$
$$= 4(4x - 1)(4x + 1)$$

55. $9x^2 - 15x + 4 = (3x - 4)(3x - 1)$

Quick Quiz 8.3

1. $3x + 5y = -1 \quad (1)$
$-5x + 4y = -23 \quad (2)$

Multiply (1) by 5, (2) by 3 to eliminate x.

$$15x + 25y = -5$$
$$\underline{-15x + 12y = -69}$$
$$37y = -74$$
$$y = -2$$

Substitute −2 for y in (1).

$$3x + 5(-2) = -1$$
$$x = 3$$

(3, −2)

Check:

$$3(3) + 5(-2)) \stackrel{?}{=} -1$$
$$-1 = -1$$
$$-5(3) + 4(-2)) \stackrel{?}{=} -23$$
$$-23 = -2$$

2. $-7x + 3y = -31 \quad (1)$
$4x + 6y = 10 \quad (2)$

Multiply (1) by −2 to eliminate y.

$$14x - 6y = 62$$
$$\underline{4x + 6y = 10}$$
$$18x = 72$$
$$x = 4$$

Substitute 4 for x in (2).

$$4(4) + 6y = 10$$
$$6y = -6$$
$$y = -1$$

(4, −1)

Check:

$$-7(4) + 3(-1)) \stackrel{?}{=} -31$$
$$-31 = -31$$
$$4(4) + 6(-1)) \stackrel{?}{=} 10$$
$$10 = 10$$

3. $\frac{1}{3}x + y = \frac{8}{3} \quad (1)$
$\frac{4}{5}x - \frac{2}{5}y = \frac{18}{5} \quad (2)$

Multiply (2) by 5, (1) by 2 to eliminate y.

$$\frac{2}{3}x + 2y = \frac{16}{3}$$
$$\underline{4x - 2y = 18}$$
$$\frac{14}{3}x = \frac{70}{3}$$
$$x = 5$$

Substitute 5 for x in (2).

$$\frac{4}{5}(5) - \frac{2}{5}y = \frac{18}{5}$$
$$20 - 2y = 18$$
$$y = 1$$

(5, 1)

Check:

$$\frac{1}{3}(5) + 1) \stackrel{?}{=} \frac{8}{3}$$
$$\frac{8}{3} = \frac{8}{3}$$
$$\frac{4}{5}(5) - \frac{2}{5}(1)) \stackrel{?}{=} \frac{18}{5}$$
$$\frac{18}{5} = \frac{18}{5}$$

4. Answers may vary. Possible solution: Multiplying both equations by 100 will eliminate decimals. Or, multiply the first equation by 10 and the second equation by 100.

How Am I Doing? Sections 8.1–8.3

1. $2x - y = -8$
$-x + 2y = 10$

Graph both equations on the same coordinate system. The lines intersect at (−2, 4)

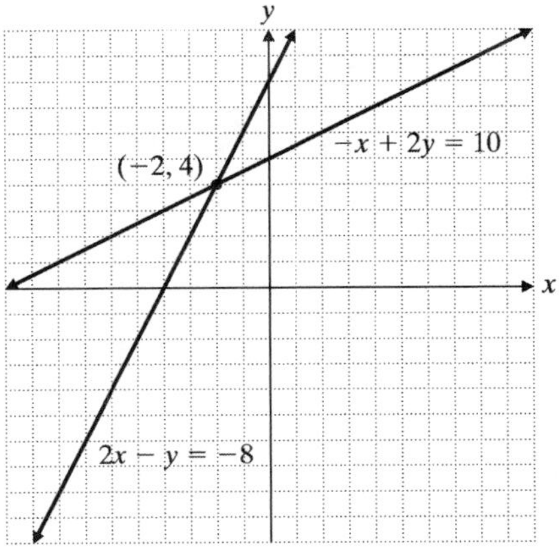

2. $-2x+y=4$
$x-3y=3$
Graph both equations on the same coordinate system. The lines intersect at (–3, –2).

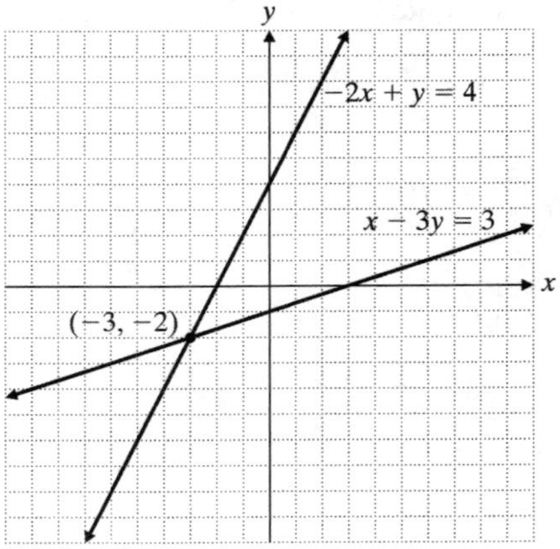

3. $5x+y=26$ (1)
$3x+2y=10$ (2)
Solve (1) for y.
$y=-5x+26$ (3)
Substitute $-5x+26$ for y in (2).
$3x+2(-5x+26)=10$
$3x-10x+52=10$
$-7x+52=10$
$-7x=-42$
$x=6$
Substitute 6 for x in (3).
$y=-5(6)+26=-4$
(6, –4)

4. $3x-5y=10$ (1)
$x+4y=-8$ (2)
Solve (2) for x.
$x=-8-4y$ (3)
Substitute $-8-4y$ for x in (1).
$3(-8-4y)-5y=10$
$-24-12y-5y=10$
$-17y=34$
$y=-2$
Substitute –2 for y in (3).
$x=-8-4(-2)$
$x=0$
(0, –2)

5. $2x+3y=5$ (1)
$\frac{1}{2}x+\frac{1}{6}y=-\frac{1}{2}$ (2)
Multiply (2) by 6.
$3x+y=-3$
Solve for y.
$y=-3x-3$ (3)
Substitute $-3x-3$ for y in (1).
$2x+3(-3x-3)=5$
$2x-9x-9=5$
$-7x-9=5$
$-7x=14$
$x=-2$
Substitute –2 for x in (3).
$y=-3(-2)-3$
$y=3$
(–2, 3)

6. $7x-5y-17=0$ (1)
$x+8y+15=0$ (2)
Solve (2) for x.
$x=-8y-15$ (3)
Substitute $-8y-15$ for x in (1).
$7(-8y-15)-5y-17=0$
$-56y-105-5y-17=0$
$-61y-122=0$
$-61y=122$
$y=-2$
Substitute –2 for y in (3).
$x=-8(-2)-15=1$
(1, –2)

7. $3x-2y=19$ (1)
$x+y=8$ (2)
Multiply (2) by 2.
$$\begin{aligned}3x-2y&=19\\ \underline{2x+2y}&\underline{=16}\\ 5x\quad&=35\\ x&=7\end{aligned}$$
Substitute 7 for x in (2).
$7+y=8$
$y=1$
(7, 1)

8. $3x+10y=-4$ (1)
$2x-8y=-10$ (2)
Multiply (1) by –2, (2) by 3 to eliminate x.
$$\begin{aligned}-6x-20y&=8\\ \underline{6x-24y}&\underline{=-30}\\ -44y&=-22\\ y&=\frac{1}{2}\end{aligned}$$
Substitute $\frac{1}{2}$ for y in (1).

$$3x + 10\left(\frac{1}{2}\right) = -4$$
$$3x + 5 = -4$$
$$x = -3$$
$$\left(-3, \frac{1}{2}\right)$$

9. $\frac{2}{3}x - \frac{3}{4}y = 3$ (1)
$\frac{1}{9}x + \frac{1}{12}y = -2$ (2)
Multiply (1) by 12 and (2) by 36.
$8x - 9y = 36$ (3)
$4x + 3y = -72$ (4)
Multiply (4) by 3.
$$\begin{array}{l} 8x - 9y = 36 \\ \underline{12x + 9y = -216} \\ 20x \quad\quad = -180 \\ \quad\quad x = -9 \end{array}$$
Substitute −9 for x in (1).
$$\frac{2}{3}(-9) - \frac{3}{4}y = 3$$
$$-6 - \frac{3}{4}y = 3$$
$$-\frac{3}{4}y = 9$$
$$y = -12$$
(−9, −12)

10. $0.6x - 0.5y = 2.1$ (1)
$0.4x - 0.3y = 1.3$ (2)
Multiply (1) and (2) by 10.
$6x - 5y = 21$ (3)
$4x - 3y = 13$ (4)
Multiply (3) by 2 and (4) by −3.
$$\begin{array}{l} 12x - 10y = 42 \\ \underline{-12x + 9y = -39} \\ \quad\quad -y = 3 \\ \quad\quad y = -3 \end{array}$$
Substitute −3 for y in (1).
$$0.6x - 0.5(-3) = 2.1$$
$$0.6x + 1.5 = 2.1$$
$$0.6x = 0.6$$
$$x = 1$$
(1, −3)

8.4 Exercises

1. If there is no solution to a system of linear equations, the graphs of the equations are <u>parallel lines</u>. Solving the system algebraically, you will obtain an equation that is <u>inconsistent</u> with known facts.

3. If there is exactly one solution, the graphs of the equations <u>intersect</u>. This system is said to be <u>independent</u> and <u>consistent</u>.

5. $-2x - 3y = 15$ (1)
$5x + 2y = 1$ (2)
Multiply (1) by 2 and (2) by 3.
$$\begin{array}{l} -4x - 6y = 30 \\ \underline{15x + 6y = 3} \\ 11x \quad\quad = 33 \\ \quad\quad x = 3 \end{array}$$
Substitute 3 for x in (2).
$$5(3) + 2y = 1$$
$$2y = -14$$
$$y = -7$$
(3, −7)

7. $3x - 4y = 2$ (1)
$-5x + 6y = -12$ (2)
Multiply (1) by 5, (2) by 3 to eliminate x.
$$\begin{array}{l} 15x - 20y = 10 \\ \underline{-15x + 18y = -36} \\ \quad\quad -2y = -26 \\ \quad\quad y = 13 \end{array}$$
Substitute 13 for y into (1).
$$3x - 4(13) = 2$$
$$3x - 52 = 2$$
$$x = 18$$
(18, 13)

9. $2x - 4y = 5$ (1)
$-4x + 8y = 9$ (2)
Multiply (1) by 2.
$$\begin{array}{l} 4x - 8y = 10 \\ \underline{-4x + 8y = 9} \\ \quad\quad 0 = 19 \end{array}$$
There is no solution.

11. $-5x+2y=2$ (1)
$15x-6y=-6$ (2)
Multiply (1) by 3.
$$\begin{aligned} -15x+6y&=6\\ 15x-6y&=-6\\ \hline 0&=0 \end{aligned}$$
There is an infinite number of solutions.

13. $5x-3y=13$ (1)
$7x+2y=43$ (2)
Multiply (1) by 2 and (2) by 3.
$$\begin{aligned} 10x-6y&=26\\ 21x+6y&=129\\ \hline 31x\qquad&=155\\ x&=5 \end{aligned}$$
Substitute 5 for x in (1).
$$\begin{aligned} 5(5)-3y&=13\\ 25-3y&=13\\ -3y&=-12\\ y&=4 \end{aligned}$$
(5, 4)

15. $3x-2y=70$ (1)
$0.6x+0.5y=50$ (2)
Multiply (1) by −2 and (2) by 10.
$$\begin{aligned} -6x+4y&=-140\\ 6x+5y&=500\\ \hline 9y&=360\\ y&=40 \end{aligned}$$
Substitute 40 for y in (1).
$$\begin{aligned} 3x-2(40)&=70\\ 3x&=150\\ x&=50 \end{aligned}$$
(50, 40)

17. $0.2x-0.3y=0.1$ (1)
$-0.5x+0.8y=0$ (2)
Multiply (1) by 50, (2) by 20 to eliminate x.
$$\begin{aligned} 10x-15y&=5\\ -10x+16y&=0\\ \hline y&=5 \end{aligned}$$
Substitute 5 for y into (1).
$$\begin{aligned} 0.2x-0.3(5)&=0.1\\ 0.2x-1.5&=0.1\\ x&=8 \end{aligned}$$
(8, 5)

19. $\frac{4}{3}x+\frac{1}{2}y=1$ (1)
$\frac{1}{3}x-y=-\frac{1}{2}$ (2)
Multiply (1) by 6 and (2) by −24.
$$\begin{aligned} 8x+3y&=6\\ -8x+24y&=12\\ \hline 27y&=18\\ y&=\frac{2}{3} \end{aligned}$$
Substitute $\frac{2}{3}$ for y in (2).
$$\begin{aligned} \frac{1}{3}x-\frac{2}{3}&=-\frac{1}{2}\\ \frac{1}{3}x&=\frac{1}{6}\\ x&=\frac{1}{2} \end{aligned}$$
$\left(\frac{1}{2}, \frac{2}{3}\right)$

21. $\frac{2}{3}x+\frac{1}{6}y=2$ (1)
$\frac{1}{4}x-\frac{1}{2}y=-\frac{3}{4}$ (2)
Multiply (1) by 12 and (2) by 4.
$$\begin{aligned} 8x+2y&=24\\ x-2y&=-3\\ \hline 9x\qquad&=21\\ x&=\frac{7}{3} \end{aligned}$$
Substitute $\frac{7}{3}$ for x in (1).
$$\begin{aligned} \frac{2}{3}\left(\frac{7}{3}\right)+\frac{1}{6}y&=2\\ \frac{1}{6}y&=\frac{4}{9}\\ y&=\frac{8}{3} \end{aligned}$$
$\left(\frac{7}{3}, \frac{8}{3}\right)$

23. $4x+3y=-2$ (1)
$5x-2y=9$ (2)
Multiply (1) by 2 and (2) by 3.
$$\begin{array}{r} 8x+6y=-4 \\ \underline{15x-6y=27} \\ 23x \quad =23 \\ x=1 \end{array}$$
Substitute 1 for x in (1).
$$\begin{aligned} 4(1)+3y&=-2 \\ 3y&=-6 \\ y&=-2 \end{aligned}$$
(1, –2)

25. $\frac{2}{3}x+y=1$ (1)
$\frac{3}{4}x+\frac{1}{2}y=\frac{3}{4}$ (2)
Multiply (1) by 3 and (2) by 4.
$2x+3y=3$ (3)
$3x+2y=3$ (4)
Multiply (3) by 3 and (2) by –2.
$$\begin{array}{r} 6x+9y=9 \\ \underline{-6x-4y=-6} \\ 5y=3 \\ y=\frac{3}{5} \end{array}$$
Substitute $\frac{3}{5}$ for y in (1).
$$\begin{aligned} \frac{2}{3}x+\frac{3}{5}&=1 \\ 10x+9&=15 \\ 10x&=6 \\ x&=\frac{3}{5} \end{aligned}$$
$\left(\frac{3}{5}, \frac{3}{5}\right)$

27. $x-2y+2=0$ (1)
$3(x-2y+1)=15$
$3x-6y+3=15$ (2)
Multiply (1) by –3.
$$\begin{array}{r} -3x+6y-6=0 \\ \underline{3x-6y+3=15} \\ -3=15 \end{array}$$
No solution; the system is inconsistent.

29. $2(a+3)=b+1$
$3(a-b)=a+1$
Expand.
$2a+6=b+1$
$3a-3b=a+1$
Rearrange.
$$\begin{array}{rl} 2a-b=-5 & (1) \\ \underline{-2a+3b=-1} & (2) \\ 2b=-6 & \\ b=-3 & \end{array}$$
Substitute –3 for b in (1).
$$\begin{aligned} 2a-(-3)&=-5 \\ 2a&=-8 \\ a&=-4 \end{aligned}$$
(–4, –3)

Cumulative Review

31. $$\begin{aligned} 8x+15\left(\frac{2}{3}x-\frac{2}{5}\right)&=-10 \\ 8x+10x-6&=-10 \\ 18x&=-4 \\ x&=-\frac{2}{9} \end{aligned}$$

32. $$\begin{aligned} 0.2(x+0.3)+x&=0.3(x+0.5) \\ 0.2x+0.06+x&=0.3x+0.15 \\ 1.2x+0.06&=0.3x+0.15 \\ 0.9x&=0.09 \\ x&=0.1 \end{aligned}$$

33. x = federal income tax owed
x = 782.50 + 0.15(27,600 – 7825)
x = \$3748.75
She had to pay \$3748.75.

34. x = federal income tax owed
x = 4386.25 + 0.25(32,900 – 31,850)
x = \$4648.75
Amount paid was \$5033.
Refund = \$5033 – \$4648.75 = \$384.25
It was more. The refund was \$384.25.

Quick Quiz 8.4

1. $12-x=3(x+y)$ (1)
$3x+14=4x+5(x+y)$ (2)
Expand the equations.
$12-x=3x+3y$
$3x+14=4x+5x+5y$
Rearrange.

$-4x - 3y = -12$ (3)
$-6x - 5y = -14$ (4)
Multiply (3) by −5, (4) by 3 to eliminate y.
$$\begin{array}{r} 20x + 15y = 60 \\ -18x - 15y = -42 \\ \hline 2x \quad\quad = 18 \\ x = 9 \end{array}$$
Substitute 9 for x in (4).
$-6(9) - 5y = -14$
$-54 - 5y = -14$
$y = -8$
(9, −8)
Check:
$-4(9) - 3(-8) \stackrel{?}{=} -12$
$-12 = -12$
$-6(9) - 5(-8) \stackrel{?}{=} -14$
$-14 = -14$

2. $-4x + 6y = 2$ (1)
$12y = 8x + 12$ (2)
Rearrange.
$-4x + 6y = 2$ (3)
$-8x + 12y = 12$ (4)
Multiply (1) by −2 to eliminate y.
$$\begin{array}{r} 8x - 12y = -4 \\ -8x + 12y = 12 \\ \hline 0 + 0 = 8 \end{array}$$
Inconsistent system of equations; no solution

3. $\frac{1}{2}x - \frac{1}{4}y = 1$ (1)
$\frac{4}{3}x + 4y = 12$ (2)
Multiply (1) by −8, (2) by 3 to eliminate x.
$$\begin{array}{r} -4x + 2y = -8 \quad (3) \\ 4x + 12y = 36 \quad (4) \\ \hline 14y = 28 \\ y = 2 \end{array}$$
Substitute 2 for y in (4).
$4x + 12(2) = 36$
$4x = 36 - 24$
$x = 3$
(3, 2)
Check:
$-4(3) + 2(2) \stackrel{?}{=} -8$
$-8 = -8$
$4(3) + 12(2) \stackrel{?}{=} 36$
$36 = 36$

4. Answers may vary. Possible solution: Multiply (1) by 10, (2) by −48. Then add equations to eliminate fractions and x.

8.5 Exercises

1. x = number of first class tickets
y = number of coach tickets
$x + y = 14$ (1)
$1150x + 250y = 10,700$ (2)
Solve (1) for x.
$x = 14 - y$ (3)
Substitute $14 - y$ for x in (2).
$1150(14 - y) + 250y = 10,700$
$16,100 - 1150y + 250y = 10,700$
$16,100 - 900y = 10,700$
$-900y = -5400$
$y = 6$
Substitute 6 for y in (3).
$x = 14 - 6$
$x = 8$
She bought 8 first class tickets and 6 coach tickets.

3. x = number of etchings sold at \$35
y = number of etchings sold at \$40
$35x + 40y = 455$ (1)
$x + y = 12$ (2)
Solve (2) for y.
$y = -x + 12$ (3)
Substitute $-x + 12$ for y in (1).
$35x + 40(-x + 12) = 455$
$35x - 40x + 480 = 455$
$-5x = -25$
$x = 5$
Substitute 5 for x in (3).
$y = -5 + 12 = 7$
He sold 5 etchings at \$35.
He sold 7 etchings at \$40.

5. x = hours on day shift
y = hours on night shift
$x + y = 23$ (1)
$13.50x + 16.50y = 352.50$ (2)
Solve (1) for x.
$x = 23 - y$ (3)
Substitute $23 - y$ for x in (2).
$13.50(23 - y) + 16.50y = 352.50$
$310.50 - 13.50y + 16.50y = 352.50$
$310.50 + 3y = 352.50$
$3y = 42$
$y = 14$
Substitute 14 for y in (3).

$x = 23 - 14$
$x = 9$
She spent 9 hours on day shift and 14 hours on night shift.

7. L = original length
W = original width
$2L + 2W = 38 \quad (1)$
$2(2L - 6) + 2(W + 5) = 56$
$4L - 12 + 2W + 10 = 56$
$4L + 2W = 58 \quad (2)$
Multiply (1) by −1.
$$\begin{array}{l} -2L - 2W = -38 \\ \underline{4L + 2W = 58} \\ 2L \quad\quad = 20 \\ L = 10 \end{array}$$
Substitute 10 for L in (1).
$2(10) + 2W = 38$
$2W = 18$
$W = 9$
original length = 10 feet
original width = 9 feet
New length = 2(10) − 6 = 14 feet
New width = 9 + 5 = 14 feet

9. x = quarts before
y = quarts added
$x + y = 16 \quad (1)$
$0.50x + 0.80y = 0.65(16) \quad (2)$
Multiply (1) by −5 and (2) by 10.
$$\begin{array}{l} -5x - 5y = -80 \\ \underline{5x + 8y = 104} \\ 3y = 24 \\ y = 8 \end{array}$$
Substitute 8 for y in (1).
$x + 8 = 16$
$x = 8$
8 quarts of 50% antifreeze solution were in the radiator originally. She added 8 quarts of 80% antifreeze solution.

11. x = weight of nuts
y = weight of raisins
$x + y = 50 \quad (1)$
$2.00x + 1.50y = 50(1.80) \quad (2)$
Multiply (1) by −2.
$$\begin{array}{l} -2x - 2y = -100 \\ \underline{2x + 1.5y = 90} \\ -0.5y = -10 \\ y = 20 \end{array}$$
Substitute 20 for y in (1).
$x + 20 = 50$
$x = 30$
They should buy 30 lb of nuts and 20 lb of raisins.

13. w = speed of wind
r = airspeed of airplane in still air
$$\text{speed} = \frac{\text{distance}}{\text{time}}$$
$$\begin{array}{l} r + w = \frac{3000}{5} = 600 \quad (1) \\ \underline{r - w = \frac{3000}{6} = 500 \quad (2)} \\ 2r \quad = 1100 \\ r \quad = 550 \end{array}$$
Substitute 550 for r in (1).
$550 + w = 600$
$w = 50$
The plane's speed in still air is 550 kilometers per hour. The windspeed is 50 kilometers per hour.

15. x = number of \$10 bills for Jewelry
y = number of \$20 bills for Jewelry
Total amount in \$10 and \$20 bills
= 320 − 25 − 55
= 240
$10x + 20y = 240 \quad (1)$
$10(2x - 2) + 20(y - 4) = 240$
$20x - 20 + 20y - 80 = 240$
$20x + 20y = 340 \quad (2)$
Multiply (1) by −1.
$$\begin{array}{l} -10x - 20y = -240 \\ \underline{20x + 20y = 340} \\ 10x \quad\quad = 100 \\ x = 10 \end{array}$$
Substitute 10 for x in (1).
$10(10) + 20y = 240$
$100 + 20y = 240$
$20y = 140$
$y = 7$
Jewelry: 10-\$10 bills and 7-\$20 bills
Cosmetics:
$2(10) - 2 = 18$ \$10 bills
7 − 4 = 3 \$20 bills

17. x = price of oranges
y = price of apples
$7x + 5y = 8.78 \quad (1)$
$3x + 8y = 8.39 \quad (2)$
Multiply (1) by 3 and (2) by −7.

$$\begin{aligned} 21x+15y &= 26.34 \\ -21x-56y &= -58.73 \\ \hline -41y &= -32.39 \\ y &= 0.79 \end{aligned}$$

Substitute 0.79 for y in (1).

$$\begin{aligned} 7x+5(0.79) &= 8.78 \\ 7x+3.95 &= 8.78 \\ 7x &= 4.83 \\ x &= 0.69 \end{aligned}$$

Oranges are \$0.69 per pound.
Apples are \$0.79 per pound.

19. x = general admission tickets
y = student admission tickets

$$\begin{aligned} x+y &= 342 \quad (1) \\ 8x+5.5y &= 2211 \quad (2) \end{aligned}$$

Solve (1) for x.
$x = 342 - y$
Substitute $342 - y$ for x into (2).

$$\begin{aligned} 8(342-y)+5.5y &= 2211 \\ 2736-8y+5.5y &= 2211 \\ y &= 210 \end{aligned}$$

Substitute 210 for y into (1).

$$\begin{aligned} x+210 &= 342 \\ x &= 132 \end{aligned}$$

132 general admission tickets and 210 student admission tickets were sold.

21.

$$\begin{aligned} 40x-y &= -5467 \quad (1) \\ 53x-2y &= -11,447 \quad (2) \end{aligned}$$

Multiply (1) by −2 to eliminate y.

$$\begin{aligned} -80x+2y &= 10,934 \\ 53x-2y &= -11,447 \\ \hline -27x \qquad &= -513 \\ x &= 19 \end{aligned}$$

Substitute 19 for x in (1).

$$\begin{aligned} 40(19)-y &= -5467 \\ y &= 6227 \end{aligned}$$

year = 2005 + 19 = 2024
population = 6,227,000
The two states' population will be the same at 6,227,000 in 2024.

Cumulative Review

23.

$$\begin{aligned} &\frac{2}{x-1}+\frac{3}{x+1}+1 \\ &= \frac{2}{x-1}\cdot\frac{x+1}{x+1}+\frac{3}{x+1}\cdot\frac{x-1}{x-1}+1\cdot\frac{(x+1)(x-1)}{(x+1)(x-1)} \\ &= \frac{2(x+1)+3(x-1)+(x+1)(x-1)}{(x+1)(x-1)} \\ &= \frac{2x+2+3x-3+x^2-x+x-1}{(x+1)(x-1)} \\ &= \frac{x^2+5x-2}{x^2-1} \end{aligned}$$

24. When $x = -2$ and $y = 3$,

$$\begin{aligned} 2x^2-3y+4y^2 &= 2(-2)^2-3(3)+4(3)^2 \\ &= 8-9+36 \\ &= 35 \end{aligned}$$

Quick Quiz 8.5

1. x = people who purchased student tickets
y = people who purchased general admission tickets

$$\begin{aligned} x+y &= 18,500 \quad (1) \\ 9x+15y &= 217,500 \quad (2) \end{aligned}$$

Solve (1) for x.
$x = 18,500 - y$
Substitute $(18,500 - y)$ for x in (2).

$$\begin{aligned} 9(18,500-y)+15y &= 217,500 \\ 166,500-9y+15y &= 217,500 \\ y &= 8500 \end{aligned}$$

Substitute 8500 for y in (1).

$$\begin{aligned} x+8500 &= 18,500 \\ x &= 10,000 \end{aligned}$$

10,000 people purchased student tickets.
8500 people purchased general admission tickets.

2. x = price of PC
y = price of iMac

$$\begin{aligned} 8x+12y &= 20,800 \quad (1) \\ 12x+7y &= 18,000 \quad (2) \end{aligned}$$

Multiply (1) by −3, (2) by 2 to eliminate x.

$$\begin{aligned} -24x-36y &= -62,400 \\ 24x+14y &= 36,000 \\ \hline -22y &= -26,400 \\ y &= 1200 \end{aligned}$$

Substitute 1200 for y in (1).

$$\begin{aligned} 8x+12(1200) &= 20,800 \\ x &= 800 \end{aligned}$$

Each Dell PC was \$800. Each Apple iMac was \$1200.

3. x = speed of plane in still air
y = speed of jet stream
$(x+y)3=1500$ (1)
$(x-y)5=1500$ (2)
Expand the equations.
$3x+3y=1500$ (3)
$5x-5y=1500$ (4)
Multiply (3) by 5, (4) by 3 to eliminate y.

$$\begin{array}{l} 15x+15y=7500 \\ \underline{15x-15y=4500} \\ 30x \qquad =12,000 \\ \qquad x=400 \end{array}$$

Substitute 400 for x in (3).

$$\begin{aligned} 3(400)+3y&=1500 \\ 1200+3y&=1500 \\ y&=100 \end{aligned}$$

Speed of plane in still air was 400 mph. Speed of jet stream was 100 mph.

4. Answers may vary. Possible solution:
The system will consist of two equations in terms of x and y. x will equal the number of gallons of 50% pure juice. y will equal the number of gallons of 30% pure juice.
The first equation states that the sum of x and y will equal 500. The second equation states that each amount of the mixture times its decimal fraction of juice will sum to equal the total mixture quantity times its decimal fraction of juice.

Putting Your Skills to Work

1. Store A: $\frac{1}{5}(400+400)=\frac{1}{5}(800)=160$
Total $= \$400+\$400-\$160=\$640$

Store B: $\frac{1}{5}(500)=100$
Total $= \$500+\$500-\$100=\$900$
Store A has the better buy.

2. **a.** about $30 · 0.10 = $3

b. about $30 · 0.20 = $6

c. about $30 · 0.30 = $9

3. **a.** about $60 · 0.20 = $12

b. about $60 · 0.25 = $15

c. about $60 · 0.30 = $18

4. $(52+36+2\times15)\times0.2 \approx (50+40+30)\times0.2$
$=120\times0.2$
$=24$
She would save about $24.

5. 1st savings: $95 × 0.10 = $9.50
New price: $95 – $9.50 = $85.50
2nd savings: $85.50 × 0.2 = $17.10
Total savings ≈ $10 + $17 = $27

6. A: (6 × 40) × 0.80 = $192
B: (4 × 30) + (2 × 30) × 0.9 = 120 + 54 = $174
Store B is offering the better buy.

7. Answers may vary.

Chapter 8 Review Problems

1. $2x+3y=0$
$-x+3y=9$
Graph both equations on the same coordinate system. The lines intersect at (–3, 2).

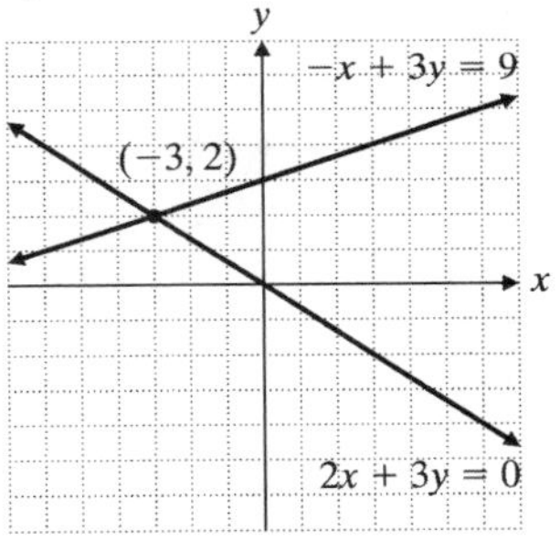

2. $-3x+y=-2$
$-2x-y=-8$
Graph both equations on the same coordinate system. The lines intersect at (2, 4).

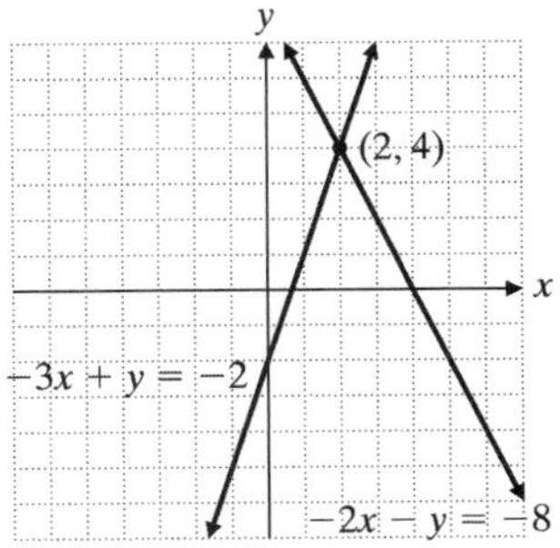

3. $2x - y = 6$
$6x + 3y = 6$
Graph both equations on the same coordinate system. The lines intersect at (2, −2).

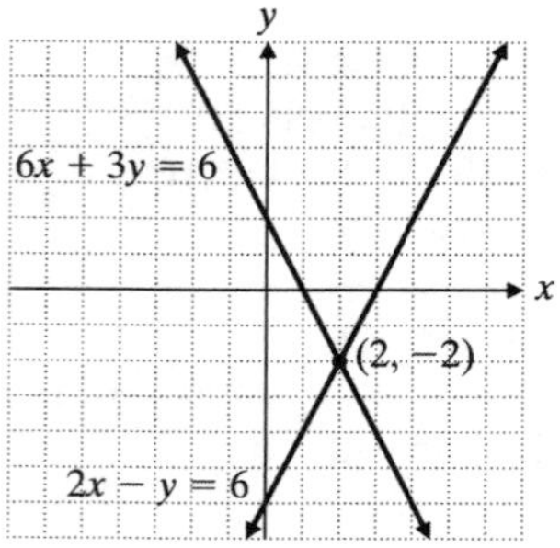

4. $2x - y = 1$
$3x + y = -6$
Graph both equations on the same coordinate system. The lines intersect at (−1, −3).

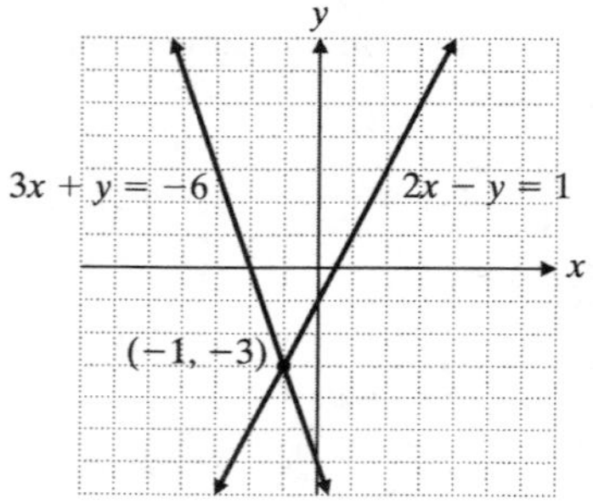

5. $x + y = 6$ (1)
$-2x + y = -3$ (2)
Solve (2) for *y*.
$y = 2x - 3$ (3)
Substitute $2x - 3$ for *y* in (1).
$x + (2x - 3) = 6$
$x + 2x - 3 = 6$
$3x = 9$
$x = 3$
Substitute 3 for *x* in (3).
$y = 2(3) - 3 = 3$
(3, 3)
Check:
$3 + 3 \stackrel{?}{=} 6$
$6 = 6$
$-2(3) + 3 \stackrel{?}{=} -3$
$-3 = -3$

6. $x + 3y = 18$ (1)
$2x + y = 11$ (2)
Solve (2) for *y*.
$y = -2x + 11$ (3)
Substitute $-2x + 11$ for *y* in (1).
$x + 3(-2x + 11) = 18$
$x - 6x + 33 = 18$
$-5x = -15$
$x = 3$
Substitute 3 for *x* in (3).
$y = -2(3) + 11 = 5$
(3, 5)
Check:
$3 + 3(5) \stackrel{?}{=} 18$
$18 = 18$
$2(3) + 5 \stackrel{?}{=} 11$
$11 = 11$

7. $3x - 2y = 3$ (1)
$x - \frac{1}{3}y = 8$
$3x - y = 24$ (2)
Solve (2) for *y*.
$y = 3x - 24$ (3)
Substitute $3x - 24$ for *y* in (1).
$3x - 2(3x - 24) = 3$
$3x - 6x + 48 = 3$
$-3x = -45$
$x = 15$
Substitute 15 for *x* in (3).
$y = 3(15) - 24 = 21$
(15, 21)
Check:
$15 - \frac{1}{3}(21) \stackrel{?}{=} 8$
$15 - 7 \stackrel{?}{=} 6$
$8 = 8$
$3(15) - 2(21) \stackrel{?}{=} 3$
$45 - 42 \stackrel{?}{=} 3$
$3 = 3$

8. $0.5x + y = 16$ (1)
$4x - 2y = 8$ (2)
Solve (1) for *y*.
$y = -0.5x + 16$ (3)
Substitute $-0.5x + 16$ for *y* in (2).
$4x - 2(-0.5x + 16) = 8$
$4x + x - 32 = 8$
$5x - 32 = 8$
$5x = 40$
$x = 8$
Substitute 8 for *x* in (3).
$y = -0.5(8) + 16$
$y = 12$
(8, 12)

Check:

$4(8)-2(12) \stackrel{?}{=} 8$

$8=8$

$0.5(8)+12 \stackrel{?}{=} 16$

$4+12 \stackrel{?}{=} 16$

$16=16$

9. $6x-2y=10$ (1)

$2x+3y=7$ (2)

Multiply (1) by 3 and (2) by 2.

$$\begin{array}{r} 18x-6y=30 \\ \underline{4x+6y=14} \\ 22x \quad\quad =44 \end{array}$$

$x=2$

Substitute 2 for x in (2).

$2(2)+3y=7$

$3y=3$

$y=1$

(2, 1)

Check:

$6(2)-2(1) \stackrel{?}{=} 10$

$10=10$

$2(2)+3(1) \stackrel{?}{=} 7$

$7=7$

10. $4x-5y=-4$ (1)

$-3x+2y=3$ (2)

Multiply (1) by 2 and (2) by 5.

$$\begin{array}{r} 8x-10y=-8 \\ \underline{-15x+10y=15} \\ -7x \quad\quad =7 \end{array}$$

$x=-1$

Substitute -1 for x in (2).

$-3(-1)+2y=3$

$2y=0$

$y=0$

(−1, 0)

Check:

$4(-1)-5(0) \stackrel{?}{=} -4$

$-4=-4$

$-3(-1)+2(0) \stackrel{?}{=} 3$

$3=3$

11. $5x-4y=2$ (1)

$-3x+10y=-5$ (2)

Multiply (1) by 3, (2) by 5 to eliminate x.

$$\begin{array}{r} 15x-12y=6 \\ \underline{-15x+50y=-25} \\ 38y=-19 \end{array}$$

$y=-\frac{1}{2}$

Substitute $-\frac{1}{2}$ for y in (1).

$5x-4\left(-\frac{1}{2}\right)=2$

$5x+2=2$

$x=0$

$\left(0, -\frac{1}{2}\right)$

Check:

$-3(0)+10\left(-\frac{1}{2}\right) \stackrel{?}{=} -5$

$-5=-5$

$5(0)-4\left(-\frac{1}{2}\right) \stackrel{?}{=} 2$

$2=2$

12. $9x+2y=-5$ (1)

$12x-y=8$ (2)

Multiply (2) by 2 to eliminate y.

$$\begin{array}{r} 9x+2y=-5 \\ \underline{24x-2y=16} \\ 33x \quad\quad =11 \end{array}$$

$x=\frac{1}{3}$

Substitute $\frac{1}{3}$ for x in (1).

$9\left(\frac{1}{3}\right)+2y=-5$

$3+2y=-5$

$2y=-8$

$y=-4$

$\left(\frac{1}{3}, -4\right)$

Check:

$12\left(\frac{1}{3}\right)+4 \stackrel{?}{=} 8$

$4+4 \stackrel{?}{=} 8$

$8=8$

$9\left(\frac{1}{3}\right)+2(-4) \stackrel{?}{=} -5$

$3+(-8) \stackrel{?}{=} -5$

$-5=-5$

13. $6x - y = 33$ (1)
$6x + 7y = 9$ (2)
Multiply (1) by -1 to eliminate x.
$$\begin{aligned} -6x + y &= -33 \\ 6x + 7y &= 9 \\ \hline 8y &= -24 \\ y &= -3 \end{aligned}$$
Substitute -3 for y in (1).
$$\begin{aligned} 6x + 3 &= 33 \\ 6x &= 30 \\ x &= 5 \end{aligned}$$
$(5, -3)$

14. $2x + 5y = 29$ (1)
$-3x + 10y = -26$ (2)
Multiply (1) by -2 to eliminate y.
$$\begin{aligned} -4x - 10y &= -58 \\ -3x + 10y &= -26 \\ \hline -7x \quad &= -84 \\ x &= 12 \end{aligned}$$
Substitute 12 for x in (1).
$$\begin{aligned} 2(12) + 5y &= 29 \\ 24 + 5y &= 29 \\ y &= 1 \end{aligned}$$
$(12, 1)$

15. $7x + 3y = 2$ (1)
$-8x - 7y = 2$ (2)
Multiply (1) by 7 and (2) by 3.
$$\begin{aligned} 49x + 21y &= 14 \\ -24x - 21y &= 6 \\ \hline 25x \quad &= 20 \\ x &= \frac{4}{5} \end{aligned}$$
Substitute $\frac{4}{5}$ for x in (1).
$$\begin{aligned} 7\left(\frac{4}{5}\right) + 3y &= 2 \\ 3y &= -\frac{18}{5} \\ y &= -\frac{6}{5} \end{aligned}$$
$\left(\frac{4}{5}, -\frac{6}{5}\right)$

16. $7x + 5y = 7$ (1)
$-3x - 4y = 1$ (2)
Multiply (1) by 4 and (2) by 5.
$$\begin{aligned} 28x + 20y &= 28 \\ -15x - 20y &= 5 \\ \hline 13x \quad &= 33 \\ x &= \frac{33}{13} \end{aligned}$$
Substitute $\frac{33}{13}$ for x in (1).
$$\begin{aligned} 7\left(\frac{33}{13}\right) + 5y &= 7 \\ \frac{231}{13} + 5y &= 7 \\ 5y &= -\frac{140}{13} \\ y &= -\frac{28}{13} \end{aligned}$$
$\left(\frac{33}{13}, -\frac{28}{13}\right)$

17. $4x + 5y = 4$ (1)
$-5x - 8y = 2$ (2)
Multiply (1) by 5 and (2) by 4.
$$\begin{aligned} 20x + 25y &= 20 \\ -20x - 32y &= 8 \\ \hline -7y &= 28 \\ y &= -4 \end{aligned}$$
Substitute -4 for y in (1).
$$\begin{aligned} 4x + 5(-4) &= 4 \\ 4x - 20 &= 4 \\ 4x &= 24 \\ x &= 6 \end{aligned}$$
$(6, -4)$

18. $7x - 2y = 1$ (1)
$-5x + 3y = -7$ (2)
Multiply (1) by 3 and (2) by 2.
$$\begin{aligned} 21x - 6y &= 3 \\ -10x + 6y &= -14 \\ \hline 11x \quad &= -11 \\ x &= -1 \end{aligned}$$
Substitute -1 for x in (1).
$$\begin{aligned} 7(-1) - 2y &= 1 \\ -7 - 2y &= 1 \\ -2y &= 8 \\ y &= -4 \end{aligned}$$
$(-1, -4)$

19. $2x-4y=6$ (1)
$-3x+6y=7$ (2)
Multiply (1) by 3 and (2) by 2.
$$\begin{array}{r} 6x-12y=18 \\ \underline{-6x+12y=14} \\ 0=32 \end{array}$$
No solution, inconsistent system

20. $4x-7y=8$ (1)
$5x+9y=81$ (2)
Multiply (1) by 9 and (2) by 7.
$$\begin{array}{r} 36x-63y=72 \\ \underline{35x+63y=567} \\ 71x \qquad =639 \\ x=9 \end{array}$$
Substitute 9 for x in (2).
$$\begin{array}{r} 5(9)+9y=81 \\ 9y=36 \\ y=4 \end{array}$$
(9, 4)

21. $2x-9y=0$ (1)
$3x+5=6y$ (2)
Rearrange (2).
$3x-6y=-5$ (3)
Multiply (1) by 3 and (3) by −2.
$$\begin{array}{r} 6x-27y=0 \\ \underline{-6x+12y=10} \\ -15y=10 \\ y=-\frac{2}{3} \end{array}$$
Substitute $-\frac{2}{3}$ for y in (1).
$$\begin{array}{r} 2x-9\left(-\frac{2}{3}\right)=0 \\ 2x=-6 \\ x=-3 \end{array}$$
$\left(-3, -\frac{2}{3}\right)$

22. $2x+10y=1$ (1)
$-4x-20y=-2$ (2)
Multiply (1) by 2.
$$\begin{array}{r} 4x+20y=2 \\ \underline{-4x-20y=-2} \\ 0=0 \end{array}$$
Infinite number of solutions, dependent equations

23. $5x-8y=3x+12$ (1)
$7x+y=6y-4$ (2)
Rearrange and simplify.
$2x-8y=12$ (3)
$7x-5y=-4$ (4)
Multiply (3) by 7 and (4) by −2.
$$\begin{array}{r} 14x-56y=84 \\ \underline{-14x+10y=8} \\ -46y=92 \\ y=-2 \end{array}$$
Substitute −2 for y in (3).
$$\begin{array}{r} 2x-8(-2)=12 \\ 2x=-4 \\ x=-2 \end{array}$$
(−2, −2)

24. $1+x-y=y+4$
$4(x-y)=3-x$
Rearrange and simplify.
$x-2y=3$ (1)
$5x-4y=3$ (2)
Multiply (1) by −5.
$$\begin{array}{r} -5x+10y=-15 \\ \underline{5x-4y=3} \\ 6y=-12 \\ y=-2 \end{array}$$
Substitute −2 for y in (1).
$$\begin{array}{r} x-2(-2)=3 \\ x=-1 \end{array}$$
(−1, −2)

25. $3x+y=9$ (1)
$x-2y=10$ (2)
Multiply (1) by 2.
$$\begin{array}{r} 6x+2y=18 \\ \underline{x-2y=10} \\ 7x \qquad =28 \\ x=4 \end{array}$$
Substitute 4 for x in (1).
$$\begin{array}{r} 3(4)+y=9 \\ y=-3 \end{array}$$
(4, −3)

26. $x+12y=0$ (1)
$3x-5y=-2$ (2)
Multiply (1) by −3.

$$\begin{array}{r} -3x-36y=0 \\ 3x-5y=-2 \\ \hline -41y=-2 \\ y=\dfrac{2}{41} \end{array}$$

Substitute $\frac{2}{41}$ for y in (1).

$$x+12\left(\frac{2}{41}\right)=0$$
$$x=-\frac{24}{41}$$

$\left(-\frac{24}{41}, \frac{2}{41}\right)$

27. $2(x+3)=y+4$
$4x-2y=-4$
Expand, rearrange and simplify.
$2x-y=-2$ (1)
$4x-2y=-4$ (2)
Multiply (1) by -2.

$$\begin{array}{r} -4x+2y=4 \\ 4x-2y=-4 \\ \hline 0=0 \end{array}$$

Infinite number of solutions, dependent equations

28. $x+y=3000$ (1)
$x-2y=-120$ (2)
Multiply (2) by -1.

$$\begin{array}{r} x+y=3000 \\ -x+2y=120 \\ \hline 3y=3120 \\ y=1040 \end{array}$$

Substitute 1040 for y in (1).
$x+1040=3000$
$x=1960$
(1960, 1040)

29. $5x+4y+3=23$
$8x-3y-4=75$
Rearrange and simplify.
$5x+4y=20$ (1)
$8x-3y=79$ (2)
Multiply (1) by 3 and (2) by 4.

$$\begin{array}{r} 15x+12y=60 \\ 32x-12y=316 \\ \hline 47x \qquad =376 \\ x=8 \end{array}$$

Substitute 8 for x in (1).
$5(8)+4y=20$
$4y=-20$
$y=-5$
$(8, -5)$

30. $4x-3y+1=6$ (1)
$5x+8y+2=-74$ (2)
Rearrange and simplify.
$4x-3y=5$ (3)
$5x+8y=-76$ (4)
Multiply (3) by 8 and (4) by 3.

$$\begin{array}{r} 32x-24y=40 \\ 15x+24y=-228 \\ \hline 47x=-188 \\ x=-4 \end{array}$$

Substitute -4 for x in (2).
$5(-4)+8y+2=-74$
$8y=-56$
$y=-7$
$(-4, -7)$

31. $\frac{2x}{3}-\frac{3y}{4}=\frac{7}{12}$ (1)
$8x+5y=9$ (2)
Multiply (1) by -12.

$$\begin{array}{r} -8x+9y=-7 \\ 8x+5y=9 \\ \hline 14y=2 \\ y=\dfrac{1}{7} \end{array}$$

Substitute $\frac{1}{7}$ for y in (2).

$$8x+5\left(\frac{1}{7}\right)=9$$
$$8x=\frac{58}{7}$$
$$x=\frac{29}{28}$$

$\left(\frac{29}{28}, \frac{1}{7}\right)$

32. $\frac{x}{2}+\frac{y}{5}=4$ (1)
$\frac{x}{3}+\frac{y}{5}=\frac{10}{3}$ (2)
Multiply (1) by 10 and (2) by -15.

$$\begin{aligned} 5x+2y&=40\\ -5x-3y&=-50\\ \hline -y&=-10\\ y&=10 \end{aligned}$$

Substitute 10 for y in (1).

$$\frac{x}{2}+\frac{10}{5}=4$$
$$\frac{x}{2}=2$$
$$x=4$$

(4, 10)

33. $\frac{1}{5}a+\frac{1}{2}b=6 \quad (1)$

$\frac{3}{5}a-\frac{1}{2}b=2 \quad (2)$

$$\begin{aligned} \frac{4}{5}a &= 8\\ a&=10 \end{aligned}$$

Substitute 10 for a in (1).

$$\frac{1}{5}(10)+\frac{1}{2}b=6$$
$$\frac{1}{2}b=4$$
$$b=8$$

(10, 8)

34. $\frac{2}{3}a+\frac{3}{5}b=-17 \quad (1)$

$\frac{1}{2}a-\frac{1}{3}b=-1 \quad (2)$

Multiply (1) by 30 and (2) by 54.

$$\begin{aligned} 20a+18b&=-510\\ 27a-18b&=-54\\ \hline 47a&=-564\\ a&=-12 \end{aligned}$$

Substitute −12 for a in (2).

$$\frac{1}{2}(-12)-\frac{1}{3}b=-1$$
$$-\frac{1}{3}b=5$$
$$b=-15$$

(−12, −15)

35. $4.8+0.6m=0.9n$

$0.6m-0.9n=-4.8$

$6m-9n=-48 \quad (1)$

$0.2m-0.3n=1.6$

$2m-3n=16 \quad (2)$

Multiply (2) by −3.

$$\begin{aligned} 6m-9n&=-48\\ -6m+9n&=-48\\ \hline 0+0&=-96 \end{aligned}$$

No solution; inconsistent system

36. $8.4-0.8m=0.4n$

$-0.8m-0.4n=-8.4$

$-8m-4n=-84 \quad (1)$

$0.2m+0.1n=2.1$

$2m+n=21 \quad (2)$

Multiply (2) by 4.

$$\begin{aligned} -8m-4n&=-84\\ 8m+4n&=84\\ \hline 0&=0 \end{aligned}$$

Infinite number of solutions; dependent equations

37. $6s-4t=5 \quad (1)$

$4s-5t=-2 \quad (2)$

Multiply (1) by −5, (2) by 4 to eliminate y.

$$\begin{aligned} -30s+20t&=-25\\ 16s-20t&=-8\\ \hline -14s&=-33\\ s&=\frac{33}{14} \end{aligned}$$

Substitute $\left(\frac{33}{14}\right)$ for s in (1).

$$6\left(\frac{33}{14}\right)-4t=5$$
$$-4t=\frac{-128}{14}$$
$$t=\frac{32}{14}$$
$$t=\frac{16}{7}$$

$\left(\frac{33}{14}, \frac{16}{7}\right)$

38. $10s+3t=4 \quad (1)$

$4s-2t=-5 \quad (2)$

Multiply (1) by 2, (2) by 3 to eliminate t.

$$\begin{aligned} 20s+6t&=8\\ 12s-6t&=-15\\ \hline 32s&=-7\\ s&=-\frac{7}{32} \end{aligned}$$

Substitute $-\frac{7}{32}$ for s in (1).

$$10\left(-\frac{7}{32}\right)+3t=4$$
$$-\frac{70}{32}+3t=4$$
$$3t=\frac{198}{32}$$
$$t=\frac{66}{32}$$
$$t=\frac{33}{16}$$

$\left(-\frac{7}{32}, \frac{33}{16}\right)$

39. $3(x+2)=-2-(x+3y)$
$3(x+y)=3-2(y-1)$
Expand, rearrange and simplify.
$4x+3y=-8$ (1)
$3x+5y=5$ (2)
Multiply (1) by 3 and (2) by –4.

$$\begin{array}{r} 12x+9y=-24 \\ \underline{-12x-20y=-20} \\ -11y=-44 \\ y=4 \end{array}$$

Substitute 4 for y in (1).

$$4x+3(4)=-8$$
$$4x=-20$$
$$x=-5$$

(–5, 4)

40. $13-x=3(x+y)+1$
$14+2x=5(x+y)+3x$
Expand, rearrange and simplify.
$4x+3y=12$ (1)
$6x+5y=14$ (2)
Multiply (1) by –3 and (2) by 2.

$$\begin{array}{r} -12x-9y=-36 \\ \underline{12x+10y=28} \\ y=-8 \end{array}$$

Substitute –8 for y in (1).

$$4x+3(-8)=12$$
$$4x=36$$
$$x=9$$

(9, –8)

41. $0.2b=1.4-0.3a$
$0.1b+0.6=0.5a$
Rearrange.
$0.3a+0.2b=1.4$ (1)
$0.5a-0.1b=0.6$ (2)
Multiply (1) by 10 and (2) by 20.

$$\begin{array}{r} 3a+2b=14 \\ \underline{10a-2b=12} \\ 13a \quad\quad =26 \\ a=2 \end{array}$$

Substitute 2 for a in (1).

$$0.3(2)+0.2b=1.4$$
$$0.2b=0.8$$
$$b=4$$

(2, 4)

42. $0.3a=1.1-0.2b$
$0.3b=0.4a-0.9$
Rearrange and simplify.
$0.3a+0.2b=1.1$ (1)
$0.4a-0.3b=0.9$ (2)
Multiply (1) by 30 and (2) by 20.

$$\begin{array}{r} 9a+6b=33 \\ \underline{8a-6b=18} \\ 17a \quad\quad =51 \\ a=3 \end{array}$$

Substitute 3 for a in (1).

$$0.3(3)+0.2b=1.1$$
$$0.2b=0.2$$
$$b=1$$

(3, 1)

43. $\frac{b}{5}=\frac{2}{5}-\frac{a-3}{2}$
$4(a-b)=3b-2(a-2)$
Multiply the first equation by 10, then expand, rearrange, and simplify both equations.
$5a+2b=19$ (1)
$6a-7b=4$ (2)
Multiply (1) by 7 and (2) by 2.

$$\begin{array}{r} 35a+14b=133 \\ \underline{12a-14b=8} \\ 47a \quad\quad =141 \\ a=3 \end{array}$$

Substitute 3 for a in (1).

$$5(3)+2b=19$$
$$2b=4$$
$$b=2$$

(3, 2)

44. $9(b + 4) = 2(2a + 5b)$
Expand and simplify.
$4a + b = 36$ (1)
$\frac{b}{5} + \frac{a}{2} = \frac{18}{5}$
Rearrange and multiply by 10.
$5a + 2b = 36$ (2)
Multiply (1) by −2.
$-8a - 2b = -72$
$5a + 2b = 36$
$-3a = -36$
$a = 12$
Substitute 12 for a in (1).
$4(12) + b = 36$
$b = -12$
$(12, -12)$

45. x = number of adults
y = number of students
$x + y = 186$ (1)
$20x + 12y = 3240$ (2)
Solve (1) for x.
$x = 186 - y$ (3)
Substitute $186 - y$ for x in (2).
$20(186 - y) + 12y = 3240$
$3720 - 20y + 12y = 3240$
$y = 60$
Substitute 60 for y in (3).
$x = 186 - 60$
$x = 126$
126 adults and 60 students visited the museum.

46. x = number of 2-pound bags
y = number of 5-pound bags
$2x + 5y = 252$ (1)
$y = 2x$ (2)
Substitute $2x$ for y in (1).
$2x + 5(2x) = 252$
$2x + 10x = 252$
$12x = 252$
$x = 21$
Substitute 21 for x in (2).
$y = 2(21) = 42$
He prepared 21 2-lb bags and 42 5-lb bags.

47. r = airspeed
w = wind speed
$\text{speed} = \frac{\text{distance}}{\text{time}}$

$r + w = \frac{1500}{5}$ (1)
$r - w = \frac{1500}{6}$ (2)
$2r = 550$
$r = 275$
Substitute 275 for r in (1).
$275 + w = 300$
$w = 25$
Speed in still air = 275 mph
Wind speed = 25 mph

48. x = liters of 20%
y = liters of 30%
$x + y = 40$ (1)
$0.2x + 0.3y = 0.25(40)$ (2)
Solve (1) for x.
$x = 40 - y$
Substitute $40 - y$ for x in (2).
$0.2(40 - y) + 0.3y = 10$
$8 - 0.2y + 0.3y = 10$
$0.1y = 2$
$y = 20$
Substitute 20 for y in (1).
$x + 20 = 40$
$x = 20$
He should mix 20 liters of 20% and 20 liters of 30%.

49. x = Balsam firs
y = Norwegian pines
$x + y = 79$ (1)
$23x + 28y = 1942$ (2)
Solve (1) for x.
$x = 79 - y$ (3)
Substitute $(79 - y)$ for x in (2).
$23(79 - y) + 28y = 1942$
$1817 - 23 + 28y = 1942$
$y = 25$
Substitute 25 for y in (3).
$x = 79 - 25$
$x = 54$
The troop sold 54 Balsam firs and 25 Norwegian pines.

50. x = tons of 15% salt mixture
y = tons of 30% salt mixture
$x + y = 24$ (1)
$0.15x + 0.30y = 0.25(24)$
Multiply by 100 and simplify.
$15x + 30y = 600$ (2)
Solve (1) for x.

$x = 24 - y \quad (3)$
Substitute $24 - y$ for x in (2).
$15(24 - y) + 30y = 600$
$360 - 15y + 30y = 600$
$15y = 240$
$y = 16$
Substitute 16 for y in (3).
$x = 24 - 16 = 8$
He should use 8 tons of 15% salt mixture and 16 tons of 30% salt mixture.

51. r = speed of boat in still water
c = speed of current
$r + c = 23 \quad (1)$
$r - c = 15 \quad (2)$
$2r = 38$
$r = 19$
Substitute 19 for r in (1).
$19 + c = 23$
$c = 4$
Speed of boat = 19 km/hr
Speed of current = 4 km/hr

52. x = cost of copper per gram
y = hourly labor rate
$5x + 3y = 27 \quad (1)$
$4x + 5y = 32 \quad (2)$
Multiply (1) by 5 and (2) by −3.
$25x + 15y = 135$
$-12x - 15y = -96$
$13x = 39$
$x = 3$
Substitute 3 for x in (1).
$5(3) + 3y = 27$
$3y = 12$
$y = 4$
It costs \$3 per gram of copper. The hourly labor rate is \$4.

53. x = Level Outfield seats
y = Upper Skydeck seats
$x + y = 25 \quad (1)$
$23x + 9y = 379 \quad (2)$
Solve (1) for x.
$x = 25 - y \quad (3)$
Substitute $(25 - y)$ for x in (2).
$23(25 - y) + 9y = 379$
$575 - 23y + 9y = 379$
$y = 14$
Substitute 14 for y in (3).
$x = 25 - 14$
$x = 11$
They bought 11 Level Outfield seats and 14 Upper Skydeck seats.

54. x = Field Level Bases seats
y = Field Level Infield seats
$x + y = 22 \quad (1)$
$41x + 57y = 1062 \quad (2)$
Solve (1) for x.
$x = 22 - y \quad (3)$
Substitute $(22 - y)$ for x in (2).
$41(22 - y) + 57y = 1062$
$902 - 41y + 57y = 1062$
$y = 10$
Substitute 10 for y in (3).
$x = 22 - 10$
$x = 12$
They bought 12 Field Level Bases seats and 10 Field Level Infield seats.

How Am I Doing? Chapter 8 Test

1. $3x - y = -5 \quad (1)$
$-2x + 5y = -14 \quad (2)$
Solve (1) for y.
$y = 3x + 5 \quad (3)$
Substitute $3x + 5$ for y in (2).
$-2x + 5(3x + 5) = -14$
$-2x + 15x + 25 = -14$
$13x = -39$
$x = -3$
Substitute −3 for x in (1).
$3(-3) - y = -5$
$-y = 4$
$y = -4$
$(-3, -4)$

2. $3x + 4y = 7 \quad (1)$
$2x + 3y = 6 \quad (2)$
Multiply (1) by 2 and (2) by −3.
$6x + 8y = 14$
$-6x - 9y = -18$
$-y = -4$
$y = 4$
Substitute 4 for y in (2).
$2x + 3(4) = 6$
$2x = -6$
$x = -3$
$(-3, 4)$

3. $2x - y = 4$
$4x + y = 2$
Graph both equations on the same coordinate system. The lines intersect at (1, −2).

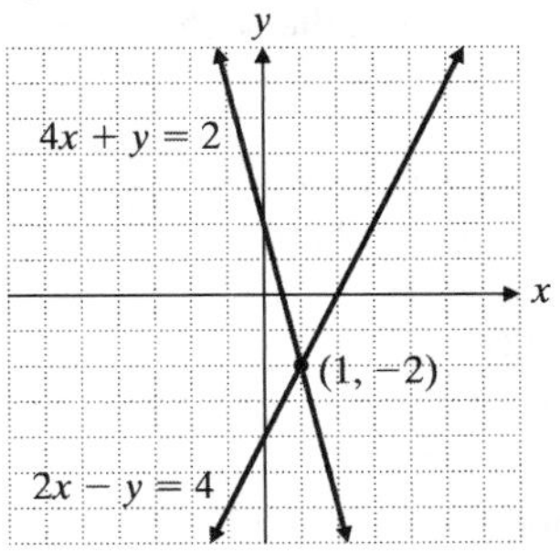

4. $x + 3y = 12 \quad (1)$
$2x - 4y = 4 \quad (2)$
Multiply (1) by −2.
$$\begin{array}{r} -2x - 6y = -24 \\ 2x - 4y = 4 \\ \hline -10y = -20 \\ y = 2 \end{array}$$
Substitute 2 for y in (1).
$x + 3(2) = 12$
$x = 6$
(6, 2)

5. $2x - y = 5 \quad (1)$
$-x + 3y = 5 \quad (2)$
Multiply (1) by 3.
$$\begin{array}{r} 6x - 3y = 15 \\ -x + 3y = 5 \\ \hline 5x \qquad = 20 \\ x = 4 \end{array}$$
Substitute 4 for x in (1).
$2(4) - y = 5$
$-y = -3$
$y = 3$
(4, 3)

6. $2x + 3y = 13 \quad (1)$
$3x - 5y = 10 \quad (2)$
Multiply (1) by 5 and (2) by 3.
$$\begin{array}{r} 10x + 15y = 65 \\ 9x - 15y = 30 \\ \hline 19x \qquad = 95 \\ x = 5 \end{array}$$
Substitute 5 for x in (1).
$2(5) + 3y = 13$
$3y = 3$
$y = 1$
(5, 1)

7. $\frac{2}{3}x - \frac{1}{5}y = 2 \quad (1)$
$\frac{4}{3}x + 4y = 4 \quad (2)$
Multiply (1) by 15 and (2) by 3.
$10x - 3y = 30 \quad (3)$
$4x + 12y = 12 \quad (4)$
Multiply (3) by 4.
$$\begin{array}{r} 40x - 12y = 120 \\ 4x + 12y = 12 \\ \hline 44x \qquad = 132 \\ x = 3 \end{array}$$
Substitute 3 for x in (2).
$\frac{4}{3}(3) + 4y = 4$
$4y = 0$
$y = 0$
(3, 0)

8. $3x - 6y = 5 \quad (1)$
$-\frac{1}{2}x + y = \frac{7}{2} \quad (2)$
Multiply (2) by 6.
$$\begin{array}{r} 3x - 6y = 5 \quad (1) \\ -3x + 6y = 21 \quad (3) \\ \hline 0 = 26 \end{array}$$
No solution, inconsistent system

9. $5x - 2 = y \quad (1)$
$10x = 4 + 2y \quad (2)$
Substitute $5x - 2$ for y in (2).
$10x = 4 + 2(5x - 2)$
$10x = 4 + 10x - 4$
$0 = 0$
Infinite number of solutions, dependent equations

10. $0.3x + 0.2y = 0 \quad (1)$
$1.0x + 0.5y = -0.5 \quad (2)$
Multiply (1) by −50 and (2) by 20.
$$\begin{array}{r} -15x - 10y = 0 \quad (3) \\ 20x + 10y = -10 \quad (4) \\ \hline 5x \qquad = -10 \\ x = -2 \end{array}$$
Substitute −2 for x in (4).

$20(-2)+10y=-10$
$10y=30$
$y=3$
$(-2, 3)$

11. $2(x+y)=2(1-y)$ (1)
$5(-x+y)=2(23-y)$ (2)
Expand, rearrange, and simplify.
$2x+4y=2$ (3)
$-5x+7y=46$ (4)
Multiply (3) by 5 and (4) by 2.
$10x+20y=10$
$\underline{-10x+14y=92}$
$34y=102$
$y=3$
Substitute 3 for y in (3).
$2x+4(3)=2$
$2x=-10$
$x=-5$
$(-5, 3)$

12. $3(x-y)=12+2y$ (1)
$8(y+1)=6x-7$ (2)
Expand, rearrange, and simplify.
$3x-5y=12$ (3)
$-6x+8y=-15$ (4)
Multiply (3) by 2.
$6x-10y=24$
$\underline{-6x+8y=-15}$
$-2y=9$
$y=-4.5$
Substitute -4.5 for y in (3).
$3x-5(-4.5)=12$
$3x+22.5=12$
$3x=-10.5$
$x=-3.5$
$(-3.3, -4.5)$ or $\left(-\frac{7}{2}, -\frac{9}{2}\right)$

13. x = first number
y = second number
$2x+3y=1$ (1)
$3x+2y=9$ (2)
Multiply (1) by 3 and (2) by -2.
$6x+9y=3$
$\underline{-6x-4y=-19}$
$5y=-15$
$y=-3$
Substitute -3 for y in (1).
$2x+3(-3)=1$
$2x-9=1$
$2x=10$
$x=5$
The first number is 5 and the second number is -3.

14. x = the number of \$8 tickets
y = the number of \$12 tickets
$x+y=30{,}500$ (1)
$8x+12y=308{,}000$ (2)
Solve (1) for y.
$y=30{,}500-x$ (3)
Substitute $30{,}500-x$ for y in (2).
$8x+12(30{,}500-x)=308{,}000$
$8x+366{,}000-12x=308{,}000$
$-4x=-58{,}000$
$x=14{,}500$
Substitute 14,500 for x in (3).
$y=30{,}500-14{,}500=16{,}000$
14,500 of the \$8 tickets and 16,000 of the \$12 tickets were sold.

15. x = cost of a shirt
y = cost of a pair of slacks
$5x+3y=172$ (1)
$3x+4y=156$ (2)
Multiply (1) by 4 and (2) by -3.
$20x+12y=688$
$\underline{-9x-12y=-468}$
$11x=220$
$x=20$
Substitute 20 for x in (1).
$5(20)+3y=172$
$3y=72$
$y=24$
A shirt costs \$20. A pair of slacks costs \$24.

16. x = number of booklets
y = cost
$y=0.25x+200$ (1)
$y=0.2x+250$ (2)
Substitute $0.25x+200$ for y in (2).
$0.25x+200=0.2x+250$
$0.05x+200=250$
$0.05x=50$
$x=1000$
Substitute 1000 for x in (1).
$y=0.25(1000)+200$
$y=450$
Order 1000 booklets at a cost of \$450.

17. r = airspeed
w = wind speed

$$r + w = \frac{2000}{4} \quad (1)$$
$$r - w = \frac{2000}{5} \quad (2)$$
$$\overline{2r \quad = 900}$$
$$r = 450$$

Substitute 450 for r in (1).

$$450 + w = 500$$
$$w = 50$$

Speed of plane in still air = 450 km/h
Wind speed = 50 km/h

Cumulative Test for Chapters 0–8

1. 15% of 42.75 = 0.15(42.75) = \$6.41

2. $\dfrac{-2^4(2^2)}{2^3} = \dfrac{-16(4)}{8} = -8$

3. $(3x^2 - 2x + 1)(5x - 3)$
$= (3x^2 - 2x + 1)(5x) - (3x^2 - 2x + 1)(3)$
$= 15x^3 - 10x^2 + 5x - 9x^2 + 6x - 3$
$= 15x^3 - 19x^2 + 11x - 3$

4. $\dfrac{x^2 - 7x + 12}{x^2 - 9} \div \dfrac{x^2 - 8x + 16}{1}$
$= \dfrac{x^2 - 7x + 12}{x^2 - 9} \cdot \dfrac{1}{x^2 - 8x + 16}$
$= \dfrac{(x-4)(x-3)}{(x+3)(x-3)} \cdot \dfrac{1}{(x-4)(x-4)}$
$= \dfrac{1}{(x+3)(x-4)}$

5. $-3(x+2) < 4(x-2)$
$-3x - 6 < 4x - 8$
$-7x < -2$
$\dfrac{-7x}{-7} > \dfrac{-2}{-7}$
$x > \dfrac{2}{7}$

6. $6x^2 + x - 2 = 0$
$(2x-1)(3x+2) = 0$

$2x - 1 = 0$	$3x + 2 = 0$
$2x = 1$	$3x = -2$
$x = \dfrac{1}{2}$	$x = -\dfrac{2}{3}$

7. 75% of x = 30
$0.75x = 30$
$x = 40$
75% of 40 is 30.

8. $\dfrac{1}{10}x + \dfrac{1}{3} = \dfrac{2}{5}x - \dfrac{1}{6}$
$30\left(\dfrac{1}{10}x\right) + 30\left(\dfrac{1}{3}\right) = 30\left(\dfrac{2}{5}\right)x - 30\left(\dfrac{1}{6}\right)$
$3x + 10 = 12x - 5$
$10 = 9x - 5$
$15 = 9x$
$\dfrac{5}{3} = x$

9. $10t^2 - 25t^3 + 15t^4 = 15t^4 - 25t^3 + 10t^2$
$= 5t^2(3t^2 - 5t + 2)$
$= 5t^2(3t - 2)(t - 1)$

10. $4x = 2(12 - 2x)$
$4x = 24 - 4x$
$8x = 24$
$x = 3$

11. $2x + y = 8 \quad (1)$
$3x + 4y = -8 \quad (2)$
Multiply (1) by −4.

$$-8x - 4y = -32$$
$$3x + 4y = -8$$
$$\overline{-5x \quad = -40}$$
$$x = 8$$

Substitute 8 for x in (1).
$2(8) + y = 8$
$y = -8$
(8, −8)

12. $\dfrac{1}{3}x - \dfrac{1}{2}y = 4$
$\dfrac{3}{8}x - \dfrac{1}{4}y = 7$
Clear the fractions.
$2x - 3y = 24 \quad (1)$
$3x - 2y = 56 \quad (2)$
Multiply (1) by −3 and (2) by 2.

$$-6x + 9y = -72$$
$$6x - 4y = 112$$
$$\overline{5y = 40}$$
$$y = 8$$

Substitute 8 for y in (1).

$$2x - 3(8) = 24$$
$$2x = 48$$
$$x = 24$$
(24, 8)

13. $1.3x - 0.7y = 0.4$ (1)
$-3.9x + 2.1y = -1.2$ (2)
Multiply (1) by 30 and (2) by 10.
$$39x - 21y = 12$$
$$-39x + 21y = -12$$
$$0 = 0$$
Infinite number of solutions, dependent system

14. $10x - 5y = 45$ (1)
$3x - 8y = 7$ (2)
Multiply (1) by −8 and (2) by 5.
$$-80x + 40y = -360$$
$$15x - 40y = 35$$
$$-65x = -325$$
$$x = 5$$
Substitute 5 for x in (2).
$$3(5) - 8y = 7$$
$$-8y = -8$$
$$y = 1$$
(5, 1)

15. Parallel lines represent an inconsistent system for which there is no solution.

16. Let x = amount of water, which is 0% acid solution.
$$0x + 0.80(60) = 0.5(x + 60)$$
$$48 = 0.5x + 30$$
$$18 = 0.5x$$
$$36 = x$$
Add 36 kilograms of water to an 80% acid solution to produce a 50% solution.

17. r = speed of boat in still water
s = speed of the stream
$$r - s = \frac{6}{3} \quad (1)$$
$$r + s = \frac{6}{1.5} \quad (2)$$
Multiply (1) by −1.
$$-r + s = -2$$
$$r + s = 4$$
$$2s = 2$$
$$s = 1$$
Speed of the stream is 1 mph.

18. x = first printer's rate
y = second printer's rate
$5x + 5y = 15{,}000$ (1)
$7x + 2y = 15{,}000$ (2)
Multiply (1) by 2 and (2) by −5.
$$10x + 10y = 30{,}000$$
$$-35x - 10y = -75{,}000$$
$$-25x = -45{,}000$$
$$x = 1800$$
Substitute 1800 for x in (1).
$$5(1800) + 5y = 15{,}000$$
$$9000 + 5y = 15{,}000$$
$$5y = 6000$$
$$y = 1200$$
The printer that broke prints 1200 labels/hr. The other printer prints 1800 labels/hr.

Chapter 9

9.1 Exercises

1. The principal square root of N, where $N \ge 0$, is a nonnegative number a that has the property $a^2 = N$.

3. No, since $(0.3)(0.3) = 0.09$.

5. $\pm\sqrt{9} = \pm 3$ since $(3)^2 = 9$ and $(-3)^2 = 9$.

7. $\pm\sqrt{49} = \pm 7$ since $(7)^2 = 49$ and $(-7)^2 = 49$.

9. $\sqrt{16} = 4$ since $4^2 = 16$.

11. $\sqrt{81} = 9$ since $9^2 = 81$.

13. $-\sqrt{36} = -6$ since $6^2 = 36$.

15. $\sqrt{0.81} = 0.9$ since $0.9^2 = 0.81$.

17. $\sqrt{\frac{36}{121}} = \frac{6}{11}$ since $\left(\frac{6}{11}\right)^2 = \frac{36}{121}$.

19. $\sqrt{\frac{49}{64}} = \frac{7}{8}$ since $\left(\frac{7}{8}\right)^2 = \frac{49}{64}$.

21. $\sqrt{625} = \sqrt{25^2} = 25$

23. $-\sqrt{10,000} = -100$ since $100^2 = 10,000$.

25. $\sqrt{169} = 13$ since $13^2 = 169$.

27. $-\sqrt{\frac{1}{64}} = -\frac{1}{8}$ since $\left(\frac{1}{8}\right)^2 = \frac{1}{64}$.

29. $\sqrt{\frac{9}{16}} = \frac{3}{4}$ since $\left(\frac{3}{4}\right)^2 = \frac{9}{16}$.

31. $\sqrt{0.0036} = 0.06$ since $(0.06)^2 = 0.0036$.

33. $\sqrt{19,600} = \sqrt{140^2} = 140$

35. $-\sqrt{289} = -17$ since $(17)^2 = 289$.

37. $\sqrt{27} \approx 5.196$

39. $\sqrt{74} \approx 8.602$

41. $-\sqrt{183} \approx -13.528$

43. $-\sqrt{195} \approx -13.964$

45. $\sqrt{169} = 13$ feet
The wire is 13 feet long.

47. If $a = 900$, $t = \frac{1}{4}\sqrt{a} = \frac{1}{4}\sqrt{900} = \frac{1}{4}(30) = 7.5$.
It will take the object 7.5 seconds to fall 900 feet.

49. $\sqrt[3]{27} = 3$ since $3^3 = 27$.

51. $\sqrt[3]{-64} = \sqrt[3]{(-4)^3} = -4$

53. $\sqrt[4]{81} = 3$ since $3^4 = 81$.

55. $\sqrt[5]{243} = 3$ since $3^5 = 243$.

57. No; a fourth power of a real number must be nonnegative.

Cumulative Review

59. $3x + 2y = 8$ (1)
$7x - 3y = 11$ (2)
Multiply (1) by 3 and (2) by 2.

$$\begin{aligned} 9x + 6y &= 24 \\ 14x - 6y &= 22 \\ \hline 23x \quad &= 46 \\ x &= 2 \end{aligned}$$

Substitute 2 for x in (1).

$$\begin{aligned} 3(2) + 2y &= 8 \\ 2y &= 2 \\ y &= 1 \end{aligned}$$

(2, 1)

60. $2x - 3y = 1$ (1)
$-8x + 12y = 4$ (2)
Multiply (1) by 4.

$$\begin{aligned} 8x - 12y &= 4 \\ -8x + 12y &= 4 \\ \hline 0 &= 8 \end{aligned}$$

No solution, inconsistent system of equations

61. $\frac{900-813}{813} \cdot 100\% = 10.7\%$

The percent increase from 2006–2007 to 2007–2008 was 10.7%.

$\frac{x-900}{900} \cdot 100\% = 10.7\%$

$x = 996.30$

The average household expenditure for 2008–2009 would be \$996.30.

62. $\frac{1841-1466}{1466} \cdot 100\% = 25.6\%$

The percent of increase from 2006–2007 to 2007–2008 was 25.6%.

$\frac{x-1841}{1841} \cdot 100\% = 25.6\%$

$x = 2312.30$

The average household expenditure for 2008–2009 would be \$2312.30.

Quick Quiz 9.1

1. $\sqrt{169} = 13$ since $13^2 = 169$.

2. $\sqrt{\frac{25}{64}} = \frac{5}{8}$ since $\left(\frac{5}{8}\right)^2 = \frac{25}{64}$.

3. $\sqrt{0.49} = 0.7$ since $(0.7)^2 = 0.49$.

4. Answers may vary. Possible solution: Start by factoring the radicand by 10^2, or 100 leaving a more manageable $10\sqrt{324}$. For the remaining radicand guess logically what its square root might be. For example, $15^2 = 225$ is too small, but $20^2 = 400$ is too big. But $18^2 = 324$.

Therefore $10\sqrt{324} = 10\sqrt{18^2} = 180$.

9.2 Exercises

1. Yes; $\sqrt{6^2} = \sqrt{6 \cdot 6} = \sqrt{36} = 6$

$\left(\sqrt{6}\right)^2 = \left(\sqrt{6}\right)\left(\sqrt{6}\right) = \sqrt{6 \cdot 6} = \sqrt{36} = 6$

3. No; $\sqrt{(-9)^2} = \sqrt{(-9)(-9)} = \sqrt{81} = 9$ but $-\sqrt{9^2} = -\sqrt{81} = -9$. The two expressions are not the same.

5. $\sqrt{8^2} = 8$

7. $\sqrt{18^4} = \sqrt{(18^2)^2} = 18^2$

9. $\sqrt{9^6} = \sqrt{(9^3)^2} = 9^3$

11. $\sqrt{33^8} = \sqrt{(33^4)^2} = 33^4$

13. $\sqrt{5^{140}} = \sqrt{(5^{70})^2} = 5^{70}$

15. $\sqrt{x^{12}} = \sqrt{(x^6)^2} = x^6$

17. $\sqrt{t^{18}} = \sqrt{(t^9)^2} = t^9$

19. $\sqrt{y^{26}} = \sqrt{(y^{13})^2} = y^{13}$

21. $\sqrt{36x^8} = \sqrt{6^2(x^4)^2} = 6x^4$

23. $\sqrt{144x^2} = \sqrt{144}\sqrt{x^2} = 12x$

25. $\sqrt{x^6y^4} = \sqrt{(x^3)^2(y^2)^2} = x^3y^2$

27. $\sqrt{16x^2y^{20}} = \sqrt{4^2 \cdot x^2 \cdot (y^{10})^2} = 4xy^{10}$

29. $\sqrt{100a^{10}b^6} = \sqrt{(10a^5b^3)^2} = 10a^5b^3$

31. $\sqrt{24} = \sqrt{4 \cdot 6} = 2\sqrt{6}$

33. $\sqrt{45} = \sqrt{9 \cdot 5} = 3\sqrt{5}$

35. $\sqrt{18} = \sqrt{9 \cdot 2} = 3\sqrt{2}$

37. $\sqrt{72} = \sqrt{36 \cdot 2} = 6\sqrt{2}$

39. $\sqrt{90} = \sqrt{9 \cdot 10} = 3\sqrt{10}$

41. $\sqrt{128} = \sqrt{64 \cdot 2} = 8\sqrt{2}$

43. $\sqrt{8x^3} = \sqrt{4 \cdot 2x^2 \cdot x} = 2x\sqrt{2x}$

45.
$$\begin{aligned}\sqrt{27w^5} &= \sqrt{9 \cdot 3 \cdot w^4 \cdot w} \\ &= \sqrt{3^2 \cdot 3(w^2)^2 \cdot w} \\ &= 3w^2\sqrt{3w}\end{aligned}$$

47. $\sqrt{32z^9} = \sqrt{16 \cdot 2 \cdot z^8 \cdot z} = 4z^4\sqrt{2z}$

49. $\sqrt{28x^5y^7} = \sqrt{4 \cdot 7 \cdot x^4 \cdot x \cdot y^6 \cdot y} = 2x^2y^3\sqrt{7xy}$

51. $\sqrt{48y^3w} = \sqrt{16 \cdot 3 \cdot y^2 \cdot y \cdot w} = 4y\sqrt{3yw}$

53. $\sqrt{75x^2y^3} = \sqrt{25 \cdot 3 \cdot x^2 \cdot y^2 \cdot y} = 5xy\sqrt{3y}$

55. $\sqrt{64x^{10}} = \sqrt{(8x^5)^2} = 8x^5$

57. $\sqrt{y^7} = \sqrt{y^6 \cdot y} = y^3\sqrt{y}$

59. $\sqrt{x^5y^{11}} = \sqrt{x^4 \cdot x \cdot y^{10} \cdot y} = x^2y^5\sqrt{xy}$

61. $\sqrt{135x^5y^7} = \sqrt{9 \cdot 15 \cdot x^4 \cdot x \cdot y^6 \cdot y} = 3x^2y^3\sqrt{15xy}$

63.
$$\begin{aligned}\sqrt{72a^2b^4c^5} &= \sqrt{36 \cdot 2 \cdot a^2 \cdot b^4 \cdot c \cdot c^4}\\ &= \sqrt{2c \cdot 36 \cdot a^2 \cdot b^4 \cdot c^4}\\ &= \sqrt{2c \cdot (6ab^2c^2)^2}\\ &= 6ab^2c^2\sqrt{2c}\end{aligned}$$

65.
$$\begin{aligned}\sqrt{169a^3b^8c^5} &= \sqrt{169 \cdot a^2 \cdot a \cdot b^8 \cdot c^4 \cdot c}\\ &= 13ab^4c^2\sqrt{ac}\end{aligned}$$

67. $\sqrt{(x+5)^2} = x+5$

69. $\sqrt{16x^2+8x+1} = \sqrt{(4x+1)^2} = 4x+1$

71.
$$\begin{aligned}\sqrt{x^2y^2+14xy^2+49y^2} &= \sqrt{y^2(x^2+14x+49)}\\ &= \sqrt{y^2(x+7)^2}\\ &= y(x+7)\end{aligned}$$

Cumulative Review

73. $3x - 2y = 14$ (1)
$y = x - 5$ (2)
$3x - 2(x-5) = 14$ Substitute $x-5$ for y into (1).
$3x - 2x + 10 = 14$
$x = 4$
$y = 4 - 5 = -1$ Substitute 4 for x into (2).
$(4, -1)$

74.
$$\frac{1}{2}(x-4) = \frac{1}{3}x - 5$$
$$\frac{x}{2} - \frac{4}{2} = \frac{1}{3}x - 5$$
Multiply both sides of the equation by 6 to eliminate fractions.
$$\begin{aligned}3x - 12 &= 2x - 30\\ 3x - 2x &= -30 + 12\\ x &= -18\end{aligned}$$

75. 2010–2020

Rate of increase $= \dfrac{142-115}{10} = 2.7$

Use in 2033 = 142 + 13(2.7) = 177.1

An estimated 177.1 trillion ft^3 of energy will be consumed in 2033.

76. Russian natural gas production:

$2010 = 0.26(115) = 29.9$ trillion ft^3

$2015 = 0.29(128) = 37.12$ trillion ft^3

Difference $= 37.12 - 29.9$ trillion ft^3
$= 7.22$ trillion ft^3

Russia will produce 7.22 trillion ft^3 more natural gas in 2015 than in 2010.

77. $x = \dfrac{142-115}{115} \cdot 100\% = 23.5\%$

The expected percent increase is about 23.5%.

Quick Quiz 9.2

1. $\sqrt{120} = \sqrt{4 \cdot 30} = \sqrt{(2)^2 \cdot 30} = 2\sqrt{30}$

2.
$$\begin{aligned}\sqrt{81x^3y^4} &= \sqrt{81 \cdot x \cdot x^2 \cdot y^4}\\ &= \sqrt{x \cdot 81x^2y^4}\\ &= \sqrt{x(9xy^2)^2}\\ &= 9xy^2\sqrt{x}\end{aligned}$$

3.
$$\begin{aligned}\sqrt{135x^8y^5} &= \sqrt{9 \cdot 15 \cdot x^8 \cdot y \cdot y^4}\\ &= \sqrt{15y(3x^4y^2)^2}\\ &= 3x^4y^2\sqrt{15y}\end{aligned}$$

4. Answers may vary. Possible solution: Factor the radicand into variables with the largest possible even exponents. All variables with even exponents may then be removed from the radicand by halving the exponent, leaving only the variables with exponents of 1 in the radicand.

9.3 Exercises

1. (1) Simplify each radical term.
(2) Combine like radicals.

3. $3\sqrt{5}-\sqrt{5}+4\sqrt{5}=(3-1+4)\sqrt{5}=6\sqrt{5}$

5. $\sqrt{2}+8\sqrt{3}-5\sqrt{3}+4\sqrt{2}=(1+4)\sqrt{2}+(8-5)\sqrt{3}$
$=5\sqrt{2}+3\sqrt{3}$

7. $\sqrt{5}+\sqrt{45}=\sqrt{5}+\sqrt{9\cdot 5}=\sqrt{5}+3\sqrt{5}=4\sqrt{5}$

9. $\sqrt{80}+4\sqrt{20}=\sqrt{16\cdot 5}+4\sqrt{4\cdot 5}$
$=4\sqrt{5}+8\sqrt{5}$
$=12\sqrt{5}$

11. $3\sqrt{8}-5\sqrt{2}=3\sqrt{4\cdot 2}-5\sqrt{2}$
$=6\sqrt{2}-5\sqrt{2}$
$=\sqrt{2}$

13. $-\sqrt{2}+\sqrt{18}+\sqrt{98}=-\sqrt{2}+\sqrt{9\cdot 2}+\sqrt{49\cdot 2}$
$=-\sqrt{2}+3\sqrt{2}+7\sqrt{2}$
$=9\sqrt{2}$

15. $\sqrt{25}+\sqrt{72}+3\sqrt{12}=\sqrt{25}+\sqrt{36\cdot 2}+3\sqrt{4\cdot 3}$
$=5+6\sqrt{2}+6\sqrt{3}$

17. $3\sqrt{20}-\sqrt{50}+\sqrt{18}=3\sqrt{4\cdot 5}-\sqrt{25\cdot 2}+\sqrt{9\cdot 2}$
$=6\sqrt{5}-5\sqrt{2}+3\sqrt{2}$
$=6\sqrt{5}-2\sqrt{2}$

19. $8\sqrt{2x}+\sqrt{50x}=8\sqrt{2x}+\sqrt{25\cdot 2x}$
$=8\sqrt{2x}+5\sqrt{2x}$
$=13\sqrt{2x}$

21. $1.2\sqrt{3x}-0.5\sqrt{12x}=1.2\sqrt{3x}-0.5\sqrt{4\cdot 3x}$
$=1.2\sqrt{3x}-\sqrt{3x}$
$=0.2\sqrt{3x}$

23. $\sqrt{20y}+2\sqrt{45y}-\sqrt{5y}=\sqrt{4\cdot 5y}+2\sqrt{9\cdot 5y}-\sqrt{5y}$
$=2\sqrt{5y}+6\sqrt{5y}-\sqrt{5y}$
$=7\sqrt{5y}$

25. $\sqrt{50}-3\sqrt{8}=\sqrt{25\cdot 2}-3\sqrt{4\cdot 2}$
$=5\sqrt{2}-6\sqrt{2}$
$=-\sqrt{2}$

27. $3\sqrt{28x}-5x\sqrt{63x}=3\sqrt{4\cdot 7x}-5x\sqrt{9\cdot 7x}$
$=6\sqrt{7x}-15x\sqrt{7x}$

29. $5\sqrt{8x^3}-3x\sqrt{50x}=5\sqrt{4\cdot 2xx^2}-3x\sqrt{25\cdot 2x}$
$=10x\sqrt{2x}-15x\sqrt{2x}$
$=-5x\sqrt{2x}$

31. $5x\sqrt{48}-2x\sqrt{75}=5x\sqrt{16\cdot 3}-2x\sqrt{25\cdot 3}$
$=5x(4)\sqrt{3}-2x(5)\sqrt{3}$
$=20x\sqrt{3}-10x\sqrt{3}$
$=10x\sqrt{3}$

33. $2\sqrt{6y^3}-2y\sqrt{54}=2\sqrt{6yy^2}-2y\sqrt{9\cdot 6}$
$=2y\sqrt{6y}-6y\sqrt{6}$

35. $5x\sqrt{8x}-24\sqrt{50x^3}=5x\sqrt{4\cdot 2x}-24\sqrt{25\cdot 2xx^2}$
$=10x\sqrt{2x}-120x\sqrt{2x}$
$=-110x\sqrt{2x}$

37. $P=2\left(\sqrt{5}+2\sqrt{3}\right)+2\left(3\sqrt{5}+\sqrt{3}\right)$
$=2\sqrt{5}+4\sqrt{3}+6\sqrt{5}+2\sqrt{3}$
$=8\sqrt{5}+6\sqrt{3}$
The perimeter is $\left(8\sqrt{5}+6\sqrt{3}\right)$ miles.

Cumulative Review

39. Let x = number of \$8.99 games and
y = number of \$12.49 games.
$x+y=12 \quad (1)$
$8.99x+12.49y=121.88 \quad (2)$
Solve (1) for y.
$y=12-x$
Substitute $12-x$ for y in (2).
$8.99x+12.49(12-x)=121.88$
$8.99x+149.88-12.49x=121.88$
$-3.5x=-28$
$x=8$
Substitute 8 for x in (1).
$8+y=12$
$y=4$
She bought 8 games at \$8.99 and 4 games at \$12.49.

40. Let x = amount invested at 8%. Then $20{,}000 - x$ = amount invested at 4%.

$$0.08x + 0.04(20{,}000 - x) = 1340$$
$$0.08x + 800 - 0.04x = 1340$$
$$0.04x = 540$$
$$x = 13{,}500$$

$20{,}000 - x = 6500$

He invested \$13,500 at 8% in the certificate of deposit and \$6500 at 4% in the money market fund.

Quick Quiz 9.3

1. $\sqrt{98} + 3\sqrt{8} = \sqrt{49 \cdot 2} + 3\sqrt{4 \cdot 2}$
$= 7\sqrt{2} + 6\sqrt{2}$
$= 13\sqrt{2}$

2. $5\sqrt{3} + 2\sqrt{75} - 3\sqrt{48} = 5\sqrt{3} + 2\sqrt{25 \cdot 3} - 3\sqrt{16 \cdot 3}$
$= 5\sqrt{3} + 10\sqrt{3} - 12\sqrt{3}$
$= 3\sqrt{3}$

3. $\sqrt{9a} + \sqrt{8a} + \sqrt{49a} + \sqrt{32a}$
$= \sqrt{9a} + \sqrt{2 \cdot 4a} + \sqrt{7 \cdot 7a} + \sqrt{16 \cdot 2a}$
$= 3\sqrt{a} + 2\sqrt{2a} + 7\sqrt{a} + 4\sqrt{2a}$
$= 10\sqrt{a} + 6\sqrt{2a}$

4. Answers may vary. Possible solution: Factor $4x^2$ out of the radicand of the first term and 49 out of the second. Combining the two terms results in $41x\sqrt{2x}$.

9.4 Exercises

1. $\sqrt{3}\sqrt{10} = \sqrt{3 \cdot 10} = \sqrt{30}$

3. $\sqrt{2}\sqrt{22} = \sqrt{44} = \sqrt{4 \cdot 11} = 2\sqrt{11}$

5. $\sqrt{18}\sqrt{3} = \sqrt{18 \cdot 3} = \sqrt{54} = \sqrt{9 \cdot 6} = 3\sqrt{6}$

7. $\left(4\sqrt{5}\right)\left(3\sqrt{2}\right) = 4 \cdot 3\sqrt{5 \cdot 2} = 12\sqrt{10}$

9. $\sqrt{5}\sqrt{10a} = \sqrt{50a} = \sqrt{25 \cdot 2a} = 5\sqrt{2a}$

11. $\left(3\sqrt{10x}\right)\left(2\sqrt{5x}\right) = 6\sqrt{50x^2}$
$= 6\sqrt{25 \cdot 2x^2}$
$= 30x\sqrt{2}$

13. $\left(-4\sqrt{8a}\right)\left(-2\sqrt{3a}\right) = 8\sqrt{24a^2}$
$= 8\sqrt{4 \cdot 6a^2}$
$= 16a\sqrt{6}$

15. $\left(-3\sqrt{ab}\right)\left(2\sqrt{b}\right) = -6\sqrt{ab^2} = -6b\sqrt{a}$

17. $\sqrt{5}\left(\sqrt{2} + \sqrt{3}\right) = \sqrt{5}\left(\sqrt{2}\right) + \sqrt{5}\left(\sqrt{13}\right)$
$= \sqrt{10} + \sqrt{15}$

19. $\sqrt{3}\left(2\sqrt{6} + 5\sqrt{15}\right) = 2\sqrt{18} + 5\sqrt{45}$
$= 2\sqrt{9 \cdot 2} + 5\sqrt{9 \cdot 5}$
$= 6\sqrt{2} + 15\sqrt{5}$

21. $2\sqrt{x}\left(\sqrt{x} - 8\sqrt{5}\right) = 2\sqrt{x^2} - 16\sqrt{5x} = 2x - 16\sqrt{5x}$

23. $\sqrt{6}\left(\sqrt{2} - 3\sqrt{6} + 2\sqrt{10}\right) = \sqrt{12} - 3\sqrt{36} + 2\sqrt{60}$
$= \sqrt{4 \cdot 3} - 3 \cdot 6 + 2\sqrt{4 \cdot 15}$
$= 2\sqrt{3} - 18 + 4\sqrt{15}$

25. $\left(2\sqrt{3} + \sqrt{6}\right)\left(\sqrt{3} - 2\sqrt{6}\right)$
$= 2\sqrt{9} - 4\sqrt{18} + \sqrt{18} - 2\sqrt{36}$
$= 6 - 3\sqrt{9 \cdot 2} - 12$
$= -6 - 9\sqrt{2}$

27. $\left(5 + 3\sqrt{2}\right)\left(3 + \sqrt{2}\right) = 15 + 5\sqrt{2} + 9\sqrt{2} + 3\sqrt{4}$
$= 15 + 14\sqrt{2} + 6$
$= 21 + 14\sqrt{2}$

29. $\left(2\sqrt{7} - 3\sqrt{3}\right)\left(\sqrt{7} + \sqrt{3}\right)$
$= 2\sqrt{49} + 2\sqrt{21} - 3\sqrt{21} - 3\sqrt{9}$
$= 14 - \sqrt{21} - 9$
$= 5 - \sqrt{21}$

31. $\left(\sqrt{3} + 2\sqrt{6}\right)\left(2\sqrt{3} - \sqrt{6}\right)$
$= 2\sqrt{9} - \sqrt{18} + 4\sqrt{18} - 2\sqrt{36}$
$= 6 + 3\sqrt{18} - 12$
$= -6 + 3\sqrt{9 \cdot 2}$
$= -6 + 9\sqrt{2}$

33. $\left(3\sqrt{7}-\sqrt{8}\right)\left(\sqrt{8}+2\sqrt{7}\right)$
$= 3\sqrt{56}+6\sqrt{49}-\sqrt{64}-2\sqrt{56}$
$= \sqrt{56}+42-8$
$= \sqrt{4\cdot 14}+34$
$= 2\sqrt{14}+34$

35. $\left(2\sqrt{6}-1\right)^2 = \left(2\sqrt{6}-1\right)\left(2\sqrt{6}-1\right)$
$= 4\sqrt{36}-2\sqrt{6}-2\sqrt{6}+1$
$= 24-4\sqrt{6}+1$
$= 25-4\sqrt{6}$

37. $\left(\sqrt{3}+5\sqrt{2}\right)^2 = \sqrt{9}+2\left(5\sqrt{6}\right)+25\sqrt{4}$
$= 3+10\sqrt{6}+50$
$= 53+10\sqrt{6}$

39. $\left(3\sqrt{5}-\sqrt{3}\right)^2 = 9\sqrt{25}-6\sqrt{15}+\sqrt{9}$
$= 45-6\sqrt{15}+3$
$= 48-6\sqrt{15}$

41. $\left(3\sqrt{5}\right)\left(2\sqrt{8}\right) = 6\sqrt{40} = 6\sqrt{4\cdot 10} = 12\sqrt{10}$

43. $\sqrt{7}\left(\sqrt{2}+3\sqrt{14}-\sqrt{6}\right) = \sqrt{14}+3\sqrt{98}-\sqrt{42}$
$= \sqrt{14}+3\sqrt{49\cdot 2}-\sqrt{42}$
$= \sqrt{14}+21\sqrt{2}-\sqrt{42}$

45. $\left(5\sqrt{2}-3\sqrt{7}\right)\left(3\sqrt{2}+\sqrt{7}\right)$
$= 15\sqrt{4}+5\sqrt{14}-9\sqrt{14}-3\sqrt{49}$
$= 30-4\sqrt{14}-21$
$= 9-4\sqrt{14}$

47. $2\sqrt{a}\left(3\sqrt{b}+\sqrt{ab}-2\sqrt{a}\right)$
$= 6\sqrt{ab}+2\sqrt{a^2b}-4\sqrt{a^2}$
$= 6\sqrt{ab}+2a\sqrt{b}-4a$

49. $\left(3x\sqrt{y}+\sqrt{5}\right)\left(3x\sqrt{y}-\sqrt{5}\right) = \left(3x\sqrt{y}\right)^2-\left(\sqrt{5}\right)^2$
$= 9x^2y-5$

51. $A = \frac{1}{2}bh,\ b = \sqrt{288},\ h = \sqrt{2}$
$A = \frac{1}{2}\sqrt{288}\cdot\sqrt{2} = \frac{1}{2}\left(12\sqrt{2}\right)\left(\sqrt{2}\right) = 12$
The area is 12 square feet.

Cumulative Review

53. $64a^2-25b^2 = (8a)^2-(5b)^2 = (8a-5b)(8a+5b)$

54. $4x^2-20x+25 = (2x-5)(2x-5) = (2x-5)^2$

55. $m = ak$
$m = \left(\frac{6076}{5280}\right)k$
$m = \left(\frac{1519}{1320}\right)k$

Destroyer is going $m = \left(\frac{1519}{1320}\right)(35) \approx 40.3$ mph.

56. $280\times 0.15 = 42$
They will save $42 per year.
$\frac{350}{42} = 8\frac{1}{3}$

It will take $8\frac{1}{3}$ years.

Quick Quiz 9.4

1. $\left(2\sqrt{x}\right)\left(4\sqrt{y}\right)\left(3\sqrt{xy}\right) = 2\cdot 4\cdot 3\sqrt{x\cdot y\cdot xy} = 24xy$

2. $\sqrt{3}\left(\sqrt{8}+2\sqrt{3}-4\sqrt{5}\right) = \sqrt{24}+2(3)-4\sqrt{15}$
$= \sqrt{4\cdot 6}+6-4\sqrt{15}$
$= 2\sqrt{6}+6-4\sqrt{15}$

3. $\left(4\sqrt{2}-5\sqrt{3}\right)\left(2\sqrt{2}+3\sqrt{3}\right)$
$= 8(2)+12\sqrt{6}-10\sqrt{6}-15(3)$
$= 16+2\sqrt{6}-45$
$= -29+2\sqrt{6}$

4. Answers may vary. Possible solution:
Expand and use FOIL method to multiply terms.
Combine terms with equal radicands.

How Am I Doing? Sections 9.1–9.4

1. $\sqrt{49} = 7$ since $7^2 = 49$.

2. $\sqrt{\frac{25}{64}} = \frac{5}{8}$ since $\left(\frac{5}{8}\right)^2 = \frac{25}{64}$.

3. $\sqrt{81} = 9$ since $9^2 = 81$.

4. $-\sqrt{100} = -10$ since $10^2 = 100$.

5. $\sqrt{5} \approx 2.236$

6. $-\sqrt{121} = -11$ since $11^2 = 121$.

7. $\sqrt{0.04} = 0.2$ since $0.2^2 = 0.04$.

8. $\sqrt{0.36} = 0.6$ since $0.6^2 = 0.36$.

9. $\sqrt{x^4} = x^2$

10. $\sqrt{25x^4y^6} = 5x^2y^3$

11. $\sqrt{50} = \sqrt{25 \cdot 2} = 5\sqrt{2}$

12. $\sqrt{x^5} = \sqrt{x^4 \cdot x} = x^2\sqrt{x}$

13. $\sqrt{144a^2b^3} = \sqrt{144 \cdot a^2 \cdot b^2 \cdot b} = 12ab\sqrt{b}$

14. $\sqrt{32x^5} = \sqrt{16 \cdot 2 \cdot x^4 \cdot x} = 4x^2\sqrt{2x}$

15.
$$\begin{aligned}\sqrt{98} + \sqrt{128} &= \sqrt{49 \cdot 2} + \sqrt{64 \cdot 2}\\ &= 7\sqrt{2} + 8\sqrt{2}\\ &= 15\sqrt{2}\end{aligned}$$

16.
$$\begin{aligned}3\sqrt{2} - \sqrt{8} + \sqrt{18} &= 3\sqrt{2} - \sqrt{4 \cdot 2} + \sqrt{9 \cdot 2}\\ &= 3\sqrt{2} - 2\sqrt{2} + 3\sqrt{2}\\ &= 4\sqrt{2}\end{aligned}$$

17.
$$\begin{aligned}2\sqrt{28} + 3\sqrt{63} - \sqrt{49} &= 2\sqrt{4 \cdot 7} + 3\sqrt{9 \cdot 7} - \sqrt{49}\\ &= 4\sqrt{7} + 9\sqrt{7} - 7\\ &= 13\sqrt{7} - 7\end{aligned}$$

18.
$$\begin{aligned}&3\sqrt{8} + \sqrt{12} + \sqrt{50} - 4\sqrt{75}\\ &= 3\sqrt{4 \cdot 2} + \sqrt{4 \cdot 3} + \sqrt{25 \cdot 2} - 4\sqrt{25 \cdot 3}\\ &= 6\sqrt{2} + 2\sqrt{3} + 5\sqrt{2} - 20\sqrt{3}\\ &= 11\sqrt{2} - 18\sqrt{3}\end{aligned}$$

19.
$$\begin{aligned}5\sqrt{2x} - 7\sqrt{8x} &= 5\sqrt{2x} - 7\sqrt{4 \cdot 2x}\\ &= 5\sqrt{2x} - 14\sqrt{2x}\\ &= -9\sqrt{2x}\end{aligned}$$

20.
$$\begin{aligned}-4x\sqrt{75} - 2x\sqrt{27} &= -4x\sqrt{25 \cdot 3} - 2x\sqrt{9 \cdot 3}\\ &= -20x\sqrt{3} - 6x\sqrt{3}\\ &= -26x\sqrt{3}\end{aligned}$$

21. $\left(5\sqrt{2}\right)\left(3\sqrt{15}\right) = 5 \cdot 3 \cdot \sqrt{2 \cdot 15} = 15\sqrt{30}$

22.
$$\begin{aligned}\left(4x\sqrt{6}\right)\left(x\sqrt{10}\right) &= 4x^2\sqrt{60}\\ &= 4x^2\sqrt{4 \cdot 15}\\ &= 8x^2\sqrt{15}\end{aligned}$$

23.
$$\begin{aligned}\sqrt{2}\left(\sqrt{6} + 3\sqrt{2}\right) &= \sqrt{12} + 3\sqrt{4}\\ &= \sqrt{4 \cdot 3} + 3 \cdot 2\\ &= 2\sqrt{3} + 6\end{aligned}$$

24.
$$\begin{aligned}\left(\sqrt{7} - 3\right)^2 &= \left(\sqrt{7} - 3\right)\left(\sqrt{7} - 3\right)\\ &= 7 - 3\sqrt{7} - 3\sqrt{7} + 9\\ &= 16 - 6\sqrt{7}\end{aligned}$$

25.
$$\begin{aligned}\left(\sqrt{11} - \sqrt{10}\right)\left(\sqrt{11} + \sqrt{10}\right) &= \sqrt{121} - \sqrt{100}\\ &= 11 - 10\\ &= 1\end{aligned}$$

26.
$$\begin{aligned}\left(2\sqrt{3} - \sqrt{5}\right)^2 &= 4\sqrt{9} - 2\left(2\sqrt{3}\right)\left(\sqrt{5}\right) + \sqrt{25}\\ &= 12 - 4\sqrt{15} + 5\\ &= 17 - 4\sqrt{15}\end{aligned}$$

27.
$$\begin{aligned}\left(\sqrt{x} + 5\right)\left(\sqrt{x} - 2\right) &= x - 2\sqrt{x} + 5\sqrt{x} - 10\\ &= x + 3\sqrt{x} - 10\end{aligned}$$

28.
$$\begin{aligned}&\left(3\sqrt{7} + 4\sqrt{3}\right)\left(5\sqrt{7} - 2\sqrt{3}\right)\\ &= 15\sqrt{49} - 6\sqrt{21} + 20\sqrt{21} - 8\sqrt{9}\\ &= 105 + 14\sqrt{21} - 24\\ &= 81 + 14\sqrt{21}\end{aligned}$$

9.5 Exercises

1. $\frac{\sqrt{12}}{\sqrt{3}} = \sqrt{\frac{12}{3}} = \sqrt{4} = 2$

3. $\frac{\sqrt{7}}{\sqrt{63}} = \sqrt{\frac{7}{63}} = \sqrt{\frac{1}{9}} = \frac{1}{3}$

5. $\frac{\sqrt{98}}{\sqrt{2}} = \sqrt{\frac{98}{2}} = \sqrt{49} = 7$

7. $\frac{\sqrt{6}}{\sqrt{x^4}} = \frac{\sqrt{6}}{x^2}$

9. $\frac{\sqrt{18}}{\sqrt{a^4}} = \frac{\sqrt{9 \cdot 2}}{\sqrt{(a^2)^2}} = \frac{3\sqrt{2}}{a^2}$

11. $\frac{3}{\sqrt{7}} = \frac{3}{\sqrt{7}} \cdot \frac{\sqrt{7}}{\sqrt{7}} = \frac{3\sqrt{7}}{\sqrt{49}} = \frac{3\sqrt{7}}{7}$

13. $\frac{8}{\sqrt{15}} = \frac{8}{\sqrt{15}} \cdot \frac{\sqrt{15}}{\sqrt{15}} = \frac{8\sqrt{15}}{15}$

15. $\frac{x\sqrt{x}}{\sqrt{2}} = \frac{x\sqrt{x}}{\sqrt{2}} \cdot \frac{\sqrt{2}}{\sqrt{2}} = \frac{x\sqrt{2x}}{\sqrt{4}} = \frac{x\sqrt{2x}}{2}$

17. $\frac{\sqrt{8}}{\sqrt{x}} = \frac{\sqrt{4 \cdot 2}}{\sqrt{x}} \cdot \frac{\sqrt{x}}{\sqrt{x}} = \frac{2\sqrt{2x}}{\sqrt{x^2}} = \frac{2\sqrt{2x}}{x}$

19. $\frac{6}{\sqrt{28}} = \frac{6}{\sqrt{28}} \cdot \frac{\sqrt{28}}{\sqrt{28}}$

$= \frac{6\sqrt{28}}{28}$

$= \frac{6\sqrt{4 \cdot 7}}{28}$

$= \frac{12\sqrt{7}}{28}$

$= \frac{3\sqrt{7}}{7}$

21. $\frac{6}{\sqrt{a}} = \frac{6}{\sqrt{a}} \cdot \frac{\sqrt{a}}{\sqrt{a}} = \frac{6\sqrt{a}}{a}$

23. $\frac{x}{\sqrt{2x^5}} = \frac{x}{\sqrt{2 \cdot x \cdot x^4}}$

$= \frac{x}{x^2\sqrt{2x}}$

$= \frac{1}{x\sqrt{2x}} \cdot \frac{\sqrt{2x}}{\sqrt{2x}}$

$= \frac{\sqrt{2x}}{x\sqrt{4x^2}}$

$= \frac{\sqrt{2x}}{2x^2}$

25. $\frac{\sqrt{18}}{\sqrt{2x^3}} = \sqrt{\frac{18}{2x^3}}$

$= \sqrt{\frac{9}{x^3}}$

$= \frac{3}{x\sqrt{x}}$

$= \frac{3}{x\sqrt{x}} \cdot \frac{\sqrt{x}}{\sqrt{x}}$

$= \frac{3\sqrt{x}}{x^2}$

27. $\sqrt{\frac{3}{5}} = \frac{\sqrt{3}}{\sqrt{5}} = \frac{\sqrt{3}}{\sqrt{5}} \cdot \frac{\sqrt{5}}{\sqrt{5}} = \frac{\sqrt{15}}{\sqrt{25}} = \frac{\sqrt{5}}{5}$

29. $\sqrt{\frac{10}{21}} = \frac{\sqrt{10}}{\sqrt{21}} = \frac{\sqrt{10}}{\sqrt{21}} \cdot \frac{\sqrt{21}}{\sqrt{21}} = \frac{\sqrt{210}}{21}$

31. $\frac{9}{\sqrt{32x}} = \frac{9}{\sqrt{16 \cdot 2x}}$

$= \frac{9}{4\sqrt{2x}} \cdot \frac{\sqrt{2x}}{\sqrt{2x}}$

$= \frac{9\sqrt{2x}}{4\sqrt{4x^2}}$

$= \frac{9\sqrt{2x}}{8x}$

33. $\frac{4}{\sqrt{3}-1} = \frac{4}{\sqrt{3}-1} \cdot \frac{\sqrt{3}+1}{\sqrt{3}+1}$
$= \frac{4\sqrt{3}+4}{\sqrt{9}-1}$
$= \frac{4\sqrt{3}+4}{3-1}$
$= \frac{4\left(\sqrt{3}+1\right)}{2}$
$= 2\sqrt{3}+2$

35. $\frac{8}{\sqrt{10}+\sqrt{2}} = \frac{8}{\sqrt{10}+\sqrt{2}} \cdot \frac{\sqrt{10}-\sqrt{2}}{\sqrt{10}-\sqrt{2}}$
$= \frac{8\sqrt{10}-8\sqrt{2}}{10-2}$
$= \frac{8\left(\sqrt{10}-\sqrt{2}\right)}{8}$
$= \sqrt{10}-\sqrt{2}$

37. $\frac{\sqrt{6}}{\sqrt{6}-\sqrt{3}} = \frac{\sqrt{6}}{\sqrt{6}-\sqrt{3}} \cdot \frac{\sqrt{6}+\sqrt{3}}{\sqrt{6}+\sqrt{3}}$
$= \frac{\sqrt{36}+\sqrt{18}}{\left(\sqrt{6}\right)^2-\left(\sqrt{3}\right)^2}$
$= \frac{6+\sqrt{9\cdot 2}}{6-3}$
$= \frac{6+3\sqrt{2}}{3}$
$= 2+\sqrt{2}$

39. $\frac{\sqrt{7}}{\sqrt{8}+\sqrt{7}} = \frac{\sqrt{7}}{\sqrt{8}+\sqrt{7}} \cdot \frac{\sqrt{8}-\sqrt{7}}{\sqrt{8}-\sqrt{7}}$
$= \frac{\sqrt{56}-\sqrt{49}}{\left(\sqrt{8}\right)^2-\left(\sqrt{7}\right)^2}$
$= \frac{\sqrt{4\cdot 14}-7}{8-7}$
$= 2\sqrt{14}-7$

41. $\frac{3x}{2\sqrt{2}-\sqrt{5}} = \frac{3x}{2\sqrt{2}-\sqrt{5}} \cdot \frac{2\sqrt{2}+\sqrt{5}}{2\sqrt{2}+\sqrt{5}}$
$= \frac{6x\sqrt{2}+3x\sqrt{5}}{\left(2\sqrt{2}\right)^2-\left(\sqrt{5}\right)^2}$
$= \frac{3x\left(2\sqrt{2}+\sqrt{5}\right)}{8-5}$
$= x\left(2\sqrt{2}+\sqrt{5}\right)$

43. $\frac{\sqrt{x}}{\sqrt{6}+\sqrt{2}} = \frac{\sqrt{x}}{\sqrt{6}+\sqrt{2}} \cdot \frac{\sqrt{6}-\sqrt{2}}{\sqrt{6}-\sqrt{2}}$
$= \frac{\sqrt{6x}-\sqrt{2x}}{\sqrt{36}-\sqrt{4}}$
$= \frac{\sqrt{6x}-\sqrt{2x}}{4}$

45. $\frac{\sqrt{10}-\sqrt{3}}{\sqrt{10}+\sqrt{3}} = \frac{\sqrt{10}-\sqrt{3}}{\sqrt{10}+\sqrt{3}} \cdot \frac{\sqrt{10}-\sqrt{3}}{\sqrt{10}-\sqrt{3}}$
$= \frac{10-2\sqrt{30}+3}{10-3}$
$= \frac{13-2\sqrt{30}}{7}$

47. $\frac{4\sqrt{7}+3}{\sqrt{5}-\sqrt{2}} = \frac{4\sqrt{7}+3}{\sqrt{5}-\sqrt{2}} \cdot \frac{\sqrt{5}+\sqrt{2}}{\sqrt{5}+\sqrt{2}}$
$= \frac{4\sqrt{35}+4\sqrt{14}+3\sqrt{5}+3\sqrt{2}}{\sqrt{25}-\sqrt{4}}$
$= \frac{4\sqrt{35}+4\sqrt{14}+3\sqrt{5}+3\sqrt{2}}{3}$

49. $\frac{4\sqrt{3}+2}{\sqrt{8}-\sqrt{6}} = \frac{4\sqrt{3}+2}{\sqrt{8}-\sqrt{6}} \cdot \frac{\sqrt{8}+\sqrt{6}}{\sqrt{8}+\sqrt{6}}$
$= \frac{4\sqrt{24}+4\sqrt{18}+2\sqrt{8}+2\sqrt{6}}{\left(\sqrt{8}\right)^2-\left(\sqrt{6}\right)^2}$
$= \frac{8\sqrt{6}+12\sqrt{2}+4\sqrt{2}+2\sqrt{6}}{8-6}$
$= \frac{10\sqrt{6}+16\sqrt{2}}{2}$
$= 5\sqrt{6}+8\sqrt{2}$

51. $\frac{x-25}{\sqrt{x}+5}=\frac{x-25}{\sqrt{x}+5}\cdot\frac{\sqrt{x}-5}{\sqrt{x}-5}$
$=\frac{(x-25)\left(\sqrt{x}-5\right)}{x-25}$
$=\sqrt{x}-5$

53. $s=\sqrt{\frac{3V}{h}}$

a. $s=\frac{\sqrt{3V}}{\sqrt{h}}\cdot\frac{\sqrt{h}}{\sqrt{h}}=\frac{\sqrt{3Vh}}{h}$

b. $s=\frac{\sqrt{3Vh}}{h}=\frac{\sqrt{3\cdot 36\cdot 12}}{12}=\frac{36}{12}=3$
$s=3$ in.

55. $x=$ width, $l=\sqrt{5}+2$
$A=lw$
$3=\left(\sqrt{5}+2\right)x$
$x=\frac{3}{\sqrt{5}+2}$
$=\frac{3}{\sqrt{5}+2}\cdot\frac{\sqrt{5}-2}{\sqrt{5}-2}$
$=\frac{3\sqrt{5}-6}{\sqrt{25}-4}$
$=\frac{3\sqrt{5}-6}{5-4}$
$=3\sqrt{5}-6$
The width should be $\left(3\sqrt{5}-6\right)$ meters.

Cumulative Review

57. $-2(x-5)=8x-3(x+1)+18$
$-2x+10=8x-3x-3+18$
$-2x=5x+18-10$
$-7x=8$
$x\approx-0.71$

58. $a^2-b\div a+a^3-b \qquad a=-1,\ b=4$
$=(-1)^2-4\div(-1)+(-1)^3-4$
$=1+4-1-4$
$=0$

Quick Quiz 9.5

1. $\frac{3}{\sqrt{10}}=\frac{3}{\sqrt{10}}\cdot\frac{\sqrt{10}}{\sqrt{10}}=\frac{3\sqrt{10}}{10}$

2. $\frac{5x}{3\sqrt{2}+\sqrt{5}}=\frac{5x}{3\sqrt{2}+\sqrt{5}}\cdot\frac{3\sqrt{2}-\sqrt{5}}{3\sqrt{2}-\sqrt{5}}$
$=\frac{15x\sqrt{2}-5x\sqrt{5}}{9(2)-5}$
$=\frac{15x\sqrt{2}-5x\sqrt{5}}{13}$

3. $\frac{\sqrt{6}+\sqrt{5}}{\sqrt{6}-\sqrt{5}}=\frac{\sqrt{6}+\sqrt{5}}{\sqrt{6}-\sqrt{5}}\cdot\frac{\sqrt{6}+\sqrt{5}}{\sqrt{6}+\sqrt{5}}$
$=\frac{6+2\sqrt{30}+5}{6-5}$
$=11+2\sqrt{30}$

4. Answers may vary. Possible solution: Rationalize the denominator by multiplying both the numerator and the denominator by the conjugate of the denominator.

9.6 Exercises

1. $c^2=5^2+12^2$
$c^2=25+144$
$c^2=169$
$c=\sqrt{169}=13$

3. $9^2=5^2+b^2$
$81=25+b^2$
$56=b^2$
$\sqrt{56}=b$
$\sqrt{4\cdot 14}=b$
$2\sqrt{14}=b$

5. $a=7, b=7$
$7^2+7^2=c^2$
$49+49=c^2$
$98=c^2$
$c=\sqrt{98}$
$c=\sqrt{49\cdot 2}$
$c=7\sqrt{2}$

7. $a=\sqrt{14},\ b=7$

$c^2=\left(\sqrt{14}\right)^2+7^2=14+49=63$

$c=\sqrt{63}=\sqrt{9\cdot 7}=3\sqrt{7}$

9. $c=15, b=13$

$a^2+b^2=c^2$

$a^2+13^2=15^2$

$a^2=225-169=56$

$a=\sqrt{56}$

$a=2\sqrt{14}$

11. $c=\sqrt{82}, a=5$

$5^2+b^2=\left(\sqrt{82}\right)^2$

$25+b^2=82$

$b^2=57$

$b=\sqrt{57}$

13. $c=13.8, b=9.42$

$a^2+b^2=c^2$

$a^2+(9.42)^2=(13.8)^2$

$a=10.08$

15. $c^2=8^2+5^2$

$c^2=64+25=89$

$c=\sqrt{89}$

$c\approx 9.4$

The guy rope is 9.4 feet.

17. $a^2+b^2=c^2$

$a^2+20^2=100^2$

$a^2+400=10,000$

$a^2=9600$

$a\approx 98.0$

The kite is 98 feet.

19. Bottom cable:

$c^2=(54)^2+(130)^2=2916+16,900=19,816$

$c=\sqrt{19,816}\approx 140.77$

The bottom cable is about 140.77 feet long.

Top cable:

$c^2=(54)^2+(135)^2=2916+18,225=21,141$

$c=\sqrt{21,141}\approx 145.40$

The top cable is about 145.40 feet long.

21. $\sqrt{x+2}=3$

$\left(\sqrt{x+2}\right)^2=3^2$

$x+2=9$

$x=7$

Check: $\sqrt{7+2}\stackrel{?}{=}3$

$\sqrt{9}\stackrel{?}{=}3$

$3=3$

23. $\sqrt{2x+7}=5$

$\left(\sqrt{2x+7}\right)^2=5^2$

$2x+7=25$

$2x=18$

$x=9$

Check: $\sqrt{2\cdot 9+7}\stackrel{?}{=}5$

$\sqrt{18+7}\stackrel{?}{=}5$

$\sqrt{25}\stackrel{?}{=}5$

$5=5$

25. $\sqrt{2x+2}=\sqrt{3x-5}$

$\left(\sqrt{2x+2}\right)^2=\left(\sqrt{3x-5}\right)^2$

$2x+2=3x-5$

$2=x-5$

$7=x$

Check: $\sqrt{2(7)+2}\stackrel{?}{=}\sqrt{3(7)-5}$

$\sqrt{16}\stackrel{?}{=}\sqrt{16}$

$4=4$

27. $\sqrt{2x}-5=4$

$\sqrt{2x}=9$

$\left(\sqrt{2x}\right)^2=9^2$

$2x=81$

$x=\frac{81}{2}$

Check: $\sqrt{2\left(\frac{81}{2}\right)}-5\stackrel{?}{=}4$

$\sqrt{81}-5\stackrel{?}{=}4$

$9-5\stackrel{?}{=}4$

$4=4$

29.
$$\sqrt{3x+10} = x$$
$$3x+10 = x^2$$
$$x^2-3x-10=0$$
$$(x-5)(x+2)=0$$
$$x-5=0 \quad \text{or} \quad x+2=0$$
$$x=5 \qquad x=-2$$
Check:
$$\sqrt{3\cdot 5+10} \stackrel{?}{=} 5$$
$$\sqrt{15+10} \stackrel{?}{=} 5$$
$$\sqrt{25} \stackrel{?}{=} 5$$
$$5=5$$
$$\sqrt{3\cdot(-2)+10} \stackrel{?}{=} -2$$
$$\sqrt{4} \stackrel{?}{=} -2$$
$$2 \neq -2$$
$x = 5$ only

31.
$$\sqrt{5y+1} = y+1$$
$$\left(\sqrt{5y+1}\right)^2 = (y+1)^2$$
$$5y+1 = y^2+2y+1$$
$$y^2-3y=0$$
$$y(y-3)=0$$
$$y=0,\ y-3=0$$
$$y=3$$
Check: $\sqrt{5(0)+1} \stackrel{?}{=} 0+1$ $\qquad \sqrt{5(3)+1} \stackrel{?}{=} 3+1$
$$\sqrt{1} \stackrel{?}{=} 1 \qquad \sqrt{16} \stackrel{?}{=} 4$$
$$1=1 \qquad 4=4$$
$y = 0, 3$

33.
$$\sqrt{x+3} = 3x-1$$
$$\left(\sqrt{x+3}\right)^2 = (3x-1)^2$$
$$x+3 = 9x^2-6x+1$$
$$9x^2-7x-2=0$$
$$(9x+2)(x-1)=0$$
$$9x+2=0 \qquad x-1=0$$
$$9x=-2 \qquad x=1$$
$$x=-\frac{2}{9}$$
Check:
$$\sqrt{-\frac{2}{9}+3} \stackrel{?}{=} 3\left(-\frac{2}{9}\right)-1$$
$$\sqrt{\frac{25}{9}} \stackrel{?}{=} -\frac{5}{3}$$
$$\frac{5}{3} \neq -\frac{5}{3} \text{ doesn't check}$$
$$\sqrt{1+3} \stackrel{?}{=} 3(1)-1$$
$$\sqrt{4} \stackrel{?}{=} 2$$
$$2=2$$
The only solution is $x = 1$.

35.
$$\sqrt{3y+1}-y=1$$
$$\sqrt{3y+1} = y+1$$
$$\left(\sqrt{3y+1}\right)^2 = (y+1)^2$$
$$3y+1 = y^2+2y+1$$
$$y^2-y=0$$
$$y(y-1)=0$$
$$y=0 \qquad y-1=0$$
$$y=1$$
Check: $\sqrt{3(0)+1}-0 \stackrel{?}{=} 1$ $\qquad \sqrt{3(1)+1}-1 \stackrel{?}{=} 1$
$$\sqrt{1}-0 \stackrel{?}{=} 1 \qquad \sqrt{4}-1 \stackrel{?}{=} 1$$
$$1=1 \qquad 2-1 \stackrel{?}{=} 1$$
$$1=1$$
$y = 0, 1$

37.
$$\sqrt{6y+1}-3y=y$$
$$\sqrt{6y+1} = 4y$$
$$\left(\sqrt{6y+1}\right)^2 = (4y)^2$$
$$6y+1 = 16y^2$$
$$0 = 16y^2-6y-1$$
$$0 = (8y+1)(2y-1)$$
$$8y+1=0 \qquad 2y-1=0$$
$$y=-\frac{1}{8} \qquad y=\frac{1}{2}$$
Check: $\sqrt{6\left(-\frac{1}{8}\right)+1}-3\left(-\frac{1}{8}\right) \stackrel{?}{=} -\frac{1}{8}$
$$\sqrt{\frac{1}{4}}+\frac{3}{8} \stackrel{?}{=} -\frac{1}{8}$$
$$\frac{1}{2}+\frac{3}{8} \stackrel{?}{=} -\frac{1}{8}$$
$$\frac{7}{8} \neq -\frac{1}{8}$$

$$\sqrt{6\left(\frac{1}{2}\right)+1}-3\left(\frac{1}{2}\right) \stackrel{?}{=} \frac{1}{2}$$
$$\sqrt{4}-\frac{3}{2} \stackrel{?}{=} \frac{1}{2}$$
$$2-\frac{3}{2} \stackrel{?}{=} \frac{1}{2}$$
$$\frac{1}{2}=\frac{1}{2}$$

$y=\frac{1}{2}$ only

39. $\sqrt{12x+1}=2\sqrt{6x}-1$

$$\left(\sqrt{12x+1}\right)^2=\left(2\sqrt{6x}-1\right)^2$$
$$12x+1=4\sqrt{36x^2}-2\left(2\sqrt{6x}\right)+1$$
$$12x+1=24x-4\sqrt{6x}+1$$
$$-12x+1=-4\sqrt{6x}+1$$
$$-12x=-4\sqrt{6x}$$
$$3x=\sqrt{6x}$$
$$(3x)^2=\left(\sqrt{6x}\right)^2$$
$$9x^2=6x$$
$$9x^2-6x=0$$
$$3x(3x-2)=0$$

$3x=0$, $x=0$ $\qquad$ $3x-2=0$, $3x=2$, $x=\frac{2}{3}$

Check: $\sqrt{12(0)+1} \stackrel{?}{=} 2\sqrt{6(0)}-1$
$$\sqrt{1} \stackrel{?}{=} -1$$
$$1 \neq -1$$
$$\sqrt{12\left(\frac{2}{3}\right)+1} \stackrel{?}{=} 2\sqrt{6\left(\frac{2}{3}\right)}-1$$
$$\sqrt{9} \stackrel{?}{=} 2(2)-1$$
$$3=3$$

$x=\frac{2}{3}$ only

Cumulative Review

41. $\frac{5x}{x-4}=5+\frac{4x}{x-4}$

$$(x-4)\left(\frac{5x}{x-4}\right)=5(x-4)+(x-4)\left(\frac{4x}{x-4}\right)$$
$$5x=5x-20+4x$$
$$5x=9x-20$$
$$-4x=-20$$
$$x=5$$

42. $(7x-2)^2=49x^2-28x+4$

43. $f(x)=2x^2-3x+6$

$f(-2)=2(-2)^2-3(-2)+6=2(4)+6+6=20$

44. $x^2+y^2=9$ is a relation. There exist two different ordered pairs, for example (0, 3) and (0, −3), that have the same x-coordinate.

Quick Quiz 9.6

1. $a^2+b^2=c^2$

$$\left(\sqrt{3}\right)^2+(5)^2=c^2$$
$$3+25=c^2$$
$$\sqrt{28}=c$$
$$\sqrt{4\cdot 7}=c$$
$$2\sqrt{7}=c$$

2. $8-\sqrt{5x-6}=0$

$$\sqrt{5x-6}=8$$
$$\left(\sqrt{5x-6}\right)^2=8^2$$
$$5x-6=64$$
$$5x=70$$
$$x=14$$

Check: $8-\sqrt{5(14)-6} \stackrel{?}{=} 0$
$$0=0 \text{ True}$$

3. $\sqrt{6x+1}=x-1$

$$\left(\sqrt{6x+1}\right)^2=(x-1)^2$$
$$6x+1=x^2-2x+1$$
$$x^2-8x=0$$
$$x(x-8)=0$$
$$x=8 \text{ or } x=0$$

Check: $x = 8$

$\sqrt{6(8)+1} \stackrel{?}{=} 8-1$

$7 = 7$ True

Check: $x = 0$

$\sqrt{6(0)+1} \stackrel{?}{=} 0-1$

$1 = -1$ False

$x = 8$ only

4. Answers may vary. Possible solution: Substitute found values into the original equation to test the logic of each value.

9.7 Exercises

1. $y = kx,\ y = 9,\ x = 2$

$9 = 2k \Rightarrow k = \frac{9}{2}$

$y = \frac{9}{2}x,\ x = 16$

$y = \frac{9}{2}(16) = 72$

3. $y = kx^3,\ y = 12,\ x = 2$

$12 = k(2)^3 \Rightarrow k = \frac{12}{8} = \frac{3}{2}$

$y = \frac{3}{2}x^3,\ x = 7$

$y = \frac{3}{2}(7)^3 = \frac{3}{2}(343) = \frac{1029}{2}$

5. $y = kx^2,\ y = 900,\ x = 25$

$900 = k(25)^2 \Rightarrow k = \frac{900}{625} = \frac{36}{25}$

$y = \frac{36}{25}x^2,\ x = 30$

$y = \frac{36}{25}(30)^2 = 1296$

7. $c = kw$

$100 = k20$

$5 = k$

$c = 5w,\ w = 45$

$c = 5 \cdot 45$

$c = 225$

They will have 225 calories.

9. T = time, s = length of side

$T = ks^3,\ T = 7,\ s = 2.0$

$7 = k(2.0)^3 \Rightarrow k = \frac{7}{8}$

$T = \frac{7}{8}s^3,\ s = 4.0$

$T = \frac{7}{8}(4.0)^3 = 56$

It will take 56 minutes.

11. $y = \frac{k}{x},\ y = 18,\ x = 5$

$18 = \frac{k}{5} \Rightarrow k = 90$

$y = \frac{90}{x},\ x = 8$

$y = \frac{90}{8} = \frac{45}{4}$

13. $y = \frac{k}{x},\ y = \frac{1}{3},\ x = 12$

$\frac{1}{3} = \frac{k}{12}$

$k = 4$

$y = \frac{4}{x},\ x = 4$

$y = \frac{4}{x}$

$y = \frac{4}{4}$

$y = 1$

15. $y = \frac{k}{x^2},\ y = 30,\ x = 2$

$30 = \frac{k}{(2)^2} \Rightarrow k = 120$

$y = \frac{120}{x^2},\ x = 9$

$y = \frac{120}{(9)^2} = \frac{120}{81} = \frac{40}{27}$

17. T = time, F = temperature

$T = \frac{k}{F}$

$2.3 = \frac{k}{60}$

$138 = k$

$T = \frac{138}{F}$, $F = 40$

$T = \frac{138}{40}$

$T = 3.45$

It will take 3.45 minutes.

19. W = weight, D = distance from center

$W = \frac{k}{D^2}$, $W = 1000$, $D = 4000$

$1000 = \frac{k}{(4000)^2} \Rightarrow k = 16{,}000{,}000{,}000$

$W = \frac{16{,}000{,}000{,}000}{D^2}$, $D = 6000$

$W = \frac{16{,}000{,}000{,}000}{(6000)^2} = \frac{4000}{9} = 444\frac{4}{9}$

It will weigh $444\frac{4}{9}$ pounds.

21. $d = \frac{15}{11}\sqrt{h}$

a. $h = 121$; $d = \frac{15}{11}\sqrt{121} = 15$ km

$h = 36$; $d = \frac{15}{11}\sqrt{36} \approx 8.2$ km

$h = 4$; $d = \frac{15}{11}\sqrt{4} \approx 2.7$ km

b. $h = 81$; $d = \frac{15}{11}\sqrt{81} \approx 12.3$ km

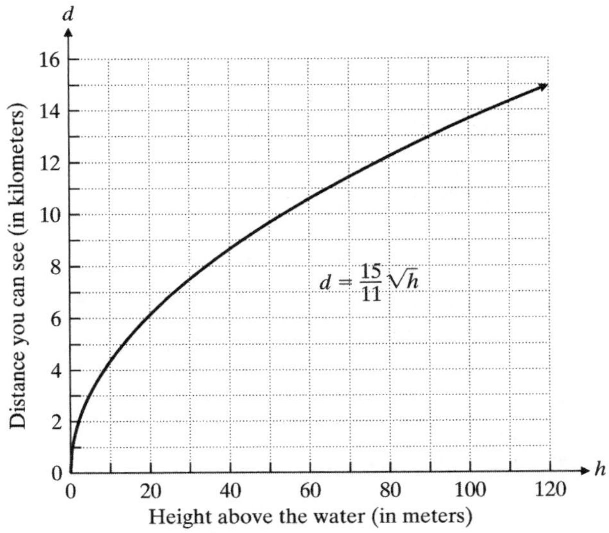

23. $t^2 = kd^3$

a. $1^2 = k(1)^3$

$1 = k$

b. Pluto: $t^2 = 1(39.5)^3$

$t^2 = 61{,}629.875$

$t = \sqrt{61{,}629.875} \approx 248.25$

The length is approximately 248.25 Earth years.

Cumulative Review

25. $\frac{80{,}000}{320} = \frac{120{,}000}{320+a}$

LCD = $320(320 + a)$

$320(320+a)\cdot\frac{80{,}000}{320} = 320(320+a)\cdot\frac{120{,}000}{320+a}$

$80{,}000(320+a) = 120{,}000(320)$

$2(320+a) = 3(320)$

$640 + 2a = 960$

$2a = 320$

$a = 160$

26.

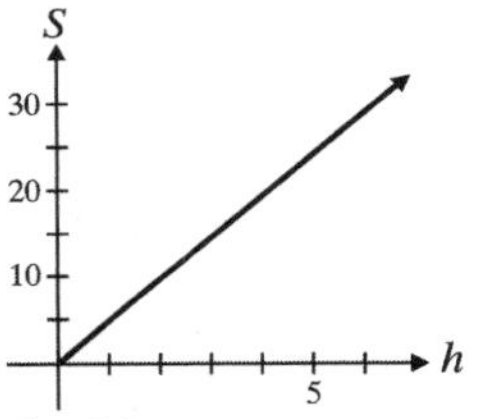

$S = kh$

$S = 4.80h$

Quick Quiz 9.7

1. i = intensity in lumens

x = distance in inches

k = unitless constant

$i = \frac{1}{kx^2}$

If $i = 45$ when $x = 2$:

$k = \frac{1}{x^2 i} = \frac{1}{(2)^2(45)} = \frac{1}{180}$

If $i = 5$ and $k = \frac{1}{180}$

$i = \frac{1}{kx^2}$ or

$x = \sqrt{\frac{1}{ki}} = \sqrt{\frac{1}{\left(\frac{1}{180}\right)5}}$

$x = 6$

It is 6 inches from the light source.

2. c = calories
w = weight in grams
k = unitless constant
$c = wk$
When $w = 40$, $c = 150$

$k = \frac{c}{w} = \frac{150}{40} = \frac{15}{4}$

If $w = 75$, and $k = \frac{15}{4}$

$c = wk = 75\left(\frac{15}{4}\right) = 281.25$

There are 281.25 calories in a bar weighing 75 g.

3. p = perimeter
a = area
k = constant

$a = kp^2$

When $a = 6.4$, $p = 8$

$k = \frac{a}{p^2} = \frac{6.4}{8^2} = 0.1$

When $p = 20$, $k = 0.1$

$a = kp^2 = 0.1(20)^2 = 40$

The area is 40 square centimeters.

4. Answers may vary. Possible solution:
t = time
d = distance
k = constant

$d = kt^2$

When $d = 64$, $t = 2$.
Substitute the values for d and t into the equation, and solve for k.

$k = \frac{d}{t^2} = \frac{64}{2^2} = 16$

Putting Your Skills to Work

1. $4 \times \frac{2}{3} = \frac{4}{1} \times \frac{2}{3} = \frac{8}{3}$

Lucy prepares $\frac{8}{3}$ cups of rice each week.

2. $\frac{21 \text{ ounces}}{1 \text{ cup}} \times \frac{8}{3} \text{ cups} = 56 \text{ ounces}$

Lucy's family consumes 56 ounces of rice each week.

3. $\frac{1 \text{ lb}}{16 \text{ oz}} \times 56 \text{ oz} = \frac{7}{2} \text{ or } 3\frac{1}{2} \text{ lb}$

Lucy's family eats $3\frac{1}{2}$ pounds each week.

4. a. $15.73 - 9.25 = 6.48$

$\frac{6.48}{25} \approx 0.26$

It increased by \$0.26/lb.

b. $\frac{9.25}{25} = 0.37; \frac{0.26}{0.37} \approx 0.70$

It is a 70% increase.

5. $\frac{15.73}{25} = 0.6292; 0.6292(3.5) \approx 2.20$

The cost in today's prices is \$2.20 per week.

6. Answers will vary.

7. Answers will vary.

Chapter 9 Review Problems

1. $\sqrt{36} = 6$ since $6^2 = 36$.

2. $\sqrt{25} = 5$ since $5^2 = 25$.

3. $\sqrt{169} = 13$ since $13^2 = 169$.

4. $\sqrt{196} = 14$ since $14^2 = 196$.

5. $-\sqrt{81} = -9$ since $9^2 = 81$.

6. $\sqrt{64} = 8$ since $8^2 = 64$.

7. $\sqrt{49} = 7$ since $7^2 = 49$.

8. $-\sqrt{121} = -11$ since $11^2 = 121$.

9. $-\sqrt{144} = -12$ since $12^2 = 144$.

10. $\sqrt{289} = 17$ since $17^2 = 289$.

11. $\sqrt{0.04} = 0.2$ since $0.2^2 = 0.04$.

12. $\sqrt{0.49} = 0.7$ since $0.7^2 = 0.49$.

13. $\sqrt{\frac{1}{100}} = \frac{1}{10}$ since $\left(\frac{1}{10}\right)^2 = \frac{1}{100}$.

14. $\sqrt{\frac{64}{81}} = \frac{8}{9}$ since $\left(\frac{8}{9}\right)^2 = \frac{64}{81}$.

15. $\sqrt{105} \approx 10.247$

16. $\sqrt{198} \approx 14.071$

17. $\sqrt{77} = 8.775$

18. $\sqrt{88} \approx 9.381$

19. $\sqrt{27} = \sqrt{9 \cdot 3} = 3\sqrt{3}$

20. $\sqrt{52} = \sqrt{4 \cdot 13} = 2\sqrt{13}$

21. $\sqrt{28} = \sqrt{4 \cdot 7} = 2\sqrt{7}$

22. $\sqrt{125} = \sqrt{25 \cdot 5} = 5\sqrt{5}$

23. $\sqrt{40} = \sqrt{4 \cdot 10} = 2\sqrt{10}$

24. $\sqrt{80} = \sqrt{16 \cdot 5} = 4\sqrt{5}$

25. $\sqrt{x^8} = \sqrt{(x^4)^2} = x^4$

26. $\sqrt{y^{10}} = \sqrt{(y^5)^2} = y^5$

27. $\sqrt{x^5 y^6} = \sqrt{x^4 \cdot x \cdot y^6} = x^2 y^3 \sqrt{x}$

28. $\sqrt{a^3 b^4} = \sqrt{a^2 \cdot a \cdot (b^2)^2} = ab^2\sqrt{a}$

29. $\sqrt{16x^3 y^5} = \sqrt{16 \cdot x^2 \cdot x \cdot y^4 \cdot y} = 4xy^2\sqrt{xy}$

30. $\sqrt{98x^4 y^6} = \sqrt{49 \cdot 2 \cdot (x^2)^2 \cdot (y^3)^2} = 7x^2 y^3 \sqrt{2}$

31. $\sqrt{12x^5} = \sqrt{4 \cdot 3 \cdot x^4 \cdot x} = 2x^2\sqrt{3x}$

32. $\sqrt{72x^9} = \sqrt{36 \cdot 2 \cdot x^8 \cdot x} = 6x^4\sqrt{2x}$

33. $\sqrt{48x^8} = \sqrt{16 \cdot 3 \cdot x^8} = 4x^4\sqrt{3}$

34. $\sqrt{125x^{12}} = \sqrt{25 \cdot 5 \cdot (x^6)^2} = 5x^6\sqrt{5}$

35. $\sqrt{120a^3 b^4 c^5} = \sqrt{4 \cdot 30 \cdot a^2 \cdot a \cdot b^4 \cdot c^4 \cdot c}$
$= 2ab^2c^2\sqrt{30ac}$

36. $\sqrt{121a^6 b^4 c} = \sqrt{11^2 \cdot (a^3)^2 \cdot (b^2)^2 \cdot c} = 11a^3b^2\sqrt{c}$

37. $\sqrt{56x^7 y^9} = \sqrt{4 \cdot 14 \cdot x^6 \cdot x \cdot y^8 \cdot y} = 2x^3y^4\sqrt{14xy}$

38. $\sqrt{99x^{13} y^7} = \sqrt{9 \cdot 11 \cdot (x^6)^2 \cdot x \cdot (y^3)^2 \cdot y}$
$= 3x^6y^3\sqrt{11xy}$

39. $\sqrt{5} - \sqrt{20} + \sqrt{80} = \sqrt{5} - \sqrt{4 \cdot 5} + \sqrt{16 \cdot 5}$
$= \sqrt{5} - 2\sqrt{5} + 4\sqrt{5}$
$= 3\sqrt{5}$

40. $5\sqrt{6} - \sqrt{24} + 2\sqrt{54} = 5\sqrt{6} - \sqrt{4 \cdot 6} + 2\sqrt{9 \cdot 6}$
$= 5\sqrt{6} - 2\sqrt{6} + 6\sqrt{6}$
$= 9\sqrt{6}$

41. $x\sqrt{3} + 3x\sqrt{3} + \sqrt{27x^2} = 4x\sqrt{3} + \sqrt{9 \cdot 3x^2}$
$= 4x\sqrt{3} + 3x\sqrt{3}$
$= 7x\sqrt{3}$

42. $a\sqrt{2} + \sqrt{12a^2} + a\sqrt{98} = a\sqrt{2} + 2a\sqrt{3} + 7a\sqrt{2}$
$= 8a\sqrt{2} + 2a\sqrt{3}$

43. $5\sqrt{5} - 6\sqrt{20} + 2\sqrt{10} = 5\sqrt{5} - 6\sqrt{4 \cdot 5} + 2\sqrt{10}$
$= 5\sqrt{5} - 12\sqrt{5} + 2\sqrt{10}$
$= -7\sqrt{5} + 2\sqrt{10}$

44. $2\sqrt{40} - 2\sqrt{90} + 3\sqrt{28}$
$= 2\sqrt{4 \cdot 10} - 2\sqrt{9 \cdot 10} + 3\sqrt{4 \cdot 7}$
$= 4\sqrt{10} - 6\sqrt{10} + 6\sqrt{7}$
$= -2\sqrt{10} + 6\sqrt{7}$

45. $\left(2\sqrt{x}\right)\left(3\sqrt{x^3}\right) = 6\sqrt{x^4} = 6x^2$

46. $\left(-5\sqrt{a}\right)\left(2\sqrt{ab}\right) = -10\sqrt{a^2 b} = -10a\sqrt{b}$

47. $\left(\sqrt{6a^2}\right)\left(\sqrt{6b^5}\right) = 6\sqrt{a^2 b^4 b} = 6ab^2\sqrt{b}$

48. $\left(-4x\sqrt{x}\right)\left(2x\sqrt{x}\right)=-8x^2\sqrt{x^2}=-8x^3$

49. $\sqrt{6}\left(\sqrt{24}-5\sqrt{6}\right)=\sqrt{6\cdot 6\cdot 4}-5\sqrt{6\cdot 6}$
$=12-30$
$=-18$

50. $\sqrt{3}\left(\sqrt{48}-8\sqrt{3}\right)=\sqrt{3\cdot 3\cdot 16}-8\sqrt{3\cdot 3}$
$=12-24$
$=-12$

51. $\sqrt{2}\left(\sqrt{5}-\sqrt{3}-2\sqrt{2}\right)=\sqrt{10}-\sqrt{6}-2\sqrt{4}$
$=\sqrt{10}-\sqrt{6}-4$

52. $\sqrt{5}\left(\sqrt{6}-2\sqrt{5}+\sqrt{10}\right)=\sqrt{30}-2(5)+\sqrt{50}$
$=\sqrt{30}-10+5\sqrt{2}$

53. $\left(\sqrt{11}+2\right)\left(2\sqrt{11}-1\right)=2\sqrt{121}-\sqrt{11}+4\sqrt{11}-2$
$=22+3\sqrt{11}-2$
$=20+3\sqrt{11}$

54. $\left(\sqrt{10}+3\right)\left(3\sqrt{10}-1\right)=3(10)-\sqrt{10}+9\sqrt{10}-3$
$=30+8\sqrt{10}-3$
$=27+8\sqrt{10}$

55. $\left(2+3\sqrt{6}\right)\left(4-2\sqrt{3}\right)=8-4\sqrt{3}+12\sqrt{6}-6\sqrt{18}$
$=8-4\sqrt{3}+12\sqrt{6}-18\sqrt{2}$

56. $\left(5-\sqrt{2}\right)\left(3-\sqrt{12}\right)=\left(5-\sqrt{2}\right)\left(3-2\sqrt{3}\right)$
$=15-10\sqrt{3}-3\sqrt{2}+2\sqrt{6}$

57. $\left(2\sqrt{3}+3\sqrt{6}\right)^2$
$=\left(2\sqrt{3}\right)^2+2\left(2\sqrt{3}\right)\left(3\sqrt{6}\right)+\left(3\sqrt{6}\right)^2$
$=4\sqrt{9}+2(6)\sqrt{18}+9\sqrt{36}$
$=12+36\sqrt{2}+54$
$=66+36\sqrt{2}$

58. $\left(5\sqrt{2}-2\sqrt{6}\right)^2$
$=\left(5\sqrt{2}\right)^2-2\left(5\sqrt{2}\right)\left(2\sqrt{6}\right)+\left(2\sqrt{6}\right)^2$
$=25(2)-20\sqrt{12}+4(6)$
$=50-40\sqrt{3}+24$
$=74-40\sqrt{3}$

59. $\dfrac{1}{\sqrt{3x}}=\dfrac{1}{\sqrt{3x}}\cdot\dfrac{\sqrt{3x}}{\sqrt{3x}}=\dfrac{\sqrt{3x}}{3x}$

60. $\dfrac{2y}{\sqrt{5}}=\dfrac{2y}{\sqrt{5}}\cdot\dfrac{\sqrt{5}}{\sqrt{5}}=\dfrac{2y\sqrt{5}}{5}$

61. $\dfrac{x^2y}{\sqrt{8}}=\dfrac{x^2y}{2\sqrt{2}}=\dfrac{x^2y}{2\sqrt{2}}\cdot\dfrac{\sqrt{2}}{\sqrt{2}}=\dfrac{x^2y\sqrt{2}}{4}$

62. $\dfrac{3ab}{\sqrt{2b}}=\dfrac{3ab}{\sqrt{2b}}\cdot\dfrac{\sqrt{2b}}{\sqrt{2b}}=\dfrac{3ab\sqrt{2b}}{2b}=\dfrac{3a\sqrt{2b}}{2}$

63. $\sqrt{\dfrac{5}{6}}=\dfrac{\sqrt{5}}{\sqrt{6}}=\dfrac{\sqrt{5}}{\sqrt{6}}\cdot\dfrac{\sqrt{6}}{\sqrt{6}}=\dfrac{\sqrt{30}}{\sqrt{36}}=\dfrac{\sqrt{30}}{6}$

64. $\sqrt{\dfrac{6}{11}}=\dfrac{\sqrt{6}}{\sqrt{11}}=\dfrac{\sqrt{6}}{\sqrt{11}}\cdot\dfrac{\sqrt{11}}{\sqrt{11}}=\dfrac{\sqrt{66}}{\sqrt{121}}=\dfrac{\sqrt{66}}{11}$

65. $\dfrac{\sqrt{x^3}}{\sqrt{5x}}=\dfrac{\sqrt{x^3}}{\sqrt{5x}}\cdot\dfrac{\sqrt{5x}}{\sqrt{5x}}=\dfrac{\sqrt{5x^4}}{5x}=\dfrac{x^2\sqrt{5}}{5x}=\dfrac{x\sqrt{5}}{5}$

66. $\dfrac{\sqrt{a^5}}{\sqrt{6a}}=\dfrac{\sqrt{a^5}}{\sqrt{6a}}\cdot\dfrac{\sqrt{6a}}{\sqrt{6a}}=\dfrac{\sqrt{6a^6}}{6a}=\dfrac{a^3\sqrt{6}}{6a}=\dfrac{a^2\sqrt{6}}{6}$

67. $\dfrac{3}{\sqrt{5}+\sqrt{2}}=\dfrac{3}{\sqrt{5}+\sqrt{2}}\cdot\dfrac{\sqrt{5}-\sqrt{2}}{\sqrt{5}-\sqrt{2}}$
$=\dfrac{3\left(\sqrt{5}-\sqrt{2}\right)}{\left(\sqrt{5}\right)^2-\left(\sqrt{2}\right)^2}$
$=\dfrac{3\left(\sqrt{5}-\sqrt{2}\right)}{3}$
$=\sqrt{5}-\sqrt{2}$

68. $\dfrac{6}{\sqrt{6}-\sqrt{3}}=\dfrac{2}{\sqrt{6}-\sqrt{3}}\cdot\dfrac{\sqrt{6}+\sqrt{3}}{\sqrt{6}+\sqrt{3}}$
$=\dfrac{2\left(\sqrt{6}+\sqrt{3}\right)}{6-3}$
$=\dfrac{2\left(\sqrt{6}+\sqrt{3}\right)}{3}$

69. $\frac{1-\sqrt{5}}{2+\sqrt{5}} = \frac{1-\sqrt{5}}{2+\sqrt{5}} \cdot \frac{2-\sqrt{5}}{2-\sqrt{5}}$
$= \frac{2-3\sqrt{5}+5}{2^2-\left(\sqrt{5}\right)^2}$
$= \frac{7-3\sqrt{5}}{-1}$
$= -7+3\sqrt{5}$

70. $\frac{1-\sqrt{3}}{3+\sqrt{3}} = \frac{1-\sqrt{3}}{3+\sqrt{3}} \cdot \frac{3-\sqrt{3}}{3-\sqrt{3}}$
$= \frac{3-4\sqrt{3}+3}{9-3}$
$= \frac{6-4\sqrt{3}}{6}$
$= \frac{2\left(3-2\sqrt{3}\right)}{6}$
$= \frac{3-2\sqrt{3}}{3}$

71. $c^2 = a^2+b^2,\ a=5,\ b=8$
$c^2 = 5^2+8^2$
$c^2 = 25+64$
$c^2 = 89$
$c = \sqrt{89}$

72. $a^2+b^2=c^2,\ c=\sqrt{11},\ b=3$
$a^2+3^2 = \left(\sqrt{11}\right)^2$
$a^2+9=11$
$a^2=2$
$a=\sqrt{2}$

73. $c^2=a^2+b^2,\ c=5,\ a=3.5$
$5^2=(3.5)^2+b^2$
$25=12.25+b^2$
$12.75=b^2$
$\sqrt{12\frac{3}{4}}=b$
$\sqrt{\frac{51}{4}}=b$
$\frac{\sqrt{51}}{2}=b$

74. $a^2+b^2=c^2$
$(2.400)^2+(2.000)^2=c^2$
$5.760+4.000=c^2$
$9.760=c^2$
$3.124=c$

75. $x^2=18^2+24^2$
$x^2=324+576$
$x^2=900$
$x=30$
It is 30 meters from his feet to the top of the pole.

76. $a^2+b^2=c^2$
$a = 143$
$b = 120$
$(143)^2+(120)^2=c^2$
$\sqrt{34849}=c$
$186.7=c$
Kingman is 186.7 miles from Fountain Hills.

77. $a^2+b^2=c^2$ (1)
$a=b$ (2)
$a^2+a^2=c^2$
$2a^2=c^2$
$a=\sqrt{\frac{c^2}{2}}$
$a=\sqrt{\frac{7200}{2}}$
$a=60$
Each side of the square is 60 ft.

78. $\sqrt{x-4}=5$
$\left(\sqrt{x-4}\right)^2=5^2$
$x-4=25$
$x=29$
Check: $\sqrt{29-4} \stackrel{?}{=} 5$
$\sqrt{25} \stackrel{?}{=} 5$
$5=5$

79.
$$\sqrt{5x-1}=8$$
$$\left(\sqrt{5x-1}\right)^2=8^2$$
$$5x-1=64$$
$$5x=65$$
$$x=13$$
Check: $\sqrt{5\cdot 13-1}\stackrel{?}{=}8$
$$\sqrt{64}\stackrel{?}{=}8$$
$$8=8$$

80.
$$\sqrt{1-5x}=\sqrt{9-x}$$
$$\left(\sqrt{1-5x}\right)^2=\left(\sqrt{9-x}\right)^2$$
$$1-5x=9-x$$
$$-4x=8$$
$$x=-2$$
Check: $\sqrt{1-5(-2)}\stackrel{?}{=}\sqrt{9-(-2)}$
$$\sqrt{11}=\sqrt{11}$$

81.
$$\sqrt{-5+2x}=\sqrt{1+x}$$
$$\left(\sqrt{-5+2x}\right)^2=\left(\sqrt{1+x}\right)^2$$
$$-5+2x=1+x$$
$$x=6$$
Check: $\sqrt{-5+2(6)}\stackrel{?}{=}\sqrt{1+6}$
$$\sqrt{-5+12}\stackrel{?}{=}\sqrt{7}$$
$$\sqrt{7}=\sqrt{7}$$

82.
$$\sqrt{10x+9}=-1+2x$$
$$\left(\sqrt{10x+9}\right)^2=(-1+2x)^2$$
$$10x+9=1-4x+4x^2$$
$$0=4x^2-14x-8$$
$$0=2(2x^2-7x-4)$$
$$0=2(2x+1)(x-4)$$
$2x+1=0$ $\quad$ $x-4=0$

$x=-\frac{1}{2}$ $\quad$ $x=4$

Check: $x=-\frac{1}{2}$
$$\sqrt{10\left(-\frac{1}{2}\right)+9}\stackrel{?}{=}-1+2\left(-\frac{1}{2}\right)$$
$$\sqrt{4}\stackrel{?}{=}-2$$
$$2\neq -2$$

Check: $x=4$
$$\sqrt{10(4)+9}\stackrel{?}{=}-1+2(4)$$
$$\sqrt{49}\stackrel{?}{=}7$$
$$7=7$$
$x=4$ only

83.
$$\sqrt{2x-5}=10-x$$
$$\left(\sqrt{2x-5}\right)^2=(10-x)^2$$
$$2x-5=100-20x+x^2$$
$$0=x^2-22x+105$$
$$0=(x-7)(x-15)$$
$x-7=0$ $\quad$ $x-15=0$

$x=7$ $\quad$ $x=15$

Check: $x=7$
$$\sqrt{2(7)-5}\stackrel{?}{=}10-7$$
$$\sqrt{9}\stackrel{?}{=}3$$
$$3=3$$
Check: $x=15$
$$\sqrt{2(15)-5}\stackrel{?}{=}10-15$$
$$\sqrt{25}\stackrel{?}{=}-5$$
$$5\neq -5$$
$x=7$ only

84.
$$6-\sqrt{5x-1}=x+1$$
$$-\sqrt{5x-1}=x-5$$
$$\left(-\sqrt{5x-1}\right)^2=(x-5)^2$$
$$5x-1=x^2-10x+25$$
$$0=x^2-15x+26$$
$$0=(x-13)(x-2)$$
$x-13=0$ $\quad$ $x-2=0$

$x=13$ $\quad$ $x=2$

Check: $x=13$
$$6-\sqrt{5(13)-1}\stackrel{?}{=}13+1$$
$$6-\sqrt{64}\stackrel{?}{=}14$$
$$6-8\neq 14$$
Check: $x=2$
$$6-\sqrt{5(2)-1}\stackrel{?}{=}2+1$$
$$6-\sqrt{9}\stackrel{?}{=}3$$
$$6-3=3$$
$x=2$ only

85. $4x+\sqrt{x+2}=5x-4$
$\sqrt{x+2}=x-4$
$\left(\sqrt{x+2}\right)^2=(x-4)^2$
$x+2=x^2-8x+16$
$0=x^2-9x+14$
$0=(x-2)(x-7)$
$x-2=0 \qquad x-7=0$
$x=2 \qquad x=7$
Check: $x = 2$
$4(2)+\sqrt{2+2}\stackrel{?}{=}5(2)-4$
$8+\sqrt{4}\stackrel{?}{=}10-4$
$10\neq 6$
Check: $x = 7$
$4(7)+\sqrt{7+2}\stackrel{?}{=}5(7)-4$
$28+\sqrt{9}\stackrel{?}{=}31$
$31=31$
$x = 7$ only

86. $y=kx^2$
When $y = 27$, $x = 3$, so
$k=\frac{y}{x^2}=\frac{27}{3^2}=3$
When $x = 4$, $k = 3$.
$y=kx^2$
$y=3(4)^2$
$y=48$

87. $y=\frac{1}{k\sqrt{x}}$
When $y = 2$, $x = 25$ so
$k=\frac{1}{y\sqrt{x}}=\frac{1}{2\sqrt{25}}=\frac{1}{10}$
When $x = 100$, $k=\frac{1}{10}$.
$y=\frac{1}{k\sqrt{x}}$
$y=\frac{1}{\frac{1}{10}\sqrt{100}}$
$y=1$

88. $y=\frac{k}{x^3}$, $y=4$, $x=2$
$4=\frac{k}{2^3}$
$4=\frac{k}{8}$
$32=k$
$y=\frac{32}{x^3}$, $x=4$
$y=\frac{32}{x^3}=\frac{32}{4^3}=\frac{32}{64}=\frac{1}{2}$

89. $I=\frac{k}{d^2}$, $I=45$, $d=4$
$45=\frac{k}{4^2}\Rightarrow k=720$
$I=\frac{720}{d^2}$, $d=12$
$I=\frac{720}{12^2}=5$
The intensity is 5 lumens.

90. l = length of skid mark, s = speed of car
$l=ks^2$
$40=k(30)^2$
$40=900k$
$\frac{2}{45}=k$
Find l when $s = 55$.
$l=\frac{2}{45}s^2=\frac{2}{45}(55)^2=134$
The skid marks will be approximately 134 feet.

91. H = horsepower, s = speed
$H=ks^3$, $s=s_m$, $H=H_m$
$H_1=k(s_m)^3\Rightarrow k=\frac{H_1}{(s_m)^3}$
$H_1=\frac{H_m}{s_m^3}s^3$, $s=2s_m$
$H=\frac{H_m}{s_m^3}(2s_m)^3=8H_m$
The horsepower is 8 times as much.

92. $\sqrt{\frac{36}{121}}=\frac{6}{11}$ since $\left(\frac{6}{11}\right)^2=\frac{36}{121}$.

93. $\sqrt{0.0004} = 0.02$ since $(0.02)^2 = 0.0004$.

94. $\sqrt{98x^6y^3} = \sqrt{49 \cdot 2 \cdot x^6 \cdot y^2 \cdot y} = 7x^3y\sqrt{2y}$

95. $3\sqrt{27} - 2\sqrt{75} + \sqrt{48} = 3\sqrt{9 \cdot 3} - 2\sqrt{25 \cdot 3} + \sqrt{16 \cdot 3}$
$= 9\sqrt{3} - 10\sqrt{3} + 4\sqrt{3}$
$= 3\sqrt{3}$

96. $\left(\sqrt{3} - 2\sqrt{5}\right)\left(2\sqrt{5} + 3\sqrt{2}\right)$
$= 2\sqrt{15} + 3\sqrt{6} - 4\sqrt{25} - 6\sqrt{10}$
$= 2\sqrt{15} + 3\sqrt{6} - 20 - 6\sqrt{10}$

97. $\left(4\sqrt{3} + 3\right)^2 = \left(4\sqrt{3} + 3\right)\left(4\sqrt{3} + 3\right)$
$= 16(3) + 2 \cdot 12\sqrt{3} + 9$
$= 48 + 24\sqrt{3} + 9$
$= 57 + 24\sqrt{3}$

98. $\frac{5}{\sqrt{12}} = \frac{5}{\sqrt{4 \cdot 3}} = \frac{5}{2\sqrt{3}} = \frac{5}{2\sqrt{3}} \cdot \frac{\sqrt{3}}{\sqrt{3}} = \frac{5\sqrt{3}}{6}$

99. $\frac{\sqrt{3} + \sqrt{6}}{2\sqrt{3} + \sqrt{2}} = \frac{\sqrt{3} + \sqrt{6}}{2\sqrt{3} + \sqrt{2}} \cdot \frac{2\sqrt{3} - \sqrt{2}}{2\sqrt{3} - \sqrt{2}}$
$= \frac{2\sqrt{9} - \sqrt{6} + 2\sqrt{18} - \sqrt{12}}{4\sqrt{9} - \sqrt{4}}$
$= \frac{6 - \sqrt{6} + 2\sqrt{9 \cdot 2} - \sqrt{4 \cdot 3}}{12 - 2}$
$= \frac{6 - \sqrt{6} + 6\sqrt{2} - 2\sqrt{3}}{10}$

100. $\sqrt{2x - 3} = 9$
$\left(\sqrt{2x - 3}\right)^2 = 9^2$
$2x - 3 = 81$
$2x = 84$
$x = 42$

101. $\sqrt{10x + 5} = 2x + 1$
$\left(\sqrt{10x + 5}\right)^2 = (2x + 1)^2$
$10x + 5 = 4x^2 + 4x + 1$
$0 = 4x^2 - 6x - 4$
$0 = 2(2x^2 - 3x - 2)$
$0 = 2(2x + 1)(x - 2)$
$2x + 1 = 0 \qquad x - 2 = 0$
$2x = -1 \qquad x = 2$
$x = -\frac{1}{2}$

How Am I Doing? Chapter 9 Test

1. $\sqrt{121} = 11$ since $11^2 = 121$.

2. $\sqrt{\frac{9}{100}} = \frac{3}{10}$ since $\left(\frac{3}{10}\right)^2 = \frac{9}{100}$.

3. $\sqrt{48x^2y^7} = \sqrt{3 \cdot 16 \cdot x^2 \cdot y^6 \cdot y} = 4x^3\sqrt{3y}$

4. $\sqrt{100x^3yz^4} = \sqrt{100 \cdot x^2 \cdot x \cdot y \cdot (z^2)^2} = 10xz^2\sqrt{xy}$

5. $8\sqrt{3} + 5\sqrt{27} - 5\sqrt{48} = 8\sqrt{3} + 5\sqrt{9 \cdot 3} - 5\sqrt{16 \cdot 3}$
$= 8\sqrt{3} + 15\sqrt{3} - 20\sqrt{3}$
$= 3\sqrt{3}$

6. $\sqrt{4a} + \sqrt{8a} + \sqrt{36a} + \sqrt{18a}$
$= 2\sqrt{a} + 2\sqrt{2a} + 6\sqrt{a} + 3\sqrt{2a}$
$= 8\sqrt{a} + 5\sqrt{2a}$

7. $\left(2\sqrt{a}\right)\left(3\sqrt{b}\right)\left(2\sqrt{ab}\right) = 12\sqrt{a^2b^2} = 12ab$

8. $\sqrt{5}\left(\sqrt{10} + 2\sqrt{3} - 3\sqrt{5}\right) = \sqrt{50} + 2\sqrt{15} - 3\sqrt{25}$
$= \sqrt{25 \cdot 2} + 2\sqrt{15} - 3 \cdot 5$
$= 5\sqrt{2} + 2\sqrt{15} - 15$

9. $\left(2\sqrt{3} + 5\right)^2 = 4\sqrt{9} + 2\left(2\sqrt{3}\right)(5) + 25$
$= 12 + 20\sqrt{3} + 25$
$= 37 + 20\sqrt{3}$

10. $\left(4\sqrt{2}-\sqrt{5}\right)\left(3\sqrt{2}+\sqrt{5}\right)$
$= 12(2)+4\sqrt{10}-3\sqrt{10}-5$
$= 24+\sqrt{10}-5$
$= 19+\sqrt{10}$

11. $\sqrt{\frac{x}{5}} = \frac{\sqrt{x}}{\sqrt{5}} = \frac{\sqrt{x}}{\sqrt{5}} \cdot \frac{\sqrt{5}}{\sqrt{5}} = \frac{\sqrt{5x}}{5}$

12. $\frac{3}{\sqrt{12}} = \frac{3}{2\sqrt{3}} = \frac{3}{2\sqrt{3}} \cdot \frac{\sqrt{3}}{\sqrt{3}} = \frac{3\sqrt{3}}{2(3)} = \frac{\sqrt{3}}{2}$

13. $\frac{\sqrt{3}+4}{5+\sqrt{3}} = \frac{\sqrt{3}+4}{5+\sqrt{3}} \cdot \frac{5-\sqrt{3}}{5-\sqrt{3}}$
$= \frac{5\sqrt{3}-\sqrt{9}+20-4\sqrt{3}}{25-\sqrt{9}}$
$= \frac{\sqrt{3}-3+20}{25-3}$
$= \frac{17+\sqrt{3}}{22}$

14. $\frac{3a}{\sqrt{5}+\sqrt{2}} = \frac{3a}{\sqrt{5}+\sqrt{2}} \cdot \frac{\sqrt{5}-\sqrt{2}}{\sqrt{5}-\sqrt{2}}$
$= \frac{3a\left(\sqrt{5}-\sqrt{2}\right)}{5-2}$
$= \frac{3a\left(\sqrt{5}-\sqrt{2}\right)}{3}$
$= a\left(\sqrt{5}-\sqrt{2}\right)$

15. $\sqrt{156} \approx 12.49$

16. $c^2 = a^2 + b^2$
$13^2 = 6^2 + x^2$
$169 = 36 + x^2$
$133 = x^2$
$\sqrt{133} = x$

17. $x^2 = \left(3\sqrt{2}\right)^2 + 3^2 = 9\sqrt{4}+9 = 27$
$x = \sqrt{27} = \sqrt{9 \cdot 3} = 3\sqrt{3}$

18. $6-\sqrt{2x+1} = 0$
$6 = \sqrt{2x+1}$
$6^2 = \left(\sqrt{2x+1}\right)^2$
$36 = 2x+1$
$35 = 2x$
$\frac{35}{2} = x$

Check: $6-\sqrt{2\left(\frac{35}{2}\right)+1} \stackrel{?}{=} 0$
$6-\sqrt{36} \stackrel{?}{=} 0$
$6-6 = 0$

19. $x = 5+\sqrt{x+7}$
$x-5 = \sqrt{x+7}$
$(x-5)^2 = \left(\sqrt{x+7}\right)^2$
$x^2-10x+25 = x+7$
$x^2-11x+18 = 0$
$(x-2)(x-9) = 0$
$x-2 = 0$ $\quad x-9 = 0$
$x = 2$ $\quad x = 9$

Check: $2 \stackrel{?}{=} 5+\sqrt{2+7}$ $\quad 9 \stackrel{?}{=} 5+\sqrt{9+7}$
$2 \stackrel{?}{=} 5+3$ $\quad 9 \stackrel{?}{=} 5+4$
$2 \neq 8$ $\quad 9 = 9$

2 is an extraneous solution.
$x = 9$

20. i = illumination
d = distance
$i = \frac{k}{d^2}$
$12 = \frac{k}{8^2}$
$k = 768$
$i = \frac{768}{d^2}$, $i = 3$
$3 = \frac{768}{d^2}$
$d^2 = 256$
$d = 16$
The distance from the light source is 16 inches.

21. $c = ks$, $c = 23.40$, $s = 780$
$23.40 = k(780)$
$0.03 = k$
$c = 0.03s$, $s = 2859$
$c = 0.03(2859)$
$c = 85.77$
He will earn \$85.77.

22. A = area, p = perimeter
$A = kp^2$
$6.93 = k(12)^2$
$6.93 = 144k$
$0.048125 = k$
$A = 0.048125p^2 = 0.048125(21)^2 \approx 21.2$
The area is 21.2 cm^2.

Cumulative Test Chapters 0–9

1. $\frac{28}{42} = \frac{2\cdot 2\cdot 7}{2\cdot 3\cdot 7} = \frac{2}{3}$

2. $5\frac{1}{3} + 1\frac{3}{4} = \frac{16}{3} + \frac{7}{4}$
$= \frac{16}{3}\cdot\frac{4}{4} + \frac{7}{4}\cdot\frac{3}{3}$
$= \frac{64}{12} + \frac{21}{12}$
$= \frac{85}{12}$
$= 7\frac{1}{12}$

3. $6\frac{1}{4} \div 6\frac{2}{3} = \frac{25}{4} \div \frac{20}{3} = \frac{25}{4}\cdot\frac{3}{20} = \frac{5(5)}{4}\cdot\frac{3}{4(5)} = \frac{15}{16}$

4. $0.07\% = 0.0007$

5. $-11-16+8+4-13+31 = -27+8+4-13+31$
$= -19+4-13+31$
$= -15-13+31$
$= -28+31$
$= 3$

6. $(-3)^2\cdot 4 - 8\div 2 - (3-2)^3 = 9\cdot 4 - 8\div 2 - 1^3$
$= 9\cdot 4 - 8\div 2 - 1$
$= 36-4-1$
$= 31$

7. $4a^2bc^3 - 2abc^2 - 7a^2bc^3 = -3a^2bc^3 - 2abc^2$

8. $(x-1)(x-4)(x+4)$
$= (x^2-4x-x+4)(x+4)$
$= (x^2-5x+4)(x+4)$
$= x^3+4x^2-5x^2-20x+4x+16$
$= x^3-x^2-16x+16$

9. $(4x-5)^2 = (4x)^2 - 2(4x)(5) + 5^2$
$= 16x^2 - 40x + 25$

10. $(3x-11)(3x+11) = (3x)^2 - 11^2 = 9x^2 - 121$

11. $\frac{4x}{x+2} + \frac{4}{x+2} = \frac{4x+4}{x+2}$

12. $\frac{x^2-5x+6}{x^2-5x-6} \div \frac{x^2-4}{x^2+2x+1}$
$= \frac{x^2-5x+6}{x^2-5x-6}\cdot\frac{x^2+2x+1}{x^2-4}$
$= \frac{(x-2)(x-3)}{(x-6)(x+1)}\cdot\frac{(x+1)^2}{(x-2)(x+2)}$
$= \frac{(x-3)(x+1)}{(x-6)(x+2)}$

13. $\sqrt{98x^5y^6} = \sqrt{49\cdot 2\cdot x^4\cdot x\cdot y^6} = 7x^2y^3\sqrt{2x}$

14. $\sqrt{50} - \sqrt{98} + \sqrt{162} = \sqrt{25\cdot 2} - \sqrt{49\cdot 2} + \sqrt{81\cdot 2}$
$= 5\sqrt{2} - 7\sqrt{2} + 9\sqrt{2}$
$= 7\sqrt{2}$

15. $\left(\sqrt{6}-\sqrt{3}\right)^2 = \sqrt{36} + 2\left(-\sqrt{3}\right)\left(\sqrt{6}\right) + \sqrt{9}$
$= 6 - 2\sqrt{18} + 3$
$= 9 - 2\sqrt{9\cdot 2}$
$= 9 - 6\sqrt{2}$

16. $\frac{\sqrt{3}+\sqrt{2}}{\sqrt{3}-\sqrt{2}} = \frac{\sqrt{3}+\sqrt{2}}{\sqrt{3}-\sqrt{2}}\cdot\frac{\sqrt{3}+\sqrt{2}}{\sqrt{3}+\sqrt{2}}$
$= \frac{3+2\sqrt{6}+2}{3-2}$
$= 5 + 2\sqrt{6}$

17. $\left(3\sqrt{6}-\sqrt{2}\right)\left(\sqrt{6}+4\sqrt{2}\right)$
$=3\sqrt{36}+12\sqrt{12}-\sqrt{12}-4\sqrt{4}$
$=18+11\sqrt{12}-8$
$=10+11\sqrt{4\cdot 3}$
$=10+22\sqrt{3}$

18. $\frac{2}{2-\sqrt{5}}=\frac{2}{2-\sqrt{5}}\cdot\frac{2+\sqrt{5}}{2+\sqrt{5}}$
$=\frac{4+2\sqrt{5}}{4-5}$
$=\frac{4+2\sqrt{5}}{-1}$
$=-4-2\sqrt{5}$

19. $-\sqrt{25}=-5$ since $5^2=25$.

20. $\sqrt{-16}$ is not a real number.

21. $c^2=a^2+b^2$, $a=4$, $b=\sqrt{7}$
$c^2=4^2+\left(\sqrt{7}\right)^2=16+7=23$
$c=\sqrt{23}$

22. $a^2+b^2=c^2$
$19^2+b^2=21^2$
$361+b^2=441$
$b^2=80$
$b=4\sqrt{5}$

23. $\sqrt{4x-3}=x$
$\left(\sqrt{4x-3}\right)^2=x^2$
$4x-3=x^2$
$x^2-4x+3=0$
$(x-3)(x-1)=0$
$x-3=0$ $\quad$ $x-1=0$
$x=3$ $\quad$ $x=1$

24. $\sqrt{6y+13}-1=y$
$\sqrt{6y+13}=y+1$
$\left(\sqrt{6y+13}\right)^2=(y+1)^2$
$\left(\sqrt{6y+13}\right)^2=y^2+2y+1$
$y^2-4y-12=0$
$(y-6)(y+2)=0$
$y-6=0$ $\quad$ $y+2=0$
$y=6$ $\quad$ $y=-2$
Check:
$\sqrt{6\cdot 6+13}-1\stackrel{?}{=}6$ $\quad$ $\sqrt{6(-2)+13}-1\stackrel{?}{=}-2$
$\sqrt{49}-1\stackrel{?}{=}6$ $\quad$ $\sqrt{1}-1\stackrel{?}{=}-2$
$7-1\stackrel{?}{=}6$ $\quad$ $0\neq -2$
$6=6$
$y=6$ only

25. $y=\frac{k}{x}$, $y=4$, $x=5$
$4=\frac{k}{5}\Rightarrow k=20$
$y=\frac{20}{x}$, $x=2$
$y=\frac{20}{2}=10$

26. A = surface area, r = radius
$A=kr^2$
$1256=k(2)^2$
$1256=4k$
$314=k$
$A=314r^2=314(0.5)^2=314(0.25)=78.5$
The surface area is 78.5 m^2.

Chapter 10

10.1 Exercises

1. $x^2 + 8x + 7 = 0$
$a = 1, b = 8, c = 7$

3. $8x^2 - 11x = 0$
$a = 8, b = -11, c = 0$

5. $x^2 + 15x - 7 = 12x + 8$
$x^2 + 3x - 15 = 0$
$a = 1, b = 3, c = -15$

7. $27x^2 - 9x = 0$
$9x(3x - 1) = 0$
$9x = 0 \quad 3x - 1 = 0$
$x = 0 \quad x = \frac{1}{3}$

9. $4x^2 - x = -9x$
$4x^2 + 8x = 0$
$4x(x + 2) = 0$
$4x = 0 \quad x + 2 = 0$
$x = 0 \quad x = -2$

11. $11x^2 - 13x = 8x - 3x^2$
$14x^2 - 21x = 0$
$7x(2x - 3) = 0$
$7x = 0 \quad 2x - 3 = 0$
$x = 0 \quad x = \frac{3}{2}$

13. $x^2 - 3x - 28 = 0$
$(x + 4)(x - 7) = 0$
$x + 4 = 0 \quad x - 7 = 0$
$x = -4 \quad x = 7$

15. $2x^2 + 11x - 6 = 0$
$(2x - 1)(x + 6) = 0$
$2x - 1 = 0 \quad x + 6 = 0$
$2x = 1 \quad x = -6$
$x = \frac{1}{2}$

17. $3x^2 - 16x + 5 = 0$
$(3x - 1)(x - 5) = 0$
$3x - 1 = 0 \quad x - 5 = 0$
$3x = 1 \quad x = 5$
$x = \frac{1}{3}$

19. $x^2 = 5x + 14$
$x^2 - 5x - 14 = 0$
$(x - 7)(x + 2) = 0$
$x - 7 = 0 \quad x + 2 = 0$
$x = 7 \quad x = -2$

21. $12x^2 + 17x + 6 = 0$
$(4x + 3)(3x + 2) = 0$
$4x + 3 = 0 \quad 3x + 2 = 0$
$4x = -3 \quad 3x = -2$
$x = -\frac{3}{4} \quad x = -\frac{2}{3}$

23. $a^2 + 9a + 39 = 7 - 3a$
$a^2 + 12a + 32 = 0$
$(a + 8)(a + 4) = 0$
$a + 8 = 0 \quad a + 4 = 0$
$a = -8 \quad a = -4$

25. $6y^2 = -7y + 3$
$6y^2 + 7y - 3 = 0$
$(3y - 1)(2y + 3) = 0$
$3y - 1 = 0 \quad 2y + 3 = 0$
$3y = 1 \quad 2y = -3$
$y = \frac{1}{3} \quad y = -\frac{3}{2}$

27. $25x^2 - 60x + 36 = 0$
$(5x - 6)(5x - 6) = 0$
$5x - 6 = 0$
$5x = 6$
$x = \frac{6}{5}$

29. $(x-5)(x+2)=-10$
$x^2-3x-10=-10$
$x^2-3x=0$
$x(x-3)=0$
$x=0 \qquad x-3=0$
$\qquad x=3$

31. $3x^2+8x-10=-6x-10$
$3x^2+14x=0$
$x(3x+14)=0$
$x=0 \qquad 3x+14=0$
$3x=-14$
$x=-\frac{14}{3}$

33. $y(y-7)=2(5-2y)$
$y^2-7y=10-4y$
$y^2-3y=10$
$y^2-3y-10=0$
$(y+2)(y-5)=0$
$y+2=0 \qquad y-5=0$
$y=-2 \qquad y=5$

35. $(x-6)(x-4)=3$
$x^2-10x+24=3$
$x^2-10x+21=0$
$(x-3)(x-7)=0$
$x-3=0 \qquad x-7=0$
$x=3 \qquad x=7$

37. $2x(x+3)=(3x+1)(x+1)$
$2x^2+6x=3x^2+4x+1$
$0=x^2-2x+1$
$0=(x-1)(x-1)$
$0=x-1$
$1=x$

39. $x+\frac{8}{x}=6$
LCD = x
$x^2+8=6x$
$x^2-6x+8=0$
$(x-2)(x-4)=0$
$x-2=0 \qquad x-4=0$
$x=2 \qquad x=4$
Check: $2+\frac{8}{2}\stackrel{?}{=}6 \qquad 4+\frac{8}{4}\stackrel{?}{=}6$
$6=6 \qquad 6=6$

41. $\frac{x}{7}-\frac{2}{x}=-\frac{5}{7}$
LCD = $7x$
$x^2-14=-5x$
$x^2+5x-14=0$
$(x+7)(x-2)=0$
$x+7=0 \qquad x-2=0$
$x=-7 \qquad x=2$
Check: $\frac{-7}{7}-\frac{2}{-7}\stackrel{?}{=}-\frac{5}{7} \qquad \frac{2}{7}-\frac{2}{2}\stackrel{?}{=}-\frac{5}{7}$
$\frac{-7}{7}+\frac{2}{7}\stackrel{?}{=}-\frac{5}{7} \qquad -\frac{5}{7}=-\frac{5}{7}$
$-\frac{5}{7}=-\frac{5}{7}$

43. $\frac{3y+5}{2}=\frac{-3}{y-2}$
LCD = $2(y-2)$
$(3y+5)(y-2)=-3(2)$
$3y^2-6y+5y-10=-6$
$3y^2-y-4=0$
$(3y-4)(y+1)=0$
$3y-4=0 \qquad y+1=0$
$3y=4 \qquad y=-1$
$y=\frac{4}{3}$
Check: $\frac{3\left(\frac{4}{3}\right)+5}{2}\stackrel{?}{=}\frac{-3}{\frac{4}{3}-2} \qquad \frac{3(-1)+5}{2}\stackrel{?}{=}\frac{-3}{-1-2}$
$\frac{4+5}{2}\stackrel{?}{=}\frac{-3}{-\frac{2}{3}} \qquad 1=1$
$\frac{9}{2}=\frac{9}{2}$

45. $\frac{4}{x}+\frac{3}{x+5}=2$
LCD = $x(x+5)$
$(x+5)(4)+3x=2x(x+5)$
$4x+20+3x=2x^2+10x$
$2x^2+3x-20=0$
$(2x-5)(x+4)=0$

$2x-5=0$ $\quad$ $x+4=0$

$x=\frac{5}{2}$ $\quad$ $x=-4$

Check: $\frac{4}{\frac{5}{2}}+\frac{3}{\frac{5}{2}+5}\stackrel{?}{=}2$ $\quad$ $\frac{4}{-4}+\frac{3}{-4+5}\stackrel{?}{=}2$

$\frac{8}{5}+\frac{6}{15}\stackrel{?}{=}2$ $\quad$ $-1+3\stackrel{?}{=}2$

$2=2$ $\quad$ $2=2$

47. $\frac{24}{x^2-4}=1+\frac{2x-6}{x-2}$

LCD = $(x+2)(x-2)$

$24=(x^2-4)(1)+(x+2)(2x-6)$

$24=x^2-4+2x^2-2x-12$

$0=3x^2-2x-40$

$0=(3x+10)(x-4)$

$3x+10=0$ $\quad$ $x-4=0$

$x=-\frac{10}{3}$ $\quad$ $x=4$

Check:

$\frac{24}{\left(-\frac{10}{3}\right)^2-4}\stackrel{?}{=}1+\frac{2\left(-\frac{10}{3}\right)-6}{-\frac{10}{3}-2}$

$\frac{27}{8}=\frac{27}{8}$

$\frac{24}{4^2-4}\stackrel{?}{=}1+\frac{2(4)-6}{4-2}$

$2=2$

49. You can always factor out x.

51. $ax^2-7x+c=0$

$x=-\frac{3}{2}\Rightarrow\frac{9}{4}a+\frac{21}{2}+c=0 \quad (1)$

$x=6\Rightarrow 36a-42+c=0 \quad (2)$

Multiply (1) by −1 and add it to (2).

$\frac{135}{4}a-\frac{105}{2}=0$

$a=\frac{14}{9}$

Substitute $\frac{14}{9}$ for a in (2).

$36\left(\frac{14}{9}\right)-42+c=0$

$c=-14$

$a=\frac{14}{9},\ c=-14$

For Exercises 53–57, use $R=-2.5n^2+80n-280$.

53. $320=-2.5n^2+80n-280$

$0=-2.5n^2+80n-600$

$0=-2.5(n^2-32n+240)$

$0=-2.5(n-12)(n-20)$

$n-12=0$ $\quad$ $n-20=0$

$n=12$ $\quad$ $n=20$

They may produce 1200 or 2000 labels.

55. $\frac{12+20}{2}=16$, average = 1600 labels

$R=-2.5(16)^2+80(16)-280=360$

The revenue for producing 1600 labels is $360, which is more than producing either 1200 or 2000 labels.

57. $R=-2.5(15)^2+80(15)-280=\$357.5$

$R=-2.5(17)^2+80(17)-280=\$357.5$

It appears that the maximum revenue is at 1600 labels.

Cumulative Review

59. $\frac{\frac{1}{x}+\frac{3}{x-2}}{\frac{4}{x^2-4}}=\frac{\frac{1}{x}+\frac{3}{x-2}}{\frac{4}{(x+2)(x-2)}}\cdot\frac{x(x+2)(x-2)}{x(x+2)(x-2)}$

$=\frac{(x+2)(x-2)+3x(x+2)}{4x}$

$=\frac{x^2-4+3x^2+6x}{4x}$

$=\frac{4x^2+6x-4}{4x}$

$=\frac{2(2x^2+3x-2)}{2(2x)}$

$=\frac{2x^2-3x-2}{2x}$

$=\frac{(2x-1)(x+2)}{2x}$

60. $\frac{2x}{3x-5}+\frac{2}{3x^2-11x+10}$
$=\frac{2x}{3x-5}\cdot\frac{x-2}{x-2}+\frac{2}{(3x-5)(x-2)}$
$=\frac{2x^2-4x+2}{(3x-5)(x-2)}$
$=\frac{2x^2-4x+2}{3x^2-11+10}$

Quick Quiz 10.1

1. $6x^2-x=15$
$6x^2-x-15=0$
$(3x-5)(2x+3)=0$
$3x-5=0$ $\quad$ $2x+3=0$
$3x=5$ $\quad$ $2x=-3$
$x=\frac{5}{3}$ $\quad$ $x=-\frac{3}{2}$

2. $x^2=2(7x-24)$
$x^2=14x-48$
$x^2-14x+48=0$
$(x-8)(x-6)=0$
$x-8=0$ $\quad$ $x-6=0$
$x=8$ $\quad$ $x=6$

3. $x^2=2-\left(\frac{7}{3}\right)x$
LCD = 3
$3x^2=6-7x$
$3x^2+7x-6=0$
$(3x-2)(x+3)=0$
$3x-2=0$ $\quad$ $x+3=0$
$3x=2$ $\quad$ $x=-3$
$x=\frac{2}{3}$

4. Answers may vary. Possible solution: Multiply each term by the LCD. Simplify and move all terms to the left side of the equation.

10.2 Exercises

1. $x^2=64$
$x=\pm\sqrt{64}$
$x=\pm 8$

3. $x^2=98$
$x=\pm\sqrt{98}$
$x=\pm 7\sqrt{2}$

5. $x^2-28=0$
$x^2=28$
$x=\pm\sqrt{28}$
$x=\pm 2\sqrt{7}$

7. $5x^2=45$
$x^2=9$
$x=\pm\sqrt{9}$
$x=\pm 3$

9. $6x^2=120$
$x^2=20$
$x=\pm\sqrt{20}$
$x=\pm 2\sqrt{5}$

11. $3x^2-375=0$
$3x^2=375$
$x^2=125$
$x=\pm\sqrt{125}$
$x=\pm 5\sqrt{5}$

13. $5x^2+13=73$
$5x^2=60$
$x^2=12$
$x=\pm\sqrt{12}$
$x=\pm 2\sqrt{3}$

15. $13x^2+17=82$
$13x^2=65$
$x^2=5$
$x=\pm\sqrt{5}$

17. $(x-7)^2=16$
$x-7=\pm 4$
$x=7\pm 4$
$x=7+4=11$
$x=7-4=3$

19. $(x+4)^2 = 6$
$x+4 = \pm\sqrt{6}$
$x = -4 \pm \sqrt{6}$

21. $(2x+5)^2 = 2$
$2x+5 = \pm\sqrt{2}$
$2x = -5 \pm \sqrt{2}$
$x = \dfrac{-5 \pm \sqrt{2}}{2}$

23. $(3x-1)^2 = 7$
$3x-1 = \pm\sqrt{7}$
$3x = 1 \pm \sqrt{7}$
$x = \dfrac{1 \pm \sqrt{7}}{3}$

25. $(5x+1)^2 = 18$
$5x+1 = \pm\sqrt{18}$
$5x+1 = \pm 3\sqrt{2}$
$5x = -1 \pm 3\sqrt{2}$
$x = \dfrac{-1 \pm 3\sqrt{2}}{5}$

27. $(4x-5)^2 = 54$
$4x-5 = \pm\sqrt{54}$
$4x = 5 \pm 3\sqrt{6}$
$x = \dfrac{5 \pm 3\sqrt{6}}{4}$

29. $x^2 - 6x = 11, \left[\frac{1}{2}(6)\right]^2 = 3^2 = 9$
$x^2 - 6x + 9 = 11 + 9$
$(x-3)^2 = 20$
$x-3 = \pm\sqrt{20}$
$x = 3 \pm 2\sqrt{5}$

31. $x^2 + 8x - 20 = 0$
$x^2 + 8x = 20$
$x^2 + 8x + 16 = 20 + 16$
$(x+4)^2 = 36$
$x+4 = \pm\sqrt{36}$
$x+4 = \pm 6$
$x = -4 \pm 6$
$x = -4 + 6 = 2$
$x = -4 - 6 = -10$

33. $x^2 - 12x - 5 = 0$
$x^2 - 12x = 5, \left[\frac{1}{2}(-12)\right]^2 = 6^2 = 36$
$x^2 - 12x + 36 = 5 + 36$
$(x-6)^2 = 41$
$x-6 = \pm\sqrt{41}$
$x = 6 \pm \sqrt{41}$

35. $x^2 - 7x = 0$
$x^2 - 7x + \dfrac{49}{4} = \dfrac{49}{4}$
$\left(x - \dfrac{7}{2}\right)^2 = \dfrac{49}{4}$
$x - \dfrac{7}{2} = \pm\sqrt{\dfrac{49}{4}}$
$x - \dfrac{7}{2} = \pm\dfrac{7}{2}$
$x = \dfrac{7}{2} \pm \dfrac{7}{2}$

$x = \dfrac{7}{2} + \dfrac{7}{2} = 7$
$x = \dfrac{7}{2} - \dfrac{7}{2} = 0$

37. $5x^2 - 25x = 0$

$x^2 - 5x = 0$

$x^2 - 5x + \frac{25}{4} = \frac{25}{4}$

$\left(x - \frac{5}{2}\right)^2 = \frac{25}{4}$

$x - \frac{5}{2} = \pm\sqrt{\frac{25}{4}}$

$x - \frac{5}{2} = \pm\frac{5}{2}$

$x = \frac{5}{2} \pm \frac{5}{2}$

$x = \frac{5}{2} + \frac{5}{2} = 5$

$x = \frac{5}{2} - \frac{5}{2} = 0$

39. $2x^2 - 7x = 9$

$x^2 - \frac{7}{2}x = \frac{9}{2}, \left[\frac{1}{2}\left(-\frac{7}{2}\right)\right]^2 = \frac{49}{16}$

$x^2 - \frac{7}{2}x + \frac{49}{16} = \frac{9}{2} + \frac{49}{16}$

$\left(x - \frac{7}{4}\right)^2 = \frac{121}{16}$

$x - \frac{7}{4} = \pm\sqrt{\frac{121}{16}}$

$x = \frac{7}{4} \pm \frac{11}{4}$

$x = \frac{7}{4} - \frac{11}{4}$ $\quad$ $x = \frac{7}{4} + \frac{11}{4}$

$x = -\frac{4}{4} = -1$ $\quad$ $x = \frac{18}{4} = \frac{9}{2}$

Cumulative Review

41. $3a - 5b = 8$ (1)

$5a - 7b = 8$ (2)

Multiply (1) by 5 and (2) by −3.

$15a - 25b = 40$

$-15a + 21b = -24$

$-4b = 16$

$b = -4$

Substitute −4 for b in (1).

$3a - 5(-4) = 8$

$3a + 20 = 8$

$3a = -12$

$a = -4$

$a = -4, b = -4$

42. $\frac{x}{x+3} - \frac{2}{x} = \frac{x-2}{x^2+3x}$

$x(x+3)\frac{x}{x+3} - x(x+3)\frac{2}{x} = x(x+3)\frac{x-2}{x(x+3)}$

$x^2 - 2(x+3) = x - 2$

$x^2 - 2x - 6 = x - 2$

$x^2 - 3x - 4 = 0$

$(x-4)(x+1) = 0$

$x - 4 = 0$ $\quad$ $x + 1 = 0$

$x = 4$ $\quad$ $x = -1$

43. $P = \frac{F}{A}, A = LW$

$F = 16, L = 8, W = 0.01$

$P = \frac{16}{8(0.01)} = \frac{16}{0.08} = 200$

The pressure is 200 pounds per square inch.

44. Total defective = 13 + 9 − 6 = 16

Six had both kinds of defects, seven had only defects in workmanship, and three had only defects in materials.

Quick Quiz 10.2

1. $4x^2 + 13 = 37$

$4x^2 = 24$

$x^2 = 6$

$x = \pm\sqrt{6}$

2. $(3x-2)^2 = 40$

$3x - 2 = \pm 2\sqrt{10}$

$3x = 2 \pm 2\sqrt{10}$

$x = \frac{2 \pm 2\sqrt{10}}{3}$

3. $x^2 + 8x - 10 = 0$

$$x^2 + 8x = 10$$
$$x^2 + 8x + 16 = 10 + 16$$
$$(x+4)^2 = 26$$
$$x + 4 = \pm\sqrt{26}$$
$$x = -4 \pm \sqrt{26}$$

4. Answers may vary. Possible solution:
(1) Subtract 2 from both sides of the equation.
(2) Add the square of quantity ten divided by two to both sides of the equation: $\left(\frac{10}{2}\right)^2 = 25$
(3) Combine like terms on both sides of the equation.

10.3 Exercises

Use $x = \frac{-b \pm \sqrt{b^2 - 4ac}}{2a}$ in Exercises 1–33.

1. $3x^2 + 4x - 7 = 0$
$a = 3, b = 4, c = -7$
$\sqrt{b^2 - 4ac} = \sqrt{16 + 84} = \sqrt{100} = 10$
There are two rational roots.

3. $4x^2 = -5x + 6$, no
$4x^2 + 5x - 6 = 0$
$a = 4, b = 5, c = -6$

5. $x^2 + 3x - 10 = 0$
$a = 1, b = 3, c = -10$

$$x = \frac{-3 \pm \sqrt{(3)^2 - 4(1)(-10)}}{2(1)}$$
$$= \frac{-3 \pm \sqrt{49}}{2}$$
$$= \frac{-3 \pm 7}{2}$$
$$x = \frac{-3+7}{2} = 2$$
$$x = \frac{-3-7}{2} = -5$$

7. $x^2 - 3x - 8 = 0$
$a = 1, b = -3, c = -8$

$$x = \frac{-(-3) \pm \sqrt{(-3)^2 - 4(1)(-8)}}{2(1)}$$
$$= \frac{3 \pm \sqrt{9 + 32}}{2}$$
$$= \frac{3 \pm \sqrt{41}}{2}$$

9. $4x^2 + 7x - 2 = 0$
$a = 4, b = 7, c = -2$

$$x = \frac{-7 \pm \sqrt{(7)^2 - 4(4)(-2)}}{2(4)} = \frac{-7 \pm \sqrt{81}}{8} = \frac{-7 \pm 9}{8}$$
$$x = \frac{-7-9}{8} = -2$$
$$x = \frac{-7+9}{8} = \frac{1}{4}$$

11. $2x^2 = 3x + 20$
$2x^2 - 3x - 20 = 0$
$a = 2, b = -3, c = -20$

$$x = \frac{-(-3) \pm \sqrt{(-3)^2 - 4(2)(-20)}}{2(2)}$$
$$x = \frac{3 \pm \sqrt{169}}{4}$$
$$x = \frac{3 \pm 13}{4}$$
$$x = \frac{3+13}{4} = 4$$
$$x = \frac{3-13}{4} = -\frac{5}{2}$$

13. $6x^2 - 3x = 1$
$6x^2 - 3x - 1 = 0$
$a = 6, b = -3, c = -1$

$$x = \frac{-(-3) \pm \sqrt{(-3)^2 - 4(6)(-1)}}{2(6)}$$
$$= \frac{3 \pm \sqrt{9 + 24}}{12}$$
$$= \frac{3 \pm \sqrt{33}}{12}$$

15. $x + \frac{3}{2} = 3x^2$

$3x^2 - x - \frac{3}{2} = 0$

$6x^2 - 2x - 3 = 0$

$a = 6, b = -2, c = -3$

$$x = \frac{-(-2) \pm \sqrt{(-2)^2 - 4(6)(-3)}}{2(6)}$$
$$= \frac{2 \pm \sqrt{76}}{12}$$
$$= \frac{2 \pm 2\sqrt{19}}{12}$$
$$= \frac{1 \pm \sqrt{19}}{6}$$

17. $\frac{x}{2} + \frac{5}{x} = \frac{7}{2}$

$x^2 + 10 = 7x$

$x^2 - 7x + 10 = 0$

$a = 1, b = -7, c = 10$

$$x = \frac{-(-7) \pm \sqrt{(-7)^2 - 4(1)(10)}}{2(1)}$$
$$= \frac{7 \pm \sqrt{49 - 40}}{2}$$
$$= \frac{7 \pm \sqrt{9}}{2}$$
$$= \frac{7 \pm 3}{2}$$
$$x = \frac{7+3}{2} = 5$$
$$x = \frac{7-3}{2} = 2$$

19. $5x^2 + 6x + 2 = 0$

$a = 5, b = 6, c = 2$

$$x = \frac{-(6) \pm \sqrt{(6)^2 - 4(5)(2)}}{2(5)} = \frac{-6 \pm \sqrt{-4}}{10}$$

No real solution

21. $4y^2 = 10y - 3$

$4y^2 - 10y + 3 = 0$

$a = 4, b = -10, c = 3$

$$y = \frac{-(-10) \pm \sqrt{(-10)^2 - 4(4)(3)}}{2(4)}$$
$$y = \frac{10 \pm \sqrt{52}}{8}$$
$$y = \frac{10 \pm 2\sqrt{13}}{8}$$
$$y = \frac{5 \pm \sqrt{13}}{4}$$

23. $\frac{d^2}{2} + \frac{5d}{6} - 2 = 0$

$3d^2 + 5d - 12 = 0$

$a = 3, b = 5, c = -12$

$$d = \frac{-5 \pm \sqrt{5^2 - 4(3)(-12)}}{2(3)}$$
$$= \frac{-5 \pm \sqrt{169}}{6}$$
$$= \frac{-5 \pm 13}{6}$$
$$d = \frac{-5-13}{6} = -3$$
$$d = \frac{-5+13}{6} = \frac{4}{3}$$

25. $4x^2 + 8x + 11 = 0$

$a = 4, b = 8, c = 11$

$$x = \frac{-8 \pm \sqrt{(8)^2 - 4(4)(11)}}{2(4)}$$
$$x = \frac{-8 \pm \sqrt{-24}}{8}$$

$\sqrt{-24}$ is not real. There is no real solution.

27. $x^2 + 12x + 36 = 0$

$a = 1, b = 12, c = 36$

$$x = \frac{-12 \pm \sqrt{(12)^2 - 4(1)(36)}}{2(1)}$$
$$x = \frac{-12 \pm \sqrt{0}}{2}$$
$$x = \frac{-12}{2}$$
$$x = -6$$

29. $x^2 + 5x - 2 = 0$
$a = 1, b = 5, c = -2$
$$x = \frac{-(5) \pm \sqrt{(5)^2 - 4(1)(-2)}}{2(1)} = \frac{-5 \pm \sqrt{33}}{2}$$
$x = 0.372, -5.372$

31. $2x^2 - 7x - 5 = 0$
$a = 2, b = -7, c = -5$
$$x = \frac{-(-7) \pm \sqrt{(-7)^2 - 4(2)(-5)}}{2(2)}$$
$$x = \frac{7 \pm \sqrt{89}}{4}$$
$$x = \frac{7 + \sqrt{89}}{4} \approx 4.108$$
$$x = \frac{7 - \sqrt{89}}{4} \approx -0.608$$

33. $5x^2 + 10x + 1 = 0$
$a = 5, b = 10, c = 1$
$$x = \frac{-(10) \pm \sqrt{(10)^2 - 4(5)(1)}}{2(5)}$$
$$= \frac{-10 \pm \sqrt{80}}{10}$$
$$= \frac{-10 \pm 4\sqrt{5}}{10}$$
$$= \frac{-5 \pm 2\sqrt{5}}{5}$$
$x = -0.106, -1.894$

35. $6x^2 - 13x + 6 = 0$
$(3x - 2)(2x - 3) = 0$
$3x - 2 = 0$ $\quad$ $2x - 3 = 0$
$3x = 2$ $\quad$ $2x = 3$
$x = \frac{2}{3}$ $\quad$ $x = \frac{3}{2}$

37. $3(x^2 + 1) = 10x$
$3x^2 + 3 = 10x$
$3x^2 - 10x + 3 = 0$
$(3x - 1)(x - 3) = 0$
$3x - 1 = 0$ $\quad$ $x - 3 = 0$
$3x = 1$ $\quad$ $x = 3$
$x = \frac{1}{3}$

39. $(t + 5)(t - 3) = 7$
$t^2 + 2t - 15 = 7$
$t^2 + 2t = 22$
$t^2 + 2t + 1 = 22 + 1$
$(t + 1)^2 = 23$
$t + 1 = \pm\sqrt{23}$
$t = -1 \pm \sqrt{23}$

41. $y^2 - \frac{2}{5}y = 2$
$5y^2 - 2y = 10$
$5y^2 - 2y - 10 = 0$
$a = 5, b = -2, c = -10$
$$y = \frac{-(-2) \pm \sqrt{(-2)^2 - 4(5)(-10)}}{2(5)}$$
$$= \frac{2 \pm \sqrt{204}}{10}$$
$$= \frac{2 \pm 2\sqrt{51}}{10}$$
$$= \frac{1 \pm \sqrt{51}}{5}$$

43. $3x^2 - 13 = 0$
$3x^2 = 13$
$x^2 = \frac{13}{3}$
$$x = \pm\sqrt{\frac{13}{3}} = \pm\frac{\sqrt{13}}{\sqrt{3}} = \pm\frac{\sqrt{39}}{3}$$

45. $x(x - 2) = 7$
$x^2 - 2x = 7$
$x^2 - 2x - 7 = 0$
$a = 1, b = -2, c = -7$
$$x = \frac{-(-2) \pm \sqrt{(-2)^2 - 4(1)(-7)}}{2(1)}$$
$$= \frac{2 \pm \sqrt{32}}{2}$$
$$= \frac{2 \pm 4\sqrt{2}}{2}$$
$$= 1 \pm 2\sqrt{2}$$

47. Total area − Pool area = Tile area

$$(2x+30)(2x+20)-20(30)=216$$
$$4x^2+100x+600-600=216$$
$$4x^2+100x-216=0$$
$$x^2+25x-54=0$$
$$(x+27)(x-2)=0$$
$$x+27=0 \qquad x-2=0$$
$$x=-27 \qquad x=2$$

Width can't be negative. Width is 2 feet.

Cumulative Review

49.
$$\begin{array}{r} x^2+5x+2 \\ x+3\overline{)\,x^3+8x^2+17x+6} \\ \underline{x^3+3x^2} \\ 5x^2+17x \\ \underline{5x^2+15x} \\ 2x+6 \\ \underline{2x+6} \end{array}$$

$$(x^3+8x^2+17x+6)\div(x+3)=x^2+5x+2$$

50.
$$2p=3gf-7p+20$$
$$9p=3gf+20$$
$$9p-20=3gf$$
$$\frac{9p-20}{3f}=g$$

Quick Quiz 10.3

1. $9x^2+6x-1=0$

$a=9,\ b=6,\ c=-1$

$$x=\frac{-6\pm\sqrt{(6)^2-4(9)(-1)}}{2(9)}$$
$$x=\frac{-6\pm\sqrt{72}}{18}$$
$$x=\frac{-6\pm6\sqrt{2}}{18}$$
$$x=\frac{-1\pm\sqrt{2}}{3}$$

2. $4x^2-4x=19$

$4x^2-4x-19=0$

$a=4,\ b=-4,\ c=-19$

$$x=\frac{-(-4)\pm\sqrt{(-4)^2-4(4)(-19)}}{2(4)}$$
$$x=\frac{4\pm\sqrt{320}}{8}$$
$$x=\frac{4\pm8\sqrt{5}}{8}$$
$$x=\frac{1\pm2\sqrt{5}}{2}$$

3.
$$2+x=3x(x+1)$$
$$2+x=3x^2+3x$$
$$3x^2+2x-2=0$$

$a=3,\ b=2,\ c=-2$

$$x=\frac{-2\pm\sqrt{(2)^2-4(3)(-2)}}{2(3)}$$
$$x=\frac{-2\pm\sqrt{28}}{6}$$
$$x=\frac{-2\pm2\sqrt{7}}{6}$$
$$x=\frac{-1\pm\sqrt{7}}{3}$$

4. Answers may vary. Possible solution: Attempt to solve using the quadratic formula. If the resulting radicand is negative there exist no real solutions.

How Am I Doing? Sections 10.1–10.3

1. $x^2-13x-48=0$

$(x-16)(x+3)=0$

$$x-16=0 \qquad x+3=0$$
$$x=16 \qquad x=-3$$

2. $5x^2+7x=14x$

$5x^2-7x=0$

$x(5x-7)=0$

$$5x-7=0 \qquad x=0$$
$$5x=7$$
$$x=\frac{7}{5}$$

3. $5x^2 = 22x - 8$

$5x^2 - 22x + 8 = 0$

$(5x-2)(x-4) = 0$

$5x - 2 = 0 \qquad x - 4 = 0$

$5x = 2 \qquad x = 4$

$x = \frac{2}{5}$

4. $-x + 1 = 6x^2$

$6x^2 + x - 1 = 0$

$(3x-1)(2x+1) = 0$

$3x - 1 = 0 \qquad 2x + 1 = 0$

$3x = 1 \qquad 2x = -1$

$x = \frac{1}{3} \qquad x = -\frac{1}{2}$

5. $x(x+9) = 4(x+6)$

$x^2 + 9x = 4x + 24$

$x^2 + 5x - 24 = 0$

$(x+8)(x-3) = 0$

$x + 8 = 0 \qquad x - 3 = 0$

$x = -8 \qquad x = 3$

6. $\frac{5}{x+2} = \frac{2x-1}{5}$

$25 = (x+2)(2x-1)$

$25 = 2x^2 + 3x - 2$

$0 = 2x^2 + 3x - 27$

$0 = (2x+9)(x-3)$

$2x + 9 = 0 \qquad x - 3 = 0$

$2x = -9 \qquad x = 3$

$x = -\frac{9}{2}$

7. $x^2 - 18 = 0$

$x^2 = 18$

$x = \pm\sqrt{18}$

$x = \pm 3\sqrt{2}$

8. $3x^2 + 1 = 76$

$3x^2 = 75$

$x^2 = 25$

$x = \pm\sqrt{25}$

$x = \pm 5$

9. $x^2 + 6x + 1 = 0$

$a = 1, b = 6, c = 1$

$x = \frac{-6 \pm \sqrt{(6)^2 - 4(1)(1)}}{2(1)}$

$x = \frac{-6 \pm \sqrt{32}}{2}$

$x = \frac{-6 \pm 4\sqrt{2}}{2}$

$x = -3 \pm 2\sqrt{2}$

10. $2x^2 + 3x - 7 = 0$

$2x^2 + 3x = 7$

$x^2 + \frac{3}{2}x = \frac{7}{2}$

$x^2 + \frac{3}{2}x + \frac{9}{16} = \frac{7}{2} + \frac{9}{16}$

$\left(x + \frac{3}{4}\right)^2 = \frac{65}{16}$

$x + \frac{3}{4} = \pm\sqrt{\frac{65}{16}}$

$x = -\frac{3}{4} \pm \frac{\sqrt{65}}{4}$

$x = \frac{-3 \pm \sqrt{65}}{4}$

11. $2x^2 + 4x - 5 = 0$

$a = 2, b = 4, c = -5$

$x = \frac{-4 \pm \sqrt{(4)^2 - 4(2)(-5)}}{2(2)}$

$x = \frac{-4 \pm \sqrt{56}}{4}$

$x = \frac{-4 \pm 2\sqrt{14}}{4}$

$x = \frac{-2 \pm \sqrt{14}}{2}$

12. $4x^2 = x+5$

$4x^2 - x - 5 = 0$

$a = 4, b = -1, c = -5$

$x = \frac{-(-1) \pm \sqrt{(-1)^2 - 4(4)(-5)}}{2(4)}$

$x = \frac{1 \pm \sqrt{81}}{8}$

$x = \frac{1 \pm 9}{8}$

$x = \frac{1-9}{8} = -1$

$x = \frac{1+9}{8} = \frac{5}{4}$

13. $3x^2 + 8x + 1 = 0$

$a = 3, b = 8, c = 1$

$x = \frac{-8 \pm \sqrt{(8)^2 - 4(3)(1)}}{2(3)}$

$x = \frac{-8 \pm \sqrt{52}}{6}$

$x = \frac{-8 \pm 2\sqrt{13}}{6}$

$x = \frac{-4 \pm \sqrt{13}}{3}$

14. $5x^2 + 3 = 4x$

$5x^2 - 4x + 3 = 0$

$a = 5, b = -4, c = 3$

$x = \frac{-(-4) \pm \sqrt{(-4)^2 - 4(5)(3)}}{2(5)}$

$x = \frac{4 \pm \sqrt{-44}}{10}$

$\sqrt{-44}$ is not a real number. There is no real solution.

10.4 Exercises

1. It may have 2, 1, or none.

3. If $b = 0$, then the parabola always has a vertex at $(0, c)$ that is on the y-axis.

5. $y = x^2 + 2$

x	y
−2	6
−1	3
0	2
1	3
2	6

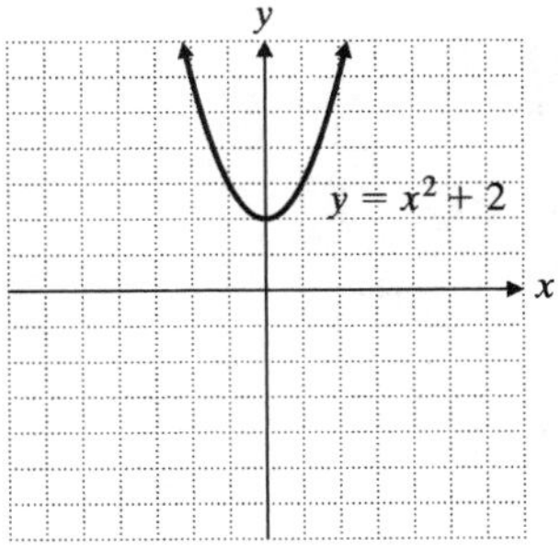

7. $y = -\frac{1}{3}x^2$

x	y
−6	−12
−3	−3
0	0
3	−3
6	−12

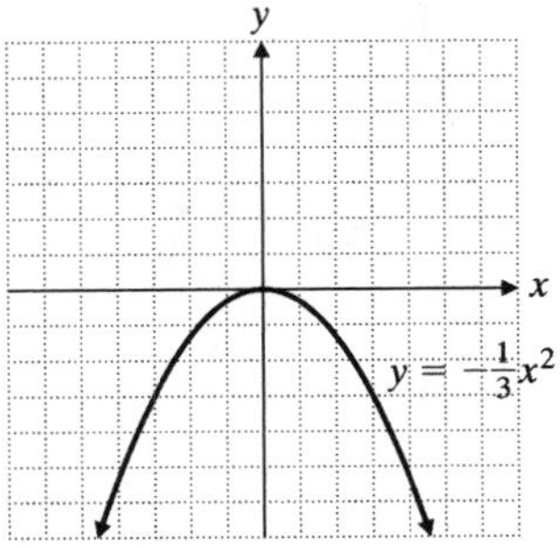

9. $y = 3x^2 - 1$

x	y
–2	11
–1	2
0	–1
1	2
2	11

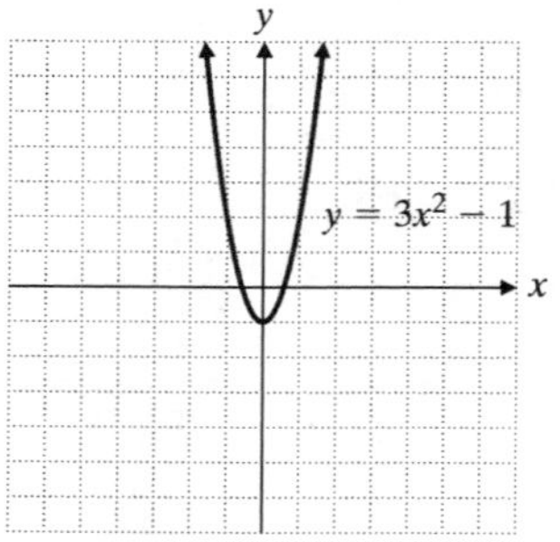

11. $y = (x - 2)^2$

x	y
0	4
1	1
2	0
3	1
4	4

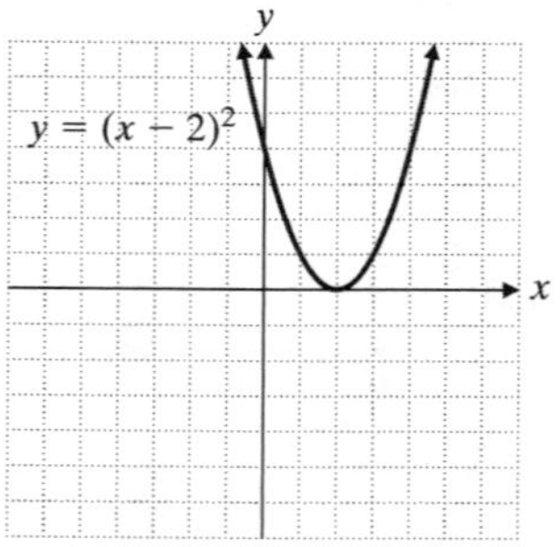

13. $y = -\frac{1}{2}x^2 + 4$

x	y
–4	–4
–2	2
0	4
2	2
4	–4

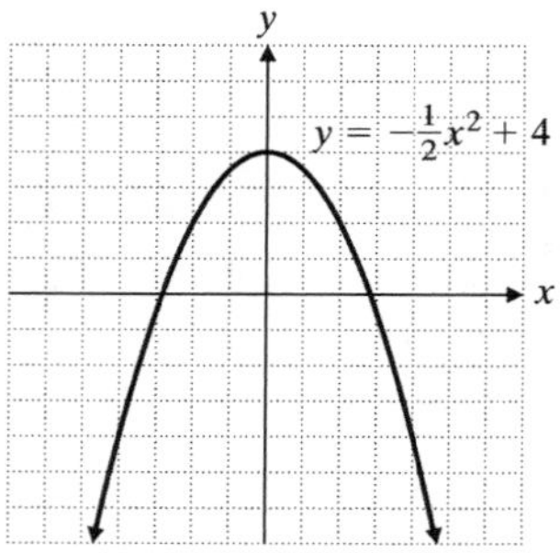

15. $y = \frac{1}{2}(x - 3)^2$

x	y
–1	8
1	2
3	0
5	2
7	8

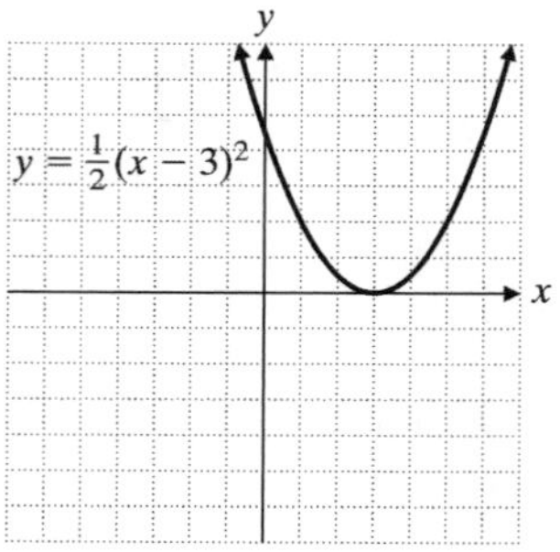

17. $y = x^2 + 4x,\ a = 1, b = 4$

$a > 0$: opens up

Vertex: $x = -\dfrac{b}{2a} = -\dfrac{4}{2(1)} = -2$

$y = (-2)^2 + 4(-2) = -4$

$V(-2, -4)$

x	y
–4	0
–3	–3
–2	–4
–1	–3
0	0

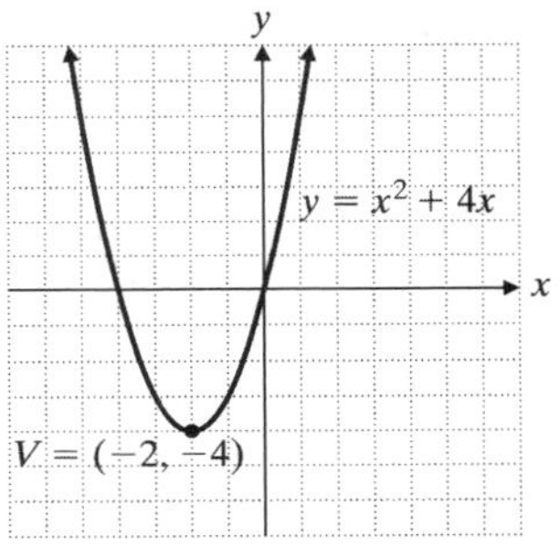

19. $y = -x^2 - 4x,\ a = -1, b = -4$

$a < 0$: opens down

Vertex: $x = -\dfrac{b}{2a} = -\dfrac{(-4)}{2(-1)} = -2$

$y = -(-2)^2 - 4(-2) = 4$

$V(-2, 4)$

x	y
–4	0
–3	3
–2	4
–1	3
0	0

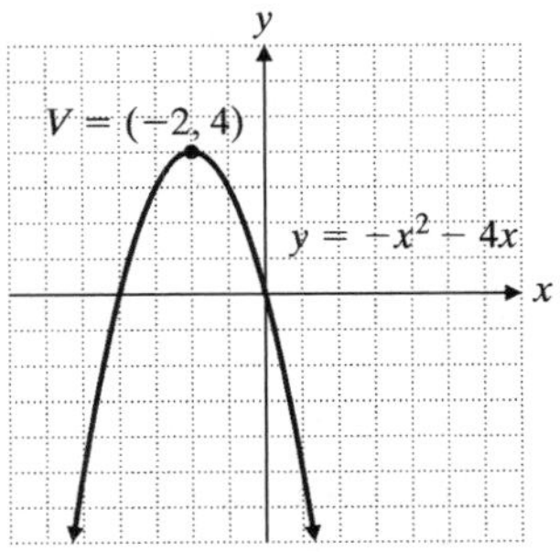

21. $y = -2x^2 + 8x,\ a = -2, b = 8$

$a < 0$: opens down

Vertex: $x = -\dfrac{b}{2a} = -\dfrac{8}{2(-2)} = 2$

$y = -2(2)^2 + 8(2) = 8$

$V(2, 8)$

x	y
–1	–10
0	0
1	8
4	0
5	–10

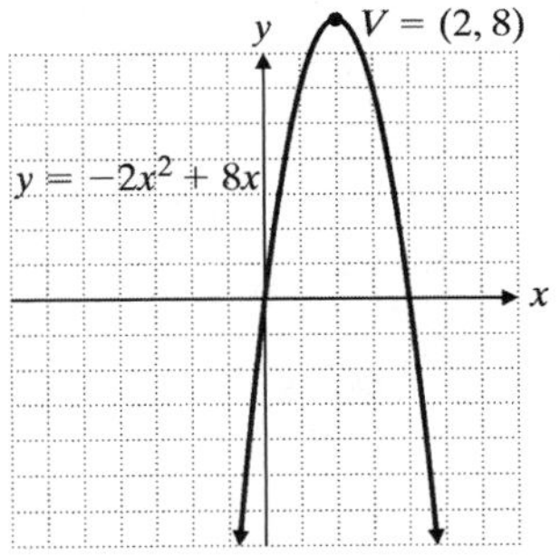

23. $y = x^2 + 2x - 3,\ a = 1, b = 2$

Vertex: $x = -\dfrac{b}{2a} = -\dfrac{2}{2(1)} = -1$

$y = (-1)^2 + 2(-1) - 3 = -4$

$V(-1, -4)$

y-intercept: $y = 0^2 + 2(0) - 3 = -3$

x-intercepts: $0 = x^2 + 2x - 3$

$0 = (x+3)(x-1)$

$x = 1, -3$

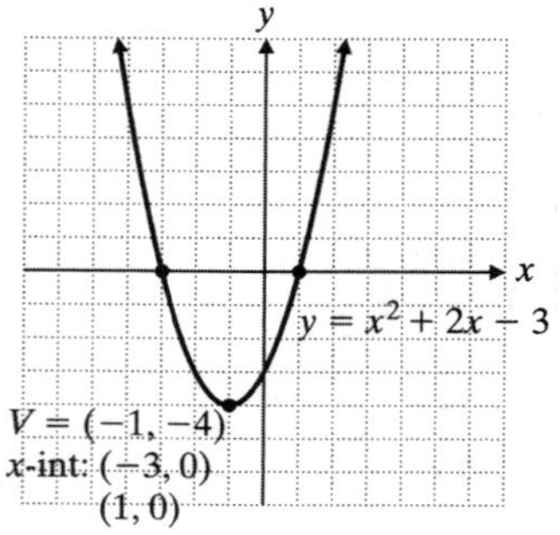

25. $y = -x^2 - 6x - 5,\ a = -1,\ b = -6$

Vertex: $x = -\dfrac{b}{2a} = -\dfrac{-6}{2(-1)} = -3$

$y = -(-3)^2 - 6(-3) - 5 = 4$

$V(-3, 4)$

y-intercept: $y = -0^2 - 6(0) - 5 = -5$

x-intercepts: $0 = -x^2 - 6x - 5$

$0 = -1(x^2 + 6x + 5)$

$0 = -1(x+5)(x+1)$

$x = -5, -1$

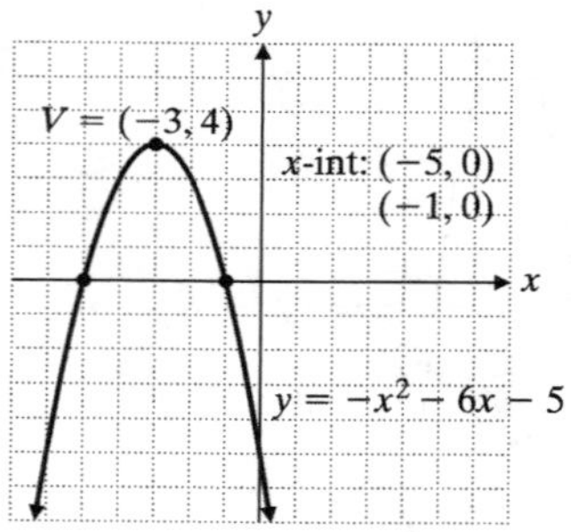

27. $y = x^2 + 6x + 9,\ a = 1,\ b = 6$

Vertex: $x = -\dfrac{b}{2a} = -\dfrac{6}{2(1)} = -3$

$y = (-3)^2 + 6(-3) + 9 = 0$

$V(-3, 0)$

y-intercept: $y = 0^2 + 6(0) + 9 = 9$

x-intercepts: $0 = x^2 + 6x + 9$

$0 = (x+3)^2$

$x = -3$

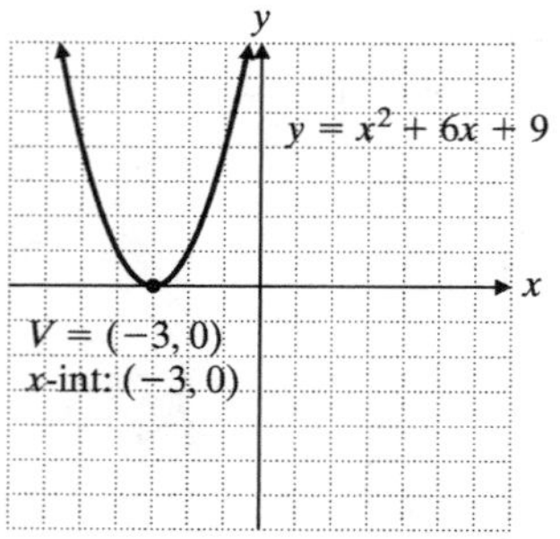

29. $h = -4.9t^2 + 39.2t + 4,\ a = -4.9,\ b = 39.2$

Vertex: $t = -\dfrac{39.2}{2(-4.9)} = 4$

$h = -4.9(4)^2 + 39.2(4) + 4 = 82.4$

$V(4, 82.4)$

h-intercept: $h = 4$

t-intercept: $0 = -4.9t^2 + 39.2t + 4$

$t \approx -0.1, 8.1$

a.

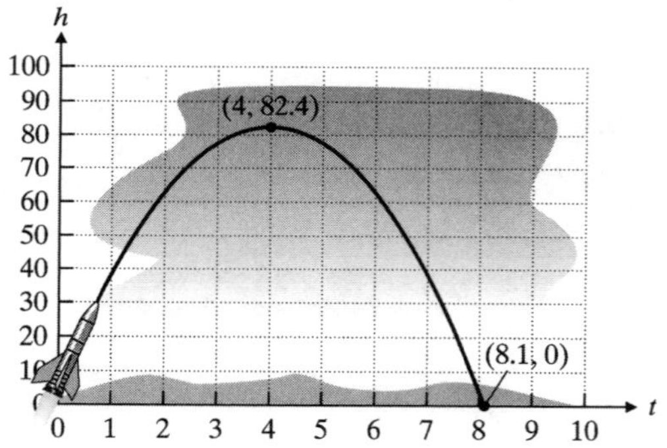

b. $t = 2$

$h = -4.9(2)^2 + 39.2(2) + 4 = 62.8$

It is 62.8 m high.

c. $V(4, 82.4)$

The maximum height is 82.4 m.

d. The x-intercept is 8.1. After 8.1 sec it will strike the earth.

31. $N = 9x - x^2$, $a = -1$, $b = 9$

a. Vertex: $x = -\dfrac{9}{2(-1)} = 4.5$

$N = 9(4.5) - (4.5)^2 = 20.25$

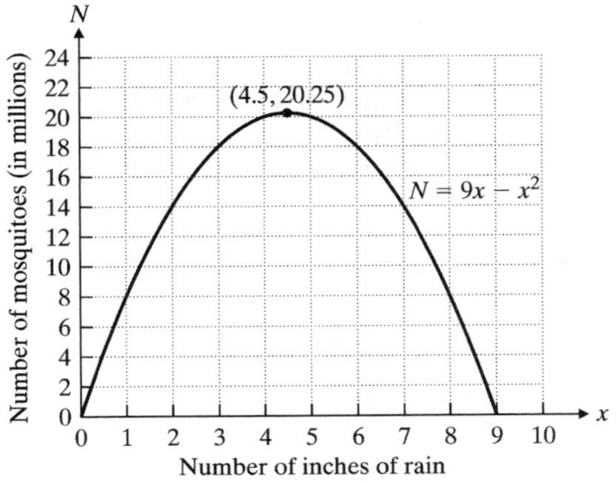

b. Maximum is at the vertex: 4.5 inches

c. Maximum is at the vertex: 20.25 million

33. $y = -301x^2 - 167x + 1528$

Vertex: $x = -\dfrac{b}{2a} = -\dfrac{-167}{2(-301)} \approx -0.277$

$y = -301(-0.277)^2 - 167(-0.277) + 1528$
≈ 1551.164

$V(-0.277, 1551.164)$
y-intercept: (0, 1528)
x-intercepts:

$$x = \frac{167 \pm \sqrt{(-167)^2 - 4(-301)(1528)}}{2(-301)}$$
$= -2.548, 1.993$

(−2.548, 0) and (1.993, 0)

Cumulative Review

34. $y = \dfrac{k}{x}$, $y = \dfrac{1}{3}$, $x = 21$

$\dfrac{1}{3} = \dfrac{k}{21} \Rightarrow k = 7$

$y = \dfrac{7}{x}$, $x = 7$

$y = \dfrac{7}{7} = 1$

35. $y = kx^2$, $y = 12$, $x = 2$

$12 = k(2)^2 \Rightarrow k = 3$

$y = 3x^2$, $x = 5$

$y = 3(5)^2 = 75$

36. $t = \dfrac{d}{r}$

First half: $t = \dfrac{30(3)\text{ ft}}{20\text{ ft/s}} = 4.5\text{ s}$

Second half: $t = \dfrac{30(3)\text{ ft}}{24\text{ ft/s}} = 3.75\text{ s}$

Total time = 4.5 + 3.75 = 8.25 s

Quick Quiz 10.4

1. $y = x^2 - 6x + 5$

$x = -\dfrac{b}{2a} = \dfrac{-(-6)}{2(1)} = 3$

$y = 3^2 - 6(3) + 5 = -4$

The vertex of the curve is (3, −4).

2. $x^2 - 6x + 5 = 0$

$(x-5)(x-1) = 0$

$x - 5 = 0$ $\quad$ $x - 1 = 0$

$x = 5$ $\quad$ $x = 1$

The x-intercepts are (5, 0) and (1, 0).

3. $y = x^2 - 6x + 5$

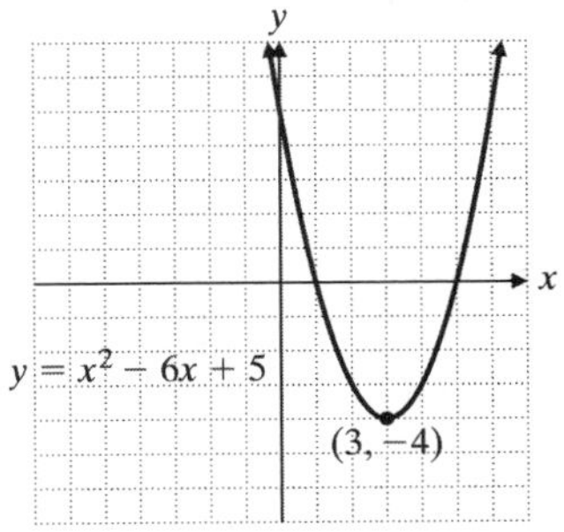

4. Answers may vary. Possible solution:
The x-coordinate of the vertex of the equation $y = 4x^2 + 16x - 2$ is found by using $x = \dfrac{-b}{2a}$ when the equation is in the form $y = ax^2 + bx + c$. After the x-coordinate has been found, it may be substituted back into the original equation to find the y-coordinate.

10.5 Exercises

1. $A = \frac{1}{2}ba$, x = altitude, $x - 5$ = base

$$88 = \frac{1}{2}(x-5)(x)$$
$$176 = x^2 - 5x$$
$$0 = x^2 - 5x - 176$$
$$0 = (x-16)(x+11)$$
$$x - 16 = 0 \qquad x + 11 = 0$$
$$x = 16 \qquad x = -11$$

Length can't be negative.
$x - 5 = 16 - 5 = 11$
The altitude is 16 centimeters. The base is 11 centimeters.

3.
$$(x+4)(x+9) = 266$$
$$x^2 + 9x + 4x + 36 = 266$$
$$x^2 + 13x - 230 = 0$$
$$(x-10)(x+23) = 0$$
$$x - 10 = 0 \qquad x + 23 = 0$$
$$x = 10 \qquad x = -23$$

dimension cannot be negative

The old garden dimensions are 10 ft × 10 ft.

5. s = original number of people
c = cost per person
$$s \cdot c = 420 \quad (1)$$
$$(s+7)(c-5) = 420 \quad (2)$$
Substitute $\frac{420}{s}$ for c in (2).
$$(s+7)\left(\frac{420}{s} - 5\right) = 420$$
$$(s+7)(420 - 5s) = 420s$$
$$(s+7)(84 - s) = 84s$$
$$s^2 + 7s - 588 = 0$$
$$(s+28)(s-21) = 0$$
$s = -28$, $s = 21$
Number of people can't be negative. Original number of people was 21.

7. s = number of members
c = cost
$$s \cdot c = 400 \quad (1)$$
$$(s+20)(c-1) = 400 \quad (2)$$
Substitute $\frac{400}{s}$ for c in (2).
$$(s+20)\left(\frac{400}{s} - 1\right) = 400$$
$$(s+20)(400 - s) = 400s$$
$$s^2 - 380s - 8000 = -400s$$
$$s^2 + 20s - 8000 = 0$$
$$(s+100)(s-80) = 0$$
$$s + 100 = 0 \qquad s - 80 = 0$$
$$s = -100 \qquad s = 80$$
Number of members can't be negative.
$s + 20 = 100$
80 members were expected, 100 members attended.

9. x = width, y = length
Length: $2x + y = 100$ (1)
Area: $xy = 1250$ (2)
Substitute $\frac{1250}{y}$ for x in (1).
$$2\left(\frac{1250}{y}\right) + y = 100$$
$$2500 + y^2 = 100y$$
$$y^2 - 100y + 2500 = 0$$
$$(y-50)(y-50) = 0$$
$$y - 50 = 0$$
$$y = 50$$
$$x = \frac{1250}{y} = \frac{1250}{50} = 25$$
The length along the building is 50 feet, width is 25 feet.

11. x = actual speed
$x + 200$ = proposed speed
Actual test: time $= \frac{2400}{x}$
Proposed test: time $= \frac{2400}{x+200}$
$$\frac{2400}{x+200} = \frac{2400}{x} - 1$$
$$2400x = 2400(x+200) - x(x+200)$$
$$2400x = 2400x + 480{,}000 - x^2 - 200x$$
$$0 = -x^2 - 200x + 480{,}000$$
$$0 = -(x+800)(x-600)$$
$$x = -800 \qquad x = 600$$
Speed can't be negative. The actual speed was 600 mph.

13. Let $x = 0$.
$y = -2.5x^2 + 22.5x + 50$
$y = -2.5(0) + 22.5(0) + 50$
50 Quick Print Centers were in operation in 1998.

15. 2000, $x = 2$
$y = -2.5(2)^2 + 22.5(2) + 50 = 85$
2002, $x = 4$
$y = -2.5(4)^2 + 22.5(4) + 50 = 100$
Difference between 2002 and 2000
$100 - 85 = 15$
15 more Quick Print Centers in operation in 2002 than in 2000.

17. $100 = -2.5x^2 + 22.5x + 50$
Rearrange.
$2.5x^2 - 22.5x + 50 = 0$
$a = 2,\ b = -22.5,\ c = 50$

$$x = \frac{-(-22.5) \pm \sqrt{(-22.5)^2 - 4(2.5)(50)}}{2(2.5)}$$

$$x = \frac{22.5 \pm \sqrt{106.25}}{5}$$

$x = 4, 5$
year = 1998 + 4 = 2002
year = 1998 + 5 = 2003
During 2002 and 2003 there were 100 Quick Print Centers.

19. $y = -0.018x^2 + 1.336x,\ y = 0$

$0 = -0.018x^2 + 1.336x$
$0 = x(-0.018x + 1.336)$
$x = 0 \qquad -0.018x + 1.336 = 0$
$-0.018x = -1.336$
$x \approx 74.2$
Big Bertha could send a shell about 74 miles.

Cumulative Review

21. $3x + 11 \ge 9x - 4$
$-6x \ge -15$
$x \le \frac{5}{2}$

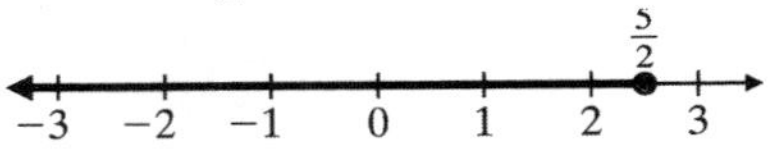

22.

$$\frac{\sqrt{3}}{\sqrt{5}+2} = \frac{\sqrt{3}}{\sqrt{5}+2} \cdot \frac{\sqrt{5}-2}{\sqrt{5}-2}$$
$$= \frac{\sqrt{15} - 2\sqrt{3}}{\left(\sqrt{5}\right)^2 - (2)^2}$$
$$= \frac{\sqrt{15} - 2\sqrt{3}}{5-4}$$
$$= \sqrt{15} - 2\sqrt{3}$$

23. x = length of section 1 and of section 2
l = length of section 3

Volume of 3 = $\left(\frac{2}{3}\right)$volume of 1

$$2(2)l = \frac{2}{3}(2)(2)x$$
$$l = \frac{2}{3}x$$

Total length = $x + x + \frac{2}{3}x = 3$

$$\frac{8}{3}x = 3$$
$$x = \frac{9}{8}$$
$$l = \frac{2}{3}\left(\frac{9}{8}\right) = \frac{3}{4}$$

Section 3 is 0.75 foot long.

Quick Quiz 10.5

1. other leg = x
one leg = $x + 7$

$$a^2 + b^2 = c^2$$
$$(x+7)^2 + x^2 = (13)^2$$
$$x^2 + 14x + 49 + x^2 = 169$$
$$2x^2 + 14x - 120 = 0$$
$$2(x^2 + 7x - 60) = 0$$
$$2(x+12)(x-5) = 0$$

$x + 12 = 0 \qquad x - 5 = 0$
$x = -12 \qquad x = 5$
$x + 7 = 12$

$x = 5$ only because length cannot be negative.
The lengths of legs are 5 and 12 yards.

2. $S = -5t^2 + vt + h$
$0 = -5t^2 + 25t + 30$
$0 = -5(t^2 - 5t - 6)$
$0 = -5(t-6)(t+1)$
$t - 6 = 0 \qquad t + 1 = 0$
$t = 6 \qquad t = -1$
negative value not possible here
The ball hits the ground after 6 seconds.

3. $t = \frac{-b}{2a} = \frac{-25}{2(-5)} = 2.5$ seconds
$S = -5(2.5)^2 + 25(2.5) + 30 = 61.25$
61.25 meters is the maximum height of the ball.

4. Answers may vary. Possible solution:
Set x = original number
d = original due per person
$x \cdot d = 750$ (1)
$(x-4)(d+6) = 750$ (2)
Two equations with two unknowns can be solved by substitution.

Putting Your Skills to Work

1. $72° - 68° = 4°$
$4 \times 0.02 \times 205 = 16.4$
They could expect to save $16.40 per month.

2. $84° - 78° = 6°$
$6 \times 0.02 \times 205 = 24.6$
They could expect to save $24.60 per month.

3. $\frac{5°}{1 \text{ mo}} \times \frac{12 \text{ mo}}{1} \times \frac{\$205}{1 \text{ mo}} \times \frac{0.02}{1°} = \$246$
They could expect to save $246 per year.

4. Answers will vary.

5. Answers will vary.

6. $2.95 × 100 = $295
$4.45 × 100 = $445
$445 – $295 = $150
The difference in cost is $150.

7. $4.45 × 100 × 5 = $2225
He can expect to pay $2225.

8. 4° lower: $\frac{4}{72} = 0.08$ or 8% savings
8% of 2225 = 0.08 × 2225 = 178
He will save $178.

9. 8° lower: $\frac{8}{72} = 0.16$ or 16% savings
16% of 2225 = 0.16 × 2225 = 356
He will save $356.

10. $445 = what percent of $2225?
$\frac{445}{2225} = 20\%$; 10° lower
He needs to get it at 72° – 10° = 62°.

Chapter 10 Review Problems

1. $6x^2 = 5x - 8$
$6x^2 - 5x + 8 = 0$
$a = 6, b = -5, c = 8$

2. $9x^2 + 3x = -5x^2 + 16$
$14x^2 + 3x - 16 = 0$
$a = 14, b = 3, c = -16$

3. $(x+3)(2x-3) = x^2 + 5$
$2x^2 - 3x + 6x - 9 = x^2 + 5$
$x^2 + 3x - 14 = 0$
$a = 1, b = 3, c = -14$

4. $3x(4x+1) = -x(5-x)$
$12x^2 + 3x = -5x + x^2$
$11x^2 + 8x = 0$
$a = 11, b = 8, c = 0$

5. $\frac{3}{x^2} - \frac{6}{x} - 2 = 0$
$x^2 \cdot \frac{3}{x^2} - x^2 \cdot \frac{6}{x} - x^2 \cdot 2 = 0$
$3 - 6x - 2x^2 = 0$
$-2x^2 - 6x + 3 = 0$
$a = -2, b = -6, c = 3$

6.
$$\frac{x}{x+2}-\frac{3}{x-2}=5$$
$$(x+2)(x-2)\frac{x}{x+2}-(x+2)(x-2)\frac{3}{x-2}=(x+2)(x-2)5$$
$$x(x-2)-3(x+2)=5(x+2)(x-2)$$
$$x^2-2x-3x-6=5x^2-20$$
$$-4x^2-5x+14=0$$
$$4x^2+5x-14=0$$

$a=4$, $b=5$, $c=-14$

7. $x^2+26x+25=0$

$(x+25)(x+1)=0$

$x+25=0 \qquad x+1=0$

$x=-25 \qquad x=-1$

8. $x^2+16x+64=0$

$(x+8)(x+8)=0$

$x+8=0$

$x=-8$

9. $x^2+6x-55=0$

$(x+11)(x-5)=0$

$x+11=0 \qquad x-5=0$

$x=-11 \qquad x=5$

10. $4x^2-16x+15=0$

$(2x-3)(2x-5)=0$

$2x-3=0 \qquad 2x-5=0$

$2x=3 \qquad 2x=5$

$x=\frac{3}{2} \qquad x=\frac{5}{2}$

11. $9x^2-24x+16=0$

$(3x-4)(3x-4)=0$

$3x-4=0$

$3x=4$

$x=\frac{4}{3}$

12. $15x^2=26x-8$

$15x^2-26x+8=0$

$(5x-2)(3x-4)=0$

$5x-2=0 \qquad 3x-4=0$

$5x=2 \qquad 3x=4$

$x=\frac{2}{5} \qquad x=\frac{4}{3}$

13. $6x^2 = 13x - 5$

$6x^2 - 13x + 5 = 0$

$(3x - 5)(2x - 1) = 0$

$3x - 5 = 0 \quad\quad 2x - 1 = 0$

$3x = 5 \quad\quad 2x = 1$

$x = \frac{5}{3} \quad\quad x = \frac{1}{2}$

14. $4x^2 = -5x + 9$

$4x^2 + 5x - 9 = 0$

$(x - 1)(4x + 9) = 0$

$x - 1 = 0 \quad\quad 4x + 9 = 0$

$x = 1 \quad\quad 4x = -9$

$x = -\frac{9}{4}$

15. $x^2 + \frac{1}{6}x - 2 = 0$

$6x^2 + x - 12 = 0$

$(2x + 3)(3x - 4) = 0$

$2x + 3 = 0 \quad\quad 3x - 4 = 0$

$2x = -3 \quad\quad 3x = 4$

$x = -\frac{3}{2} \quad\quad x = \frac{4}{3}$

16. $\frac{1}{2}x^2 = \frac{3}{4}x - \frac{1}{4}$

$2x^2 = 3x - 1$

$2x^2 - 3x + 1 = 0$

$(2x - 1)(x - 1) = 0$

$2x - 1 = 0 \quad\quad x - 1 = 0$

$2x = 1 \quad\quad x = 1$

$x = \frac{1}{2}$

17. $x^2 + \frac{2}{15}x = \frac{1}{15}$

$15x^2 + 2x = 1$

$15x^2 + 2x - 1 = 0$

$(5x - 1)(3x + 1) = 0$

$5x - 1 = 0 \quad\quad 3x + 1 = 0$

$5x = 1 \quad\quad 3x = -1$

$x = \frac{1}{5} \quad\quad x = -\frac{1}{3}$

18. $1 + \frac{13}{12x} - \frac{1}{3x^2} = 0$

$12x^2 + 13x - 4 = 0$

$(3x + 4)(4x - 1) = 0$

$2x + 4 = 0 \quad\quad 4x - 1 = 0$

$3x = -4 \quad\quad 4x = 1$

$x = -\frac{4}{3} \quad\quad x = \frac{1}{4}$

19. $x^2 + 9x + 5 = x + 5$

$x^2 + 8x = 0$

$x(x + 8) = 0$

$x = 0 \quad\quad x + 8 = 0$

$x = -8$

20. $x^2 - 2x + 7 = 7 + x$

$x^2 - 3x = 0$

$x(x - 3) = 0$

$x = 0 \quad\quad x - 3 = 0$

$x = 3$

21. $1 + \frac{2}{3x + 4} = \frac{3}{3x + 2}$

$(3x + 4)(3x + 2) + 2(3x + 2) = 3(3x + 4)$

$9x^2 + 18x + 8 + 6x + 4 = 9x + 12$

$9x^2 + 15x = 0$

$3x(3x + 5) = 0$

$3x = 0 \quad\quad 3x + 5 = 0$

$x = 0 \quad\quad 3x = -5$

$x = -\frac{5}{3}$

22. $2 - \frac{5}{x + 1} = \frac{3}{x - 1}$

$2(x + 1)(x - 1) - 5(x - 1) = 3(x + 1)$

$2x^2 - 2 - 5x + 5 = 3x + 3$

$2x^2 - 8x = 0$

$2x(x - 4) = 0$

$2x = 0 \quad\quad x - 4 = 0$

$x = 0 \quad\quad x = 4$

23.
$$5+\frac{24}{2-x}=\frac{24}{2+x}$$
$$5(2-x)(2+x)+24(2+x)=24(2-x)$$
$$20-5x^2+48+24x=48-24x$$
$$-5x^2+48x+20=0$$
$$5x^2-48x-20=0$$
$$(5x+2)(x-10)=0$$
$5x+2=0$ or $x-10=0$

$5x=-2$, $x=10$

$x=-\frac{2}{5}$

24.
$$\frac{4}{9}=\frac{5x}{3}-x^2$$
$$4=15x-9x^2$$
$$9x^2-15x+4=0$$
$$(3x-1)(3x-4)=0$$
$3x-1=0$ or $3x-4=0$

$3x=1$, $3x=4$

$x=\frac{1}{3}$, $x=\frac{4}{3}$

25. $x^2-8=41$
$$x^2=49$$
$$x=\pm\sqrt{49}$$
$$x=\pm 7$$

26. $x^2+11=92$
$$x^2=81$$
$$x=\pm\sqrt{81}$$
$$x=\pm 9$$

27. $x^2-5=17$
$$x^2=22$$
$$x=\pm\sqrt{22}$$

28. $x^2+11=50$
$$x^2=39$$
$$x=\pm\sqrt{39}$$

29. $2x^2-1=15$
$$2x^2=16$$
$$x^2=8$$
$$x=\pm\sqrt{8}$$
$$x=\pm 2\sqrt{2}$$

30. $3x^2+4=154$
$$3x^2=150$$
$$x^2=50$$
$$x=\pm\sqrt{50}$$
$$x=\pm 5\sqrt{2}$$

31. $3x^2+6=60$
$$3x^2=54$$
$$x^2=18$$
$$x=\pm\sqrt{18}$$
$$x=\pm 3\sqrt{2}$$

32. $2x^2-5=43$
$$2x^2=48$$
$$x^2=24$$
$$x=\pm\sqrt{24}$$
$$x=\pm 2\sqrt{6}$$

33. $(x-4)^2=7$
$$x-4=\pm\sqrt{7}$$
$$x=4\pm\sqrt{7}$$

34. $(x-2)^2=3$
$$x-2=\pm\sqrt{3}$$
$$x=2\pm\sqrt{3}$$

35. $(3x+2)^2=28$
$$3x+2=\pm\sqrt{28}$$
$$3x=-2\pm\sqrt{28}$$
$$3x=-2\pm 2\sqrt{7}$$
$$x=\frac{-2\pm 2\sqrt{7}}{3}$$

36. $(2x-1)^2=32$
$$2x-1=\pm\sqrt{32}$$
$$2x-1=\pm 4\sqrt{2}$$
$$2x=1\pm 4\sqrt{2}$$
$$x=\frac{1\pm 4\sqrt{2}}{2}$$

37. $x^2+8x+7=0$

$$x^2+8x=-7$$
$$x^2+8x+16=-7+16$$
$$(x+4)^2=9$$
$$x+4=\pm 3$$
$$x=-4\pm 3$$
$$x=-1,\ -7$$

38. $x^2+14x+33=0$

$$x^2+14x=-33$$
$$x^2+14x+49=-33+49$$
$$(x+7)^2=16$$
$$x+7=\pm 4$$
$$x=-7\pm 4$$
$$x=-3,\ -11$$

39. $2x^2-8x-90=0$

$$2x^2-8x=90$$
$$x^2-4x=45,\ \left[\frac{1}{2}(-4)\right]^2=(-2)^2=4$$
$$x^2-4x+4=45+4$$
$$(x-2)^2=49$$
$$x-2=\pm\sqrt{49}$$
$$x=2\pm 7$$

$x=2+7$ $\quad$ $x=2-7$

$x=9$ $\quad$ $x=-5$

40. $-5x^2+30x-35=0$

$$x^2-6x+7=0$$
$$x^2-6x=-7$$
$$x^2-6x+9=-7+9$$
$$(x-3)^2=2$$
$$x-3=\pm\sqrt{2}$$
$$x=3\pm\sqrt{2}$$

41. $3x^2+6x-6=0$

$$3x^2+6x=6$$
$$x^2+2x=2,\ \left[\frac{1}{2}(2)\right]^2=1^2=1$$
$$x^2+2x+1=2+1$$
$$(x+1)^2=3$$
$$x+1=\pm\sqrt{3}$$
$$x=-1\pm\sqrt{3}$$

42. $2x^2+10x-3=0$

$$2x^2+10x=3$$
$$x^2+5x=\frac{3}{2},\ \left[\frac{1}{2}(5)\right]^2=\left(\frac{5}{2}\right)^2=\frac{25}{4}$$
$$x^2+5x+\frac{25}{4}=\frac{3}{2}+\frac{25}{4}$$
$$\left(x+\frac{5}{2}\right)^2=\frac{31}{4}$$
$$x+\frac{5}{2}=\pm\sqrt{\frac{31}{4}}$$
$$x=-\frac{5}{2}\pm\frac{\sqrt{31}}{2}$$
$$x=\frac{-5\pm\sqrt{31}}{2}$$

Use $x=\dfrac{-b\pm\sqrt{b^2-4ac}}{2a}$ in Exercises 43–50.

43. $x^2+4x-6=0$

$a=1,\ b=4,\ c=-6$

$$x=\frac{-4\pm\sqrt{(4)^2-4(1)(-6)}}{2(1)}$$
$$=\frac{-4\pm\sqrt{40}}{2}$$
$$=\frac{-4\pm 2\sqrt{10}}{2}$$
$$x=-2\pm\sqrt{10}$$

44. $x^2+4x-8=0$

$a=1,\ b=4,\ c=-8$

$$x=\frac{-4\pm\sqrt{(4)^2-4(1)(-8)}}{2(1)}$$
$$=\frac{-4\pm\sqrt{48}}{2}$$
$$=\frac{-4\pm 4\sqrt{3}}{2}$$
$$x=-2\pm 2\sqrt{3}$$

45. $2x^2-7x+4=0$

$a=2,\ b=-7,\ c=4$

$$x=\frac{-(-7)\pm\sqrt{(-7)^2-4(2)(4)}}{2(2)}$$
$$x=\frac{7\pm\sqrt{17}}{4}$$

46. $2x^2+5x-6=0$
$a=2,\ b=5,\ c=-6$
$$x=\frac{-(5)\pm\sqrt{(5)^2-4(2)(-6)}}{2(2)}$$
$$x=\frac{-5\pm\sqrt{73}}{4}$$

47. $3x^2-8x-4=0$
$a=3,\ b=-8,\ c=-4$
$$x=\frac{-(-8)\pm\sqrt{(-8)^2-4(3)(-4)}}{2(3)}$$
$$x=\frac{8\pm\sqrt{112}}{6}$$
$$x=\frac{8\pm4\sqrt{7}}{6}$$
$$x=\frac{4\pm2\sqrt{7}}{3}$$

48. $4x^2-2x-11=0$
$a=4,\ b=-2,\ c=-11$
$$x=\frac{-(-2)\pm\sqrt{(-2)^2-4(4)(-11)}}{2(4)}$$
$$x=\frac{2\pm\sqrt{180}}{8}$$
$$x=\frac{2\pm6\sqrt{5}}{8}$$
$$x=\frac{1\pm3\sqrt{5}}{4}$$

49. $5x^2-x=2$
$5x^2-x-2=0$
$a=5,\ b=-1,\ c=-2$
$$x=\frac{-(-1)\pm\sqrt{(-1)^2-4(5)(-2)}}{2(5)}$$
$$x=\frac{1\pm\sqrt{41}}{10}$$

50. $3x^2+3x=1$
$3x^2+3x-1=0$
$a=3,\ b=3,\ c=-1$
$$x=\frac{-3\pm\sqrt{(3)^2-4(3)(-1)}}{2(3)}$$
$$x=\frac{-3\pm\sqrt{21}}{6}$$

51. $2x^2-9x+10=0$
$(2x-5)(x-2)=0$
$2x-5=0$ $\quad$ $x-2=0$
$x=\frac{5}{2}$ $\quad$ $x=2$

52. $4x^2-4x-3=0$
$(2x+1)(2x-3)=0$
$2x+1=0$ $\quad$ $2x-3=0$
$x=-\frac{1}{2}$ $\quad$ $x=\frac{3}{2}$

53. $25x^2+10x+1=0$
$(5x+1)^2=0$
$5x+1=0$
$x=-\frac{1}{5}$

54. $2x^2-11+12=0$
$(x-4)(2x-3)=0$
$x-4=0$ $\quad$ $2x-3=0$
$x=4$ $\quad$ $x=\frac{3}{2}$

55. $3x^2-6x+2=0$
$a=3,\ b=-6,\ c=2$
$$x=\frac{-(-6)\pm\sqrt{(-6)^2-4(3)(2)}}{2(3)}$$
$$=\frac{6\pm\sqrt{12}}{6}$$
$$=\frac{6\pm2\sqrt{3}}{6}$$
$$=\frac{3\pm\sqrt{3}}{3}$$

56. $5x^2-7x=8$
$5x^2-7x-8=0$
$a=5,\ b=-7,\ c=-8$
$$x=\frac{-(-7)\pm\sqrt{(-7)^2-4(5)(-8)}}{2(5)}=\frac{7\pm\sqrt{209}}{10}$$

57. $4x^2 + 4x = x^2 + 5$

$3x^2 + 4x - 5 = 0$

$a = 3, b = 4, c = -5$

$$x = \frac{-4 \pm \sqrt{4^2 - 4(3)(-5)}}{2(3)} = \frac{-4 \pm \sqrt{76}}{6} = \frac{-4 \pm 2\sqrt{19}}{6} = \frac{-2 \pm \sqrt{19}}{3}$$

58. $5x^2 + 7x + 1 = 0$

$a = 5, b = 7, c = 1$

$$x = \frac{-(7) \pm \sqrt{(7)^2 - 4(5)(1)}}{2(5)} = \frac{-7 \pm \sqrt{29}}{10}$$

59. $x^2 = 9x + 3$

$x^2 - 9x - 3 = 0$

$a = 1, b = -9, c = -3$

$$x = \frac{-(-9) \pm \sqrt{(-9)^2 - 4(1)(-3)}}{2(1)}$$

$$x = \frac{9 \pm \sqrt{93}}{2}$$

60. $2x^2 = 9x + 5$

$2x^2 - 9x - 5 = 0$

$a = 2, b = -9, c = -5$

$$x = \frac{-(-9) \pm \sqrt{(-9)^2 - 4(2)(-5)}}{2(2)}$$

$$x = \frac{9 \pm \sqrt{121}}{4}$$

$$x = \frac{9 \pm 11}{4}$$

$$x = 5, -\frac{1}{2}$$

61. $-20x^2 - 15x = -8x^2 - 18$

$-12x^2 - 15x + 18 = 0$

$a = -12, b = -15, c = 18$

$$x = \frac{-(-15) \pm \sqrt{(-15)^2 - 4(-12)(18)}}{2(-12)}$$

$$x = \frac{15 \pm \sqrt{1089}}{-24}$$

$$x = \frac{15 \pm 33}{-24}$$

$$x = -2, \frac{3}{4}$$

62. $2x^2 - 1 = 35$

$2x^2 = 36$

$x^2 = 18$

$x = \pm\sqrt{18} = \pm 3\sqrt{2}$

63. $\frac{(y-2)^2}{20} + 3 + y = 0$

$(y-2)^2 + 60 + 20y = 0$

$y^2 - 4y + 4 + 60 + 20y = 0$

$y^2 + 16y + 64 = 0$

$(y+8)^2 = 0$

$y + 8 = 0$

$y = -8$

64. $\frac{(y+2)^2}{5} + 2y = -9$

$(y+2)^2 + 10y = -45$

$y^2 + 4y + 4 + 10y + 45 = 0$

$y^2 + 14y + 49 = 0$

$(y+7)^2 = 0$

$y + 7 = 0$

$y = -7$

65. $3x^2+1=6-8x$

$3x^2+8x-5=0$

$a=3,\ b=8,\ c=-5$

$$x=\frac{-8\pm\sqrt{8^2-4(3)(-5)}}{2(3)}=\frac{-8\pm\sqrt{124}}{6}=\frac{-8\pm2\sqrt{31}}{6}=\frac{-4\pm\sqrt{31}}{3}$$

66. $4x^2+11x=5x+1$

$4x^2+6x-1=0$

$$x=\frac{-(6)\pm\sqrt{(6)^2-4(4)(-1)}}{2(4)}$$

$$x=\frac{-6\pm\sqrt{52}}{8}$$

$$x=\frac{-6\pm2\sqrt{13}}{8}$$

$$x=\frac{-3\pm\sqrt{13}}{4}$$

67. $8y(y+1)=7y+9$

$8y^2+8y=7y+9$

$8y^2+y-9=0$

$a=8,\ b=1,\ c=-9$

$$y=\frac{-1\pm\sqrt{(1)^2-4(8)(-9)}}{2(8)}$$

$$y=\frac{-1\pm\sqrt{289}}{16}$$

$$y=\frac{-1\pm17}{16}$$

$$y=1,\ -\frac{9}{8}$$

68. $\frac{3y-2}{4}=\frac{y^2-2}{y}$

$y(3y-2)=4(y^2-2)$

$3y^2-2y=4y^2-8$

$0=y^2+2y-8$

$0=(y+4)(y-2)$

$y+4=0$ $\quad$ $y-2=0$

$y=-4$ $\quad$ $y=2$

69. $\frac{y^2+5}{2y}=\frac{2y-1}{3}$

$3(y^2+5)=2y(2y-1)$

$3y^2+15=4y^2-2y$

$0=y^2-2y-15$

$0=(y-5)(y+3)$

$y-5=0$ $\quad$ $y+3=0$

$y=5$ $\quad$ $y=-3$

70. $\frac{5x^2}{2}=x-\frac{7x^2}{2}$

$5x^2=2x-7x^2$

$12x^2-2x=0$

$2x(6x-1)=0$

$2x=0$ $\quad$ $6x-1=0$

$x=0$ $\quad$ $x=\frac{1}{6}$

71. $y=2x^2$

x	y
−2	8
−1	2
0	0
1	2
2	8

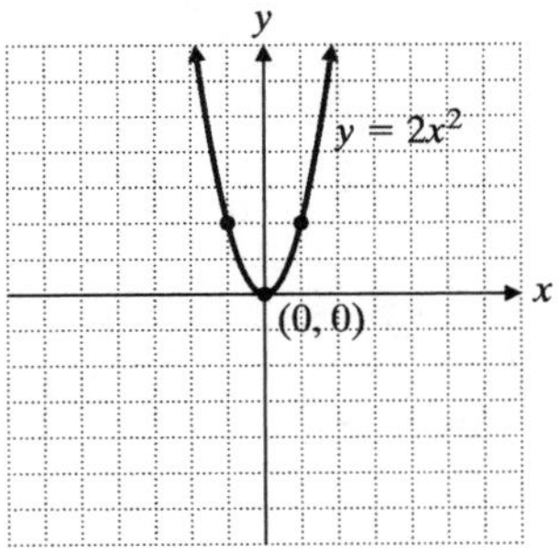

72. $y = x^2 + 4$

x	y
−2	8
−1	5
0	4
1	5
2	8

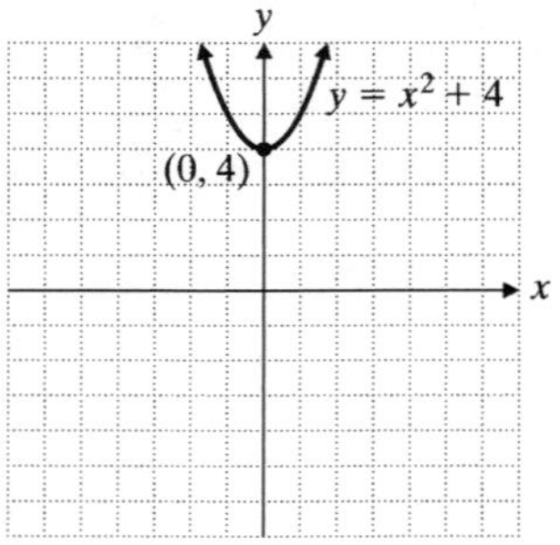

73. $y = x^2 - 3$

x	y
−2	1
−1	−2
0	−3
1	−2
2	1

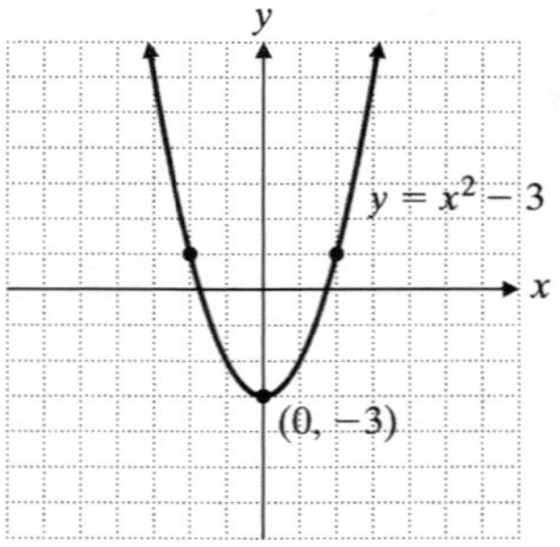

74. $y = -\frac{1}{2}x^2$

x	y
−2	−2
−1	$-\frac{1}{2}$
0	0
1	$-\frac{1}{2}$
2	−2

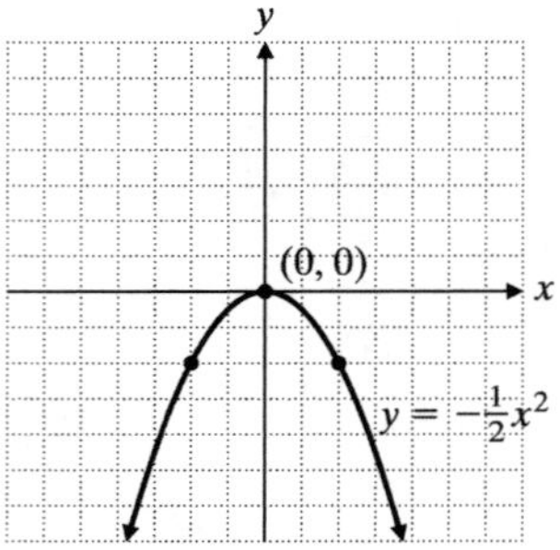

75. $y = x^2 - 3x - 4$

Vertex: $x = \frac{-b}{2a} = \frac{-(-3)}{2(1)} = \frac{3}{2}$

$$y = \left(\frac{3}{2}\right)^2 - 3\left(\frac{3}{2}\right) - 4 = -\frac{25}{4}$$

$V\left(\frac{3}{2}, -\frac{25}{4}\right)$

x	y
−1	0
0	−4
1.5	−6.25
3	−4
4	0

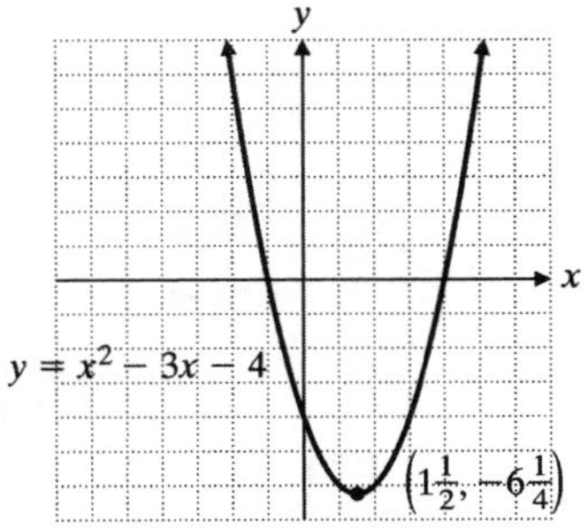

76. $y = \frac{1}{2}x^2 - 2$

x	y
−2	0
−1	−1.5
0	−2
1	−1.5
2	0

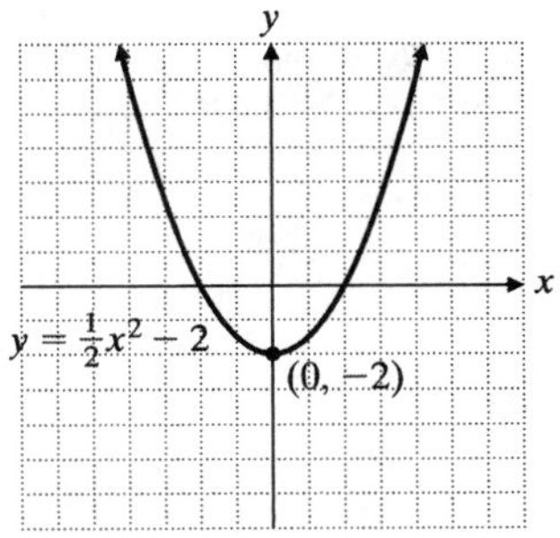

77. $y = -2x^2 + 12x - 17$

Vertex: $x = \frac{-b}{2a} = \frac{-12}{2(-2)} = 3$

$$y = -2(3)^2 + 12(3) - 17 = 1$$

$V(3, 1)$

x	y
1	−7
2	−1
3	1
4	−1
5	−7

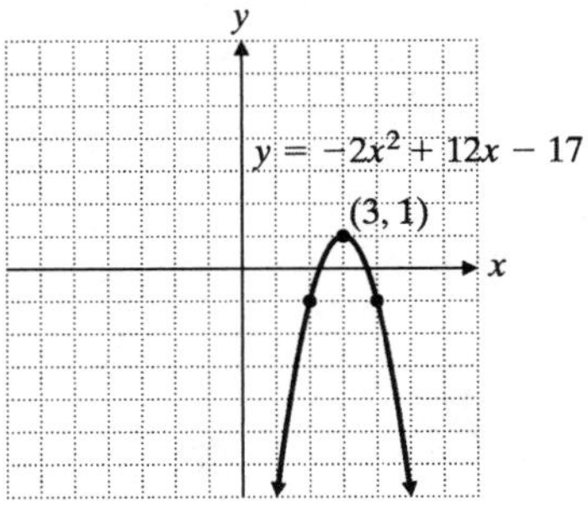

78. $y = -3x^2 - 2x + 4$

Vertex: $x = \frac{-b}{2a} = -\frac{-2}{2(-3)} = -\frac{1}{3}$

$$y = -3\left(-\frac{1}{3}\right)^2 - 2\left(-\frac{1}{3}\right) + 4 = \frac{13}{3}$$

$V\left(-\frac{1}{3}, \frac{13}{3}\right)$

x	y
−2	−4
−1	3
0	4
1	−1

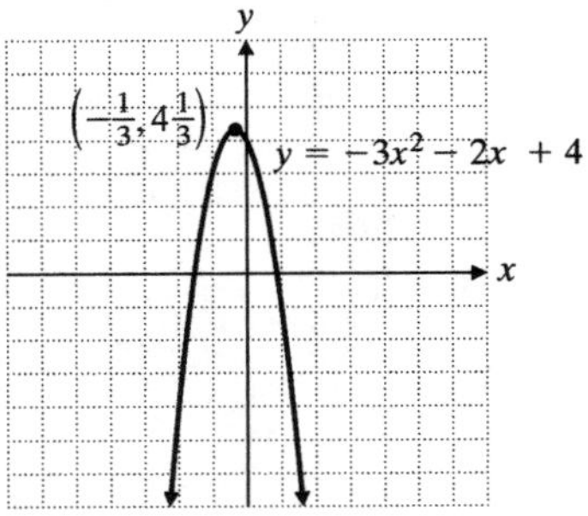

79. l = length
w = width
length: $l + 2w = 15$ (1)
area: $lw = 28$ (2)

Substitute $\frac{28}{w}$ for l in (1).

$$\frac{28}{w} + 2w = 15$$

$$28 + 2w^2 = 15w$$

Rearrange.

$$2w^2 - 15w + 28 = 0$$

$a = 2, b = -15, c = 28$

$$w = \frac{-(-15) \pm \sqrt{(-15)^2 - 4(2)(28)}}{2(2)}$$

$$w = \frac{15 \pm \sqrt{1}}{4}$$

$$w = \frac{15 \pm 1}{4} = 4, 3\frac{1}{2}$$

$w = 4$ $\quad$ $w = 3\frac{1}{2}$

$l = \frac{28}{w}$ $\quad$ $l = \frac{28}{w}$

$l = 7$ $\quad$ $l = 8$

Dimensions of pen are either 4 ft by 7 ft or $3\frac{1}{2}$ ft by 8 ft.

80. x = first leg
$x - 4$ = second leg

$$x^2 + (x-4)^2 = 20^2$$
$$x^2 + x^2 - 8x + 16 = 400$$
$$2x^2 - 8x - 384 = 0$$
$$x^2 - 4x - 192 = 0$$
$$(x-16)(x+12) = 0$$

$x - 16 = 0$ $\quad$ $x + 12 = 0$

$x = 16$ $\quad$ $x = -12$

Negative length isn't allowed.
$x - 4 = 16 - 4 = 12$
The legs are 16 cm and 12 cm.

81. x = width
$x + 6$ = length

$$x^2 + (x+6)^2 = 30^2$$
$$x^2 + x^2 + 12x + 36 = 900$$
$$2x^2 + 12x - 864 = 0$$
$$2(x^2 + 6x - 432) = 0$$
$$x^2 + 6x - 432 = 0$$
$$(x+24)(x-18) = 0$$

$x + 24 = 0$ $\quad$ $x - 18 = 0$

$x = -24$ $\quad$ $x = 18$

Negative length isn't allowed.
width = x = 18 inches
length = $x + 6 = 18 + 6 = 24$ inches

82. $h = -225t^2 + 202,500$

$$0 = -225t^2 + 202,500$$
$$0 = -225(t^2 - 900)$$
$$0 = -225(t-30)(t+30)$$

$t - 30 = 0$ $\quad$ $t + 30 = 0$

$t = 30$ $\quad$ $t = -30$

negative value, not valid

The time for the alien space probe to reach Earth from the mother ship is 30 seconds.

83. x = members last year
$x - 4$ = members this year

$$\frac{720}{x} + 6 = \frac{720}{x-4}$$
$$(x-4)720 + 6x(x-4) = 720x$$
$$720x - 2880 + 6x^2 + 24x = 720x$$
$$6x^2 - 24x - 2880 = 0$$
$$6(x^2 - 4x - 480) = 0$$
$$x^2 - 4x - 480 = 0$$
$$(x-24)(x+20) = 0$$

$x - 24 = 0$ $\quad$ $x + 20 = 0$

$x = 24$ $\quad$ $x = -20$

Negative number of members isn't allowed.
There were 24 members last year.

84. r = speed on the trip to there
$r + 15$ = speed on the trip back

$d = r \cdot t$

$\frac{d}{r} = t$

$$3.5 = \frac{90}{r} + \frac{90}{r+15}$$
$$3.5r(r+15) = 90(r+15) + 90r$$
$$3.5r^2 + 52.5r = 90r + 1350 + 90r$$
$$3.5r^2 - 127.5r - 1350 = 0$$
$$7r^2 - 255r - 2700 = 0$$
$$(7r+60)(r-45) = 0$$

$7r + 60 = 0$ $\quad$ $r - 45 = 0$

$r = -\frac{60}{7}$ $\quad$ $r = 45$

Speed must be positive.
$r + 15 = 45 + 15 = 60$
Speed going was 45 mph; speed returning was 60 mph.

Use $y = -0.05x^2 + 2.5x + 60$ in exercises 85–88.

85. $x = 2030 - 1980 = 50$

$y = -0.05(50)^2 + 2.5(50) + 60$
$= 60$

There will be 60 pizza stores.

86. $x = 2040 - 1980 = 60$

$y = -0.05(60)^2 + 2.5(60) + 60 = 30$

Difference $= 30 - 60 = -30$

There will be 30 fewer pizza stores.

87. $-0.05x^2 + 2.5x^2 + 60 = 80$

$-0.05x^2 + 2.5x - 20 = 0$

$x^2 - 50x + 400 = 0$

$(x-10)(x-40) = 0$

$x - 1 = 0$ $\quad x - 40 = 0$

$x = 10$ $\quad x = 40$

year $= 1980 + 10$ $\quad$ year $= 1980 + 40$

$= 1990$ $\quad = 2020$

During years 1990 and 2020 there will be exactly 80 pizza stores.

88. $-0.05x^2 + 2.5x + 60 = 90$

$-0.05x^2 + 2.5x - 30 = 0$

$x^2 - 50x + 600 = 0$

$(x-20)(x-30) = 0$

$x - 20 = 0$ $\quad x - 30 = 0$

$x = 20$ $\quad x = 30$

year $= 1980 + 20$ $\quad$ year $= 1980 + 30$

$= 2000$ $\quad = 2010$

During years 2000 and 2010, there will be exactly 90 stores.

How Am I Doing? Chapter 10 Test

1. $5x^2 + 7x = 4$

$5x^2 + 7x - 4 = 0$

$a = 5, b = 7, c = -4$

$$x = \frac{-7 \pm \sqrt{7^2 - 4(5)(-4)}}{2(5)} = \frac{-7 \pm \sqrt{129}}{10}$$

2. $3x^2 + 13x = 10$

$3x^2 + 13x - 10 = 0$

$(x+5)(3x-2) = 0$

$x + 5 = 0$ $\quad 3x - 2 = 0$

$x = -5$ $\quad x = \frac{2}{3}$

3. $2x^2 = 2x - 5$

$2x^2 - 2x + 5 = 0$

$a = 2, b = -2, c = 5$

$$x = \frac{-(-2) \pm \sqrt{(-2)^2 - 4(2)(5)}}{2(2)} = \frac{2 \pm \sqrt{-36}}{4}$$

There is no real solution.

4. $4x^2 - 19x = 4x - 15$

$4x^2 - 23x + 15 = 0$

$(4x-3)(x-5) = 0$

$4x - 3 = 0$ $\quad x - 5 = 0$

$4x = 3$ $\quad x = 5$

$x = \frac{3}{4}$

5. $12x^2 + 11x = 5$

$12x^2 + 11x - 5 = 0$

$(4x+5)(3x-1) = 0$

$4x + 5 = 0$ $\quad 3x - 1 = 0$

$4x = -5$ $\quad 3x = 1$

$x = -\frac{5}{4}$ $\quad x = \frac{1}{3}$

6. $18x^2 + 32 = 48x$

$18x^2 - 48x + 32 = 0$

$9x^2 - 24x + 16 = 0$

$(3x-4)^2 = 0$

$3x - 4 = 0$

$3x = 4$

$x = \frac{4}{3}$

7. $2x^2 - 11x + 3 = 5x + 3$

$2x^2 - 16x = 0$

$2x(x-8) = 0$

$2x = 0$ $\quad x - 8 = 0$

$x = 0$ $\quad x = 8$

8. $5x^2 + 7 = 52$

$5x^2 = 45$

$x^2 = 9$

$x = \pm\sqrt{9}$

$x = \pm 3$

9.

$$2x(x-6)=6-x$$
$$2x^2-12x=6-x$$
$$2x^2-11x-6=0$$
$$(2x+1)(x-6)=0$$
$$2x+1=0 \qquad x-6=0$$
$$x=-\frac{1}{2} \qquad x=6$$

10.

$$x^2-x=\frac{3}{4}$$
$$4x^2-4x=3$$
$$4x^2-4x-3=0$$
$$(2x+1)(2x-3)=0$$
$$2x+1=0 \qquad 2x-3=0$$
$$x=-\frac{1}{2} \qquad x=\frac{3}{2}$$

11. $y=3x^2-6x$

Vertex: $x=\frac{-b}{2a}=-\frac{-6}{2(3)}=1$

$$y=3(1)^2-6(1)=-3$$

$V(1, -3)$

x	y
−1	9
0	0
1	−3
2	0
3	9

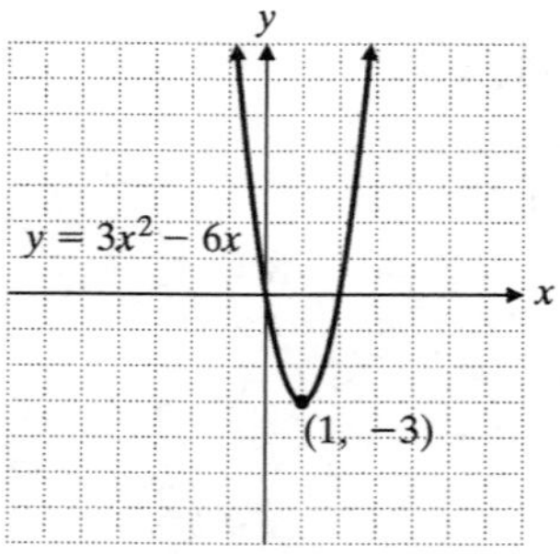

12. $y=-x^2+8x-12$

Vertex: $x=-\frac{b}{2a}=-\frac{8}{2(-1)}=4$

$$y=-(4)^2+8(4)-12=4$$

$V(4, 4)$

x	y
2	0
3	3
4	4
5	3
6	0

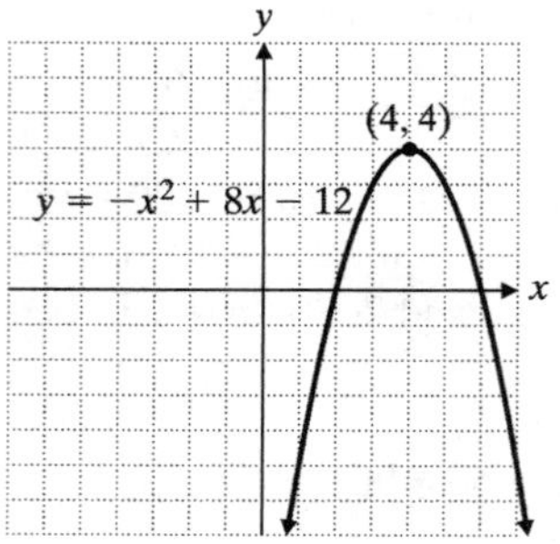

13. x = length of one leg
$x + 3$ = length of other leg

$$x^2+(x+3)^2=(15)^2$$
$$x^2+x^2+6x+9=225$$
$$2x^2+6x-216=0$$
$$x^2+3x-108=0$$
$$(x+12)(x-9)=0$$
$$x+12=0 \qquad x-9=0$$
$$x=-12 \qquad x=9$$

Length can't be negative, so $x = 9$ and $x + 3 = 12$. The lengths are 9 m and 12 m.

14. $S=-5t^2+vt+h$

$$S=-5t^2+33t+14$$
$$0=-5t^2+33t+14$$
$$0=5t^2-33t-14$$
$$0=(5t+2)(t-7)$$
$$5t+2=0 \qquad t-7=0$$
$$t=-\frac{2}{5} \qquad t=7$$

Time must be positive. It will strike the ground after 7 sec.

Practice Final Examination

1. $$\begin{aligned}-2x+3y\{7-2[x-(4x+y)]\} &= -2x+3y[7-2(x-4x-y)]\\ &= -2x+3y[7-2(-3x-y)]\\ &= -2x+3y(7+6x+2y)\\ &= -2x+21y+18xy+6y^2\end{aligned}$$

2. $2x^2-3xy-4y,\ x=-2,\ y=3$

$2(-2)^2-3(-2)(3)-4(3)=8+18-12=14$

3. $$\begin{aligned}(-3x^2y)(-6x^3y^4) &= (-3)(-6)x^{2+3}y^{1+4}\\ &= 18x^5y^5\end{aligned}$$

4. $$\begin{aligned}5x^2y-6xy+8xy-3x^2y-10xy &= (5x^2y-3x^2y)+(-6xy+8xy-10xy)\\ &= 2x^2y-8xy\end{aligned}$$

5. $$\begin{aligned}\frac{1}{2}(x+4)-\frac{2}{3}(x-7) &= 4x\\ 3(x+4)-4(x-7) &= 6(4x)\\ 3x+12-4x+28 &= 24x\\ 40 &= 25x\\ \frac{8}{5} &= x\end{aligned}$$

6. $$\begin{aligned}p &= 2a+2b\\ p-2a &= 2b\\ \frac{p-2a}{2} &= b\end{aligned}$$

7. $$\begin{aligned}5x+3-(4x-2) &\le 6x-8\\ 5x+3-4x+2 &\le 6x-8\\ -5x &\le -13\\ x &\ge \frac{13}{5}=2.6\end{aligned}$$

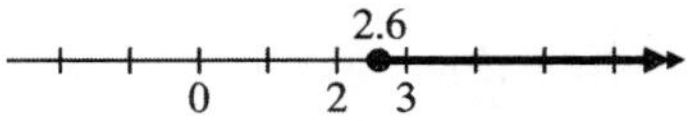

8. $$\begin{aligned}(2x+y)(x^2-3xy+2y^2) &= 2x^3-6x^2y+4xy^2+x^2y-3xy^2+2y^3\\ &= 2x^3-5x^2y+xy^2+2y^3\end{aligned}$$

9. $$\begin{aligned}4x^2-18x-10 &= 2(2x^2-9x-5)\\ &= 2(2x+1)(x-5)\end{aligned}$$

10. $$\begin{aligned}3x^3-9x^2-30x &= 3x(x^2-3x-10)\\ &= 3x(x-5)(x+2)\end{aligned}$$

11. $\dfrac{2}{x-3}-\dfrac{3}{x^2-x-6}+\dfrac{4}{x+2}=\dfrac{2}{x-3}\cdot\dfrac{x+2}{x+2}-\dfrac{3}{(x-3)(x+2)}+\dfrac{4}{x+2}\cdot\dfrac{x-3}{x-3}$

$$=\frac{2x+4-3+4x-12}{(x+2)(x-3)}$$

$$=\frac{6x-11}{(x+2)(x-3)}$$

12. $\dfrac{\frac{3}{x}+\frac{5}{2x}}{1+\frac{2}{x+2}}=\dfrac{\frac{3}{x}+\frac{5}{2x}}{1+\frac{2}{x+2}}\cdot\dfrac{2x(x+2)}{2x(x+2)}$

$$=\frac{6(x+2)+5(x+2)}{2x(x+2)+2x(2)}$$

$$=\frac{11(x+2)}{2x(x+2+2)}$$

$$=\frac{11(x+2)}{2x(x+4)}$$

13.

$$\frac{2}{x+2}=\frac{4}{x-2}+\frac{3x}{x^2-4}$$

$$(x+2)(x-2)\cdot\frac{2}{x+2}=(x+2)(x-2)\cdot\frac{4}{x-2}+(x+2)(x-2)\cdot\frac{3x}{x^2-4}$$

$$(x-2)(2)=4(x+2)+3x$$

$$2x-4=4x+8+3x$$

$$-12=5x$$

$$-\frac{12}{5}=x$$

14. $5x-2y-3=0$

$$-2y=-5x+3$$

$$y=\frac{-5}{-2}x+\frac{3}{-2}$$

$$y=\frac{5}{2}x-\frac{3}{2}$$

$m=\dfrac{5}{2}$, y-intercept $\left(0,\ -\dfrac{3}{2}\right)$

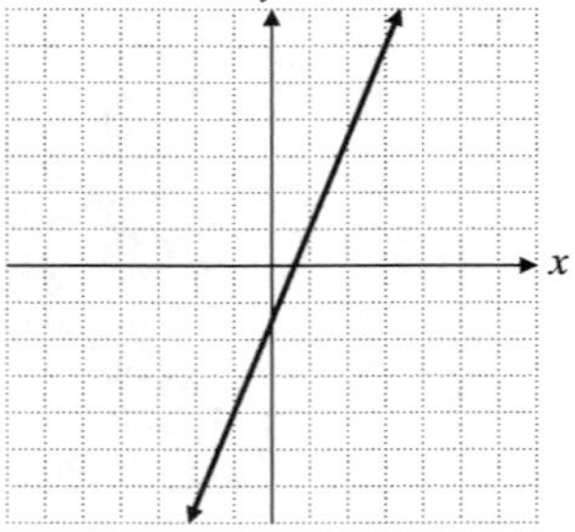

15. $m = -\frac{3}{4}$, $(-2, 5)$
$y = mx + b$
$5 = -\frac{3}{4}(-2) + b \Rightarrow b = \frac{7}{2}$
$y = -\frac{3}{4}x + \frac{7}{2}$ or $3x + 4y = 14$

16. $A = \frac{1}{2}\pi r^2 + wl$, $r = 3$, $w = 3$, $l = 8$
$A = \frac{1}{2}(3.14)(2)^2 + (3)(8) = 38.13$
The area is 38.13 in.^2

17. $2x + 7y = 4$ (1)
$-3x - 5y = 5$ (2)
Multiply (1) by 3 and (2) by 2.
$$\begin{aligned} 6x + 21y &= 12 \\ \underline{-6x - 10y} &\underline{= 10} \\ 11y &= 22 \\ y &= 2 \end{aligned}$$
Substitute 2 for y in (1).
$$\begin{aligned} 2x + 7(2) &= 4 \\ 2x + 14 &= 4 \\ 2x &= -10 \\ x &= -5 \end{aligned}$$
$(-5, 2)$

18. $a - \frac{3}{4}b = \frac{1}{4}$ (1)
$\frac{3}{2}a + \frac{1}{2}b = -\frac{9}{2}$ (2)
Multiply (1) by 4 and (2) by 6.
$$\begin{aligned} 4a - 3b &= 1 \\ \underline{9a + 3b} &\underline{= -27} \\ 13a \quad &= -26 \\ a &= -2 \end{aligned}$$
Substitute -2 for a in (1).
$$\begin{aligned} -2 - \frac{3}{4}b &= \frac{1}{4} \\ -8 - 3b &= 1 \\ -3b &= 9 \\ b &= -3 \end{aligned}$$
$(-2, -3)$

19.
$$\begin{aligned} &\sqrt{45x^3} + 2x\sqrt{20x} - 6\sqrt{5x^3} \\ &= \sqrt{9 \cdot 5 \cdot x^2 \cdot x} + 2x\sqrt{4 \cdot 5 \cdot x} - 6\sqrt{5 \cdot x^2 \cdot x} \\ &= 3x\sqrt{5x} + 4x\sqrt{5x} - 6x\sqrt{5x} \\ &= x\sqrt{5x} \end{aligned}$$

20.
$$\begin{aligned} \sqrt{6}\left(3\sqrt{2} - 2\sqrt{6} + 4\sqrt{3}\right) &= 3\sqrt{12} - 2\sqrt{36} + 4\sqrt{18} \\ &= 3\sqrt{4(3)} - 2(6) + 4\sqrt{9(2)} \\ &= 6\sqrt{3} - 12 + 12\sqrt{2} \end{aligned}$$

21.
$$\begin{aligned} \frac{\sqrt{3}+\sqrt{7}}{\sqrt{5}-\sqrt{7}} &= \frac{\sqrt{3}+\sqrt{7}}{\sqrt{5}-\sqrt{7}} \cdot \frac{\sqrt{5}+\sqrt{7}}{\sqrt{5}+\sqrt{7}} \\ &= \frac{\sqrt{15}+\sqrt{21}+\sqrt{35}+\sqrt{49}}{\left(\sqrt{5}\right)^2 - \left(\sqrt{7}\right)^2} \\ &= -\frac{7+\sqrt{15}+\sqrt{21}+\sqrt{35}}{2} \end{aligned}$$

22. $12x^2 - 5x - 2 = 0$
$(4x + 1)(3x - 2) = 0$

$4x + 1 = 0$	$3x - 2 = 0$
$4x = -1$	$3x = 2$
$x = -\frac{1}{4}$	$x = \frac{2}{3}$

23.
$$\begin{aligned} 2y^2 &= 6y - 1 \\ 2y^2 - 6y + 1 &= 0 \end{aligned}$$
$a = 2$, $b = -6$, $c = 1$
$$\begin{aligned} y &= \frac{-b \pm \sqrt{b^2 - 4ac}}{2a} \\ y &= \frac{-(-6) \pm \sqrt{(-6)^2 - 4(2)(1)}}{2(2)} \\ &= \frac{6 \pm \sqrt{28}}{4} \\ &= \frac{6 \pm 2\sqrt{7}}{4} \\ &= \frac{3 \pm \sqrt{7}}{2} \end{aligned}$$

24.
$$\begin{aligned} 4x^2 + 3 &= 19 \\ 4x^2 &= 16 \\ x^2 &= 4 \\ x &= \pm\sqrt{4} \\ x &= \pm 2 \end{aligned}$$

25. $c^2 = a^2 + b^2$
$8^2 = a^2 + 5^2$
$64 = a^2 + 25$
$39 = a^2$
$\pm\sqrt{39} = a$
The length can't be negative.
$a = \sqrt{39}$

26. x = a number
$3x + 6 = 21$
$3x = 15$
$x = 5$
The number is 5.

27. w = width
$2w - 2$ = length
$P = 2L + 2w$
$38 = 2(2w - 2) + 2w$
$38 = 4w - 4 + 2w$
$42 = 6w$
$7 = w$
$2w - 2 = 12$
The dimensions are 12 meters by 7 meters.

28. x = amount invested at 10%
y = amount invested at 14%
$x + y = 7000$ (1)
$0.10x + 0.14y = 860$ (2)
Solve (1) for x.
$x = 7000 - y$ (3)
Substitute $7000 - y$ for x in (2).
$0.10(7000 - y) + 0.14y = 860$
$700 - 0.10y + 0.14y = 860$
$0.04y = 160$
$y = 4000$
Substitute 4000 for y in (3).
$x = 7000 - 4000 = 3000$
She invested \$3000 at 10% and \$4000 at 14%.

29. x = number of general admission tickets
y = number of reserved seat tickets
$x + y = 360$ (1)
$3x + 5y = 1480$ (2)
Multiply (1) by −3 and add it to (2).
$-3x - 3y = -1080$
$3x + 5 = 1480$
$2y = 400$
$y = 200$
Substitute 200 for y in (1).
$x + 200 = 360$
$x = 160$
200 reserved-seat tickets sold and 160 general admission tickets sold.

30. $A = \frac{1}{2}bh$, x = base, $2x + 1$ = altitude

$68 = \frac{1}{2}x(2x + 1)$
$136 = 2x^2 + x$
$0 = 2x^2 + x - 136$
$0 = (2x + 17)(x - 8)$
$2x + 17 = 0$ $x - 8 = 0$
$x = -\frac{17}{2}$ $x = 8$
Length must be positive.
$2x + 1 = 2(8) + 1 = 17$
The base is 8 m and the altitude is 17 m.